Representation Theory and Beyond

CONTEMPORARY MATHEMATICS

758

Representation Theory and Beyond

Workshop and 18th International Conference on
Representations of Algebras (ICRA 2018)
August 8–17, 2018
Prague, Czech Republic

Jan Šťovíček
Jan Trlifaj
Editors

AMS AMERICAN MATHEMATICAL SOCIETY

2010 *Mathematics Subject Classification.* Primary 16Gxx, 18Dxx, 18Gxx.

Library of Congress Cataloging-in-Publication Data

Names: Workshop and International Conference on Representations of Algebras (2018: Prague, Czech Republic), author. | Šťovíček, Jan, 1980– editor. | Trlifaj, Jan, editor.
Title: Representation theory and beyond: Workshop and 18th International Conference on Representations of Algebras, August 8–17, 2018, Prague, Czech Republic / Jan Šťovíček, Jan Trlifaj, editors.
Description: Providence, Rhode Island: American Mathematical Society, [2020] | Series: Contemporary mathematics, 0271-4132; volume 758 | Includes bibliographical references.
Identifiers: LCCN 2020002000 | ISBN 9781470451318 (paperback) | ISBN 9781470456993 (ebook)
Subjects: LCSH: Representations of algebras–Congresses. | Homology theory–Congresses. | AMS: Associative rings and algebras – Representation theory of rings and algebras. | Category theory; homological algebra
Classification: LCC QA150 .W66 2018 | DDC 512–dc23
LC record available at https://lccn.loc.gov/2020002000
DOI: https://doi.org/10.1090/conm/758

∞ The paper used in this book is acid-free and falls within the guidelines
established to ensure permanence and durability.
Visit the AMS home page at `https://www.ams.org/`

10 9 8 7 6 5 4 3 2 1 25 24 23 22 21 20

Dedicated to Idun Reiten whose work has shaped modern representation theory of algebras

Contents

Preface

The Workshop and 18th International Conference on Representations of Algebras (ICRA 2018) was held in Prague, Czech Republic, from August 8–17, 2018. It continued the long tradition of conferences dedicated to representation theory of associative algebras and related areas of mathematics started at Carleton University in Canada in 1974. The immediate predecessor, ICRA 2016, was held at Syracuse University, NY, USA, and the immediate successor, ICRA 2020, will be held at ICTP in Trieste, Italy, in July 2020.[1] Besides presenting main themes of contemporary representation theory, ICRA 2018 also brought in some new tools, such as stable ∞-categories, stable derivators, and contramodules. We are grateful to all the authors who provided their contributions to this volume, as well as the referees for their assistance.

ICRA 2018 started with a four-day Workshop presenting six lecture series of three hours each. These series were delivered by Pramod Achar (Louisiana State University) who spoke on Characters of Tilting Modules for Algebraic Groups, Silvana Bazzoni (University of Padua) speaking on Contramodules and Their Application in Tilting Theory, Serge Bouc (University of Picardie - Jules Verne) on Representations of Finite Sets and Correspondences, Giovanni Cerulli Irelli (The Sapienza University of Rome) on Quiver Grassmanians, Moritz Groth (University of Bonn) on Higher Symmetries in Abstract Stable Homotopy Theories, and Michael Weymyss (University of Glasgow) who spoke on Contraction Algebras: Properties, Prophesies and Predictions. Part I of the present volume consists of expanded lecture notes for the four lecture series delivered by Bazzoni, Bouc, Cerulli Irelli, and Groth.

The 18th International Conference on Representations of Algebras spanned the week of August 13–17 and consisted of 20 morning plenary talks followed by 135 afternoon contributed talks presented in six parallel sessions. Part II of the present volume consists of research papers based on six of the plenary talks, delivered by Karin Erdmann (University of Oxford), Ivo Herzog (Ohio State University), Gustavo Jasso (University of Bonn), Bernhard Keller (Université Paris Diderot), Markus Reineke (University of Bochum), and Andrea Solotar (University of Buenos Aires).

ICRA 2018 had about 250 participants coming from 34 countries. There were 52 graduate students participating; about half of them gave a talk at the Conference, including Sota Asai (University of Nagoya) whose talk was selected by the Scientific Committee as a plenary.

[1]Note Added in Proof: Due to the COVID-19 pandemic, ICRA 2020 will be held online, with the scheduled date of November 9–27, 2020.

For more than ten years, ICRA has had the tradition of granting the ICRA Award to the best young scientists working in representation theory. In 2018, there were two awardees:

- Gustavo Jasso Ahuja from the University of Bonn, "for his fundamental contributions to τ-tilting theory introducing the original methods of reduction, g-vectors, bricks, as well as his foundational work in higher Auslander-Reiten theory," and
- Julian Külshammer from the University of Stuttgart, "for developing a new approach to quasi-hereditary algebras and Ringel duality in terms of regular bocses, with applications to exact Borel subalgebras and representation types of standard filtered categories, as well as for his study of higher Nakayama algebras."

The Awards were announced at the beginning of the Conference and presented during the Conference banquet held at the Kaiserstein Palace in Prague.

ICRA 2018 would not take place without responsible work of members of several committees:

The Scietific Committee consisted of Lidia Angeleri Hügel (University of Verona), William Crawley-Boevey (Bielefeld University), Steffen Koenig (University of Stuttgart), Henning Krause (Bielefeld University), Bernard Leclerc (University of Caen Normandy), Idun Reiten (Norwegian University of Science and Technology), Changchang Xi (Capital Normal University, Beijing), Kunio Yamagata (Tokyo University of Agriculture and Technology), and Dan Zacharia (Syracuse University).

The Organizing Committee consisted of Jan Trlifaj (Charles University, Prague - chair), Jan Šťovíček (Charles University), Miroslav Korbelář (Czech Technical University, Prague), Pavel Příhoda (Charles University), Pavel Růžička (Charles University), Jan Šaroch (Charles University), Jan Žemlička (Charles University), Alena Gollová (Czech Technical University) Pavel Pták (Czech Technical University), and Jiří Rosický (Masaryk University, Brno). The Committee appreciated the partnership of AMCA, the Academic and Medical Conference Agency (Prague), in organizing ICRA 2018. Special thanks are also due to Alexander Slávik (Charles University) and Michal Hrbek (Math. Inst., Czech. Acad. Sci.)."

The ICRA 2018 Award Committee consisted of Karin Erdmann (University of Oxford), Jan Schröer (University of Bonn), Jan Trlifaj (Charles University), Osamu Iyama (Nagoya University), and Dan Zacharia (Syracuse University).

ICRA 2018 was partially supported by a grant of the Neuron endowment fund, by project 17-23112S of the Grant Agency of the Czech Republic, by the "Support of Internationalization" project of Charles University, and by a grant of the RSJ Foundation. The Faculty of Electrical Engineering of the Czech Technical University and the Faculty of Mathematics and Physics of Charles University contributed by providing substantial discounts for rentals of the lecture halls. All of this support is gratefully appreciated.

Last but not least, this volume is dedicated to Professor Idun Reiten from the Norwegian University of Science and Technology, Trondheim, whose work has shaped modern representation theory of algebras.

Prague, January 2020 Jan Šťovíček and Jan Trlifaj

Workshop Lecture Notes

Contemporary Mathematics
Volume **758**, 2020
https://doi.org/10.1090/conm/758/15230

Contramodules and their applications to tilting theory

Silvana Bazzoni

Introduction

Contramodules were first introduced by Eilenberg and Moore in 1965 alongside comodules over coalgebras or corings, but they achieved so little success that they were forgotten for three decades. From the first beginning of the 21st century, contramodules appeared again in literature thanks mainly to work by Leonid Positselski. The motivation towards their study was in particular for their many applications in algebraic geometry. They were studied for the purposes of the semi-infinite cohomology theory and the comodule-contramodule correspondence. Recently the notion of contramodules has been applied profitably in commutative algebra and tilting theory.

The aim of these three lectures is to advertise the theory of contramodules and to capture the interest of researchers and get them involved in applications of the theory to different contexts.

Generally, contramodules are sets with infinitary additive operations of the "arity" bounded by some cardinal. Typical examples are contramodules over complete topological rings. They provide a way of having an abelian category of non-topological modules with some completeness properties over a topological ring.

More precisely, let R be a commutative complete local ring with maximal ideal m. The category of infinitely generated m-adically complete R-modules is not abelian already for $R = k[[\epsilon]]$. The category of R-contramodules is the natural abelian category into which complete R-modules are embedded. In particular, when R is the ring of p-adic integers, the abelian category of R-contramodules is the $0,1$-perpendicular category of $\mathbb{Z}[p^{-1}]$, i.e. the category of the abelian groups C such that $\mathrm{Hom}_{\mathbb{Z}}(\mathbb{Z}[p^{-1}], C) = 0 = \mathrm{Ext}^1_{\mathbb{Z}}(\mathbb{Z}[p^{-1}], C)$. Similarly for contramodules over $R = k[[\epsilon]]$.

The $k[[\epsilon]]$-contramodules form a full subcategory of the category of $k[[\epsilon]]$-modules and even a full subcategory of the category of $k[\epsilon]$ modules. This subcategory contains all the $k[[\epsilon]]$-modules M, such that $M \cong \varprojlim_n M/\epsilon^n M$, i.e. separated and complete modules, but contains also some complete and non separated modules, i.e. the modules such that the natural map $M \to \varprojlim_n M/\epsilon^n M$ is surjective but not injective.

2010 *Mathematics Subject Classification.* Primary 18E10; Secondary 13D07, 16E35.

Analogous results hold for the adic completion of a commutative Noetherian ring with respect to an arbitrary ideal.

So surprisingly, in these cases (and some other cases) the natural forgetful functors from the categories of contramodules to the related categories of modules turn out be fully faithful. That is an infinitary additive operation is uniquely determined by its finite aspects.

1. First lecture

1.1. Infinite summation operations. An elementary approach to the theory of contramodules can be achieved by introducing the notion of an s-power infinite summation operation on abelian groups.

Let C be an abelian group and s a symbol. An s-power infinite summation operation (s-power i.s.o.) on C is a map

$$\prod_{n\geq 0} C \longrightarrow C$$

$$(c_0, c_1, \ldots, c_n, \ldots) \longmapsto \sum_{n=0}^{\infty} s^n c_n,$$

satisfying three axioms:

(1) **(U)** Unitality: $\sum_{n=0}^{\infty} s^n c_n = c_0$ if $c_1 = c_2 = \ldots, c_n = \cdots = 0$.

(2) **(A)** Additivity: $\prod_{n\geq 0} C_n \longrightarrow C$ is a homomorphism of abelian groups, i.e.

$$\sum_{n=0}^{\infty} s^n(c_n + b_n) = \sum_{n=0}^{\infty} s^n c_n + \sum_{n=0}^{\infty} s^n b_n$$

$$\forall (c_0, c_1, \ldots, c_n, \ldots), (b_0, b_1, \ldots, b_n, \ldots) \in \prod_{n\geq 0} C.$$

(3) **(CA)** Contrassociativity:

$$\sum_{i=0}^{\infty} s^i \sum_{j=0}^{\infty} s^j(c_{ij}) = \sum_{n=0}^{\infty} s^n \sum_{i+j=n} c_{ij}.$$

An s-power i.s.o. on an abelian group C defines an abelian group endomorphism $s : C \to C$ by

$$sc = \sum_{n=0}^{\infty} s^n c_n, \qquad \text{where } c_1 = c \text{ and } c_i = 0 \text{ for } i \neq 1,$$

$$s^n c = \sum_{n=0}^{\infty} s^n c_n, \qquad \text{where } c_n = c \text{ and } c_i = 0 \text{ for } i \neq n.$$

EXAMPLE 1.1.1. [**11**, Example 3.1 (1), (2)]

(1) For any abelian group C, the group of formal power series $C[[z]]$ is naturally endowed with a z-power i.s.o.
(2) The group of p-adic integers J_p is naturally endowed with a p-power i.s.o.
(3) For any set X the group $\prod_{x\in X} J_p$ is endowed with a p-power i.s.o.

(4) The subgroup $C = J_p[[X]] \subset \prod_{x\in X} J_p$ consisting of families

$$(a_x)_{x\in X}, a_x \in J_p$$

converging to zero in the topology of J_p is preserved by the p-power i.s.o. on $\prod_{x\in X} J_p$.

(5) In all these cases, the infinite sum can be computed as the limit of finite partial sums in the adic topology of the group in question.

(6) The category of abelian groups with s-power i.s.o. and group homomorphisms preserving the infinite summation operations is an abelian category with products (indeed, kernels and cokernels inherit the s-power i.s.o. and for any family of groups C_i with s-power i.s.o. there is a natural s-power i.s.o. on $\prod_i C_i$.)

LEMMA 1.1.2. [**11**, Lemma 3.2] *An abelian group C endowed with an s-power i.s.o. has no non zero s-divisible subgroups, i.e. if $D \leq C$ and $sD = D$ then $D = 0$.*

PROOF. Idea of the proof. Given a sequence $(c_0, c_1, \ldots, c_n, \ldots)$ of elements of C satisfying $c_n = sc_{n+1}$, for every $n \geq 0$, consider the element $\sum_{n=0}^{\infty} s^n c_n$ and

$$\sum_{n=0}^{\infty} s^n c_n = \sum_{n=0}^{\infty} s^n sc_{n+1} = \sum_{n=0}^{\infty} s^{n+1} c_{n+1} = \sum_{n=1}^{\infty} s^n c_n,$$

hence $c_0 = 0$. To make the calculation rigorous one should use the Contraassociativity axiom (CA). □

THEOREM 1.1.3. *Let C be an abelian group. The following hold.*

(1) *([**11**, Theorem 3.3 (1)] An s-power i.s.o. on an abelian group C is uniquely determined by the endomorphism $s : C \to C$, that is, given $s : C \to C$ there exists at most one s-power i.s.o. on C restricting to s.*

(2) *An endomorphism $s : C \to C$ can be extended to an s-power i.s.o. on C if and only if for any sequence $(a_0, a_1, \ldots, a_n, \ldots)$ of elements of C the infinite system of non-homogeneous equations*

$$(*) \quad b_n - sb_{n+1} = a_n, \qquad n \geq 0,$$

admits a unique solution $(b_0, b_1, \ldots, b_n, \ldots)$ in C.

(3) *An abelian group homomorphism $f\colon C \to D$ between two abelian groups C, D endowed with s-power i.s.o., preserves the s-power i.s.o. if and only if it commutes with the s-endomorphisms on C and D.*

PROOF. (2) Let C be an abelian group with an s-power i.s.o.. By Lemma 1.1.2 C is s-reduced; thus, if a solution of the system $(*)$ exists, then it is unique.

For the existence, set

$$b_n := \sum_{i=0}^{\infty} s^i a_{n+i}$$

for all $n \geq 0$. The sequence $(b_n)_{n\geq 0}$ is a solution of the system:

$$b_n - sb_{n+1} = \sum_{i=0}^{\infty} s^i a_{n+i} - s\sum_{i=0}^{\infty} s^i a_{n+i+1} =$$

$$= \sum_{i=0}^{\infty} s^i a_{n+i} - \sum_{i=0}^{\infty} s^{i+1} a_{n+i+1} = \sum_{i=0}^{\infty} s^i a_{n+i} - \sum_{i=1}^{\infty} s^i a_{n+i} = a_n.$$

To make this calculation rigorous one should use axiom (CA) with the sequence of elements

$$a_{ij} := \begin{cases} a_{n+j+1} & \text{for } i = 1, j \geq 0 \\ 0 & \text{otherwise} \end{cases}$$

For the converse, suppose now that the system $(*)$ is uniquely solvable in C for any sequence $(a_n)_{n\geq 0}$ of elements of C. Given such a sequence, take the unique solution $(b_n)_{n\geq 0}$ of the system and set

$$\sum_{n=0}^{\infty} s^n a_n := b_0.$$

This defines an s-power i.s.o. on C. One then has to check that the three axioms are satisfied.

We have proved not only (2), but also (1), since we see that once the endomorphism $s : C \to C$ is given, the s-power i.s.o. on C, can be recovered from the solution of the system $(*)$ which is unique.

To prove (3), note that any abelian group homomorphism $f\colon C \to D$ commuting with the endomorphisms s, takes solutions of the system $(*)$ in C to solutions in D and thus it preserves the s-power i.s.o. on C and D. □

From now on in this lecture R will denote a commutative ring with unit.

Definition 1.1.4. Let R be a commutative ring, $s \in R$ a fixed element and C an R-module. Let $s\colon C \to C$ be multiplication by s. We say that C admits an s-power i.s.o. if it admits an s-power i.s.o. as an abelian group in a way that it is compatible with the action of s.

If C admits an s-power i.s.o. then for $r \in R$, $r(\sum_{n=0}^{\infty} s^n c_n) = \sum_{n=0}^{\infty} s^n (rc_n)$.

If R is a commutative ring and $s \in R$ consider the localization $R[s^{-1}]$ and the inductive system

$$R \xrightarrow{s} R \xrightarrow{s} R \to \dots$$

where the maps are multiplications by s. Then $R[s^{-1}] \cong \varinjlim_{n\geq 0} R$, and $R[s^{-1}]$ admits a presentation

$$\textbf{(a)} \quad 0 \to \bigoplus_{n=0}^{\infty} Rf_n \xrightarrow{\mu} \bigoplus_{n=0}^{\infty} Re_n \to R[s^{-1}] \to 0, \quad \mu(f_n) = e_n - se_{n+1}.$$

For every R-module C we get the following exact sequence.

$$\textbf{(b)} \quad 0 \to \operatorname{Hom}_R(R[s^{-1}], C) \to \prod_{n=0}^{\infty} C \xrightarrow{\operatorname{Hom}(\mu,C)} \prod_{n=0}^{\infty} C \to \operatorname{Ext}^1_R(R[s^{-1}], C) \to 0,$$

$$\operatorname{Hom}(\mu, C)((b_n)_{n\geq 0}) = (b_n - sb_{n+1})_{n\geq 0}.$$

The following is an immediate consequence of the above discussion.

Theorem 1.1.5. *An R-module admits an s-power i.s.o. if and only if*

$$\operatorname{Hom}_R(R[s^{-1}], C) = 0 = \operatorname{Ext}^1_R(R[s^{-1}], C).$$

Proof. From the exact sequence **(b)** we infer that the system

$$(*) \quad b_n - sb_{n+1} = a_n, n \geq 0,$$

admits a solution $(b_0, b_1, \dots, b_n, \dots)$ in $\prod_{n=0}^{\infty} C$ if and only if

$$\mathrm{Ext}^1_R(R[s^{-1}], C) = 0.$$

It admits a unique solution if moreover $\mathrm{Hom}_R(R[s^{-1}], C) = 0$. The conclusion follows by Theorem 1.1.3. □

DEFINITION 1.1.6. Let R be a commutative ring and $s \in R$ a fixed element. An R-module C is an **s-contramodule** or s-contramodule R-module, if

$$\mathrm{Hom}_R(R[s^{-1}], C) = 0 = \mathrm{Ext}^1_R(R[s^{-1}], C).$$

Denote by $R\text{–}\mathsf{Mod}_{\mathbf{s-ctr}}$ the subcategory of s-contramodules.

Thus s-contramodules are the objects of the $0,1$-perpendicular category of $R[s^{-1}]$. Since the projective dimension of $R[s^{-1}]$ is at most one, the category of s-contramodules is closed under kernels, cokernels, extensions and products in $R\text{–}\mathsf{Mod}$, that is, it is exactly embedded in $R\text{–}\mathsf{Mod}$. Moreover, if

$$0 \to F_1 \xrightarrow{f} F_0 \to R[s^{-1}] \to 0$$

is a projective resolution of $R[s^{-1}]$, then M is an s-contramodule if and only if $\mathrm{Hom}_R(f, M)$ is a bijection.

Every R-module annihilated by s is clearly an s-contramodule.

1.2. Relations between s-contramodules and s-adic completion. An R-module M over a commutative ring is s-torsion if for every $x \in M$ there is $n \in \mathbb{N}$ such that $s^n x = 0$. Thus M is s-torsion if and only if $R[s^{-1}] \otimes_R M = 0$ ($\mathrm{Tor}_1^R(R[s^{-1}], M) = 0$ always). This shows a duality between s-torsion modules and s-contramodules.

Let $\mathcal{D}(R)$ denote the derived category of the ring R and $\otimes_R^{\mathbf{L}}$ denote the total left derived functor of the tensor product functor. Then:

LEMMA 1.2.1. [**11**, Lemma 6.2]
Let $M^\bullet, N^\bullet, C^\bullet$ be complexes of R-modules.

(1) *If either $H^n(M^\bullet)$ or $H^n(N^\bullet)$ are s-torsion, for all $n \in \mathbb{Z}$, then the $H^n(M^\bullet \otimes^{\mathbf{L}}{}_R N^\bullet)$ are s-torsion for all $n \in \mathbb{Z}$.*
(2) *If either $H^n(M^\bullet)$ are s-torsion or $H^n(C^\bullet)$ are s-contramodules for all $n \in \mathbb{Z}$, then $\mathrm{Hom}_{\mathcal{D}(R)}(M^\bullet, C^\bullet[n])$ are s-contramodule, for all $n \in \mathbb{Z}$.*

An R-module C is said to be **s-complete** if the natural map to its s-adic completion

$$\lambda_{s,C} \colon C \to \varprojlim_{n \geq 1} C/s^n C,$$

is surjective. The R-module C is said to be **s-separated** if the map $\lambda_{s,C}$ is injective. The module $\varprojlim_{n\geq 1} C/s^n C$ is denoted by $\Lambda_s(C)$.

LEMMA 1.2.2. *Let C be an R-module. Then*

(1) [**11**, proof of Theorem 2.3] *C is s-complete if and only if for any sequence $(a_n)_{n\geq 0}$ of elements of C, the infinite system of linear equations:*

$$(**)\ s^n(b_n - s b_{n+1}) = s^n a_n, \quad n \geq 0$$

admits a solution $(b_n)_{n\geq 0}$ in C;
(2) *C is s-separated if, and only if, for any pair of solutions $(b'_n)_{n\geq 0}$, $(b''_n)_{n\geq 0}$ of the system $(**)$ in C, one has $b'_0 = b''_0$.*
(3) *if C is an s-contramodule, then C is s-complete.*

PROOF. (1) Recall that:

$$\varprojlim C/s^nC = \{(\bar{c}_n)_{n\geq 1} \in \prod_{n\geq 1} \frac{C}{s^nC} \mid c_{n+1} \equiv c_n \bmod s^nC, \ \forall n \geq 1\}.$$

Asking $\lambda_{s,C}$ to be surjective means that, for any sequence of elements $(\bar{c}_n)_{n\geq 1}$ in $\varprojlim C/s^nC$, there exists $c \in C$ such that $c \equiv c_n \bmod s^nC$ for all $n \geq 1$.

Now, set $a_0 := c_1$ and for $n \geq 1$ choose elements $a_n \in C$ such that $c_{n+1} = c_n + s^na_n$. Then for any $n \geq 0$ one has:

$$c_{n+1} = s^na_n + s^{n-1}a_{n-1} + \cdots + sa_1 + a_0.$$

So surjectivity becomes equivalent to ask that for any sequence $(a_n)_{n\geq 0}$ of elements of C, there exists $c \in C$ such that $c - (s^na_n + s^{n-1}a_{n-1} + \cdots + sa_1 + a_0) \in s^{n+1}C$.

Finally set $b_0 := c$. Then the statement above is equivalent to ask that for any sequence $(a_n)_{n\geq 0}$ of elements of C there exists a solution $(b_n)_{n\geq 0}$ to the system of linear equations

$$b_0 - s^{n+1}b_{n+1} = s^na_n + \cdots + sa_1 + a_0, \quad n \geq 0$$

which in turn is equivalent to the system:

$$s^n(b_n - sb_{n+1}) = s^na_n, \quad n \geq 0.$$

(2) If $\lambda_{s,C}$ is injective, then $b_0' = b_0''$, since by the argument in (1) $\lambda_{s,C}(b_0') = \lambda_{s,C}(b_0'')$. Conversely if $\lambda_{s,C}(b_0') = \lambda_{s,C}(b_0'')$ for some $b_0', b_0'' \in C$, then for any $n \geq 1$, $b_0' - b_0'' \in s^nC$, i.e. $b_0' - b_0'' = s^nb_n$. So $(b_0' - b_0'', b_1, b_2, \ldots)$ is a solution of the homogeneous version of the system $(**)$, hence by assumption it must be $b_0' - b_0'' = 0$.

(3) $\operatorname{Ext}^1_R(R[s^{-1}], C) = 0$ implies that there is a solution of the system $(*)$, hence C is s-complete. □

LEMMA 1.2.3. [**11**, Theorem 2.4]

(1) *An s-separated, s-complete R-module C is an s-contramodule.*
(2) *An s-torsion free s-contramodule R-module C is s-separated and s-complete*

PROOF. (1) The category R–$\mathsf{Mod}_{\mathbf{s-ctr}}$ is closed under kernels and infinite direct products, hence under infinite projective limits. Moreover, for an R-module C, the quotient C/s^nC is always an s-contramodule for any $n \geq 1$. So an s-separated and s-complete R-module C is an s-contramodule, since $C \cong \varprojlim_{n\geq 1} C/s^nC$.

(2) If C is s-torsion free then the systems $(*)$ and $(**)$ are equivalent. If C is moreover an s-contramodule, then the system $(*)$ has a unique solution, by Theorem 1.1.5, hence Lemma 1.2.2 (2) is satisfied. □

There are examples of s-contramodules which are not s-separated.

EXAMPLE 1.2.4. [**11**, Example 2.7 (1)] For any prime $p \in \mathbb{Z}$ let J_p be the abelian group of p-adic integers and let C be the subgroup of $\prod_{n\geq 0} J_p$ consisting of all sequences converging to zero in the p-adic topology. Let $D \subseteq C$ be the subgroup of all sequences $(p^nv_n)_{n\geq 0}$, where $v_n \in J_p$ and let $E \subseteq D$ be the subgroup of all sequences $(p^nv_n)_{n\geq 0}$, where $(v_n)_{n\geq 0}$ converges to zero in the p-adic topology. All the three groups are p-contramodules, as they are p-separated and p-complete. Thus, the quotient C/E is a p-contramodule, too. However, it is not p-separated because $\bigcap_{n\geq 1} p^n(C/E) = D/E$.

The above example allows to construct an abelian group with an s-power i.s.o which cannot be interpreted as any kind of limit of finite partial sums.

EXAMPLE 1.2.5. [**11**, Example 3.1 (3)]
In the above notations, set $B = C/E$, and let $b_n = c_n + E$, $c_n = (u_m)_{m\geq 0}$ with $u_0 = 0, u_1 = 0, ..., u_n = 1, u_{n+1} = 0, \ldots$. Then $p^n c_n \in E$, but $\sum_{n=0}^{\infty} p^n c_n \notin E$, because the sequence $1, p, p^2, \ldots, p^n \ldots$ does not have the form $p^n v_n$ with $v_n \to 0$ in J_p for $n \to \infty$.

Contrarily to the class of s-contramodules, the class of s-separated and s-complete R-modules is not abelian and it is not closed under cokernels (even under the cokernels of injective morphisms) in R–Mod, nor under extensions, see [**11**, Remark 2.6]. Similarly, the class of s-complete R-modules does not have nice closure properties, since it is not closed under extensions, even though it is closed under quotients.

1.3. The left adjoint functor Δ_s. In this subsection we will show that the category R–Mod$_{\mathbf{s-ctr}}$ is a *reflective* subcategory of R–Mod (i.e. the embedding functor has a left adjoint).

NOTATION 1.3.1. *Let R be a commutative ring, s an element of R and l_R the natural ring homomorphism $R \to R[s^{-1}]$.*

Denote by $K^\bullet(s)$ the two-term complex $R \overset{l_R}{\to} R[s^{-1}]$, concentrated in cohomological degrees -1 and 0.

Denote by $T^\bullet(s)$ the two term complex

$$\bigoplus_{n=0}^{\infty} R \overset{\psi_R^s}{\to} \bigoplus_{n=1}^{\infty} R,$$

where $\psi_R^s : (x_0, x_1, x_2, \ldots) \longmapsto (x_1 - sx_0, x_2 - sx_1, \ldots)$, concentrated in cohomological degrees 0 and 1.

LEMMA 1.3.2. [**11**, Remark 6.5] *The complex $T^\bullet(s)[1]$ is a projective resolution of the complex $K^\bullet(s)$.*

PROOF. A quasi isomorphism between $T^\bullet(s)[1]$ and $K^\bullet(s)$ is given by the following diagram

$$\begin{array}{ccc}
\bigoplus_{n=0}^{\infty} R & \xrightarrow{\psi_R^s} & \bigoplus_{n=1}^{\infty} R \\
{\scriptstyle p}\downarrow & & \downarrow{\scriptstyle f} \\
R & \xrightarrow{l_R} & R[s^{-1}]
\end{array}$$

where p is the projection on the first factor and f is defined by $f(y_1, y_2, \ldots) := -\sum_{n\geq 1} y_n/s^n$. Then $f \circ \psi_R^s = l_R \circ p$ and (p, f) is a quasi isomorphism. □

We can now define the left adjoint functor to the embedding functor

$$R\text{–Mod}_{\mathbf{s-ctr}} \overset{\iota}{\to} R\text{–Mod}.$$

Let $\mathcal{D}^b(R)$ denote the bounded derived category of R and $\mathcal{K}(R)$ the homotopy category of R.

THEOREM 1.3.3. [**11**, Theorem 6.4] *Let R be a commutative ring and $s \in R$ be an element. For any R-module C, the following R-modules are naturally isomorphic:*

(1) *The cokernel of the R-module morphism*

$$\phi^s_C := \operatorname{Hom}_R(\psi^s_R, C) : \prod_{n\geq 1} C \to \prod_{n\geq 0} C$$

defined by $\phi^s_C((c_1, c_2, c_3, \dots)) := (-sc_1, c_1 - sc_2, c_2 - sc_3, \dots)$;

(2) *the cokernel of the endomorphism of the R-module $C[[z]]$ of formal power series in one variable z with coefficients in C*

$$(z - s) : C[[z]] \to C[[z]]$$

which is the difference of the endomorphism of multiplication by z and the endomorphism of multiplication by s (induced by the multiplication by s in C);

(3) *the R-module $\operatorname{Hom}_{\mathcal{D}^b(R)}(K^\bullet(s), C[1]) = H_0(\operatorname{Hom}_R(T^\bullet(s), C))$ (denoted by $\operatorname{Ext}^1_R(K^\bullet(s), C)$).*

PROOF. (1) $\Leftrightarrow$ (2). The natural R-module isomorphisms $\prod_{n\geq 1} C \cong C[[z]] \cong \prod_{n\geq 0} C$ identifies the R-module morphism ϕ^s_C with the endomorphism $(z-s)$.

(1)$\Leftrightarrow$ (3). We have

$$\operatorname{Hom}_{\mathcal{D}^b(R)}(K^\bullet(s), C[1]) = \operatorname{Hom}_{\mathcal{K}(R)}(T^\bullet(s), C)$$

since the complex $T^\bullet(s)[1]$ is a projective resolution of $K^\bullet(s)$ by Lemma 1.3.2. Now, a cochain map $T^\bullet(s) \to C$ is an R-linear map $p : T^0 \to C$, and is homotopic to zero if, and only if it factors through ψ^s_R

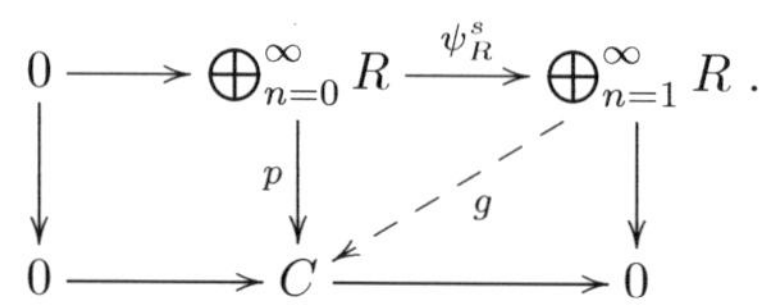

In other words

$$\operatorname{Hom}_{\mathcal{K}(R)}(T^\bullet(s), C) \cong \operatorname{Coker} \operatorname{Hom}_R(\psi^s_R, C) = \operatorname{Coker} \phi^s_C.$$

□

We will denote by $\Delta_s(C)$ the R-module described in Theorem 1.3.3.

Now we have a functor: $\Delta_s : R\text{–}\mathsf{Mod} \to R\text{–}\mathsf{Mod}$. Indeed this is a functor, since it is built using direct products and cokernels of morphisms (both functorial).

PROPOSITION 1.3.4. *For any R-module C, $\Delta_s(C)$ is an s-contramodule R-module.*

PROOF. Using the characterization of $\Delta_s(C)$ as $\operatorname{Hom}_{\mathcal{D}(R)}(K^\bullet(s), C[1])$ from Theorem 1.3.3 (3), we have that $\Delta_s(C)$ is an s-contramodule by Lemma 1.2.1 (2), since $H^{-1}(K^\bullet(s))$ ($=\operatorname{Ker} l_R$) and $H^0(K^\bullet(s))$ ($= \operatorname{Coker} l_R$) are s-torsion R-modules. □

The following theorem is of key importance and it will be used to prove that Δ_s is left adjoint to the embedding functor.

PROPOSITION 1.3.5. *For any R-module C, there exists a 5-term exact sequence of R-modules:*

$$(\mathbf{c})\, 0 \to \mathrm{Hom}_R(\mathrm{Coker}\, l_R, C) \to \mathrm{Hom}_R(R[s^{-1}], C) \to C \overset{\delta^s_C}{\to} \Delta_s(C)$$
$$\to \mathrm{Ext}^1_R(R[s^{-1}], C) \to 0.$$

where l_R is the localization map $R \to R[s^{-1}]$.

PROOF. Apply the triangulated functor $\mathrm{Hom}_{\mathcal{D}^b(R)}(-, C)$ to the distinguished triangle

$$R \to R[s^{-1}] \to K^\bullet(s) \to R[1]$$

in $\mathcal{D}^b(R\text{–}\mathsf{Mod})$ to get the long exact sequence of R-modules:

$$0 \to \mathrm{Hom}_{\mathcal{D}^b(R)}(K^\bullet(s), C) \to \mathrm{Hom}_{\mathcal{D}^b(R)}(R[s^{-1}], C) \to C \to$$

$$\to \mathrm{Hom}_{\mathcal{D}^b(R)}(K^\bullet(s)[-1], C) \cong \Delta_s(C) \to \mathrm{Ext}^1_R(R[s^{-1}], C) \to 0.$$

By Lemma 1.3.2, $\mathrm{Hom}_{\mathcal{D}^b(R)}(K^\bullet(s), C) = \mathrm{Hom}_{\mathcal{K}(R)}(T^\bullet(s)[1], C)$. A map of complexes $T^\bullet(s)[1] \to C$ is null-homotopic if and only if $f = 0$.

$$\begin{array}{ccccc} \bigoplus_{n\geq 0} R & \xrightarrow{\psi^s_R} & \bigoplus_{n\geq 1} R & \longrightarrow & 0 \\ \downarrow & & \downarrow f & & \\ 0 & \longrightarrow & C & \longrightarrow & 0 \end{array}$$

Thus:

$$\mathrm{Hom}_{\mathcal{K}(R)}(T^\bullet(s)[1], C) \cong \mathrm{Hom}_R(\mathrm{Coker}\, \psi^s_R, C) \cong \mathrm{Hom}_R(\mathrm{Coker}\, l_R, C).$$

□

COROLLARY 1.3.6. *Let C be an R-module. Then C is an s-contramodule if, and only if the adjunction morphism $C \overset{\delta^s_C}{\to} \Delta_s(C)$ is an isomorphism.*

PROOF. Immediate from Proposition 1.3.4 and the exact sequence in Proposition 1.3.5. □

PROPOSITION 1.3.7. [**11**, Theorem 6.4] $\Delta_s : R\text{–}\mathsf{Mod} \to R\text{–}\mathsf{Mod}_{\mathbf{s-ctr}}$ *is left adjoint to the embedding functor $R\text{–}\mathsf{Mod}_{\mathbf{s-ctr}} \to R\text{–}\mathsf{Mod}$.*

PROOF. We need to check that for any R-module C and any s-contramodule R-module D, there is a natural isomorphism:

$$\mathrm{Hom}_{R\text{–}\mathsf{Mod}_{\mathbf{s-ctr}}}(\Delta_s(C), D) \cong \mathrm{Hom}_R(C, D)$$

functorial both in C and D. We show the isomorphism.

Let $f : C \to D$ be an R-module morphism and consider the R-module morphism $\delta^s_C : C \to \Delta_s(C)$ given by sequence ($\mathbf{c}$). $\mathrm{Ker}\, \delta^s_C$ is an s-divisible R-module while $\mathrm{Coker}\, \delta^s_C \cong \mathrm{Ext}^1_R(R[s^{-1}], C)$ is an $R[s^{-1}]$-module.

The restriction of f to $\mathrm{Ker}\, \delta^s_C$ is zero, whence there is a map $\bar{f} : C/\mathrm{Ker}\, \delta^s_C \to D$. Considering an $R[s^{-1}]$-module presentation of $\mathrm{Coker}\, \delta^s_C$ one sees $\mathrm{Ext}^1_R(\mathrm{Coker}\, \delta^s_C, D) = 0$, since $\mathrm{Ext}^i_R(R[s^{-1}], D) = 0$, $i = 0, 1$. Applying the functor $\mathrm{Hom}_R(-, D)$ to the exact sequence

$$0 \to C/\mathrm{Ker}\, \delta^s_C \to \Delta_s(C) \to \mathrm{Coker}\, \delta^s_C \to 0$$

we obtain an isomorphism $\operatorname{Hom}_R(\Delta_s(C), D) \cong \operatorname{Hom}_R(C/\operatorname{Ker}\delta^s_C, D)$. By the diagram

$$\begin{array}{ccccccc}
\operatorname{Ker}\delta^s_C & \longrightarrow & C & \xrightarrow{\delta^s_C} & \Delta_s(C) & \longrightarrow & \operatorname{Coker}\delta^s_C \\
 & & \downarrow & & \downarrow g & & \\
 & & C/\operatorname{Ker}\delta^s_C & \xrightarrow{\overline{f}} & D & &
\end{array}$$

there is a unique map $g \in \operatorname{Hom}_R(\Delta_s(C), D)$ such that $g\delta^s_C = f$. $\square$

REMARK 1.3.8. *Δ_s is left adjoint to an exact functor, hence it sends projective modules to projective objects of $R\text{–}\mathsf{Mod}_{\mathbf{s-ctr}}$. Thus $\Delta_s(R) = P$ is a projective object of $R\text{–}\mathsf{Mod}_{\mathbf{s-ctr}}$ and it is moreover a generator since $\operatorname{Hom}_{R\text{-}\mathsf{Mod}_{\mathbf{s-ctr}}}(\Delta_s(R), C) \cong \operatorname{Hom}_R(R, C) \cong C$, for any object $C \in R\text{–}\mathsf{Mod}_{\mathbf{s-ctr}}$. Furthermore, a coproduct $P^{(X)}$ of copies of P in $R\text{–}\mathsf{Mod}_{\mathbf{s-ctr}}$ is computed as $\Delta_s(R^{(X)})$, since Δ_s preserves coproducts.*

We look for relations between the functor Δ_s and the completion functor Λ_s. Let C be an R-module. Consider the inverse systems

$$(1) \qquad \cdots \longrightarrow C/s^3C \longrightarrow C/s^2C \longrightarrow C/sC$$

and

$$(2) \qquad \cdots \longrightarrow {}_{s^3}C \longrightarrow {}_{s^2}C \longrightarrow {}_sC,$$

where for any element $r \in R$ we denote by ${}_rC \subset C$ the submodule of all the elements of C annihilated by r. The transition map ${}_{s^{n+1}}C \longrightarrow {}_{s^n}C$ acts by the multiplication with s.

The next theorem shows the relation between the functor Δ_s and the completion functor Λ.

THEOREM 1.3.9. [**11**, Lemma 6.7] *Let R be a commutative ring and $s \in R$ be an element. For any R-module C there is a natural short exact sequence of R-modules:*

$$0 \to \varprojlim\nolimits^1_{n\geq 1} {}_{s^n}C \to \Delta_s(C) \to \varprojlim\nolimits_{n\geq 1} C/s^nC = \Lambda_s(C) \to 0.$$

PROOF. The complex $K^\bullet(s)[-1]$ is the inductive limit of the system of complexes $K^\bullet_n(s) = (R \xrightarrow{s^n} R)$. For $n \geq 1$, denote by $T^\bullet_n(s)$ the subcomplex of $T^\bullet(s)$

$$\bigoplus_{i=0}^{n-1} R \xrightarrow{\psi_s} \bigoplus_{i=1}^{n} R$$

The complex $T^\bullet_n(s)$ is quasi-isomorphic to $(R \xrightarrow{s^n} R)$ (in degrees $0, 1$).

The complexes $\operatorname{Hom}_R(T^\bullet_n(s), C)$ form a projective system of complexes with surjective structure maps. By standard arguments we obtain the short exact sequence of R-modules

$$0 \to \varprojlim\nolimits^1_{n\geq 1} H_1(\operatorname{Hom}_R(T^\bullet_n(s), C)) \to H_0(\varprojlim_{n\geq 1} \operatorname{Hom}_R(T^\bullet_n(s), C))$$
$$\to \varprojlim_{n\geq 1} H_0(\operatorname{Hom}_R(T^\bullet_n(s), C)) \to 0$$

where:

- $H_1(\operatorname{Hom}_R(T^\bullet_n(s), C)) = H_1(\operatorname{Hom}_R(K^\bullet_n(s), C) = \operatorname{Ker}(C \xrightarrow{r^n} C) = {}_{s^n}C$;

- $H_0(\varprojlim_{n\geq 1} \mathrm{Hom}_R(T_n^\bullet(s), C)) = H_0(\mathrm{Hom}_R(\varinjlim_{n\geq 1} T_n^\bullet(s), C)) = H_0(\mathrm{Hom}_R(T^\bullet(s), C)) = \Delta_s(C)$;
- $H_0(\mathrm{Hom}_R(T_n^\bullet(s), C)) = H_0(\mathrm{Hom}_R(R \xrightarrow{s^n} R, C)) = \mathrm{Coker}(C \xrightarrow{s^n} C) = C/s^nC$.

□

A module C is said to be of bounded s-torsion if there exists $m \geq 1$ such that $s^n c = 0$ implies $s^m c = 0$ for every $n \geq 1$ and every $c \in C$.

REMARK 1.3.10. *Recall that an inverse system $\{M_n; f_{nk}, k \geq n\}$ of R-modules satisfies the trivial Mittag-Leffler condition if and only if for every $n \geq 1$ there is $k \geq n$ such that $M_k \xrightarrow{f_{nk}} M_n$ is the zero map.*

COROLLARY 1.3.11. *Let R be a commutative ring and $s \in R$ be an element. Let C be an R-module of bounded s-torsion. Then*

$$\varprojlim\nolimits^1_{n\geq 1} {}_{s^n}C = 0, \qquad \Delta_s(C) \cong \Lambda_s(C).$$

PROOF. If C has bounded s-torsion, the inverse system (2) satisfies the trivial Mittag-Leffler condition, hence $\varprojlim^1_n {}_{s^n}C = 0$. □

2. Second lecture

Given a ring R, one can define (left) R-modules in the following fancy way. For any set X, let $R[X]$ denote the set of all finite formal linear combinations of the elements of X with coefficients in R. We have the obvious embedding $X \to R[X]$, and the opening of parentheses map $R[R[X]] \to R[X]$ which makes the functor $X \to R[X]$ a monad on the category of sets (see the definition below). The R-modules are the algebras/modules over this monad.

Recall that a **monad** on the category of sets is a covariant functor

$$\mathbb{T}\colon \mathsf{Sets} \to \mathsf{Sets}$$

endowed with natural transformations $\epsilon\colon \mathrm{Id} \longrightarrow \mathbb{T}$ and multiplication $\phi\colon \mathbb{T}\circ\mathbb{T} \longrightarrow \mathbb{T}$ satisfying the equations of *associativity*

$$\mathbb{T}\circ\mathbb{T}\circ\mathbb{T} \rightrightarrows \mathbb{T}\circ\mathbb{T} \to \mathbb{T}, \quad \phi(\phi\circ\mathbb{T}) = \phi(\mathbb{T}\circ\phi)$$

unitality

$$\mathbb{T} \rightrightarrows \mathbb{T}\circ\mathbb{T} \to \mathbb{T}, \quad \phi(\mathbb{T}\circ\epsilon) = \phi(\epsilon\circ\mathbb{T}) = \mathrm{Id}_{\mathsf{Sets}}.$$

2.1. Topological rings ([9, Section 1]). Let $\mathfrak{R}$ be a complete, separated topological ring with a base $\mathfrak{B}$ of neighbourhoods of zero formed by open right ideals $\mathfrak{I}$.

- Given a set X, denote by $\mathfrak{R}[[X]]$ the set of all infinite formal linear combinations $\sum_{x\in X} r_x x$ of elements of X with the coefficients in $\mathfrak{R}$ such that the *X-indexed family of elements r_x converges to zero in the topology of $\mathfrak{R}$.* This means that the set $\mathfrak{R}[[X]] \subset \mathfrak{R}^X$ consists of all the infinite formal linear combinations $\sum_{x\in X} r_x x$ such that, for every open right ideal $\mathfrak{I} \subset \mathfrak{R}$, $r_x \in \mathfrak{I}$ for *all but a finite number of indices $x \in X$.*
- In other words, $\mathfrak{R}[[X]] = \varprojlim_{\mathfrak{I}\in\mathfrak{B}} (\mathfrak{R}/\mathfrak{I})[X]$. $\mathfrak{R}[[X]]$ does not depend on the choice of the basis $\mathfrak{B}$.

- The map assigning to a set X the set $\mathfrak{R}[[X]]$ extends naturally to a covariant functor $\mathbb{T}_{\mathfrak{R}}$ from the category of sets to the category of sets. Given a map of sets $f\colon X \longrightarrow Y$, one defines the induced map $\mathfrak{R}[[f]]\colon \mathfrak{R}[[X]] \longrightarrow \mathfrak{R}[[Y]]$ by the rule $\sum_{x\in X} r_x x \longmapsto \sum_{y\in Y}\big(\sum_{f(x)=y} r_x\big)y$, where the sum of elements r_x in the parentheses is understood to be the limit of finite partial sums in the topology of $\mathfrak{R}$. Such a limit is unique and exists because the topological ring $\mathfrak{R}$ is separated and complete, while the family of elements $(r_x)_{x\in X}$, and consequently its subfamily indexed by all $x \in X$ with $f(x) = y$ for a fixed $y \in Y$, converges to zero in $\mathfrak{R}$.
- The functor $\mathbb{T}_{\mathfrak{R}}$ is endowed with natural transformations

$$\epsilon\colon \mathrm{Id} \longrightarrow \mathbb{T}_{\mathfrak{R}},\ \phi\colon \mathbb{T}_{\mathfrak{R}} \circ \mathbb{T}_{\mathfrak{R}} \longrightarrow \mathbb{T}_{\mathfrak{R}}$$

satisfying the monad equations.

The monad unit $\epsilon_X : X \longrightarrow \mathfrak{R}[[X]]$ is the "point measure" map, assigning to an element $x_0 \in X$ the (finite) formal linear combination $\sum_{x\in X} r_x x \in \mathfrak{R}[[X]]$, where $r_{x_0} = 1$ and $r_x = 0$ for all $x \neq x_0$.

The monad multiplication $\phi_X : \mathfrak{R}[[\mathfrak{R}[[X]]]] \longrightarrow \mathfrak{R}[[X]]$ is the "opening of parentheses" defined by

$$\sum_{y\in\mathfrak{R}[[X]]} r_y y \mapsto \sum_{x\in X}\left(\sum_{y\in\mathfrak{R}[[X]]} r_y r_{yx}\right) x, \qquad \text{where } y = \sum_{x\in X} r_{yx} x.$$

The sum $\sum_{y\in\mathfrak{R}[[X]]} r_y r_{yx}$ converges in $\mathfrak{R}$, since the family r_y converges to zero and $\mathfrak{R}$ is complete.

(For every open right ideal $\mathfrak{I}$, one has $r_y r_{yx} \in \mathfrak{I}$ whenever $r_y \in \mathfrak{I}$; and there is only a finite set of indices y with $r_y \notin \mathfrak{I}$, because $\sum_{y\in\mathfrak{R}[[X]]} r_y y \in \mathfrak{R}[[Y]]$, where $Y = \mathfrak{R}[[X]]$)

2.2. $\mathfrak{R}$-contramodules ([**9**, Section 1]. Recall that a module over a monad $\mathbb{T}\colon \mathsf{Sets} \to \mathsf{Sets}$ is a set C endowed with a map of sets $\pi_C\colon \mathbb{T}(C) \longrightarrow C$, called the *action map* satisfying the equations of associativity

$$\mathbb{T}(\mathbb{T}(C)) \rightrightarrows \mathbb{T}(C) \to C, \quad \pi_C \circ \phi_C = \pi_C \circ \mathbb{T}(\pi_C),$$

and unitality

$$C \to \mathbb{T}(C) \to C, \quad \pi_C \circ \epsilon_C = \mathrm{id}_C\,.$$

A left $\mathfrak{R}$-contramodule is a module over the monad $\mathbb{T}_{\mathfrak{R}}\colon X \longmapsto \mathfrak{R}[[X]]$ on the category of sets. This means that a left $\mathfrak{R}$-contramodule $\mathfrak{C}$ is a set endowed with a *left contraaction map* $\pi_{\mathfrak{C}}\colon \mathfrak{R}[[\mathfrak{C}]] \longrightarrow \mathfrak{C}$ satisfying the following conditions

$$\pi_{\mathfrak{C}} \circ \epsilon_{\mathfrak{C}} = \mathrm{id}_{\mathfrak{C}} : \quad \mathfrak{C} \xrightarrow{\epsilon_{\mathfrak{C}}} \mathfrak{R}[[\mathfrak{C}]] \xrightarrow{\pi_{\mathfrak{C}}} \mathfrak{C}$$

$$\pi_{\mathfrak{C}} \circ \phi_{\mathfrak{C}} = \pi_{\mathfrak{C}} \circ \mathfrak{R}[[\pi_{\mathfrak{C}}]] : \quad \mathfrak{R}[[\mathfrak{R}[[\mathfrak{C}]]]] \overset{\phi_{\mathfrak{C}}}{\underset{\mathfrak{R}[[\pi_{\mathfrak{C}}]]}{\rightrightarrows}} \mathfrak{R}[[\mathfrak{C}]] \xrightarrow{\pi_{\mathfrak{C}}} \mathfrak{C}$$

In other words, a left $\mathfrak{R}$-contramodule $\mathfrak{C}$ can be defined as a set endowed with the following infinite summation operations.

For any family of elements r_α converging to zero in $\mathfrak{R}$ and any family of elements $c_\alpha \in \mathfrak{C}$ there is a well-defined element $\sum_\alpha r_\alpha c_\alpha \in \mathfrak{C}$ satisfying

(a) *contraassociativity*

$$\sum_\alpha r_\alpha \sum_\beta r_{\alpha\beta} c_{\alpha\beta} = \sum_{\alpha,\beta} (r_\alpha r_{\alpha\beta}) c_{\alpha\beta} \text{ if } r_\alpha \to 0 \text{ and } r_{\alpha\beta} \to 0 \text{ in } \mathfrak{R}, \forall\alpha.$$

(b) the *contraunitality*:

$$\sum_{\alpha\in A} r_\alpha c_\alpha = c_{\alpha_0}, \text{ if } A=\{\alpha_0\} \text{ and } r_{\alpha_0}=1,$$

(c) *distributivity*

$$\sum_{\alpha,\beta} r_{\alpha\beta} c_\alpha = \sum_\alpha \left(\sum_\beta r_{\alpha\beta}\right) c_\alpha \quad \text{if } r_{\alpha\beta}\to 0 \in \mathfrak{R}.$$

(d) The finite and infinite operations are compatible.

A morphism of $\mathfrak{R}$-contramodules $\mathfrak{C}$ and $\mathfrak{D}$ is a map f such that the following diagram is commutative

$$\begin{array}{ccc} \mathfrak{R}[[\mathfrak{C}]] & \xrightarrow{\mathfrak{R}[[f]]} & \mathfrak{R}[[\mathfrak{D}]] \\ \downarrow{\scriptstyle \pi_{\mathfrak{C}}} & & \downarrow{\scriptstyle \pi_{\mathfrak{D}}} \\ \mathfrak{C} & \xrightarrow[f]{} & \mathfrak{D} \end{array}$$

We denote the category of left $\mathfrak{R}$-contramodules by $\mathfrak{R}$–contra.

2.3. The category $\mathfrak{R}$–contra ([9, Section 1], [15, Section 1, 5], [16, Section 6.4], [13, Section 1]).

(1) For any left $\mathfrak{R}$-contramodule $\mathfrak{C}$, let $\mathfrak{R}[\mathfrak{C}]$ be the set of all finite formal linear combinations. The composition

$$\mathfrak{R}[\mathfrak{C}] \longrightarrow \mathfrak{R}[[\mathfrak{C}]] \xrightarrow{\pi_{\mathfrak{C}}} \mathfrak{C}$$

defines a natural structure of a left $\mathfrak{R}$-module on $\mathfrak{C}$.

• In particular, it means that all left $\mathfrak{R}$-contramodules, which were originally defined as only sets endowed with a contraaction map, are actually abelian groups, even $\mathfrak{R}$-modules.

(2) Using the above identities (a)–(d), one can define the $\mathfrak{R}$-contramodule structure on the kernel and cokernel of an $\mathfrak{R}$-contramodule morphism taken in the category of $\mathfrak{R}$-modules. Hence $\mathfrak{R}$–contra is an abelian category and the forgetful functor $\mathfrak{R}$–contra $\longrightarrow$ $\mathfrak{R}$–Mod is an exact functor.

(3) For any set X the set $\mathfrak{R}[[X]]$ has a natural structure of an $\mathfrak{R}$-contramodule: The monad multiplication $\phi_X : \mathfrak{R}[[\mathfrak{R}[[X]]]] \longrightarrow \mathfrak{R}[[X]]$ plays the rôle of the contraaction map.

(4) Let $\mathbb{T}_{\mathfrak{R}}$ be the functor sending a set X to $\mathfrak{R}[[X]]$. Then $\mathbb{T}_{\mathfrak{R}}$ is left adjoint to the forgetful functor from $\mathfrak{R}$–contra to the category of sets.

$$\text{Sets} \underset{\text{Forget}}{\overset{\mathbb{T}_{\mathfrak{R}}}{\rightleftarrows}} \mathfrak{R}\text{–contra}$$

Hence, contramodules of the form $\mathfrak{R}[[X]]$ are the free $\mathfrak{R}$-contramodules. They are projective objects in the abelian category $\mathfrak{R}$–contra, there are enough of them, and hence every projective $\mathfrak{R}$-contramodule is a direct summand of a free $\mathfrak{R}$-contramodule.

(5) For any collection of sets X_α, the free contramodule $\mathfrak{R}[[\coprod_\alpha X_\alpha]]$ generated by the disjoint union of X_α is the direct sum of the free contramodules $\mathfrak{R}[[X_\alpha]]$ in the category $\mathfrak{R}$–contra. This allows to compute, the direct sum

of $\mathfrak{R}$-contramodules, by presenting them as cokernels of morphisms of free contramodules and using the fact that infinite direct sums commute with cokernels. So infinite direct sums exist in $\mathfrak{R}$–contra.

(6) The category $\mathfrak{R}$–contra is cocomplete, with a projective generator $\mathfrak{R} = \mathbb{T}_{\mathfrak{R}}(*)$, hence it is also complete. The forgetful functor $\mathfrak{R}\text{–contra} \longrightarrow \mathfrak{R}\text{–Mod}$ preserves infinite products (but not coproducts).

(7) The category $\mathfrak{R}$–contra has the additional property that for every family of projective objects $P_\alpha \in \mathfrak{R}$–contra, the natural morphism $\coprod_\alpha P_\alpha \to \prod_\alpha P_\alpha$ is a monomorphism. This is because the property can be checked for free $\mathfrak{R}$-contramodules.

(8) The monad $\mathbb{T}_{\mathfrak{R}}$ is additive meaning that the category of $\mathbb{T}_{\mathfrak{R}}$-modules is an additive category (see [**13**, Lemma 1.1]).

(9) (See below for the terminology) Let λ^+ be the successor cardinal of the cardinality of a base of neighbourhoods of zero in $\mathfrak{R}$. The monad $\mathbb{T}_{\mathfrak{R}}$ is λ^+-accessible, meaning that it preserves λ^+-filtered colimits.

So the abelian category $\mathfrak{R}$–contra is locally λ^+-presentable with a natural λ^+-presentable projective generator, which is the free left $\mathfrak{R}$-contramodule with one generator $\mathfrak{R} = \mathfrak{R}[[*]]$.

Recall:

- Let κ be a regular cardinal. A poset is κ-directed if every subset of cardinality smaller than κ has an upper bound. A colimit of a diagram indexed by a κ-directed poset is called κ-directed colimit.
- An object C of a category $\mathcal{C}$ is called κ-presentable if $\mathrm{Hom}_{\mathcal{C}}(C, -)$ preserves κ-directed colimits.
- A category is called *locally κ-presentable* if it is cocomplete and has a set $\mathcal{A}$ of κ-presentable objects such that every object is a κ-directed colimit of objects from $\mathcal{A}$.

PROPOSITION 2.3.1. [**15**, Section 1], [**16**, Section 6.4] *Every cocomplete abelian category $\mathcal{B}$ with a projective generator P is equivalent to the category of modules over the monad*

$$\mathbb{T}_P\colon \mathsf{Sets} \to \mathsf{Sets}; \quad X \mapsto \mathrm{Hom}_{\mathcal{B}}(P, P^{(X)}).$$

PROOF. (shortly) First we describe the monad multiplication

$$\mathbb{T}_P \circ \mathbb{T}_P(X) = \mathrm{Hom}_{\mathcal{B}}(P, P^{(\mathbb{T}(X))}) \to \mathrm{Hom}_{\mathcal{B}}(P, P^{(X)}) = \mathbb{T}_P(X).$$

Note that $\mathrm{Hom}_{\mathcal{B}}(P^{(Y)}, P^{(X)})$ is computed as $\mathrm{Hom}_{\mathcal{B}}(P, P^{(X)})^Y = \mathbb{T}_P(X)^Y$.

Let $f\colon Y \to \mathbb{T}_P(X)$ be viewed as a map $h\colon P^{(Y)} \to P^{(X)}$. Then

$$\mathrm{Hom}_{\mathcal{B}}(P, h)\colon \mathrm{Hom}_{\mathcal{B}}(P, P^{(Y)}) \to \mathrm{Hom}_{\mathcal{B}}(P, P^{(X)}).$$

Letting $Y = \mathbb{T}_P(X)$ and f the identity we get the monad multiplication.

The equivalence is given by assigning to every $N \in \mathcal{B}$ the set $\mathrm{Hom}_{\mathcal{B}}(P, N)$with its natural structure as a $\mathbb{T}_P$-module.

Indeed, every $t \in \mathbb{T}_P(X) = \mathrm{Hom}_{\mathcal{B}}(P, P^{(X)})$ induces an X-ary operation on $\mathrm{Hom}_{\mathcal{B}}(P, N)$:

$$\mathrm{Hom}_{\mathcal{B}}(P, N)^X = \mathrm{Hom}_{\mathcal{B}}(P^{(X)}, N) \to \mathrm{Hom}_{\mathcal{B}}(P, N),$$

$$(f\colon P^{(X)} \to N) \mapsto f \circ t.$$

□

- The functors of infinite direct sum are not exact in $\mathfrak{R}$–contra in general; they are not exact already for the ring $\mathfrak{R} = k[[z,t]]$ of formal power series in two variables over a field k.
- When $\mathfrak{R}$–contra has global homological dimension at most 1, the infinite direct sums in $\mathfrak{R}$–contra are exact.

2.4. The category of I-contramodules for a finitely generated ideal I of a commutative ring R ([11, Section 7]).

DEFINITION 2.4.1. [**11**, Section 7] Let R be a commutative ring and $I \leq R$ be the ideal generated by a finite set of elements $s_1, \dots, s_m \in R$. An R-module C is said to be an **I-contramodule** if $\mathrm{Hom}_R(R[s_j^{-1}], C) = 0 = \mathrm{Ext}^1_R(R[s_j^{-1}], C)$ for all $1 \leq j \leq m$.

This property does not depend on the chosen set of generators, but only on the ideal I.

Properties of the category of I-contramodules

Denote by $R\text{–}\mathsf{Mod}_{\mathbf{I-ctr}}$ the full subcategory of $R\text{–}\mathsf{Mod}$ consisting of I-contramodule R-modules.

(i) I-contramodules are the objects of the 0, 1-perpendicular category of $\oplus_{j=1}^m R[s_j^{-1}]$. Since the projective dimension of $R[s_j^{-1}]$ is at most one for every j, the category of I-contramodules is closed under kernels, cokernels, extensions and products in $R\text{–}\mathsf{Mod}$, i.e. it is exactly embedded in $R\text{–}\mathsf{Mod}$.

(ii) Every R-module annihilated by I^n for some $n \geq 1$ is an I-contramodule.

(iii) Let $E = \oplus_{j=1}^m R[s_j^{-1}]$, and let $f\colon U^{-1} \to U^0$ be a two terms free resolution of E. Then $R\text{–}\mathsf{Mod}_{\mathbf{I-ctr}}$ coincides with the full subcategory $f^{\perp}$, consisting of the R-modules C such that $\mathrm{Hom}_R(f, C)$ is an isomorphism.

THEOREM 2.4.2. [**11**, Theorem 7.2], [**13**, Example 2.2 (1)] *Let R be a commutative ring and $I \leq R$ be the ideal generated by a finite set of elements $s_1, \dots, s_m \in R$. The following hold:*

(1) *The exact embedding functor $R\text{–}\mathsf{Mod}_{\mathbf{I-ctr}} \to R\text{–}\mathsf{Mod}$ has a left adjoint functor $\Delta_I\colon R\text{–}\mathsf{Mod}_{\mathbf{I-ctr}} \to R\text{–}\mathsf{Mod}$ given by $\Delta_{s_m} \dots \Delta_{s_2}\Delta_{s_1}$ (Δ_{s_i} from First Lecture).*

(2) *$\Delta_I(R) = P$ is a projective generator of $R\text{–}\mathsf{Mod}_{\mathbf{I-ctr}}$ and the coproduct of X copies of P in $R\text{–}\mathsf{Mod}_{\mathbf{I-ctr}}$ is given by $\Delta_I(R[X])$.*

(3) *The abelian category $R\text{–}\mathsf{Mod}_{\mathbf{I-ctr}}$ is equivalent to the category of modules over the additive monad $\mathbb{T}_I$ assigning to every set X the underlying set of the R-module $\Delta_I(R[X])$.*

PROOF. (1) is a generalization of the 1-element case: Δ_s is left adjoint to the exact embedding $R\text{–}\mathsf{Mod}_{\mathbf{s-ctr}} \to R\text{–}\mathsf{Mod}$.

(2) Follows from the fact that Δ_I is left adjoint to an exact functor (see also Remark 1.3.8).

(3) Let $\mathcal{B} = R\text{–}\mathsf{Mod}_{\mathbf{I-ctr}}$. By (1) and (2) $\mathcal{B}$ is a cocomplete category with a projective generator $P = \Delta_I(R)$. By Proposition 2.3.1, $\mathcal{B}$ is equivalent to the category of modules over the monad $\mathbb{T}_P$.

Now $\mathbb{T}_P(X) = \mathrm{Hom}_{\mathcal{B}}(P, P^{(X)}) \cong \mathrm{Hom}_R(R, \Delta_I(R[X])) \cong \Delta_I(R[X])$, thus $\mathbb{T}_P \cong \mathbb{T}_I$. □

The functor Δ_I and the I-adic completion functor.
Fix a finitely generated ideal I of a commutative ring R.

NOTATION 2.4.3.

(1) *For every R-module C denote by $\Lambda_I(C) = \varprojlim_n C/I^nC$ the I-adic completion of C. By* [**11**, Lemma 5.7 and Theorem 5.8] *$\Lambda_I(C)$ is an I-contramodule and it is separated and complete in the I-adic topology which coincides with the projective limit topology.*

(2) *Denote by $\mathfrak{R} = \varprojlim_n R/I^n$ the I-adic completion of the ring R endowed with the I-adic topology.*

We generalize the complex $T^\bullet(s)$, quasi isomorphic to $R \to R[s^{-1}]$ and its subcomplex $T^\bullet_n(s)$, quasi isomorphic to $R \xrightarrow{s^n} R$ from the first lecture, by setting

(3)

$$(a)\ T^\bullet(s_1, \dots, s_m) = T^\bullet(s_1) \otimes_R \cdots \otimes_R T^\bullet(s_m)$$

in cohomological degrees $0, \dots, m$, quasi isomorphic to

$$(R \to R[s_1^{-1}]) \otimes_R \cdots \otimes (R \to R[s_m^{-1}]),\ \text{and}$$

$$(b)\ T^\bullet_n(s_1, \dots, s_m) = T^\bullet_n(s_1) \otimes_R \cdots \otimes_R T^\bullet_n(s_m),$$

(4) *then, $T^\bullet(s_1, \dots, s_m) = \varinjlim_{n\geq 1} T^\bullet_n(s_1, \dots, s_m)$, and*

$$\Delta_I(C) = H^0(\mathrm{Hom}_R(T^\bullet(s_1, \dots, s_m), C).$$

PROPOSITION 2.4.4. [**11**, Lemma7.5] *In Notations* 2.4.3 *there is a natural short exact sequence of R-modules:*

$$0 \to \varprojlim\nolimits^1_{n\geq 1} H_1(\mathrm{Hom}_R(T^\bullet_n(s_1, \dots, s_m), C) \to \Delta_I(C) \to \Lambda_I(C) \to 0.$$

In particular, for every R-module C, there is a natural surjection

$$\Delta_I(C) \to \Lambda_I(C).$$

PROOF. The complexes $\mathrm{Hom}_R(T^\bullet_n(s_1, \dots, s_m), C)$ form a countable projective system with surjective maps in each degree. Hence there is a short exact sequence

$$0 \to \varprojlim\nolimits^1_{n\geq 1} H_1(\mathrm{Hom}_R(T^\bullet_n(s_1, \dots, s_m), C)) \to H_0(\varprojlim_{n\geq 1} \mathrm{Hom}_R(T^\bullet_n(s_1, \dots, s_m), C))$$

$$\to \varprojlim_{n\geq 1} H_0(\mathrm{Hom}_R(T^\bullet_n(s_1, \dots, s_m), C)) \to 0.$$

Now we have:

- $H_0(\mathrm{Hom}_R(T^\bullet_n(s_1, \dots, s_m), C)) \cong C/(s_1^n, \dots, s_m^n)C$, hence $\varprojlim_{n\geq 1} H_0(\mathrm{Hom}_R(T^\bullet_n(s_1, \dots, s_m), C)) \cong \Lambda_I(C)$,
- $H_0(\varprojlim_{n\geq 1} \mathrm{Hom}_R(T^\bullet_n(s_1, \dots, s_m), C)) \cong H_0(\mathrm{Hom}_R(T^\bullet(s_1, \dots, s_m), C)) \cong \Delta_I(C)$. □

PROPOSITION 2.4.5. [**13**, Example 2.2 (2)] *In Notations* 2.4.3, *the forgetful functor $\mathfrak{R}$–contra $\to R$–Mod is fully faithful and its image is contained in the full subcategory of I-contramodule R-modules.*

PROOF. For the claim about the image it suffices to check that the free $\mathfrak{R}$-contramodules are I-contramodule R-modules, as every $\mathfrak{R}$-contramodule is the cokernel of a morphism of free $\mathfrak{R}$-contramodules and $R\text{–}\mathsf{Mod}_{\mathbf{I-ctr}}$ is closed under cokernels. For any set X, the free $\mathfrak{R}$-contramodule $\mathfrak{R}[[X]]$ coincides with $\varprojlim_{n\geq 1} R/I^n[X]$ which is an I-contramodule by Notation 2.4.3 (1).

Fully faithfulness: the abelian category $\mathfrak{R}\text{–}\mathsf{contra}$ is the category of modules over the monad $\mathbb{T}_{\mathfrak{R}}\colon X \to \mathfrak{R}[[X]]$ and

$$\mathfrak{R}[[X]] = \varprojlim_{n\geq 1}(R/I^nR)[X] = \Lambda_I(R[X]),$$

while the abelian category $R\text{–}\mathsf{Mod}_{\mathbf{I-ctr}}$ is the category of modules over the monad $X \to \Delta_I(R[X])$, by Theorem 2.4.2 (3).

The functor $\mathfrak{R}\text{–}\mathsf{contra} \to R\text{–}\mathsf{Mod}_{\mathbf{I-ctr}}$ is induced by the morphism of monads $\Delta_I(R[X]) \to \Lambda_I(R[X])$, and surjectivity of this map for every set X implies that the forgetful functor is fully faithful.

The idea is that the forgetful functor can be seen as a "restriction functor" by means of the surjection $\Delta_I(R[X]) \to \Lambda_I(R[X])$. □

Consequence: The abelian category $\mathfrak{R}\text{–}\mathsf{contra}$ is a full subcategory of the abelian category $R\text{–}\mathsf{Mod}_{\mathbf{I-ctr}}$.

2.5. When is $\mathfrak{R}\text{–}\mathsf{contra}$ equivalent to $R\text{–}\mathsf{Mod}_{\mathbf{I-ctr}}$?

PROPOSITION 2.5.1. [**13**, Proposition 2.1] *The forgetful functor* $\mathfrak{R}\text{–}\mathsf{contra} \to R\text{–}\mathsf{Mod}_{\mathbf{I-ctr}}$ *is an equivalence of abelian categories if and only if the natural morphism*

$$\Delta_I(R[X]) \to \Lambda_I(R[X])$$

is an isomorphism for every set X.

By Proposition 2.4.4, the kernel of the natural morphism in the above statement is

$$\varprojlim\nolimits^1_{n\geq 1} H_1(\mathrm{Hom}_R(T_n^\bullet(s_1,\dots,s_m), R[X]),$$

and a sufficient condition for the kernel to be zero is that I is **weakly proregular**.

We follow the notations in [**19**]. For any sequence $\mathbf{s} = (s_1, s_2, \dots, s_m)$ of elements of R denote by $K^\bullet(\mathbf{s})$ the Koszul complex $K^\bullet(s_1, s_2, \dots, s_m)$ in degrees $-m, \dots, 0$. Moreover, for every $n \geq 1$, write $\mathbf{s}^n = (s_1^n, s_2^n, \dots, s_m^n)$ and denote by $K^\bullet(\mathbf{s}^n)$ the Koszul complex $K^\bullet(s_1^n, s_2^n, \dots, s_m^n)$. Then for every $k \geq n$, there are morphisms $K^\bullet(\mathbf{s}^k) \to K^\bullet(\mathbf{s}^n)$.

DEFINITION 2.5.2. [**18**], [**19**] An ideal $I = (s_1, \dots, s_m)$ is **weakly proregular** if the inverse systems of the Koszul cohomology modules $\{H^i(K^\bullet(\mathbf{s}^n))\}_{n\geq 1}$ are pro-zero for every $i = -m, \dots, -1$ meaning that they satisfy the trivial Mittag-Leffler condition *(see Remark* 1.3.10*)*.

- If R is noetherian every finitely generated ideal is weakly proregular.
- If $I = (s)$ then I is weakly proregular if and only if the s-torsion of R is bounded, i.e. there is $n \geq 1$ such that the s-torsion submodule of R is annihilated by s^n.

By the remarks above, we conclude that

PROPOSITION 2.5.3. *The forgetful functor* $\mathfrak{R}\text{–}\mathsf{contra} \to R\text{–}\mathsf{Mod}_{\mathbf{I-ctr}}$ *is an equivalence of abelian categories for any weakly proregular finitely generated ideal I in a commutative ring R.*

2.6. The case of a multiplicative subset S of R with p.dim $R_S \leq 1$ ([12], [13, Section 2]). Let R be a commutative ring and S a multiplicative subset of R such that the projective dimension (p.dim) of the localization R_S is at most one.

An R-module M is S-torsion if for every $x \in M$ there is $s \in S$ such that $xs = 0$ (i.e. $M \otimes_R R_S = 0$) and it is S-divisible if $Ms = M$ for every $s \in S$.

DEFINITION 2.6.1. An R-module C is said to be an **S-contramodule** if

$$\mathrm{Hom}_R(R_S, C) = 0 = \mathrm{Ext}^1_R(R_S, C).$$

Denote by $R\text{–}\mathsf{Mod}_{\mathbf{S-ctr}}$ the full subcategory of $R\text{–}\mathsf{Mod}$ consisting of S-contramodule R-modules.

- Let $K^\bullet_S$ be the complex $R \overset{l_S}{\to} R_S$ in cohomological degrees $-1, 0$.
- For every R-module C, let $\Delta_S(C) = \mathrm{Hom}_{\mathcal{D}^b(R)}(K^\bullet_S, C[1])$.
- A generalization of Lemma 1.2.1 shows that $\Delta_S(C)$ is an S-contramodule, since the modules $H^n(K^\bullet)$ are S-torsion for every $n \in \mathbb{Z}$.

PROPOSITION 2.6.2. [13, Example 2.4]

(1) *The category $R\text{–}\mathsf{Mod}_{\mathbf{S-ctr}}$ is an abelian category exactly embedded in $R\text{–}\mathsf{Mod}$.*
(2) *The functor Δ_S is left adjoint to the embedding functor $R\text{–}\mathsf{Mod}_{\mathbf{S-ctr}} \to R\text{–}\mathsf{Mod}$, hence $\Delta_S(R)$ is a projective generator of $R\text{–}\mathsf{Mod}_{\mathbf{S-ctr}}$.*
(3) *The abelian category $R\text{–}\mathsf{Mod}_{\mathbf{S-ctr}}$ is equivalent to the category of modules over the additive monad $\mathbb{T}$ assigning $\Delta_S(R[X])$ to every set X.*

PROOF. (1) follows from the assumption p.dim $R_S \leq 1$.
(2) From the triangle $R \to R_S \to K^\bullet_S \to R[1]$ we get the exact sequence

$$0 \to \mathrm{Hom}_R(\mathrm{Coker}\, l_S, C) \cong \mathrm{Hom}_{\mathcal{D}^b(R)}(K^\bullet_S, C) \to \mathrm{Hom}_R(R_S, C) \to C \overset{\delta^S_C}{\to}$$

$$\overset{\delta^S_C}{\to} \mathrm{Hom}_{\mathcal{D}^b(R)}(K^\bullet_S, C[1]) = \Delta_S(C) \to \mathrm{Ext}^1_R(R_S, C) \to 0,$$

where $\mathrm{Ker}\, \delta^S_C$ is an S-divisible R-module while $\mathrm{Coker}\, \delta^S_C \cong \mathrm{Ext}^1_R(R_S, C)$ is an R_S-module.

Then the proof continues arguing similarly to the case of the localization $R[s^{-1}]$ at a single element $s \in R$ (see Proposition 1.3.7).

(3) We argue as in Theorem 2.4.2. Let $\mathcal{B} = R\text{–}\mathsf{Mod}_{\mathbf{S-ctr}}$. By (2) $\mathcal{B}$ is a cocomplete category with a projective generator $P = \Delta_S(R)$, hence by Proposition 2.3.1 $\mathcal{B}$ is equivalent to the category of modules over the monad $\mathbb{T}_P \colon X \to \mathrm{Hom}_{\mathcal{B}}(P, P^{(X)})$.
Now $\mathrm{Hom}_{\mathcal{B}}(P, P^{(X)}) \cong \mathrm{Hom}_R(R, \Delta_S(R[X])) \cong \Delta_S(R[X])$. □

Denote by $\mathfrak{R}$ the ring $\varprojlim_{s \in S} R/sR$, the S-completion of the ring R.

(i) $\mathfrak{R}$ endowed with the **projective limit topology** is a complete, separated topological commutative ring.
(ii) For every R-module C, let $\Lambda_S(C) = \varprojlim_{s \in S} C/sC$ and let $\lambda_S(C) \colon C \to \varprojlim_{s \in S} C/sC$ the canonical map. The following hold:
   (1) $\Lambda_S(C)$ is an S-contramodule (use that C/sC is an S-contramodule for every $s \in S$ and the closure properties of $R\text{–}\mathsf{Mod}_{\mathbf{S-ctr}}$).

(2) For every R module C there is a unique R-module morphism $\beta_S(C)\colon \Delta_S(C) \to \Lambda_S(C)$ forming a commutative diagram

$$\begin{array}{ccc} C & \xrightarrow{\delta_C^S} & \Delta_S(C) \\ & {\scriptstyle\lambda_C^S}\searrow & \Big\downarrow {\scriptstyle\beta_S(C)} \\ & & \Lambda_S(C) \end{array}$$

(analogously to the 1-element case $s \in R$, see the proof of Proposition 1.3.7 and [**12**, Lemma 2.1 (b)]).

PROPOSITION 2.6.3. [**12**, Theorem 2.5] *The natural morphism*

$$\beta_S(C)\colon \Delta_S(C) \to \Lambda_S(C)$$

is an isomorphism provided that the S-torsion of C is bounded, i.e. the S-torsion submodule of C is annihilated by some element $s \in S$.

THEOREM 2.6.4. [**13**, Example 2.4]

(1) *The image of the forgetful functor $\mathfrak{R}$–$\mathsf{contra} \to R$–Mod is contained in the full subcategory R–$\mathsf{Mod}_{\mathbf{S-ctr}}$.*
(2) *The forgetful functor $\mathfrak{R}$–$\mathsf{contra} \to R$–$\mathsf{Mod}_{\mathbf{S-ctr}}$ is an equivalence of categories if and only if the natural morphism $\Delta_S(R[X]) \to \Lambda_S(R[X])$ is an isomorphism for every set X.*

PROOF. (1) It suffices to check that the free $\mathfrak{R}$-contramodules are S-contramodule R-modules, since every $\mathfrak{R}$-contramodule is the cokernel of a morphism of free $\mathfrak{R}$-contramodules and R–$\mathsf{Mod}_{\mathbf{S-ctr}}$ is closed under cokernels.

For any set X, the free $\mathfrak{R}$-contramodule $\mathfrak{R}[[X]] = \varprojlim_{s\in S} R/sR[X]$ is an S-contramodule R-module (by (ii) (1).)

(2) Follows from [**13**, Proposition 2.1]. □

Combining Proposition 2.6.3 with Theorem 2.6.4 (2) we obtain:

THEOREM 2.6.5. *If the S-torsion of R is bounded, the forgetful functor $\mathfrak{R}$–contra $\to R$–Mod is an equivalence of categories.*

3. Third lecture

3.1. n-tilting objects.

NOTATION 3.1.1. *Let $\mathcal{A}$ be an abelian category with coproducts, and $\mathcal{B}$ be an abelian category with products. For any object $T \in \mathcal{A}$ we denote by $\mathsf{Add}(T) = \mathsf{Add}_{\mathcal{A}}(T) \subset \mathcal{A}$ the class of all direct summands of the coproducts $T^{(X)}$ of copies of T in $\mathcal{A}$. For any object $W \in \mathcal{B}$ we denote by $\mathsf{Prod}(W) = \mathsf{Prod}_{\mathcal{B}}(W) \subset \mathcal{B}$ the class of all direct summands of the products W^X of copies of W in $\mathcal{B}$.*

In the sequel $\mathcal{A}$ will be a complete, cocomplete abelian category with an injective cogenerator.

DEFINITION 3.1.2. [**16**, Pages 5-6] Let $n \geq 0$. An object $T \in \mathcal{A}$ is an n-**tilting** object if

(i) the projective dimension of T is at most n, that is $\mathrm{Ext}^i_{\mathcal{A}}(T, A) = 0$ for all $A \in \mathcal{A}$ and $i > n$;

(ii) $\mathrm{Ext}^i_{\mathcal{A}}(T, T^{(X)}) = 0$ for all $i > 0$, for any set X;
(iii) every $X^\bullet \in \mathcal{D}(\mathcal{A})$ such that $\mathrm{Hom}_{\mathcal{D}(\mathcal{A})}(T, X[i]) = 0$ for all $i \in \mathbb{Z}$ is acyclic.

Assume that T satisfies (i) and (ii), let

$$\mathcal{E} = \{E \in \mathcal{A} \mid \mathrm{Ext}^i_{\mathcal{A}}(T, E) = 0, \text{ for all } i > 0\},$$

then, $\mathcal{A}_{\mathsf{inj}}$, the class of injective objects of $\mathcal{A}$, is contained in $\mathcal{E}$ and by the condition (ii), $\mathsf{Add}_{\mathcal{A}}(T) \subset \mathcal{E}$.

For each integer $m \geq 0$ and every n-tilting object T, let

$$\mathcal{L}_m = \{L \in \mathcal{A} \mid \exists \text{ an exact sequence } 0 \to L \to T^0 \to T^1 \to \cdots \to T^m \to 0\},$$

with the objects $T^i \in \mathsf{Add}(T)$, for every $i = 0, \dots, m$. The following hold ([**16**, Lemma 2.2])

- $\mathsf{Add}(T) = \mathcal{L}_0 \subset \mathcal{L}_1 \subset \mathcal{L}_2 \subset \dots$ and $\mathcal{L}_n = \mathcal{L}_{n+1} = \mathcal{L}_{n+2} = \cdots$.
- Set $\mathcal{L} = \mathcal{L}_n$,
  then $\mathrm{Ext}^i_{\mathcal{A}}(L, E) = 0$ for every $i > 0$, for every $L \in \mathcal{L}$, $E \in \mathcal{E}$.
- $\mathcal{L} \cap \mathcal{E} = \mathsf{Add}(T) \subset \mathcal{A}$.

THEOREM 3.1.3. *Assume that an object $T \in \mathcal{A}$ satisfies (i) and (ii) of Definition* 3.1.2. *Then T satisfies also (iii) if and only if every object of $\mathcal{E}$ is a quotient of an object from* $\mathsf{Add}(T)$ *in $\mathcal{A}$ if and only if every object of $\mathcal{A}$ is a quotient of an object from $\mathcal{L}$. (see* [**16**, Theorem 2.4]*).*

Assume that T is an n-tilting object in $\mathcal{A}$. Then

(1) [**16**, Theorem 2.4] *The pair of classes $(\mathcal{L}, \mathcal{E})$ in $\mathcal{A}$ is a hereditary complete cotorsion pair called the n-***tilting cotorsion pair** *associated with T.*
In particular, $\mathcal{E}$ is a coresolving class in $\mathcal{A}$, meaning that it is closed under summands, extensions and cokernels of monomorphisms in $\mathcal{A}$.
(2) [**16**, Lemma 4.1] *$\mathcal{E}$ consists of the objects E such that there is an exact sequence:*
$$T^{(I_n)} \to \dots T^{(I_1)} \to E \to 0,$$
for some sets $I_1, \dots, I_n$.
(3) [**16**, Lemma 4.1] *$\mathcal{E}$ is closed under coproducts.*

REMARK 3.1.4. *If $\mathcal{A}$ is the category A–*Mod *for an associative ring A, an n-tilting object T is exactly an n-tilting module, i.e. it satisfies (i) and (ii) and the ring A has a finite coresolution in* $\mathsf{Add}T$*:*

$$0 \to A \to T^0 \to \dots \ T^n \to 0.$$

THEOREM 3.1.5. [**16**, Theorem 1.3 and Corollary 1.4] *Let $T \in \mathcal{A}$ be an n-tilting object. Then the pair of full subcategories*

$${}^T\mathcal{D}^{\leq 0} = \{X^\bullet \in \mathcal{D}(\mathcal{A}) \mid \mathrm{Hom}_{\mathcal{D}(\mathcal{A})}(T, X^\bullet[i]) = 0, \textit{ for all } i > 0\},$$

$${}^T\mathcal{D}^{\geq 0} = \{X^\bullet \in \mathcal{D}(\mathcal{A}) \mid \mathrm{Hom}_{\mathcal{D}(\mathcal{A})}(T, X^\bullet[i]) = 0, \textit{ for all } i < 0\},$$

is a t-structure on the unbounded derived category $\mathcal{D}(\mathcal{A})$, called the **tilting** *t-***structure***.*

Moreover, the pair $({}^T\mathcal{D}^{b,\leq 0}, {}^T\mathcal{D}^{b,\geq 0})$, restriction of the tilting t-structure to the bounded derived category $\mathcal{D}^b(\mathcal{A})$ of $\mathcal{A}$, is a t-structure in $\mathcal{D}^b(\mathcal{A})$.

PROPOSITION 3.1.6. [**16**, Proposition 1.5] *Assume that T satisfies (i) and (ii) of Definition* 3.1.2 *and that $({}^T\mathcal{D}^{b,\leq 0}, {}^T\mathcal{D}^{b,\geq 0})$ is a t-structure on $\mathcal{D}^b(\mathcal{A})$. Then T satisfies also condition (iii).*

PROPOSITION 3.1.7. [**16**, Proposition 1.6] *Let*

$$\mathcal{B} = {}^T\mathcal{D}^{\leq 0} \cap {}^T\mathcal{D}^{\geq 0} = \{X^\bullet \in \mathcal{D}(\mathcal{A}) \mid \mathrm{Hom}_{\mathcal{A}}(T, X^\bullet[i]) = 0, \forall i \neq 0\}$$

be the **heart** *of the tilting t-structure.*

(1) *The n-tilting object $T \in \mathcal{A} \subseteq \mathcal{D}(\mathcal{A})$ belongs to $\mathcal{B}$ and is a projective generator of $\mathcal{B}$. $\mathcal{B}$ has coproducts and the projective objects of $\mathcal{B}$ are the summands of coproducts of copies of T.*
(2) *The subcategory $\mathcal{E} \subseteq \mathcal{A}$ can be described as $\mathcal{A} \cap \mathcal{B}$, the intersection of the hearts of the standard t-structure and the n-tilting t-structure on $\mathcal{D}(\mathcal{A})$.*

3.2. n-cotilting objects.

DEFINITION 3.2.1. [**16**, Pag 18] Let $\mathcal{B}$ be a complete, cocomplete abelian category with a projective generator $P \in \mathcal{B}$. Let $n \geq 0$. An object $W \in \mathcal{B}$ is an **n-cotilting** object if W^{op} is n-tilting in the abelian category $\mathcal{B}^{\mathrm{op}}$, that is:

(i*) the injective dimension of W is at most n, that is $\mathrm{Ext}^i_{\mathcal{B}}(B, W) = 0$ for all $B \in \mathcal{B}$ and $i > n$;
(ii*) $\mathrm{Ext}^i_{\mathcal{B}}(W^X, W) = 0$ for all $i > 0$, for any set X;
(iii*) every $Y^\bullet \in \mathcal{D}(\mathcal{B})$ such that $\mathrm{Hom}_{\mathcal{D}(\mathcal{B})}(Y, W[i]) = 0$ for all $i \in \mathbb{Z}$ is acyclic.

All the statements about an n-tilting object in $\mathcal{A}$ dualize for an n-cotilting object in $\mathcal{B}$. In particular:
Assume that W is an n-cotilting object in $\mathcal{B}$, and let

$$\mathcal{F} = \{F \in \mathcal{B} \mid \mathrm{Ext}^i_{\mathcal{B}}(F, W) = 0, \text{ for all } i > 0,\}$$

- $\mathcal{B}_{\mathsf{proj}}$, the class of projective objects of $\mathcal{B}$, is contained in $\mathcal{F}$ and by condition (ii^*), $\mathsf{Prod}_{\mathcal{B}}(W) \subset \mathcal{F}$.
- $\mathcal{F}$ consists of the objects $F \in \mathcal{B}$ such that there is a exact sequence
$$0 \to F \to W^{I_1} \to \cdots \to W^{I_n},$$
for some sets $I_1, \ldots, I_n$.
- There is a class $\mathcal{R}$ such that the pair $(\mathcal{F}, \mathcal{R})$ in $\mathcal{B}$ is a hereditary complete cotorsion pair, called the **n-cotilting cotorsion pair** associated to W.
  In particular, $\mathcal{F}$ is a resolving class in $\mathcal{B}$, meaning that it is closed under summands, extensions and kernels of epimorphisms in $\mathcal{B}$.
- The class $\mathcal{F}$ is closed under under products.

THEOREM 3.2.2. [**16**, Theorem 3.3 and Corollary 3.4] *Let $\mathcal{B}$ be a complete, cocomplete abelian category with a projective generator $P \in \mathcal{B}$ and an n-cotilting object W. Then the pair of full subcategories*

$${}^W\mathcal{D}^{\leq 0} = \{Y^\bullet \in \mathcal{D}(\mathcal{B}) \mid \mathrm{Hom}_{\mathcal{D}(\mathcal{B})}(Y^\bullet, W[i]) = 0, \textit{ for all } i < 0\},$$

$${}^W\mathcal{D}^{\geq 0} = \{Y^\bullet \in \mathcal{D}(\mathcal{B}) \mid \mathrm{Hom}_{\mathcal{D}(\mathcal{B})}(Y^\bullet, W[i]) = 0, \textit{ for all } i > 0\},$$

is a t-structure on the unbounded derived category $\mathcal{D}(\mathcal{B})$, called the **cotilting t-structure.**

Moreover, the pair $({}^W\mathcal{D}^{b,\leq 0}, {}^W\mathcal{D}^{b,\geq 0})$, restriction of the cotilting t-structure to the bounded derived category $\mathcal{D}^b(\mathcal{B})$ of $\mathcal{B}$ is a t-structure in $\mathcal{D}^b(\mathcal{B})$.

PROPOSITION 3.2.3. [**16**, Proposition 3.8] *Let W be an n-cotilting object in $\mathcal{B}$ and $\mathcal{A} = {}^W\mathcal{D}^{\leq 0} \cap {}^W\mathcal{D}^{\geq 0}$ be the* **heart** *of the cotilting t-structure on $\mathcal{D}(\mathcal{B})$.*

(1) *$W \in \mathcal{B} \subseteq \mathcal{D}(\mathcal{B})$ belongs to $\mathcal{A}$ and is an injective cogenerator of $\mathcal{A}$. $\mathcal{A}$ has products and the injective objects of $\mathcal{A}$ are the summands of products of copies of W.*

(2) *The subcategory $\mathcal{F} \subseteq \mathcal{B}$ can be described as $\mathcal{B} \cap \mathcal{A}$, the intersection of the hearts of the standard t-structure and the n-cotilting t-structure on $\mathcal{D}(\mathcal{B})$.*

3.3. The tilting-cotilting correspondence.

THEOREM 3.3.1. [**16**, Theorem 3.10 and Theorem 3.11] *Let $\mathcal{A}$ be a complete, cocomplete abelian category with an injective cogenerator W and an n-tilting object T and let $\mathcal{B} = {}^T\mathcal{D}^{\leq 0} \cap {}^T\mathcal{D}^{\geq 0}$ be the heart of the tilting t-structure.*

(1) *The object $W \in \mathcal{B} \subseteq \mathcal{D}(\mathcal{A})$ is an n-cotilting object in the abelian category $\mathcal{B}$.*

Let $\mathcal{B}$ be a complete, cocomplete abelian category with a projective generator T and an n-cotilting object W and let $\mathcal{A} = {}^W\mathcal{D}^{\leq 0} \cap {}^W\mathcal{D}^{\geq 0}$ be the heart of the cotilting t-structure.

(2) *The object $T \in \mathcal{A} \subseteq \mathcal{D}(\mathcal{B})$ is an n-tilting object in the abelian category $\mathcal{A}$.*

The previous results can be summarized as follows.

THEOREM 3.3.2. [**16**, Theorem 3.10 and 3.11], [**3**, Theorem 12.1] *There is a bijective correspondence between complete, cocomplete abelian categories $\mathcal{A}$ with an injective cogenerator J and an n-tilting object $T \in \mathcal{A}$, and complete, cocomplete abelian categories $\mathcal{B}$ with a projective generator P and an n-cotilting object $W \in \mathcal{B}$.*

(1) *The correspondence is given by a pair of adjoint functors (Φ, Ψ)*

$$\Phi\colon \mathcal{B} \longrightarrow \mathcal{A}; \quad \Psi\colon \mathcal{A} \to \mathcal{B},$$

obtained respectively from the truncation functors with respect to the cotilting t-structure on $\mathcal{D}(\mathcal{B})$ and the tilting t-structure on $\mathcal{D}(\mathcal{A})$.

(2) *The restrictions of Φ and Ψ between $\mathcal{F} \subset \mathcal{B}$ and $\mathcal{E} \subset \mathcal{A}$ are mutually inverse equivalences, as exact categories.*

(3) *Under the equivalence $\mathcal{E} \cong \mathcal{F}$, the injective cogenerator $J \in \mathcal{E} \subset \mathcal{A}$ corresponds to the n-cotilting object $W \in \mathcal{F} \subset \mathcal{B}$, and the n-tilting object $T \in \mathcal{E} \subset \mathcal{A}$ corresponds to the projective generator $P \in \mathcal{F} \subset \mathcal{B}$.*

(4) [**16**, Theorem 4.5] *The exact embedding $\mathcal{E} \hookrightarrow \mathcal{A}$ induces a triangle equivalence $\mathcal{D}(\mathcal{E}) \cong \mathcal{D}(\mathcal{A})$ and the exact embedding $\mathcal{F} \hookrightarrow \mathcal{B}$ induces a triangle equivalence $\mathcal{D}(\mathcal{F}) \cong \mathcal{D}(\mathcal{B})$. Hence there is a triangle equivalence $\mathcal{D}(\mathcal{A}) \cong \mathcal{D}(\mathcal{B})$.*

3.4. Applications to full subcategories of a module category.

PROPOSITION 3.4.1. [**16**, Theorem 7.1] *Let R be an associative ring and M a left R-module. The category $\mathsf{Add}(M)$ is equivalent to the category of projective left $\mathfrak{R}$-contramodules where $\mathfrak{R}$ is the complete, separated topological ring $\mathrm{Hom}_R(M, M)^{op}$ with a basis of neighbourhoods of zero formed by right ideals.*

PROOF. Consider the ring $\mathrm{Hom}_R(M, M)$ with the finite topology, that is the topology in which the base of neighbourhoods of zero is formed by the annihilator ideals $\mathrm{Ann}(F)$ of the finitely generated R-submodules $F \subseteq M$.

$\mathrm{Ann}(F) \cong \mathrm{Hom}_R(M/F, M)$ and $\mathrm{Hom}_R(M, M)/\mathrm{Ann}(F)$ is the set of all morphisms $F \to M$ that can be extended to morphisms $M \to M$.

A morphism $f\colon M \to M$ is given by a compatible system of R-module morphisms $F \to M$, for all the finitely generated submodules $F \subseteq M$, hence we have an isomorphism

$$\mathrm{Hom}_R(M,M) \cong \varprojlim_{F\subseteq M} \frac{\mathrm{Hom}_R(M,M)}{\mathrm{Ann}(F)}.$$

Let $\mathfrak{R} = \mathrm{Hom}_R(M,M)^{op}$. $\mathfrak{R}$ is a complete, separated topological ring with a base of the topology formed by open right ideals.

We have the two monads

$$\mathbb{T}_M\colon \mathsf{Sets} \to \mathsf{Sets}, \quad X \mapsto \mathrm{Hom}_R(M, M^{(X)})$$

$$\mathbb{T}_{\mathfrak{R}}\colon \mathsf{Sets} \to \mathsf{Sets}, \quad X \mapsto \mathfrak{R}[[X]],$$

where $\mathrm{Hom}_R(M,M^{(X)}) \subset \mathrm{Hom}_R(M,M)^X$ and $\mathfrak{R}[[X]] \subset \mathfrak{R}^X$.

We claim that $\mathbb{T}_M \cong \mathbb{T}_{\mathfrak{R}}$, that is $\mathrm{Hom}_R(M,M^{(X)}) = \mathfrak{R}[[X]]$. In fact, an X-indexed family of morphisms $g_x\colon M \to M$ corresponds to a morphism $M \to M^{(X)}$ if and only if for every finitely generated submodule F of M, $g_x(F) = 0$ for all but finitely many indices $x \in X$, that is, if and only if it converges to zero in the topology of $\mathrm{Hom}_R(M,M)$, if and only if $\{g_x\}_{x\in X} \in \mathfrak{R}[[X]]$.

The functor

$$\mathrm{Hom}_R(M,-)\colon R\text{–}\mathsf{Mod} \to \mathfrak{R}\text{–}\mathsf{contra}$$

sends $M^{(X)}$ to the free $\mathfrak{R}$-contramodule $\mathfrak{R}[[X]]$ and induces an equivalence between $\mathsf{Add}M$ and the full subcategory of projective $\mathfrak{R}$-contramodules. □

REMARK 3.4.2. *More explicitly: For any associative ring R and R-modules M, N, the group $\mathrm{Hom}_R(M,N)$ has a natural structure of a left $\mathfrak{R}$-contramodule over the topological ring $\mathfrak{R} = \mathrm{Hom}_R(M,M)^{op}$ described as follows.*

For every set X, let r_x be a family of elements of $\mathfrak{R}$ converging to zero and corresponding to a family $g_x \in \mathrm{Hom}_R(M,M)$. For every family $f_x \in \mathrm{Hom}_R(M,N)$ and $m \in M$ define $\sum_{x\in X} r_x f_x \in \mathrm{Hom}_R(M,N)$ by:

$$\left(\sum_{x\in X} r_x f_x\right)(m) = \sum_{x\in X} f_x(g_x(m)),$$

the sum on the right-hand side is finite since the family g_x converges to zero.

THEOREM 3.4.3. [**16**, Corollary 7.2 and Corollary 7.4] *Let R be an associative ring and $\mathcal{A} \subseteq R$–Mod be a full subcategory closed under coproducts. Suppose that $\mathcal{A}$ is abelian with products and an injective cogenerator. Then, for any n-tilting object $T \in \mathcal{A}$, the abelian category $\mathcal{B}$, that is the heart of the n-tilting t-structure on $\mathcal{D}(\mathcal{A})$ associated with T is equivalent to the abelian category of left contramodules $\mathfrak{R}$–contra over the topological ring $\mathfrak{R} = \mathrm{Hom}_{\mathcal{A}}(T,T)^{op}$:*

$$\mathcal{B} \cong \mathfrak{R}\text{–}\mathsf{contra}.$$

In particular, there is a derived equivalence:

$$\mathbb{R}\Psi\colon \mathcal{D}(\mathcal{A}) \leftrightarrows \mathcal{D}(\mathfrak{R}\text{–}\mathsf{contra})\colon \mathbb{L}\Phi.$$

PROOF. The abelian categories $\mathcal{B}$ and $\mathfrak{R}$–contra have enough projectives. The category of projective objects in $\mathcal{B}$ is equivalent to $\mathsf{Add}(T) \subseteq \mathcal{A} \subseteq R$–$\mathsf{Mod}$ by Proposition 3.1.7. An abelian category with enough projectives is determined by its full subcategory of projective objects and by (the proof of) Proposition 3.4.1

$\mathsf{Add}(T)$ is equivalent to the category of projective objects in $\mathfrak{R}$–contra. Hence the derived equivalence follows by Theorem 3.3.2 (4). □

3.5. Example [**3**, Sections 17 and 19]**.** Let $f\colon R \longrightarrow S$ be an injective ring epimorphism such that $\mathrm{Tor}_1^R(S,S) = 0$, the flat dimension (f.dim) of S as a right R-module is at most 1 and the projective dimension (p.dim) of S as a left R-module is at most 1. There is a short exact sequence

$$0 \to R \to S \to S/R = K \to 0$$

where K and S are R-R-bimodules. It is known (see [**1**]) that the left R-module $S \oplus K$ is a 1-tilting left R-module.

Consider the full subcategory $\mathcal{A}$ of R–Mod consisting of the left R-modules M such that

$$S \otimes_R M = 0 = \mathrm{Tor}_1^R(S, M).$$

Under our homological assumptions we have:

- $\mathcal{A}$ is closed under kernels, cokernels, extensions, and coproducts in R–Mod, hence it is exactly embedded in R–Mod.
- [**3**, Proposition 17.1] The functor $\Gamma_S = \mathrm{Tor}_1^R(K, -)$ is right adjoint to the embedding functor $\mathcal{A} \to R$–Mod.
- $X \in \mathcal{A}$ if an only if $X \cong \mathrm{Tor}_1^R(K, X)$.
- [**3**, Proposition 17.4] $\mathcal{A}$ is a Grothendieck category; if I is an injective cogenerator of R–Mod, then $\Gamma_S(I)$ is an injective cogenerator of $\mathcal{A}$.

REMARK 3.5.1. [**3**, Remark 16.9] *Note that $\mathcal{A}$ is not a torsion class in R–Mod. It is contained in the torsion class $\mathcal{T}$ consisting of the left R-modules M such that $S \otimes_R M = 0$ and the two classes $\mathcal{A}$ and $\mathcal{T}$ coincide if and only if S is flat as a right R-module. In this case the torsion class is also hereditary.*

PROPOSITION 3.5.2. [**3**, Theorem 19.1] *The object K is a 1-tilting object of the category $\mathcal{A}$.*

PROOF. $K \in \mathcal{A}$ and the left R-module $S \oplus K$ is a 1-tilting module in R–Mod, hence K satisfies conditions (i) and (ii) in $\mathcal{A}$. The tilting class in R–Mod corresponding to the 1-tilting module $S \oplus K$ is the class of S-divisible R-modules, that is the class generated by S. Let $\mathcal{E} = \mathrm{Ker}(\mathrm{Ext}^1_{\mathcal{A}}(K, -))$. An object $X \in \mathcal{A}$ is in $\mathcal{E}$ if and only if X is S-divisible. There is a left approximation $0 \to Y \to S^{(\alpha)} \oplus K^{(\alpha)} \to X \to 0$ of X in R–Mod with Y an S-divisible left R-module.

Applying the functor $\mathrm{Tor}_1^R(K, -)$ to the above sequence we get

$$\mathrm{Tor}_1^R(K, S^{(\alpha)} \oplus K^{(\alpha)}) \cong K^{(\alpha)} \to \mathrm{Tor}_1^R(K, X) \cong X \to K \otimes_R Y = 0$$

where the last term vanishes since Y is S-divisible and $K \otimes S = 0$.

Hence K satisfies condition (iii) by the characterization in Theorem 3.1.3. □

Consider the full subcategory R–$\mathsf{Mod}_{S\text{–}\mathsf{contra}}$ of R–Mod consisting of the left R-modules C such that

$$\mathrm{Hom}_R(S, C) = 0 = \mathrm{Ext}^1_R(S, C).$$

- By the assumption $\mathrm{p.dim}_R S \leq 1$, the full subcategory R–$\mathsf{Mod}_{S\text{–}\mathsf{contra}}$ is closed under kernels, cokernels extensions and products in R–Mod, hence it is exactly embedded in R–Mod.

- For every R-module C the module $\mathrm{Ext}^1_R(K, C)$ is an S-contramodule. This can be proved analogously to Lemma 1.2.1 (2) using that
$$S \otimes_R K = 0 = \mathrm{Tor}^R_1(S, K).$$
(See also [**3**, Lemma 16.7 (c)].)

PROPOSITION 3.5.3. [**3**, Proposition 17.2] *The functor*
$$\Delta_S = \mathrm{Ext}^1_R(K, -)\colon R\text{–}\mathsf{Mod} \longrightarrow R\text{–}\mathsf{Mod}_{S\text{–}\mathsf{contra}}$$
is left adjoint to the embedding functor $R\text{–}\mathsf{Mod}_{S\text{–}\mathsf{contra}} \longrightarrow R\text{–}\mathsf{Mod}$. *In particular,* $\Delta_S(R) = \mathrm{Ext}^1_R(K, R)$ *is a projective generator of* $R\text{–}\mathsf{Mod}_{S\text{–}\mathsf{contra}}$.

PROOF. The natural isomorphism $\mathrm{Hom}_{R\text{–}\mathsf{Mod}_{S\text{–}\mathsf{contra}}}(\Delta_S(C), D) \cong \mathrm{Hom}_R(C, D)$, for every R-module C and S-contramodule D is proved similarly to the case of one element localization $R[s^{-1}]$(see Proposition 1.3.7). Indeed, there is an exact sequence
$$\mathrm{Hom}_R(S, C) \to C \overset{\delta_S}{\to} \mathrm{Ext}^1_R(K, C) \to \mathrm{Ext}^1_R(S, C) \to 0$$
where $\operatorname{Ker}\delta_S$ is S-divisible and $\mathrm{Ext}^1_R(S, C)$ is an S-module. □

THEOREM 3.5.4. [**3**, Theorems 19.1, 19.2] *Let $\mathcal{A}$ and K be as above. Let $\mathfrak{R}$ be the topological ring* $\mathrm{Hom}_R(K, K)^{\mathrm{op}}$ *with the finite topology. The following hold:*

(1) *The forgetful functor* $\mathfrak{R}\text{–}\mathsf{contra} \longrightarrow R\text{–}\mathsf{Mod}$ *is fully faithful, and its essential image coincides with the full subcategory* $R\text{–}\mathsf{Mod}_{S\text{–}\mathsf{contra}}$ *of S-contramodules. In particular,* $\mathfrak{R}\text{–}\mathsf{contra} \cong R\text{–}\mathsf{Mod}_{S\text{–}\mathsf{contra}}$.
(2) *The categories $\mathcal{A}$ and* $R\text{–}\mathsf{Mod}_{S\text{–}\mathsf{contra}}$ *are connected by the 1-tilting correspondence:*
$$\mathrm{Hom}_R(K, -)\colon \mathcal{A} \leftrightarrows R\text{–}\mathsf{Mod}_{S\text{–}\mathsf{contra}} \colon K \otimes_R -.$$

PROOF. (1) The forgetful functor $\mathfrak{R}\text{–}\mathsf{contra} \longrightarrow R\text{–}\mathsf{Mod}$ sends $\mathfrak{R}[[X]]$ to $\Delta_S(R^{(X)})$ (see the next paragraph) and it preserves cokernels. The full subcategory $R\text{–}\mathsf{Mod}_{S\text{–}\mathsf{contra}} \subset R\text{–}\mathsf{Mod}$ is closed under cokernels, and every left $\mathfrak{R}$-contramodule is the cokernel of a morphism of free left $\mathfrak{R}$-contramodules, hence the image of the forgetful functor is contained in $R\text{–}\mathsf{Mod}_{S\text{–}\mathsf{contra}}$.

To show that $\mathfrak{R}\text{–}\mathsf{contra}$ is equivalent to $R\text{–}\mathsf{Mod}_{S\text{–}\mathsf{contra}}$ it is enough to show the equivalence between their subcategories of projective objects. By [**13**, Proposition 2.1] this holds provided that there is a natural isomorphism $\Delta_S(R^{(X)}) \cong \mathfrak{R}[[X]]$ for every set X. Now, from the exact sequence
$$0 \to R^{(X)} \to S^{(X)} \to K^{(X)} \to 0,$$
we have natural isomorphisms $\mathfrak{R}[[X]] \cong \mathrm{Hom}_R(K, K^{(X)}) \cong \mathrm{Ext}^1_R(K, R^{(X)}) = \Delta_S(R^{(X)})$.

(2) By Theorem 3.3.2 (1) there is a correspondence
$$\Psi\colon \mathcal{A} \leftrightarrows \mathcal{B}\colon \Phi$$
where $\mathcal{B}$ is the heart of the tilting t-structure on $\mathcal{D}(\mathcal{A})$ induced by K. By Theorem 3.4.3, $\mathcal{B} \cong \mathfrak{R}\text{–}\mathsf{contra}$, hence by part (1) we get the correspondence between $\mathcal{A}$ and $R\text{–}\mathsf{Mod}_{S\text{–}\mathsf{contra}}$. By [**16**, Corollary 7.4] the functor Ψ can be computed as $\mathrm{Hom}_R(K, -)$ and Φ is given by $K \otimes_R -$, since the forgetful functor $\mathfrak{R}\text{–}\mathsf{contra} \longrightarrow R\text{–}\mathsf{Mod}$ is fully faithful (see [**16**, Lemma 7.9].) □

3.6. Good tilting modules.

(1) If ${}_RT$ is a finitely generated n-tilting module with endomorphism ring S, then it is well known that there is a derived equivalence:

$$T \otimes^{\mathbb{L}}_S -\colon \mathcal{D}(S) \rightleftarrows \mathcal{D}(R)\colon \mathbb{R}\operatorname{Hom}_R(T,-).$$

(2) An infinitely generated n-tilting module ${}_RT$ is said to be **good** if the ring R has an $\mathsf{Add}T$-coresolution $0 \to R \to T^0 \to \cdots \to T^n \to 0$ where T^i are summands of finite direct sums of copies of T.

(For every n-tilting module T there is a set Y such that $T^{(Y)}$ is a good n-tilting module giving rise to the same tilting class).

(3) [**4**] Let ${}_RT$ be a good n-tilting module with endomorphism ring S. There is an adjunction

$$T \otimes^{\mathbb{L}}_S -\colon \mathcal{D}(S) \rightleftarrows \mathcal{D}(R)\colon \mathbb{R}\operatorname{Hom}_R(T,-),$$

where $\mathbb{R}\operatorname{Hom}_R(T,-)$ is fully faithful and induces a derived equivalence:

$$\mathcal{D}(R) \cong \frac{\mathcal{D}(S)}{\operatorname{Ker}(T \otimes^{\mathbb{L}}_S -)}$$

where $\mathcal{D}(S)/\operatorname{Ker}(T \otimes^{\mathbb{L}}_S -)$ is a Verdier quotient of $\mathcal{D}(S)$.

(4) In Proposition 3.6.4 we will see that the above derived equivalence can be stated in terms of contramodules.

Definition 3.6.1. An object M in a category with coproducts is said to be **κ-small** if every morphism $M \to M^{(X)}$ factors through $M^{(Z)} \to M^{(X)}$ for a subset Z of X of cardinality strictly smaller than κ.

Theorem 3.6.2. [**16**, Theorem 6.6] *Let $\mathcal{B}$ be a cocomplete abelian category with a κ-small projective generator P. Let Y be a set of cardinality λ such that $\lambda^+ \geq \kappa$. Let $Q = P^{(Y)}$ and $S = \operatorname{Hom}_{\mathcal{B}}(Q,Q)^{\mathrm{op}}$. Then the functor*

$$\operatorname{Hom}_{\mathcal{B}}(Q,-)\colon \mathcal{B} \to S\text{–}\mathsf{Mod}$$

is fully faithful and admits a left adjoint functor.

Proposition 3.6.3. [**16**, Theorem 7.10] *Let R be an associative ring, M a left R-module, and Y be a set of cardinality greater or equal to the minimal cardinality of a set of generators of M. Consider $L = M^{(Y)}$. Let $\mathfrak{R} = \operatorname{Hom}_R(M,M)^{\mathrm{op}}$ and $\mathfrak{S} = \operatorname{Hom}_R(L,L)^{\mathrm{op}}$.*

Then $\mathsf{Add}M \cong \mathfrak{R}_{\mathsf{proj}}$, $\mathsf{Add}L \cong \mathfrak{S}_{\mathsf{proj}}$, hence $\mathfrak{R}\text{–}\mathsf{contra} \cong \mathfrak{S}\text{–}\mathsf{contra}$ (since $\mathsf{Add}M = \mathsf{Add}L$).

Moreover, the forgetful functor

$$\mathfrak{S}\text{–}\mathsf{contra} \to \mathfrak{S}\text{–}\mathsf{Mod}$$

is fully faithful.

Proof. The functor $\operatorname{Hom}_R(M,-)\colon R\text{–}\mathsf{Mod} \to \mathfrak{R}\text{–}\mathsf{contra}$ sends $L = M^{(Y)}$ to $\mathfrak{R}[[Y]]$ (see the proof of Proposition 3.4.1). The annihilators of the finitely generated submodules of M give a basis for the topology of $\mathfrak{R}$, hence, by the choice of Y, $\mathfrak{R}$ is a κ-small projective generator of $\mathfrak{R}\text{–}\mathsf{contra}$.

Theorem 3.6.2 shows that

$$\operatorname{Hom}_{\mathfrak{R}\text{–}\mathsf{contra}}(\mathfrak{R}[[Y]],-)\colon \mathfrak{R}\text{–}\mathsf{contra} \to \mathfrak{S}\text{–}\mathsf{Mod}$$

is fully faithful. □

Application [**16**, Corollary 7.11] Let T be an n-tilting left R-module and $\mathfrak{R} = \mathrm{Hom}_R(T,T)^{\mathrm{op}}$. Let $\mathcal{B}$ be the heart of the tilting t-structure induced by T on $\mathcal{D}(R)$ and Y a set as in Proposition 3.6.3, then

$$\mathcal{B} \cong \mathfrak{R}\text{–contra} \cong \mathfrak{S}\text{–contra}$$

and the adjunction

$$\Phi\colon \mathcal{B} \longrightarrow R\text{–Mod}; \quad \Psi\colon R\text{–Mod} \to \mathcal{B},$$

from the Tilting-Cotilting correspondence is given by the restriction of the adjunction

$$T^{(Y)} \otimes_{\mathfrak{S}} -\colon \mathfrak{S}\text{–Mod} \rightleftarrows R\text{–Mod}\colon \mathrm{Hom}_R(T^{(Y)}, -).$$

As a consequence of the previous discussion we obtain:

PROPOSITION 3.6.4. [**16**, Proposition 7.13] *If $T \in R\text{–Mod}$ is a good n-tilting module and $\mathfrak{S} = \mathrm{Hom}_R(T,T)^{\mathrm{op}}$, then the forgetful functor*

$$\mathfrak{S}\text{–contra} \to \mathfrak{S}\text{–Mod}$$

induces a fully faithful functor $\mathcal{D}(\mathfrak{S}\text{–contra}) \to \mathcal{D}(\mathfrak{S})$. The functor

$$\mathbb{R}\mathrm{Hom}_R(T,-)\colon \mathcal{D}(R) \to \mathcal{D}(\mathfrak{S})$$

is fully faithful and the adjunction

$$T \otimes^{\mathbb{L}}_{\mathfrak{S}} -\colon \mathcal{D}(\mathfrak{S}) \rightleftarrows \mathcal{D}(R)\colon \mathbb{R}\mathrm{Hom}_R(T,-)$$

restricts to the equivalence

$$T \otimes^{\mathbb{L}}_{\mathfrak{S}} -\colon \mathcal{D}(\mathfrak{S}\text{–contra}) \rightleftarrows \mathcal{D}(R)\colon \mathbb{R}\mathrm{Hom}_R(T,-).$$

References

[1] Lidia Angeleri Hügel and Javier Sánchez, *Tilting modules arising from ring epimorphisms*, Algebr. Represent. Theory **14** (2011), no. 2, 217–246, DOI 10.1007/s10468-009-9186-x. MR2776783

[2] Silvana Bazzoni and Leonid Positselski, *S-almost perfect commutative rings*, J. Algebra **532** (2019), 323–356, DOI 10.1016/j.jalgebra.2019.05.018. MR3959696

[3] S. Bazzoni, L. Positselski. Contramodules over pro-perfect topological rings, the covering property in categorical tilting theory, and homological ring epimorphisms. Electronic preprint `arXiv:1807.10671 [math.CT]`.

[4] Silvana Bazzoni, Francesca Mantese, and Alberto Tonolo, *Derived equivalence induced by infinitely generated n-tilting modules*, Proc. Amer. Math. Soc. **139** (2011), no. 12, 4225–4234, DOI 10.1090/S0002-9939-2011-10900-6. MR2823068

[5] Werner Geigle and Helmut Lenzing, *Perpendicular categories with applications to representations and sheaves*, J. Algebra **144** (1991), no. 2, 273–343, DOI 10.1016/0021-8693(91)90107-J. MR1140607

[6] Rüdiger Göbel and Jan Trlifaj, *Approximations and endomorphism algebras of modules. Volume 2*, Second revised and extended edition, De Gruyter Expositions in Mathematics, vol. 41, Walter de Gruyter GmbH & Co. KG, Berlin, 2012. Predictions. MR2985654

[7] D. K. Harrison, *Infinite abelian groups and homological methods*, Ann. of Math. (2) **69** (1959), 366–391, DOI 10.2307/1970188. MR104728

[8] Leonid Positselski, *Homological algebra of semimodules and semicontramodules: Semi-infinite homological algebra of associative algebraic structures*, Instytut Matematyczny Polskiej Akademii Nauk. Monografie Matematyczne (New Series) [Mathematics Institute of the Polish Academy of Sciences. Mathematical Monographs (New Series)], vol. 70, Birkhäuser/Springer Basel AG, Basel, 2010. Appendix C in collaboration with Dmitriy Rumynin; Appendix D in collaboration with Sergey Arkhipov. MR2723021

[9] Leonid Positselski, *Weakly curved A_∞-algebras over a topological local ring* (English, with English and French summaries), Mém. Soc. Math. Fr. (N.S.) **159** (2018), vi+206. MR3939851

[10] L. Positselski. Contraherent cosheaves. Electronic preprint `arXiv:1209.2995 [math.CT]`.

[11] Leonid Positselski, *Contraadjusted modules, contramodules, and reduced cotorsion modules*, Mosc. Math. J. **17** (2017), no. 3, 385–455, DOI 10.17323/1609-4514-2017-17-3-385-455. MR3711003

[12] Leonid Positselski, *Triangulated Matlis equivalence*, J. Algebra Appl. **17** (2018), no. 4, 1850067, 44, DOI 10.1142/S0219498818500676. MR3786746

[13] L. Positselski. Abelian right perpendicular subcategories in module categories. Electronic preprint `arXiv:1705.04960 [math.CT]`.

[14] Leonid Positselski, *Flat ring epimorphisms of countable type*, Glasg. Math. J. **62** (2020), no. 2, 383–439, DOI 10.1017/s001708951900017x. MR4085049

[15] L. Positselski and J. Rosický, *Covers, envelopes, and cotorsion theories in locally presentable abelian categories and contramodule categories*, J. Algebra **483** (2017), 83–128, DOI 10.1016/j.jalgebra.2017.03.029. MR3649814

[16] L. Positselski, J. Šťovíček. The Tilting-Cotilting Correspondence. Electronic preprint `arXiv:1710.02230 [math.CT]`. To appear in Int. Math. Res. Not. Pub. online 2 July 2019.

[17] Leonid Positselski and Jan Šťovíček, ∞*-tilting theory*, Pacific J. Math. **301** (2019), no. 1, 297–334, DOI 10.2140/pjm.2019.301.297. MR4007380

[18] Peter Schenzel, *Proregular sequences, local cohomology, and completion*, Math. Scand. **92** (2003), no. 2, 161–180, DOI 10.7146/math.scand.a-14399. MR1973941

[19] Marco Porta, Liran Shaul, and Amnon Yekutieli, *On the homology of completion and torsion*, Algebr. Represent. Theory **17** (2014), no. 1, 31–67, DOI 10.1007/s10468-012-9385-8. MR3160712

Dipartimento di Matematica Pura e Applicata, Università di Padova, Via Trieste 63, 35121 Padova, Italy

Email address: `bazzoni@math.unipd.it`

Contemporary Mathematics
Volume **758**, 2020
https://doi.org/10.1090/conm/758/15231

Representations of finite sets and correspondences

Serge Bouc

Abstract. This is a report on a long term joint work with Jacques Thévenaz on *Correspondence functors*. It is an expanded version of the mini-course I gave at the workshop of the ICRA2018 conference in Prague in August 2018.

1. Introduction

This report is an expanded version of the mini-course I gave during the workshop of the ICRA2018 conference in Prague in August 2018. It it a survey of the main results obtained so far in this long term joint work with Jacques Thévenaz on *Correspondence functors*, in which we develop the representation theory of finite sets and correspondences.

The representation theory of finite sets has been considered from various points of view. Pirashvili [**Pir00**] treats the case of pointed sets and maps, while Church, Ellenberg and Farb [**CEF15**] consider the case where the morphisms are all injective maps. Putman and Sam [**PS17**] use all k-linear splittable injections between finite-rank free k-modules (where k is a commutative ring). We have chosen instead not to use any kind of maps, but rather all correspondences as morphisms, in order to get a self dual category of finite sets. At first we had no specific application in mind, our main motivation being provided by the fact that finite sets are basic objects in mathematics, and correspondences a natural generalization of maps. Still, it turns out that this functorial approach of sets and correspondences sheds a new light on the representation theory of the algebra of relations on a finite set (Theorem 11.7). The theory has many other surprising results, e.g. the fact that the finitely generated correspondence functors over a field have finite length, and that they are characterized by the exponential behaviour of the dimension of their evaluations (Theorem 7.5).

This survey is organized as follows: in Section 2, we recall some basics on representations of categories. The category of finite sets and correspondences is introduced in Section 3, as well as the algebra of essential relations on a finite set. Sections 4 and 5 describe the simple modules for this algebra, leading to a parametrization of the simple correspondence functors, introduced in Section 6. Sections 7 and 8 are concerned with finiteness properties and stability properties

2010 *Mathematics Subject Classification.* Primary 06B05, 06B15, 06D05, 06D50, 16B50, 18B05, 18B10, 18B35, 18E05.

Key words and phrases. Finite set, correspondence, functor category, simple functor, finite length, poset.

of correspondence functors. In Section 9, correspondence functors associated to finite lattices are introduced, and some of their properties are stated. This leads in Section 10 to the notion of fundamental functor associated to a finite poset. Finally, Section 11 gives a complete description of the simple correspondence functors.

2. The representation theory of categories

We first recall some standard facts from the representation theory of categories. Let $\mathcal{D}$ be a category and let X and Y be two objects of $\mathcal{D}$. We reverse the usual notation and write $\mathcal{D}(Y,X)$ for the set of all morphisms from X to Y. We assume that $\mathcal{D}$ is skeletally small. This allows us to talk about the *set* of natural transformations between two functors starting from $\mathcal{D}$.

Throughout this paper, k denotes a commutative ring.

DEFINITION 2.1. The *k-linearization* of a category $\mathcal{D}$, where k is any commutative ring, is defined as follows :

- The objects of $k\mathcal{D}$ are the objects of $\mathcal{D}$.
- For any two objects X and Y, the set of morphisms from X to Y is the free k-module $k\mathcal{D}(Y,X)$ with basis $\mathcal{D}(Y,X)$.
- The composition of morphisms in $k\mathcal{D}$ is the k-bilinear extension

$$k\mathcal{D}(Z,Y) \times k\mathcal{D}(Y,X) \longrightarrow k\mathcal{D}(Z,X)$$

  of the composition in $\mathcal{D}$.

DEFINITION 2.2. Let $\mathcal{D}$ be a category and k a commutative ring. A *k-representation* of the category $\mathcal{D}$ is a k-linear functor from $k\mathcal{D}$ to the category $k\text{-}\mathrm{Mod}$ of k-modules.

We could have defined a k-representation of $\mathcal{D}$ as a functor from $\mathcal{D}$ to $k\text{-}\mathrm{Mod}$, but it is convenient to linearize first the category $\mathcal{D}$ (just as for group representations, where one can first introduce the group algebra).

If $F : k\mathcal{D} \to k\text{-}\mathrm{Mod}$ is a k-representation of $\mathcal{D}$ and if X is an object of $\mathcal{D}$, then $F(X)$ will be called the *evaluation* of F at X. Morphisms in $k\mathcal{D}$ act on the left on the evaluations of F by setting, for every $m \in F(X)$ and for every morphism $\alpha \in k\mathcal{D}(Y,X)$,

$$\alpha \cdot m := F(\alpha)(m) \in F(Y) .$$

We often use a dot for this action of morphisms on evaluation of functors.

The category $\mathcal{F}_k(k\mathcal{D}, k\text{-}\mathrm{Mod})$ of all k-representations of $\mathcal{D}$ is an abelian category. A sequence of functors

$$0 \longrightarrow F_1 \longrightarrow F_2 \longrightarrow F_3 \longrightarrow 0$$

is exact if and only if, for every object X, the evaluation sequence

$$0 \longrightarrow F_1(X) \longrightarrow F_2(X) \longrightarrow F_3(X) \longrightarrow 0$$

is exact. Also, a k-representation of $\mathcal{D}$ is called *simple* if it is nonzero and has no proper nonzero subfunctor.

For any object X of $\mathcal{D}$, consider the representable functor $k\mathcal{D}(-,X)$ (which is a projective functor by Yoneda's lemma). Its evaluation at an object Y is the k-module $k\mathcal{D}(Y,X)$, which has a natural structure of a $(k\mathcal{D}(Y,Y), k\mathcal{D}(X,X))$-bimodule by composition.

NOTATION 2.3. Let X be an object of $\mathcal{D}$ and let V be a $k\mathcal{D}(X,X)$-module. We define

$$L_{X,V} := k\mathcal{D}(-,X) \otimes_{k\mathcal{D}(X,X)} V \,.$$

This is a k-representation of $\mathcal{D}$.

This satisfies the following adjunction property.

LEMMA 2.4. *Let $\mathcal{F} = \mathcal{F}_k(k\mathcal{D}, k\text{-Mod})$ be the category of all k-representations of $\mathcal{D}$ and let X be an object of $\mathcal{D}$.*

(a) *The functor*

$$k\mathcal{D}(X,X)\text{-Mod} \longrightarrow \mathcal{F}\,, \quad V \mapsto L_{X,V}$$

is left adjoint of the evaluation functor

$$\mathcal{F} \longrightarrow k\mathcal{D}(X,X)\text{-Mod}\,, \quad F \mapsto F(X)\,.$$

In other words, for any k-representation $F : k\mathcal{D} \to k\text{-Mod}$ and any $k\mathcal{D}(X,X)$-module V, there is a natural isomorphism

$$\mathrm{Hom}_{\mathcal{F}}(L_{X,V}, F) \cong \mathrm{Hom}_{k\mathcal{D}(X,X)}\big(V, F(X)\big)\,.$$

Moreover $L_{X,V}(X) \cong V$ as $k\mathcal{D}(X,X)$-modules. In particular, there is a k-algebra isomorphism $\mathrm{End}_{\mathcal{F}}(L_{X,V}) \cong \mathrm{End}_{k\mathcal{D}(X,X)}(V)$.

(b) *The functor $k\mathcal{D}(X,X)\text{-Mod} \longrightarrow \mathcal{F}$ is right exact. It maps projective modules to projective functors, and indecomposable modules to indecomposable functors.*

PROOF. Part (a) is straightforward and is proved in Section 2 of [**Bou96**]. Part (b) follows because this functor is left adjoint of an exact functor and satisfies the property $L_{X,V}(X) \cong V$. □

Our next result is a slight extension of the first lemma of [**Bou96**].

LEMMA 2.5. *Let X be an object of $\mathcal{D}$ and let V be a $k\mathcal{D}(X,X)$-module. For any object Y of $\mathcal{D}$, let*

$$J_{X,V}(Y) := \Big\{ \sum_i \phi_i \otimes w_i \in L_{X,V}(Y) \mid \forall \psi \in k\mathcal{D}(X,Y), \sum_i (\psi\phi_i)\cdot w_i = 0 \Big\}\,.$$

(a) *$J_{X,V}$ is the largest subfunctor of $L_{X,V}$ which vanishes at X.*

(b) *If V is a simple module, then $J_{X,V}$ is the unique maximal subfunctor of $L_{X,V}$ and $L_{X,V}/J_{X,V}$ is a simple functor.*

PROOF. The proof is sketched in Lemma 2.3 of [**BST13**] in the special case of biset functors for finite groups, but it extends without change to representations of an arbitrary category $\mathcal{D}$. □

Lemma 2.5 is our main tool for dealing with simple functors. We first fix the notation.

NOTATION 2.6. Let X be an object of $\mathcal{D}$ and let V be a $k\mathcal{D}(X,X)$-module. We define

$$S_{X,V} := L_{X,V}/J_{X,V}\,.$$

If V is a simple $k\mathcal{D}(X,X)$-module, then $S_{X,V}$ is a simple functor.

We emphasize that $L_{X,V}$ and $S_{X,V}$ are defined for any $k\mathcal{D}(X,X)$-module V and any commutative ring k. Note that we always have $J_{X,V}(X) = \{0\}$ because if $a = \sum_i \phi_i \otimes w_i \in J_{X,V}(X)$, then $a = \mathrm{id}_X \otimes (\sum_i \phi_i \cdot w_i) = 0$.

Therefore, we have isomorphisms of $k\mathcal{D}(X,X)$-modules

$$L_{X,V}(X) \cong S_{X,V}(X) \cong V\ .$$

PROPOSITION 2.7. *Let S be a simple k-representation of $\mathcal{D}$ and let Y be an object of $\mathcal{D}$ such that $S(Y) \neq 0$.*

(a) *$S(Y)$ is a simple $k\mathcal{D}(Y,Y)$-module.*
(b) *$S \cong S_{Y,S(Y)}$.*
(c) *S is generated by $S(Y)$, that is, $S(X) = k\mathcal{D}(X,Y)S(Y)$ for all objects X. More precisely, if $0 \neq u \in S(Y)$, then $S(X) = k\mathcal{D}(X,Y) \cdot u$.*

PROOF. (c) Given $0 \neq u \in S(Y)$, let $S'(X) = k\mathcal{D}(X,Y) \cdot u$ for all objects X. This clearly defines a nonzero subfunctor S' of S, so $S' = S$ by simplicity of S.

(a) This follows from (c).

(b) By the adjunction of Lemma 2.4, the identity $\mathrm{id} : S(Y) \to S(Y)$ corresponds to a non-zero morphism $\theta : L_{Y,S(Y)} \to S$, which must be surjective since S is simple. But $S_{Y,S(Y)}$ is the unique simple quotient of $L_{Y,S(Y)}$, by Lemma 2.5 and Notation 2.6, so $S \cong S_{Y,S(Y)}$. □

2.8. It should be noted that a simple k-representation S has many possible realizations $S \cong S_{Y,V}$ as above, where $V = S(Y) \neq 0$. However, if there is a notion of unique minimal object, then one can parametrize simple functors S by setting $S \cong S_{Y,V}$, where Y is the unique minimal object such that $S(Y) \neq 0$. In this case, the evaluation $V = S(Y)$ is actually a module for the quotient algebra

$$k\widehat{\mathcal{D}}(Y) = k\mathcal{D}(Y,Y)/\sum_Z k\mathcal{D}(Y,Z)k\mathcal{D}(Z,Y)$$

where Z runs through objects which are stricly smaller than Y. In the case of the category of finite sets and correspondences, this will be the motivation for considering the *algebra of essential relations* on a finite set (see Definition 3.3).

The following result establishes a link between the simple subquotients of a functor and the simple subquotients of its evaluations:

PROPOSITION 2.9. [[**BST13**] Proposition 3.5, [**BT18b**] Proposition 2.8] *Let S be a simple k-representation of $\mathcal{D}$ and let Y be an object of $\mathcal{D}$ such that $S(Y) \neq 0$. Let F be any k-representation of $\mathcal{D}$. Then the following are equivalent:*

(a) *S is isomorphic to a subquotient of F.*
(b) *The simple $k\mathcal{D}(Y,Y)$-module $S(Y)$ is isomorphic to a subquotient of the $k\mathcal{D}(Y,Y)$-module $F(Y)$.*

PROOF. It is clear that (a) implies (b). Suppose that (b) holds and let W_1, W_2 be submodules of $F(Y)$ such that $W_2 \subset W_1$ and $W_1/W_2 \cong S(Y)$. For $i \in \{1,2\}$, and for any object X of $\mathcal{D}$, set $F_i(X) = k\mathcal{D}(X,Y) \cdot W_i \subseteq F(X)$. Then $F_2 \subseteq F_1$ are subfunctors of F. Moreover $F_i(Y) = W_i$ for $i \in \{1,2\}$, and $(F_1/F_2)(Y) = W_1/W_2 \cong S(Y)$. The isomorphism $S(Y) \to (F_1/F_2)(Y)$ induces, by the adjunction of Lemma 2.4, a nonzero morphism $\theta : L_{Y,S(Y)} \to F_1/F_2$. Since $S(Y)$ is simple, the functor $L_{Y,S(Y)}$ has a unique maximal subfunctor $J_{Y,S(Y)}$, by Lemma 2.5, and

$L_{Y,S(Y)}/J_{Y,S(Y)} \cong S_{Y,S(Y)} \cong S$, by Proposition 2.7. Let $F_1' = \theta(L_{Y,S(Y)})$ and $F_2' = \theta(J_{Y,S(Y)})$. Since $\theta \neq 0$, we obtain

$$F_1'/F_2' \cong L_{Y,S(Y)}/J_{Y,S(Y)} \cong S_{Y,S(Y)} \cong S\,,$$

showing that S is isomorphic to a subquotient of F. □

(Actually, as observed by Hida and Yagita in Lemma 3.1 of [**HY14**], we have an equality $F_1' = F_1$, because both subfunctors are generated by their common evaluation at Y.)

3. Correspondences and relations

3.1. Let X and Y be finite sets. A *correspondence* from X to Y is a subset R of $Y \times X$. In case $X = Y$, we say that R is a *relation* on X.

Correspondences can be composed as follows. If $R \subseteq Y \times X$ and $S \subseteq Z \times Y$, then SR is the correspondence from X to Z defined by

$$SR = \{(z,x) \in Z \times X \mid \exists y \in Y \;\text{ such that }\; (z,y) \in S \;\text{ and }\; (y,x) \in R\}\,.$$

One can check easily that the composition of correspondences is associative. For a finite set X, the *equality relation*

$$\Delta_X = \{(x,x) \mid x \in X\} \subseteq X \times X$$

acts as an identity element for the composition of correspondences. In particular the set of all relations on X is a monoid.

DEFINITION 3.2. The *category $\mathcal{C}$ of finite sets and correspondences* is defined as follows:

- The objects of $\mathcal{C}$ are finite sets.
- if X and Y are finite sets, the set of morphisms from X to Y in $\mathcal{C}$ is the set $\mathcal{C}(Y,X)$ of correspondences from X to Y.
- The composition of morphisms in $\mathcal{C}$ is the composition of correspondences.
- The identity morphism of the finite set X is $\Delta_X \in \mathcal{C}(X,X)$.

Given a commutative ring k, we will also consider the k-linearization $k\mathcal{C}$ of $\mathcal{C}$, introduced in Section 2, called the *k-category of finite sets and correspondences*.

When X is a finite set, the endomorphism algebra $\mathcal{R}_X = k\mathcal{C}(X,X)$ of X in the category $k\mathcal{C}$ is called the *algebra of relations* on X. It is the algebra over k of the monoid of relations on X.

DEFINITION 3.3. A relation R on a finite set X is called *inessential* if there exists a finite set Y with $|Y| < |X|$ and correspondences $U \in \mathcal{C}(X,Y)$ and $V \in \mathcal{C}(Y,X)$ such that $R = UV$. A relation is called *essential* if it is not inessential.

The subset $\mathcal{I}_X$ of the algebra $\mathcal{R}_X$ consisting of k-linear combinations of inessential relations is a two sided ideal. The quotient algebra $\mathcal{E}_X = \mathcal{R}_X/\mathcal{I}_X$ is called the *algebra of essential relations* on X (or the *essential algebra*, for short).

In other words, an essential relation is a relation of maximal Schein rank, in the sense of Section 1.4 of [**Kim82**].

The algebra $\mathcal{E}_X$ has a k-basis consisting of essential relations. The product in $\mathcal{E}_X$ of two essential relations R and S on X is equal to RS if RS is essential, and to 0 otherwise.

EXAMPLE 3.4.
- Let X be a finite set with $|X| \geq 2$, and A, B be subsets of X. Then the relation $R = A \times B$ on X is inessential: if Y is a set of cardinality 1, then $R = UV$, where $U = A \times Y$ and $V = Y \times B$.
- Let $X = \{1, 2\}$, and $R = \{(1, 1), (1, 2), (2, 1)\} \in \mathcal{C}(X, X)$. Then one checks easily that R is essential, but $R^2 = X \times X$ is not.

LEMMA 3.5 ([**BT16**, Lemma 2.1]). *Let X be a finite set of cardinality n, and $R \in \mathcal{C}(X, X)$. Then R is inessential if and only if there are subsets U_i and V_i of X, for $1 \leq i \leq n-1$, such that $R = \bigcup_{i=1}^{n-1} (U_i \times V_i)$.*

3.6. Classical definitions When X and Y are finite sets and $R \in \mathcal{C}(Y, X)$, we denote by $R^{op} \in \mathcal{C}(X, Y)$ the *opposite* correspondence, defined by

$$R^{op} = \{(x, y) \in X \times Y \mid (y, x) \in R\}.$$

With this notation, the following classical properties of a relation R on a finite set X can be translated as follows:

$$\begin{aligned}
R \text{ is } \textit{reflexive} &\iff \Delta_X \subseteq R \\
R \text{ is } \textit{transitive} &\iff R^2 \subseteq R \\
R \text{ is } \textit{a preorder} &\iff \Delta_X \subseteq R = R^2 \\
R \text{ is } \textit{symmetric} &\iff R = R^{op} \\
R \text{ is } \textit{an equivalence relation} &\iff \Delta_X \subseteq R = R^{op} = R^2 \\
R \text{ is } \textit{antisymmetric} &\iff R \cap R^{op} \subseteq \Delta_X \\
R \text{ is } \textit{an order} &\iff R = R^2 \text{ and } R \cap R^{op} = \Delta_X.
\end{aligned}$$

REMARK 3.7.
- Note that unless otherwise specified, by *order* we always mean *partial order*.
- It follows from Lemma 3.5 that if R is a preorder on X which is not an order, then R is inessential.
- On the other hand, if R is an order on X and $\Delta_X \subseteq Q \subseteq R$, then Q is essential.

3.8. Let X be a finite set. For an element σ of the group Σ_X of permutations of X, we set

$$\Delta_\sigma = \{(\sigma(x), x) \mid x \in X\}.$$

We call Δ_σ the *permuted diagonal* associated to σ. One checks easily that the map $\sigma \in \Sigma_X \mapsto \Delta_\sigma \in \mathcal{C}(X, X)$ is a monoid homomorphism, and it follows in particular that Δ_σ is an essential relation on X, for any $\sigma \in \Sigma_X$.

THEOREM 3.9. [[**BT16**] Theorem 3.2] *Let X be a finite set and R be an essential relation on X. Then there exists a permutation σ of X such that $\Delta_\sigma \subseteq R$, i.e. $R = S\Delta_\sigma$, where S is a reflexive relation on X.*

PROOF. Let R be an essential relation on X. For any $a \in A$, set

$$R_a = \{x \in X \mid (x, a) \in R\} \text{ and } {}_aR = \{x \in X \mid (a, x) \in R\}.$$

For any subset A of X, define

$$R_A = \{x \in X \mid \exists a \in A \text{ such that } (x, a) \in R\} = \bigcup_{a \in A} R_a.$$

Then R decomposes as a union

$$R = \Big(\bigcup_{y\notin A}(R_y \times \{y\})\Big)\bigcup\Big(\bigcup_{x\in R_A}(\{x\}\times {}_xR)\Big).$$

Since R is essential, $\mathrm{Card}(X-A)+\mathrm{Card}(R_A)$ cannot be strictly smaller than $\mathrm{Card}(X)$, by Lemma 3.5. Therefore $\mathrm{Card}(R_A)\geq \mathrm{Card}(A)$, for all subsets A of X, that is

$$\mathrm{Card}\,\Big(\bigcup_{a\in A} R_a\Big)\geq \mathrm{Card}(A).$$

This is precisely the assumption in a theorem of Philip Hall (see Theorem 5.1.1 in [**Hal86**], or [**Hal35**] for the original version which is slightly different). The conclusion is that there exist elements $x_y\in R_y$, where y runs over X, which are all distinct. In other words $\sigma: y\mapsto x_y$ is a permutation and

$$(\sigma(y),y)=(x_y,y)\in R \quad \text{for all } y\in X.$$

This means that R contains Δ_σ, as required. □

4. The algebra of permuted orders

4.1. Let X be a finite set, and S be a reflexive relation on X. Then $\Delta_X\subseteq S$, and for any integer $m\geq 1$

$$\Delta_X\subseteq S\subseteq S^2\subseteq\ldots\subseteq S^m.$$

Since X is finite, there exists m such that $S^m=S^{m+1}$. Let $\overline{S}=S^m$ denote this limit value. The relation $\overline{S}$ is called the *transitive closure* of S. It is a preorder. There are now two cases, by Remark 3.7:

- either $\overline{S}$ is not an order, and then $\overline{S}=0$ in $\mathcal{E}_X$.
- or $\overline{S}$ is an order, and then S is essential, since $\Delta_X\subseteq S\subseteq \overline{S}$.

THEOREM 4.2. [[**BT16**] Theorem 5.3] *Let $\mathcal{N}_X$ be the subset of $\mathcal{E}_X$ consisting of k-linear combinations of elements of the form $(S-\overline{S})\Delta_\sigma$, where S is a reflexive relation on X, and $\sigma\in\Sigma_X$.*

(a) *$\mathcal{N}_X$ is a two sided nilpotent ideal of $\mathcal{E}_X$.*
(b) *The quotient algebra $\mathcal{P}_X=\mathcal{E}_X/\mathcal{N}_X$ has a k-basis consisting of elements of the form $S\Delta_\sigma$, where S is an order on X, and $\sigma\in\Sigma_X$.*

PROOF SKETCH. (a) Let $S\supseteq\Delta_X$, and $m\in\mathbb{N}-\{0\}$ such that $S^m=\overline{S}$.

- Let $Q\supseteq\Delta_X$. Then $Q(S-\overline{S})=QS-Q\overline{S}=(QS-\overline{QS})-(Q\overline{S}-\overline{Q\overline{S}})$ since $\overline{QS}=\overline{Q\overline{S}}$. Hence $Q\mathcal{N}_X\subseteq\mathcal{N}_X$, and similarly $\mathcal{N}_XQ\subseteq\mathcal{N}_X$.
- Since S and $\overline{S}$ commute, we have

$$\begin{aligned}(S-\overline{S})^m &= \sum_{i=0}^m(-1)^i\binom{m}{i}S^{m-i}\overline{S}^i\\ &= \overline{S}+\sum_{i=1}^m(-1)^i\binom{m}{i}\underbrace{S^{m-i}\overline{S}}_{\overline{S}}\\ &= \Big(\sum_{i=0}^m(-1)^i\binom{m}{i}\Big)\overline{S}=0.\end{aligned}$$

Then $\mathcal{N}_X$ is a two sided ideal of the k-algebra $\mathcal{E}_X$, linearly generated by a finite number of nilpotent elements, so it is a nilpotent ideal of $\mathcal{E}_X$ (see Lemma 5.1 of [**BT16**] for details).

(b) By Theorem 3.9, every essential relation R on X can be written $R = S\Delta_\sigma$, where S is reflexive and $\sigma \in \Sigma_X$. By definition of $\mathcal{N}_X$, we have $S\Delta_\sigma = \overline{S}\Delta_\sigma$ in $\mathcal{P}_X$, and $\overline{S} = 0$ in $\mathcal{E}_X$ if $\overline{S}$ is not an order, by Remark 3.7. Hence $\mathcal{P}_X$ is linearly generated by the elements $S\Delta_\sigma$, where S is an order on X, and one checks easily that these elements are linearly independent. □

Theorem 4.2 is particularly useful to describe the simple $\mathcal{E}_X$-modules: as $\mathcal{N}_X$ acts by zero on every such module, the simple $\mathcal{E}_X$-modules are actually simple $\mathcal{P}_X$-modules.

4.3. A relation R on a finite set X which can be written $R = S\Delta_\sigma$, where S is an order and $\sigma \in \Sigma_X$, is called a *permuted order* on X. One can show that this decomposition of R is unique (in other words, if R is a permuted order, there is a unique $\sigma \in \Sigma_X$ such that $\Delta_\sigma \subseteq R$).

In view of Assertion (b) of Theorem 4.2, the algebra $\mathcal{P}_X$ is called the *algebra of permuted orders* on X. If S and T are orders on X, and if $\sigma, \tau \in \Sigma_X$, the product $S\Delta_\sigma T\Delta_\tau$ in $\mathcal{P}_X$ is equal to $\overline{S.{}^\sigma T}\Delta_{\sigma\tau}$ if $\overline{S.{}^\sigma T}$ is an order, and to 0 otherwise, where ${}^\sigma T = \Delta_\sigma T \Delta_{\sigma^{-1}}$.

In particular $\mathcal{P}_X$ is Σ_X-graded: for $\sigma \in \Sigma_X$, the degree σ part of $\mathcal{P}_X$ is the k-submodule generated by elements $S\Delta_\sigma$, where S is an order on X. The subalgebra $\mathcal{P}_1$ has a k-basis consisting of the set $\mathcal{O}_X$ of orders on X. For $S, T \in \mathcal{O}_X$, the product of ST in $\mathcal{P}_1$ is equal to $\overline{ST} = \overline{S \cup T}$ if this is an order, and to 0 otherwise. In particular $\mathcal{P}_1$ is *commutative*. Moreover, the group Σ_X acts by conjugation on $\mathcal{O}_X$, and the algebra $\mathcal{P}_X$ identifies with the semidirect product $\mathcal{P}_1 \rtimes \Sigma_X$.

If $R \in \mathcal{O}_X$, then $R^2 = R$ in $\mathcal{P}_1$. Moreover if $R, S \in \mathcal{P}_1$, then either $RS = 0$ in $\mathcal{P}_1$, or $RS = \overline{R \cup S}$ if this is an order. We observe that if *we order $\mathcal{O}_X$ by inclusion of subsets*, then RS is either 0 or the least upper bound of R and S in the poset $\mathcal{O}_X$. So we have a k-basis $\mathcal{O}_X$ of idempotents of the commutative algebra $\mathcal{P}_1$, which is ordered and contains $1 = \Delta_X$, such that the product of any two of them is either 0 or their supremum in the poset $\mathcal{O}_X$. In such a situation, there is a standard procedure to produce a decomposition of the identity as a sum of orthogonal idempotents.

NOTATION 4.4. For $R \in \mathcal{O}_X$, let f_R be the element of $\mathcal{P}_1$ defined by

$$f_R = \sum_{\substack{S \in \mathcal{O}_X \\ S \supseteq R}} \mu_{\mathcal{O}_X}(R, S) S,$$

where $\mu_{\mathcal{O}_X}$ is the Möbius function of the poset $\mathcal{O}_X$.

THEOREM 4.5 ([**BT16**, Theorem 6.2]). (a) *The elements f_R, for $R \in \mathcal{O}_X$, are orthogonal idempotents of $\mathcal{P}_1$, and $\sum_{R \in \mathcal{O}_X} f_R = 1$.*

(b) *Moreover $\mathcal{P}_1 f_R = k f_R$, for any $R \in \mathcal{O}_X$.*

(c) *The algebra $\mathcal{P}_1$ is isomorphic to the product $\prod_{R \in \mathcal{O}_X} k f_R \cong k^{|\mathcal{O}_X|}$.*

NOTATION 4.6. For $R \in \mathcal{O}_X$, let $\Sigma_{X,R}$ denote the stabilizer of R in Σ_X, i.e.

$$\Sigma_{X,R} = \{\sigma \in \Sigma_X \mid {}^\sigma R = R\}.$$

Let moreover $e_R \in \mathcal{P}_1$ be the orbit sum

$$e_R = \sum_{\sigma \in [\Sigma_X/\Sigma_{X,R}]} f_{^\sigma R}.$$

We note that the group $\Sigma_{X,R}$ is isomorphic to the automorphism group $\mathrm{Aut}(X, R)$ of the poset (X, R). We will generally identify those two groups.

THEOREM 4.7 ([**BT16**, Lemma 7.4 and Theorem 7.5]). (a) *The elements e_R, for R in a set $[\Sigma_X \backslash \mathcal{O}_X]$ of representatives of Σ_X-orbits on $\mathcal{O}_X$, are central orthogonal idempotents of $\mathcal{P}_X$, and* $\sum_{R \in [\Sigma_X \backslash \mathcal{O}_X]} e_R = 1$.

(b) *The algebra $\mathcal{P}_X$ is isomorphic to* $\prod_{R \in [\Sigma_X \backslash \mathcal{O}_X]} \mathcal{P}_X e_R$.

(c) *For $R \in \mathcal{O}_X$, the algebra $\mathcal{P}_X e_R$ is isomorphic to the full matrix algebra $M_{|\Sigma_X : \Sigma_{X,R}|}(k\Sigma_{X,R})$.*

PROOF SKETCH. The proof relies on the fact that the set $\{\Delta_\sigma f_R \mid \sigma \in \Sigma_X,\ R \in \mathcal{O}_X\}$ is a k-basis of $\mathcal{P}_X$, and that the product of two of these basis elements is either zero or another element of the basis (see [**BT16**] for details). □

5. The simple modules for the essential algebra

5.1. In this section we deal with the simple modules for the essential algebra $\mathcal{E}_X$ and the algebra of permuted orders $\mathcal{P}_X = \mathcal{E}_X/\mathcal{N}_X$ on a finite set X. Theorems 4.2 and 4.7 allow for a description of the simple modules for these algebras.

Recall that Σ_X is the group of permutations of X. The group Σ_X acts on the set $\mathcal{O}_X$ of orders on X. For $R \in \mathcal{O}_X$, we denote by $\Sigma_{X,R}$ the stabilizer of R in Σ_X, and by f_R the idempotent associated to R in $\mathcal{P}_X$ (see Notation 4.6).

THEOREM 5.2 ([**BT16**, Theorem 7.5 and Theorem 8.1]). *Let X be a finite set.*

(a) *The surjection $\mathcal{E}_X \twoheadrightarrow \mathcal{P}_X$ induces a one to one correspondence between the simple $\mathcal{E}_X$-modules and the simple $\mathcal{P}_X$-modules.*

(b) *Let $R \in \mathcal{O}_X$. Then $\mathcal{P}_X f_R$ has a k-basis $\{\Delta_\sigma f_R \mid \sigma \in \Sigma_X\}$, so $\mathcal{P}_X f_R \cong k\Sigma_X$ as a k-module. It is an $(\mathcal{R}_X, k\Sigma_{X,R})$-bimodule, free as a right $k\Sigma_{X,R}$-module.*

(c) *The simple $\mathcal{P}_X$-modules (up to isomorphism) are the modules of the form $\mathcal{P}_X f_R \otimes_{k\Sigma_{X,R}} W$, where W is a simple $k\Sigma_{R,X}$-module (up to isomorphism).*

(d) *If k is a field with $\mathrm{char}(k) = 0$ or $\mathrm{char}(k) > |X|$, then $\mathcal{P}_X$ is semisimple, and $\mathcal{N}_X$ is equal to the Jacobson radical of $\mathcal{E}_X$.*

PROOF. (a) follows from Theorem 4.2. Now (b) and (c) follow from Theorem 4.7, and then (d) follows from the semisimplicity of all the group algebras $k\Sigma_{X,R}$, when $\mathrm{char}(k) = 0$ or $\mathrm{char}(k) > |X|$. □

As the essential algebra $\mathcal{E}_X$ is a quotient of the algebra $\mathcal{R}_X = k\mathcal{C}(X, X)$, we now get a precise description of some simple $\mathcal{R}_X$-modules.

PROPOSITION 5.3 ([**BT16**, Theorem 8.1 and Proposition 8.5]). *Let R be an order on X. If $S \in \mathcal{R}_X$, define a k-endomorphism $\beta_R(S)$ of $k\Sigma_X$ by*

$$\beta_R(S) : \sigma \in \Sigma_X \mapsto \begin{cases} \tau\sigma & \text{if } \exists \tau \in \Sigma_X,\ \Delta \subseteq \Delta_{\tau^{-1}} S \subseteq {}^\sigma R \\ 0 & \text{if there is no such } \tau. \end{cases}$$

(a) *The map $\beta_R(S)$ is well defined, and $\beta_R(S) \in \mathrm{End}_{k\Sigma_{X,R}}(k\Sigma_X)$.*

(b) *The map* $S \mapsto \beta_R(S)$ *extends to an algebra homomorphism from* $\mathcal{R}_X$ *to* $\mathrm{End}_{k\Sigma_{X,R}}(k\Sigma_X)$, *which makes* $k\Sigma_X$ *a* $(\mathcal{R}_X, k\Sigma_{X,R})$*-bimodule, isomorphic to* $\mathcal{P}_X f_R$.

(c) *For a* $k\Sigma_{X,R}$*-module* W, *let* $\Lambda_{R,W}$ *denote the* $\mathcal{R}_X$*-module* $\mathcal{P}_X f_R \otimes_{k\Sigma_{X,R}} W$. *If* W *is a simple* $k\Sigma_{X,R}$*-module, then* $\Lambda_{R,W}$ *is a simple* $\mathcal{R}_X$*-module.*

(d) *If* (R', W') *is another pair consisting of an order* R' *on* X *and a simple* $k\Sigma_{X,R'}$*-module* W', *then the* $\mathcal{R}_X$*-modules* $\Lambda_{R,W}$ *and* $\Lambda_{R',W'}$ *are isomorphic if and only if the pairs* (R, W) *and* (R', W') *are conjugate by* Σ_X.

PROOF SKETCH. (a) and (b) follow from a precise description of the action of $\mathcal{R}_X$ on $\mathcal{P}_X f_R$, using the basis $\{\Delta_\sigma f_R \mid \sigma \in \Sigma_X\}$ of $\mathcal{P}_X f_R$. Then (*c*) and (*d*) follow from Theorem 4.7. □

REMARK 5.4. If G is a finite group, a simple kG-module is actually a $(k/m)G$-module, where m is some maximal ideal of k. It follows that when dealing with simple correspondence functors over k, one could always suppose that k is a field.

EXAMPLES 5.5. Let k be a field.

- If $R = \Delta_X$, then $\Sigma_{X,R} = \Sigma_X$, and $\mathcal{R}_X$ maps surjectively to $k\Sigma_X$, by $S \mapsto \sigma$ if $S = \Delta_\sigma$ for $\sigma \in \Sigma_X$, and $S \mapsto 0$ is there is no such $\sigma \in \Sigma_X$.
- If R is a total order, then $\Sigma_{X,R} = \{1\}$, and $\mathcal{P}_X e_R \cong M_{n!}(k)$ (where $n = |X|$). In this case $k\Sigma_X$ becomes a simple $\mathcal{R}_X$-module.

REMARK 5.6. Recall that for an order R on a finite set X, the group $\Sigma_{X,R}$ is isomorphic to the group $\mathrm{Aut}(X, R)$ of automorphisms of the poset (X, R). One may ask which finite groups can occur as automorphism group of some finite poset. The answer is quite simple: every finite group can. This was first proved by Birkhoff ([**Bir46**]), and later simplified by Thornton ([**Tho72**]) and Barmak-Minian ([**BM09**])

6. Correspondence functors

DEFINITION 6.1. Let k be a commutative ring. A *correspondence functor* over k is a k-representation of the category $k\mathcal{C}$ introduced in Definition 3.2. The category of correspondence functors $\mathcal{F}_k(k\mathcal{C}, k\text{-}\mathrm{Mod})$ is denoted by $\mathcal{F}_k$.

REMARKS 6.2. The category $k\mathcal{C}$ has the following special important properties:

- The functor sending a finite set to itself and a correspondence to its opposite (introduced in **3.6**) induces a k-linear equivalence from $k\mathcal{C}$ to the opposite category $k\mathcal{C}^{op}$.
- ([[**BT18b**] Lemma 3.4] Let X and Y be finite sets such that $|X| \leq |Y|$. Let $i : X \hookrightarrow Y$ be an injective map, and set

$$i_* = \{(i(x), x) \mid x \in X\} \in \mathcal{C}(Y, X),$$
$$i^* = \{(x, i(x)) \mid x \in X\} \in \mathcal{C}(X, Y).$$

Then $i^* i_* = \Delta_X$. It follows that if $F \in \mathcal{F}_k$, then $F(i^*)F(i_*) = \mathrm{id}_{F(X)}$, so $F(X)$ is isomorphic to a direct summand of $F(Y)$. In particular $F(X) \neq 0$ implies $F(Y) \neq 0$.

EXAMPLES 6.3.
- Yoneda functors $\mathsf{Y}_{E,k} : X \mapsto k\mathcal{C}(X,E)$, where E is a fixed finite set, e.g.
  * $E=\emptyset$: then $\mathsf{Y}_{\emptyset,k}(X) = k\mathcal{C}(X,\emptyset) \cong k$, for any finite set X. The functor $\mathsf{Y}_{\emptyset,k}$ is the constant functor with value k.
  * $E=\{\bullet\}$: then $\mathsf{Y}_{\bullet,k}(X) = k\mathcal{C}(X,\bullet) \cong k(2^X)$, for any finite set X.

  Since $\mathrm{Hom}_{\mathcal{F}_k}(Y_{E,k}, M) \cong M(E)$ by the Yoneda Lemma, for any $M \in \mathcal{F}_k$, the functor $\mathsf{Y}_{E,k}$ is a *projective* object of $\mathcal{F}_k$, for any E.
- In particular $\mathrm{End}_{\mathcal{F}_k}(\mathsf{Y}_{E,k}) \cong k\mathcal{C}(E,E)$. Let R be a *preorder* on E, i.e. $R \in \mathcal{C}(E,E)$ such that $\Delta_E \subseteq R = R^2$. Then $\mathsf{Y}_{E,k}R : X \mapsto k\mathcal{C}(X,E)R$ is a projective object of $\mathcal{F}_k$.

LEMMA 6.4 ([**BT18b**, Corollary 3.5 and Corollary 3.6]). *Let X and Y be finite sets with $|X| \leq |Y|$.*

(a) *The representable functor $k\mathcal{C}(-,X)$ is isomorphic to a direct summand of the representable functor $k\mathcal{C}(-,Y)$.*
(b) *The left $k\mathcal{C}(Y,Y)$-module $k\mathcal{C}(Y,X)$ is projective.*

PROOF. (a) In the category $k\mathcal{C}$, the object X is a retract of Y, by Remark 6.2.
(b) In particular $k\mathcal{C}(Y,X)$ is isomorphic to a direct summand of the free module $k\mathcal{C}(Y,Y)$, so it is projective. □

DEFINITION 6.5. For $F \in \mathcal{F}_k$, *the dual* $F^\natural$ of F is defined by $F^\natural(X) = \mathrm{Hom}_k\big(F(X),k\big)$ for a finite set X, and by $F^\natural(S) = {}^tF(S^{op})$ for a correspondence S from X to a finite set Y.

6.6. The results of Sections 2 and 5 yield a parametrization of simple correspondence functors, as well as a rough description of their evaluations, which will be enough for stating finiteness properties of correspondence functors in the next section. We will give a more complete description of simple correspondence functors in Section 11.

Let S be a simple correspondence functor (i.e. a simple object of $\mathcal{F}_k$). If E is a set of minimal cardinality such that $V = S(E) \neq 0$, then by Proposition 2.7 and the comments in **2.8**, the functor S is isomorphic to $S_{E,V}$, and V is a simple module for the essential algebra $\mathcal{E}_E$. So there exists a unique pair (R,W) consisting of an order R on E and a simple $k\,\mathrm{Aut}(E,R)$-module W such that V is isomorphic to the module $\Lambda_{R,W} = \mathcal{P}_E f_R \otimes_{k\,\mathrm{Aut}(E,R)} W$ of Proposition 5.3.

Conversely, if (E,R,W) is a triple consisting of a finite set E, an order R on E, and a simple $k\,\mathrm{Aut}(E,R)$-module W, we denote by $S_{E,R,W}$ the simple correspondence functor $S_{E,\Lambda_{R,W}}$. The assignment $(E,R,W) \mapsto S_{E,R,W}$ is a parametrization of the simple correspondence functors over k.

THEOREM 6.7 ([**BT18b**, Theorem 4.7]). *Assume that k is a field. The set of isomorphism classes of simple correspondence functors over k is parametrized by the set of isomorphism classes of triples (E,R,W), where E is a finite set, R is an order on E, and W is a simple $k\,\mathrm{Aut}(E,R)$-module.*

EXAMPLES 6.8. Let k be a field.

- The constant functor $Y_{\emptyset,k}$ is simple, isomorphic to $S_{\emptyset,tot,k}$: here *tot* denotes the unique (total) order on the empty set, and k is the trivial module for the trivial group $\mathrm{Aut}(\emptyset, tot)$.

- The Yoneda functor $Y_{\bullet,k}$ is not simple, but splits as a direct sum of the previous functor $S_{\emptyset,tot,k}$ and the simple functor $S_{\bullet,tot,k}$, where *tot* is the unique (total) order on a set $\bullet$ of cardinality 1, and k is the trivial module for the trivial group $\mathrm{Aut}(\bullet, tot)$.

As the Yoneda functors are projective, the simple functors $S_{\emptyset,tot,k}$ and $S_{\bullet,tot,k}$ are also projective. These are two special cases of Theorem 11.2.

The following result is the first step in the description of simple correspondence functors. It gives lower and upper estimates on the dimension of their evaluations.

THEOREM 6.9 ([**BT18b**, Theorem 8.2]). *Suppose that k is a field and let $S_{E,R,W}$ be a simple correspondence functor, where E is a finite set, R is an order on E, and W is a simple $k\,\mathrm{Aut}(E,R)$-module. There exists a positive integer N and a positive real number c such that, for any finite set X of cardinality at least N, we have*

$$c|E|^{|X|} \leq \dim\big(S_{E,R,W}(X)\big) \leq \big(2^{|E|}\big)^{|X|}.$$

PROOF SKETCH. The main tool is the fact that the simple functor $S_{E,R,W}$ is a quotient of the representable functor $k\mathcal{C}(-,E)$. This readily gives the upper bound, since $\dim_k k\mathcal{C}(X,E) = 2^{|E||X|}$ for any finite set X.

The proof of the lower bound is harder. Let A be a set of representatives of the action of the group Σ_E on the set of surjections $\varphi : X \to E$. For $\varphi \in A$, set

$$\Lambda_\varphi = \{(x,e) \in X \times E \mid \big(\varphi(x), e\big) \in R\} \in \mathcal{C}(X,E).$$

Then one can show that the images under the surjection $k\mathcal{C}(X,E) \to S_{E,R,W}(X)$ of the elements Λ_φ, for $\varphi \in A$, are linearly independent. The lower bound now follows from an estimate of the cardinality of A (see [**BT18b**] for details). □

7. Finiteness properties

The notion of *finitely generated module* over an algebra splits in two different notions in the realm of correspondence functors. Recall that k is a commutative ring, and that $\mathcal{F}_k$ denotes the category of correspondence functors over k.

DEFINITION 7.1. Let $M \in \mathcal{F}_k$.

(a) Let $(E_i)_{i\in I}$ be a family of finite sets, and for each $i \in I$, let $m_i \in M(E_i)$. We say that M is *generated* by $(m_i)_{i\in I}$, and write $M = \langle m_i \rangle_{i\in I}$, if for any finite set X and any $m \in M(X)$, there exists a finite subfamily $J \subseteq I$ and elements $\alpha_j \in k\mathcal{C}(X, E_j)$, for $j \in J$, such that $m = \sum\limits_{j\in J} M(\alpha_j)(m_j)$.

(b) M is *finitely generated* if $M = \langle m_i \rangle_{i\in I}$, where I is finite.

(c) M has *bounded type* if there is a finite set E such that $M = \langle M(E) \rangle$.

EXAMPLES 7.2.
- The Yoneda functor $Y_{E,k}$ is finitely generated (by the single element Δ_E of $Y_{E,k}(E) = k\mathcal{C}(E,E)$), for any finite set E.
- When E is a finite set, and V a $k\mathcal{C}(E,E)$-module, the functor $L_{E,V}$ (see Notation 2.3) has bounded type, generated by $L_{E,V}(E) \cong V$. It is finitely generated if and only if V is finitely generated.

PROPOSITION 7.3 ([**BT18b**, Proposition 6.4]). *Let k be a commutative ring, and M be a correspondence functor over k. The following are equivalent:*

(a) *M is finitely generated.*

(b) *M is isomorphic to a quotient of a finite direct sum* $\bigoplus_{i=1}^{n} Y_{E_i,k}$.
(c) *M is isomorphic to a quotient of a finite direct sum* $(Y_{E,k})^{\oplus n}$.
(d) *There exists a finite set E and a finite subset B of $M(E)$ such that M is generated by B. In particular M has bounded type.*

PROOF. (a)$\Rightarrow$(b). If M is generated by the set $\{m_i \mid i = 1,\dots,n\}$, where $m_i \in M(E_i)$, then each m_i yields a morphism $Y_{E_i,k} \to M$, by Yoneda's lemma. The direct sum of these morphisms is surjective.

(b)$\Rightarrow$(c). If M is a quotient of a finite direct sum $\bigoplus_{i=1}^{n} Y_{E_i,k}$. If E is the largest of the sets E_i, then each $Y_{E_i,k}$ is a direct summand, hence a quotient, of $Y_{E,k}$, so M is a quotient of $(Y_{E,k})^{\oplus n}$.

(c)$\Rightarrow$(d). Let M be a quotient of $\big(k\mathcal{C}(-,E)\big)^{\oplus n}$. Since $k\mathcal{C}(-,E)$ is generated by $\Delta_E \in k\mathcal{C}(E,E)$, the functor M is generated by the images in $M(E)$ of the elements Δ_E of each of the n components of the direct sum.

(d)$\Rightarrow$(a). This is obvious. □

LEMMA 7.4 ([**BT18b**, Lemma 9.1]). *Let k be a field and let M be a finitely generated correspondence functor over k. Then M has a maximal subfunctor.*

PROOF. Since M is finitely generated, it is generated by $M(E)$ for some finite set E, by Proposition 7.3. Moreover $M(E)$ is a finitely generated $\mathcal{R}_E$-module. Let N be a maximal submodule of $M(E)$ as an $\mathcal{R}_E$-module. Then $M(E)/N$ is a simple $\mathcal{R}_E$-module. By Proposition 2.9, there exist two subfunctors $F \subseteq G \subseteq M$ such that G/F is simple, $G(E) = M(E)$, and $F(E) = N$. Since M is generated by $M(E)$ and $G(E) = M(E)$, we have $G = M$. Therefore, F is a maximal subfunctor of M. □

The following result gives a characterization of the finitely generated correspondence functors over a field. It shows that a correspondence functor over a field is finitely generated if and only if it has finite length (i.e. it admits a finite composition series). It should be noted that this equivalence is false in general for other categories of functors (e.g. biset functors).

THEOREM 7.5 ([**BT18b**, Theorem 8.7 and Theorem 9.2]). *Let k be a field, and M be a correspondence functor over k. The following conditions are equivalent:*

(a) *M is finitely generated.*
(b) *there exist positive real numbers a, b, r such that* $\dim_k M(X) \leq ab^{|X|}$ *for any finite set X with $|X| \geq r$.*
(c) *M has finite length.*

PROOF SKETCH. (a)$\Rightarrow$(b). If M is finitely generated, it is a quotient of a finite direct sum $\big(k\mathcal{C}(-,E)\big)^{\oplus n}$, by Proposition 7.3. Now $\dim_k M(X) \leq n \dim_k k\mathcal{C}(X,E)$ $= n2^{|X||E|}$, for any finite set X, so (b) holds.

The proof that (b) implies (a) is harder. We first use Theorem 6.9 to show that if $S_{E,R,W}$ is a simple subquotient of M, then the cardinality of E is bounded above by the number b appearing in (b). Then one builds a morphism π from a finite direct sum L of representable functors to M, which is surjective on evaluations at any set X of cardinality at most b. If now F is a finite set of minimal cardinality such that $\operatorname{Coker}\pi(F)$ is non-zero, then $|F| > b$. Moreover $\operatorname{Coker}\pi(F)$ is a non-zero module for the essential algebra $\mathcal{E}_F$, hence it admits a simple submodule V. By

Proposition 2.9, it follows that $\operatorname{Coker}\pi$, hence M, admits a simple subquotient isomorphic to $S_{F,V}$. But then $|F| \leq b$, a contradiction proving that π is surjective, and that (b) implies (a). So (a) and (b) are equivalent, and in particular, any subfunctor of a finitely generated correspondence functor over k is finitely generated.

(c)$\Rightarrow$(b). This follows from Theorem 6.9.

(b)$\Rightarrow$(c). By Lemma 7.4, we can build a filtration $M = M_0 \supset M_1 \supset M_2 \ldots$ in which every subquotient M_i/M_{i+1} is simple, isomorphic to S_{E_i,R_i,W_i} for some triple (E_i, R_i, W_i). Now (b) implies by Theorem 6.9 that $|E_i| \leq b$ for any i. In particular, there is only a finite number of simple quotients S_{E_i,R_i,W_i} in the filtration, up to isomorphism. But since the dimension of $M(X)$ is finite for any finite set X, each of the simple quotients S_{E_i,R_i,W_i} can only occur a finite number of times in the filtration. This shows that the filtration is finite, so M has finite length. □

The equivalence of (a) and (c) in Theorem 7.5 has been proved independently by Gitlin [**Git18**] (for an infinite field k), using results of Wiltshire-Gordon on categories of dimension 0 [**WG19**].

The following is a consequence of the adjunction of Lemma 2.4.

PROPOSITION 7.6 ([**BT18b**, Lemma 7.3]). *Let M be a correspondence functor over k, and E be a finite set.*

(a) *If M is projective in $\mathcal{F}_k$, and M is generated by $M(E)$, then $M \cong L_{F,M(F)}$ for any finite set F with $|F| \geq |E|$, and $M(F)$ is a projective $\mathcal{R}_F$-module.*
(b) *The functor $L_{E,V}$ is projective (resp. indecomposable) if and only if V is a projective (resp. indecomposable) $\mathcal{R}_E$-module.*

We introduce next the notion of *residue* of a correspondence functor at a finite set, which is an important tool to study finiteness conditions.

DEFINITION 7.7. Let M be a correspondence functor over k, and E be a finite set. The *residue* of M at E is the k-module defined by

$$\overline{M}(E) = M(E)/\sum_{|F|<|E|} k\mathcal{C}(E,F)M(F).$$

THEOREM 7.8 ([**BT18b**, Theorem 11.4 and Corollary 11.5]). *Let k be a commutative noetherian ring. Let M be a subfunctor of a correspondence functor L over k, and let E and F be finite sets.*

(a) *If L is generated by $L(F)$ and $\overline{M}(E) \neq 0$, then $|E| \leq 2^{|F|}$.*
(b) *If L is generated by $L(F)$ and $|E| \geq 2^{|F|}$, then M is generated by $M(E)$.*
(c) *If L has bounded type, then M has bounded type.*
(d) *If L is finitely generated, then M is finitely generated.*

PROOF SKETCH. (a) is proved by localization and Artin-Rees Lemma (Theorem 8.5 in [**Mat89**]). Then (*b*), (*c*), and (*d*) follow easily. □

COROLLARY 7.9. *Functors of bounded type form an abelian subcategory $\mathcal{F}_k^b$ of $\mathcal{F}_k$. Finitely generated functors form an abelian subcategory $\mathcal{F}_k^f$ of $\mathcal{F}_k^b$.*

PROPOSITION 7.10 ([**BT18b**, Proposition 6.6]). *Let k be a noetherian ring.*

(a) *For any a finitely generated correspondence functor M over k, the algebra $\operatorname{End}_{\mathcal{F}_k}(M)$ is a finitely generated k-module.*
(b) *For any two finitely generated correspondence functors M and N over k, the k-module $\operatorname{Hom}_{\mathcal{F}_k}(M,N)$ is finitely generated.*

(c) *If k is a field, then the Krull-Remak-Schmidt theorem holds for finitely generated correspondence functors over k.*

PROOF. (a) M is a quotient of a projective functor $\overset{n}{\underset{i=1}{\oplus}} k\mathcal{C}(-,E)$, so $\mathrm{End}_{\mathcal{F}_k}(M)$ is a quotient of a k-submodule of the finitely generated k-module $M_n\big(k\mathcal{C}(E,E)\big)$. (b) $\mathrm{Hom}_{\mathcal{F}_k}(M,N)$ is a direct summand of $\mathrm{End}_{\mathcal{F}_k}(M\oplus N)$. (c) Splitting $M\in\mathcal{F}_k$ amounts to splitting the identity as a sum of orthogonal idempotents in the finite dimensional k-algebra $\mathrm{End}_{\mathcal{F}_k}(M)$. □

The following three results show that when k is a field, the category $\mathcal{F}_k^f$ shares many properties of the category of modules over a finite group algebra.

THEOREM 7.11 ([**BT18b**, Theorem 10.2 and Corollary 10.3]). *Let E be a finite set.*

(a) *The representable functor $Y_{E,k}=k\mathcal{C}(-,E)$ is isomorphic to its dual.*
(b) *Let $\mathcal{R}_E=k\mathcal{C}(E,E)$ be the algebra of relations on E. Then $\mathcal{R}_E$ is a symmetric k-algebra, in the sense of* [**Bro09**]. *More precisely, let $t_E:\mathcal{R}_E\to k$ be defined by*

$$\forall R\in\mathcal{C}(E,E),\ t_E(R)=\begin{cases}1 & \text{if } R\cap\Delta_E=\emptyset,\\ 0 & \text{otherwise.}\end{cases}$$

Then t_E is a symmetrizing form for $\mathcal{R}_E$.
(c) *If k is a field (or more generally if k is self injective), any finitely generated projective correspondence functor over k is also injective. In particular $\mathcal{F}_k^f$ has infinite global dimension.*

PROOF. (b) Let $R,S\in\mathcal{C}(E,E)$. Then $t_E(RS)=1\Leftrightarrow R\subseteq(E\times E)-S^{op}$. The matrix $\big(t_E(RS)\big)_{R,S\in\mathcal{C}(E,E)}$ is the product of a permutation matrix ($S\mapsto(E\times E)-S^{op}$) with the matrix of an order ($\subseteq$), hence it is invertible (over $\mathbb{Z}$).

(a) follows from a similar argument, and (c) is an easy consequence of (a). □

THEOREM 7.12 ([**BT18b**, Theorem 10.6]). *Let k be a field, and M be a finitely generated correspondence functor over k. The following are equivalent:*

(a) *M is projective and indecomposable.*
(b) *M is projective and admits a unique maximal proper subfunctor.*
(c) *M is projective and admits a unique minimal non-zero subfunctor.*
(d) *M is injective and indecomposable.*
(e) *M is injective and admits a unique maximal proper subfunctor.*
(f) *M is injective and admits a unique minimal non-zero subfunctor.*

PROOF. We first note that if M is finitely generated, it has finite length by Theorem 7.5, hence (if it is non zero) it admits a maximal subfunctor and a minimal subfunctor.

(a)⇒(b). If M is projective and indecomposable, and generated by its evaluation at some finite set E, we can assume $M=L_{E,V}$, where V is an indecomposable projective $\mathcal{R}_E$-module, by Proposition 7.6. Since $\mathcal{R}_E$ is a finite dimensional k-algebra, the module V has a unique maximal submodule W. If N is a subfunctor of M, then $N(E)$ is a submodule of $M(E)=V$. If $N(E)=V$, then $N=M$, as M is generated by $M(E)=V$. Hence $N(E)\subseteq W$ if $N\neq M$. So if N and N' are distinct maximal subfunctors of M, then $N+N'=M$, which implies $N(E)+N'(E)=V$, a contradiction since $N(E)$ and $N'(E)$ are both contained in W.

(b)⇒(a). If M has a unique maximal subfunctor and $M = M_1 \oplus M_2$ for non zero subfunctors M_1 and M_2, then M_1 and M_2 are finitely generated by Theorem 7.5, so admit maximal subfunctors N_1 and N_2, respectively. Then $M_1 \oplus N_2$ and $N_1 \oplus M_2$ are distinct maximal subfunctors of M.

(a)⇒(d). If M is a finitely generated indecomposable projective functor, then M is injective (and indecomposable) by Theorem 7.11.

(d)⇒(a). If M is finitely generated and injective, then $M^\natural$ is projective and finitely generated (as M and $M^\natural$ both have finite length), hence injective, so $M \cong (M^\natural)^\natural$ is projective.

(a)⇒(c). For a finitely generated functor M, the duality between M and $M^\natural$ induces an order reversing bijection between the subfunctors of M and the subfunctors of $M^\natural$. If M is projective and indecomposable, then so is $M^\natural$, that is, (a) holds for $M^\natural$. Thus (b) holds for $M^\natural$ and the functor $M^\natural$ has a unique maximal subfunctor. Hence M has a unique minimal subfunctor.

(c)⇒(a). If M is projective and admits a unique minimal subfunctor, then M is also injective, and its dual $M^\natural$ is projective and admits a unique maximal subfunctor. Hence $M^\natural$ is indecomposable, so M is indecomposable. It is now clear that (e) and (f) are both equivalent to (a), (b), (c) and (d). □

THEOREM 7.13 ([**BT18b**, Theorem 10.7]). *Let k be a field.*

(a) *Let M be a finitely generated projective correspondence functor. Then*

$$M/\operatorname{Rad}(M) \cong \operatorname{Soc}(M).$$

(b) *Let M and N be finitely generated correspondence functors over k. If M is projective, then*

$$\dim_k \operatorname{Hom}_{\mathcal{F}_k}(M, N) = \dim_k \operatorname{Hom}_{\mathcal{F}_k}(N, M) < +\infty.$$

PROOF SKETCH. The proof is a generalization of the proof of the similar result for modules over a symmetric algebra ([**Ben91**] Theorem 1.6.3). It relies mainly on Theorem 7.11 (see [**BT18b**] for details). □

8. Stability

When k is noetherian, the Hom-sets and more generally the Ext-groups from correspondence functors of bounded type over k to arbitrary correspondence functors over k are detected on their evaluations at large enough finite sets. More precisely:

THEOREM 8.1 ([**BT18b**, Theorem 12.3]). *Let k be a noetherian ring, let $M, N \in \mathcal{F}_k$, and let E, F be finite sets.*

(a) *If M is generated by $M(E)$, then for $|F| \geq 2^{|E|}$, the evaluation map*

$$\operatorname{Hom}_{\mathcal{F}_k}(M, N) \to \operatorname{Hom}_{\mathcal{R}_F}\big(M(F), N(F)\big)$$

is an isomorphism.

(b) *If M has bounded type, then for any $i \in \mathbb{N}$, there exists $n_i \in \mathbb{N}$ such that if $|F| \geq n_i$, the map*

$$\operatorname{Ext}^i_{\mathcal{F}_k}(M, N) \to \operatorname{Ext}^i_{\mathcal{R}_F}\big(M(F), N(F)\big)$$

is an isomorphism.

When k is noetherian, this leads to the following alternative description of $\mathcal{F}_k^b$, up to equivalence of categories.

DEFINITION 8.2. Let $\mathcal{G}_k$ be the following category:

- the objects of $\mathcal{G}_k$ are pairs (E,U), where E is a finite set, and U is an $\mathcal{R}_E$-module.
- a morphism $(E,U) \to (F,V)$ in $\mathcal{G}_k$ is a morphism of $\mathcal{R}_E$-modules
$$U \to k\mathcal{C}(E,F) \otimes_{\mathcal{R}_F} V.$$
- the composition in $\mathcal{G}_k$ of $f : (E,U) \to (F,V)$ and $g : (F,V) \to (G,W)$ given respectively by
$$f : U \to k\mathcal{C}(E,F) \otimes_{\mathcal{R}_F} V \text{ and } g : V \to k\mathcal{C}(F,G) \otimes_{\mathcal{R}_G} W$$
is the morphism $(E,U) \to (G,W)$ given by the composition
$$U \xrightarrow{f} k\mathcal{C}(E,F) \otimes_{\mathcal{R}_F} V \xrightarrow{h} k\mathcal{C}(E,F) \otimes_{\mathcal{R}_F} k\mathcal{C}(F,G) \otimes_{\mathcal{R}_G} W \xrightarrow{l} k\mathcal{C}(E,G) \otimes_{\mathcal{R}_G} W$$
where $\begin{cases} h = \mathrm{id}_{k\mathcal{C}(E,F)} \otimes_{\mathcal{R}_F} g \\ l = \mu \otimes_{\mathcal{R}_G} \mathrm{id}_W \end{cases}$, and $\mu : k\mathcal{C}(E,F) \otimes_{\mathcal{R}_F} k\mathcal{C}(F,G) \to k\mathcal{C}(E,G)$ is the composition in $k\mathcal{C}$.
- the identity morphism of (E,U) is $U \overset{\cong}{\to} k\mathcal{C}(E,E) \otimes_{\mathcal{R}_E} U$.

THEOREM 8.3 ([**BT18b**, Theorem 12.7]). (a) *The assignment* $(E,U) \mapsto L_{E,U}$ *is a fully faithful* k*-linear functor* $\mathcal{G}_k \to \mathcal{F}_k^b$.

(b) *When* k *is noetherian, it is an equivalence of categories. In particular* $\mathcal{G}_k$ *is abelian.*

9. Correspondence functors and lattices

9.1. In this section, we consider finite lattices, and build important examples of correspondence functors from them. Recall that a lattice $(T, \vee, \wedge)$ is a poset in which any pair $\{x, y\}$ of elements admits a least upper bound $x \vee y$, called the *join* of x and y, and a largest lower bound $x \wedge y$, called the *meet* of x and y.

If T is a lattice, and T^{op} is the opposite poset, then $(T^{op}, \wedge, \vee)$ is a lattice. A lattice T is called *distributive* if $x \wedge (y \vee z) = (x \wedge y) \vee (x \wedge z)$ for any x, y, and z in T. One can show that T is distributive if and only if T^{op} is, that is, if $x \vee (y \wedge z) = (x \vee y) \wedge (x \vee z)$ for any x, y, and z in T.

A *finite* lattice T always admits a *largest element* (the join of all elements of T), denoted by 1_T, and a *smallest element* (the meet of all elements of T), denoted by 0_T.

NOTATION 9.2. Let T be a finite lattice.

- For a finite set X, we denote by $F_T(X) = k(T^X)$ the free k-module with basis the set T^X of all maps from X to T.
- For a finite set Y and a correspondence $S \in \mathcal{C}(Y,X)$, we define a k-linear map $F_T(S) : F_T(X) \to F_T(Y)$ by sending $\varphi : X \to T$ to $S\varphi : Y \to T$ defined by
$$\forall y \in Y,\ (S\varphi)(y) = \bigvee_{(y,x) \in S} \varphi(x).$$

THEOREM 9.3 ([**BT19a**, Proposition 4.2 and Theorem 4.12]). *Let* k *be a commutative ring, and* T *be finite lattice.*

(a) F_T *is a correspondence functor over* k.
(b) F_T *is* projective *in* $\mathcal{F}_k$ *if and only if* T *is* distributive.

In order to study the functorial aspects of the construction $T \mapsto F_T$, we introduce the following category $\mathcal{L}$ and its k-linearization (see Definition 2.1):

DEFINITION 9.4. • Let T and T' be finite lattices. A map $f : T \to T'$ is called a *join morphism* if $f(\bigvee_{t\in A} t) = \bigvee_{t\in A} f(t)$, for any subset A of T.

- Let $\mathcal{L}$ denote the following category:
  - the objects of $\mathcal{L}$ are the finite lattices.
  - for finite lattices T and T', the set $\mathrm{Hom}_{\mathcal{L}}(T, T')$ is the set of join morphisms from T to T'.
  - the composition of morphisms in $\mathcal{L}$ is the composition of maps, and the identity morphism of a finite lattice T is the identity map of T.
- Let $k\mathcal{L}$ be the k-linearization of $\mathcal{L}$.

Note that for finite lattices T and T', a join morphism $f : T \to T'$ always maps 0_T to $0_{T'}$ (this follows from the case $A = \emptyset$ in the definition). Conversely, if f is a map from T to T' such that $f(0_T) = 0_{T'}$ and $f(a \vee b) = f(a) \vee f(b)$ for any a and b in T, then f is a join morphism.

For our next theorem, we need some notation. Let $f : T \to T'$ be a morphism in the category $\mathcal{L}$. For a finite set X, let $F_{f,X} : F_T(X) \to F_{T'}(X)$ be the k-linear map sending the function $\varphi : X \to T$ to the function $f \circ \varphi : X \to T'$.

THEOREM 9.5 ([**BT19a**, Theorem 4.8]). (a) *Let $f : T \to T'$ be a morphism in the category $\mathcal{L}$. Then the collection of maps $F_{f,X} : F_T(X) \to F_{T'}(X)$, for all finite sets X, yields a natural transformation $F_f : F_T \to F_{T'}$ of correspondence functors.*

(b) *The assignment sending a lattice T to F_T, and a morphism $f : T \to T'$ in $\mathcal{L}$ to $F_f : F_T \to F_{T'}$, yields a functor $\mathcal{L} \to \mathcal{F}_k$. This functor extends uniquely to a k-linear functor*

$$F_? : k\mathcal{L} \longrightarrow \mathcal{F}_k \ .$$

(c) *The functor $F_?$ is fully faithful.*

The functor $F_?$ is also well behaved with the duality $F \mapsto F^\natural$ of functors introduced in Definition 6.5. To state this, we need the following notation:

NOTATION 9.6. For a morphism $f : T \to T'$ in $\mathcal{L}$, let $f^{op} : T' \to T$ be the map defined by

$$\forall t' \in T',\ \ f^{op}(t') = \bigvee_{\substack{x\in T\\ f(x)\leq t'}} x,$$

where the join is taken in T.

THEOREM 9.7 ([**BT19a**, Lemma 8.1 and Theorem 8.9]). (a) *Let T and T' be finite lattices. If $f : T \to T'$ is a morphism in $\mathcal{L}$, then $f^{op} : T'^{op} \to T^{op}$ is a morphism in $\mathcal{L}$.*

(b) *The assignment $T \mapsto T^{op}$ extends to an equivalence $\mathcal{L} \to \mathcal{L}^{op}$ and a k-linear equivalence $k\mathcal{L} \to k\mathcal{L}^{op}$.*

(c) *The functors $T \mapsto (F_T)^\natural$ and $T \mapsto F_{T^{op}}$ from $k\mathcal{L}$ to $\mathcal{F}_k^{op}$ are naturally isomorphic.*

10. Fundamental functors

10.1. Let (E,R) be a finite poset. In Proposition 5.3, we have introduced, for any $k\,\mathrm{Aut}(E,R)$-module W, an $\mathcal{R}_E$-module $\Lambda_{R,W} = \mathcal{P}_E f_R \otimes_{k\,\mathrm{Aut}(E,R)} W$. So far we have used this construction only in the case where W is simple, in order to parametrize the simple correspondence functors over k, by setting $S_{E,R,W} = S_{E,\Lambda_{R,W}}$, where $S_{E,\Lambda_{R,W}}$ is the functor $S_{X,V}$ introduced in Notation 2.6, in the case of the object $X = E$ of the category $\mathcal{D} = \mathcal{C}$, and the $k\mathcal{D}(X,X) = \mathcal{R}_E$-module $V = \Lambda_{R,W}$.

We will now consider the correspondence functor $S_{E,\Lambda_{R,W}}$ over k associated to the *free module* $W = k\,\mathrm{Aut}(E,R)$. Then $\Lambda_{R,W} = \mathcal{P}_E f_R \otimes_{k\,\mathrm{Aut}(E,R)} k\,\mathrm{Aut}(E,R) \cong \mathcal{P}_E f_R$.

Definition 10.2. Let (E,R) be a finite poset. The *fundamental correspondence functor* over k associated to (E,R) is the functor

$$\mathbb{S}_{E,R} = S_{E,\mathcal{P}_E f_R}.$$

We will see that the fundamental functors associated to finite posets allow for a precise description of all the simple correspondence functors.

Our first observation is that $\mathbb{S}_{E,R}$ is by definition a quotient of the functor $L_{E,\mathcal{P}_E f_R} = k\mathcal{C}(-,E) \otimes_{\mathcal{R}_E} \mathcal{P}_E f_R$, which is itself a quotient of the representable functor $k\mathcal{C}(-,E)$, since $L_{E,\mathcal{P}_E f_R}$ is generated by the element $\Delta_E \otimes_{\mathcal{R}_E} f_R$ of its evaluation $L_{E,\mathcal{P}_E f_R}(E) = k\mathcal{C}(E,E) \otimes_{\mathcal{R}_E} \mathcal{P}_E f_R \cong \mathcal{P}_E f_R$.

Since $R = \sum\limits_{\substack{S \in \mathcal{O}_E \\ S \supseteq R}} f_S$ in $\mathcal{P}_E$ by definition of the idempotents f_S, we have $R f_R = f_R$ in $\mathcal{P}_E$. Thus $R(\Delta_E \otimes_{\mathcal{R}_E} f_R) = R \otimes_{\mathcal{R}_E} f_R = \Delta_E \otimes_{\mathcal{R}_E} R f_R = \Delta_E \otimes_{\mathcal{R}_E} f_R$, and it follows that the direct summand $k\mathcal{C}(-,E)R$ of $k\mathcal{C}(-,E)$ maps surjectively onto $\mathbb{S}_{E,R}$, so we get a canonical surjective morphism

$$\pi_{E,R} : k\mathcal{C}(-,E)R \twoheadrightarrow \mathbb{S}_{E,R}.$$

Before going further, we need to recall some classical definitions and results on finite lattices.

Definition 10.3.
- Let T be a finite lattice. An element e of T is called *join-irreducible* if whenever e can be written $e = \bigvee\limits_{a \in A} a$ for some subset A of T, then $e \in A$. We denote by $\mathrm{Irr}(T)$ the set of join-irreducible elements of T, viewed as a full subposet of T.
- Let (E,R) be a finite poset. For $e \in E$, set

$$]\cdot, e]_{E,R} = R_e = \{x \in E \mid (x,e) \in R\}.$$

A *lower ideal* of (E,R) is a *closed downwards* subset of E, i.e. a subset A such that $]\cdot, x]_{E,R} \subseteq A$ whenever $x \in A$. Let $I_\downarrow(E,R)$ be the set of lower ideals of (E,R), ordered by inclusion of subsets.

The poset $I_\downarrow(E,R)$ is a distributive lattice, the join operation being union of subsets, and the meet operation being intersection of subsets. It is easily seen moreover that

$$\mathrm{Irr}(I_\downarrow(E,R)) = \{]\cdot, e]_{E,R} \mid e \in E\}.$$

The map $e \in E \mapsto]\cdot, e]_{E,R}$ is an isomorphism of posets from (E,R) to $\mathrm{Irr}(I_\downarrow(E,R))$. We will freely abuse notation and identify these two posets.

When T is a finite lattice, and $(E,R)=\mathrm{Irr}(T)$, the map

$$s_T: A\in I_\downarrow(E,R)\mapsto \bigvee_{a\in A} a\in T$$

is a surjective join-morphism of lattices. By standard theorems (see [**Sta12**] Theorem 3.4.1 and Proposition 3.4.2, and also [**Rom08**] Theorem 6.2), it is a bijection if and only if T is distributive, and then it is an isomorphism of lattices.

PROPOSITION 10.4 ([**BT19a**, Proposition 4.5 and Theorem 6.5]). *Let T be a finite lattice, and $(E,R)=\mathrm{Irr}(T)$.*

(a) *There is a canonical isomorphism of functors $F_{I_\downarrow(E,R)}\cong k\mathcal{C}(-,E)R^{op}$.*

(b) *The morphism $F_{I_\downarrow(E,R)}\cong k\mathcal{C}(-,E)R^{op}\xrightarrow{\pi_{R,T}}\mathbb{S}_{E,R^{op}}$ factors as*

$$F_{I_\downarrow(E,R)}\xrightarrow{F_{s_T}} F_T\xrightarrow{\Theta_T}\mathbb{S}_{E,R^{op}}.$$

DEFINITION 10.5. Let T be a finite lattice, and $(E,R)=\mathrm{Irr}(T)$.

- For $t\in T$, let $r(t)=\bigvee_{\substack{e\in E\\ e<t}} e$ and $\sigma(t)=\bigwedge_{\substack{e\in E\\ e>t}} e$. Define inductively $r^n(t)$ and $\sigma^n(t)$ by $r^n(t)=r\big(r^{n-1}(t)\big)$ and $\sigma^n(t)=\sigma\big(\sigma^{n-1}(t)\big)$.
 Then $t\geq r(t)\geq\ldots\geq r^n(t)=r^{n+1}(t)$ for some n. Set $r^\infty(t)=r^n(t)$. Similarly $t\leq\sigma(t)\leq\ldots\sigma^m(t)=\sigma^{m+1}(t)$ for some m. Set $\sigma^\infty(t)=\sigma^m(t)$.
- Let $G_T=E\sqcup\{t\in T\mid t=r^\infty\sigma^\infty(t)\}$, and $G(E,R)=G_{I_\downarrow(E,R)}$.

Note that the union in the definition of G_T is disjoint because $r(t)<t$ if (and only if) $t\in E$.

THEOREM 10.6 ([**BT18a**, Theorem 5.6]). *Let (E,R) be a finite poset, and T be a finite lattice such that $\mathrm{Irr}(T)\cong(E,R)$.*

(a) *Let $\overline{G}(E,R)$ denote the image of $G(E,R)$ by $s_T: I_\downarrow(E,R)\twoheadrightarrow T$.*
 - *The restriction of the map s_T to $G(E,R)$ is an isomorphism of posets $G(E,R)\to\overline{G}(E,R)$. Moreover $\overline{G}(E,R)=G_T$. In particular G_T only depends on the poset $\mathrm{Irr}(T)$, up to isomorphism.*
 - *For any finite set X, the set*

 $$\{\varphi: X\to T\mid E\subseteq\varphi(X)\subseteq\overline{G}(E,R)\}\subseteq F_T(X)$$

 is mapped by $\Theta_{T,X}: F_T(X)\to\mathbb{S}_{E,R^{op}}(X)$ to a k-basis of $\mathbb{S}_{E,R^{op}}(X)$.

(b) *For any finite set X, the k-module $\mathbb{S}_{E,R}(X)$ is free of rank*

$$rk_k\,\mathbb{S}_{E,R}(X)=\sum_{i=0}^{e}(-1)^i\binom{e}{i}(g-i)^{|X|},$$

where $e=|E|$ and $g=|G(E,R)|$.

EXAMPLES 10.7. (The black nodes of $I_\downarrow(E,R)$ are the join-irreducible elements)

(E,R)	$I_\downarrow(E,R)$	$G(E,R)$
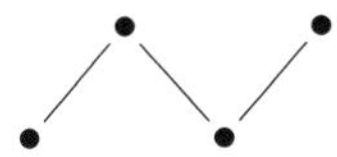	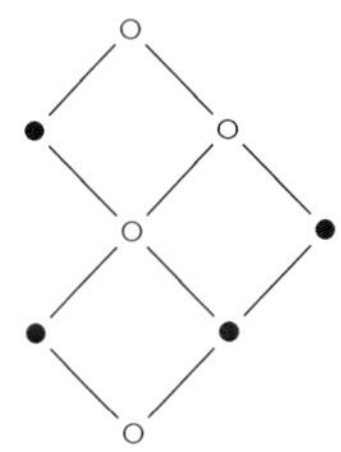	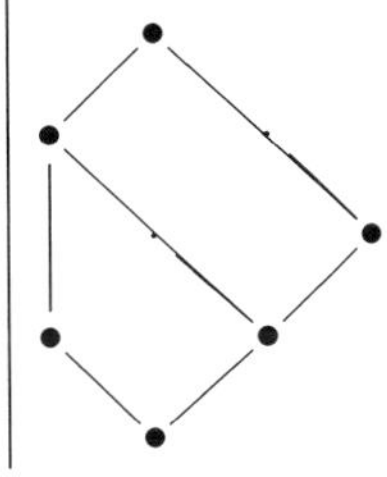

$rk_k\mathbb{S}_{E,R}(X) = 6^x - 4.5^x + 6.4^x - 4.3^x + 2^x$, where $x = |X|$.

(E,R)	$I_\downarrow(E,R)$	$G(E,R)$
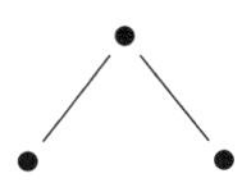	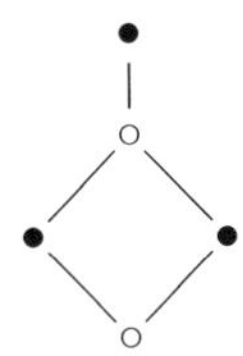	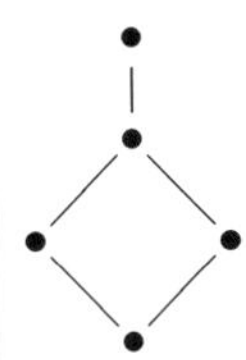

$rk_k\mathbb{S}_{E,R}(X) = 5^x - 3.4^x + 3.3^x - 2^x$, where $x = |X|$.

11. The simple correspondence functors

THEOREM 11.1 ([**BT18a**, Theorem 6.6 and Theorem 6.10]). *Let (E,R) be a finite poset, and X be a finite set.*

(a) *$\mathbb{S}_{E,R}(X)$ is a free right $k\,\mathrm{Aut}(E,R)$-module.*
(b) *Let k be a field, and W be a simple $k\,\mathrm{Aut}(E,R)$-module. Then*

$$S_{E,R,W}(X) \cong \mathbb{S}_{E,R}(X) \otimes_{k\,\mathrm{Aut}(E,R)} W.$$

In particular

$$\dim_k S_{E,R,W}(X) = \frac{\dim_k W}{|\,\mathrm{Aut}(E,R)|}\sum_{i=0}^{e}(-1)^i\binom{e}{i}(g-i)^{|X|},$$

where $e = |E|$ and $g = |G(E,R)|$.

In the case of a totally ordered poset, the previous result takes the following form. Let $n \in \mathbb{N}$. Set $\underline{n} = \{0 < 1 < \ldots < n\}$, and $[n] = \mathrm{Irr}(\underline{n}) = \underline{n} - \{0\}$. Set $\mathbb{S}_n = \mathbb{S}_{[n],tot}$, where tot is the natural total order on $[n]$. Observe that $([n],tot) \cong ([n],tot^{op})$

THEOREM 11.2 ([**BT19a**, Proposition 10.2, Theorems 11.1, 11.6 and 11.8]).

(a) *The surjection $F_{\underline{n}} \to \mathbb{S}_n$ splits. The functor $\mathbb{S}_n$ is projective.*
(b) *If X is a finite set, then $\mathbb{S}_n(X)$ is a free k-module of rank*

$$\sum_{i=0}^{n}(-1)^{n-i}\binom{n}{i}(i+1)^{|X|}.$$

(c) $F_{\underline{n}} \cong \bigoplus_{A\subseteq[n]} \mathbb{S}_{|A|} \cong \bigoplus_{j=0}^{n} (\mathbb{S}_j)^{\oplus\binom{n}{j}}.$

(d) $\mathrm{End}_{k\mathcal{L}}(\underline{n}) \cong \mathrm{End}_{\mathcal{F}_k}(F_{\underline{n}}) \cong \prod_{j=0}^{n} M_{\binom{n}{j}}(k)$.

(e) *If k is a field, then $\mathbb{S}_n$ is simple (and projective, and injective), isomorphic to $S_{[n],tot,k}$.*

The case of a total order raises the question of determining all simple projective correspondence functors. The answer is as follows.

DEFINITION 11.3.
- If X and Y are posets, let $X * Y$ denote the set $X \sqcup Y$, ordered by the orders of X and Y, and moreover $x \leq y$ for $x \in X$ and $y \in Y$.
- A *pole poset* is a poset of the form $E_1 * E_2 * \ldots E_n$, where E_i is a set of cardinality 1 or 2, ordered by equality, for $i = 1, \ldots, n$.

THEOREM 11.4 ([**BT19b**, Theorem 4.5]). *Let k be a field, and $S_{E,R,W}$ be a simple correspondence functor, where (E,R) is a finite poset, and W is a simple $k\,\mathrm{Aut}(E,R)$-module. The following conditions are equivalent:*

(a) *$S_{E,R,W}$ is projective.*
(b) *(E,R) is a pole poset, and W is a projective $k\,\mathrm{Aut}(E,R)$-module.*
(c) *Either (E,R) is totally ordered or (E,R) is a pole poset and* $\mathrm{char}(k) \neq 2$.

EXAMPLES 11.5. [[**BT18a**] Example 8.9 and Example 8.10]

- Let D be the following lattice:

$D =$ 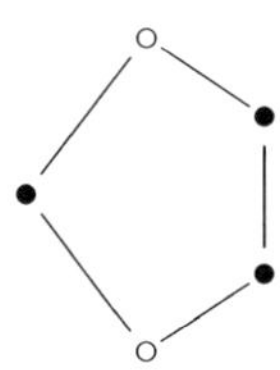

One can show that the functor F_D splits as

$$F_D \cong \mathbb{S}_0 \oplus 4\mathbb{S}_1 \oplus 4\mathbb{S}_2 \oplus \mathbb{S}_3 \oplus 2\mathbb{S}_{\bullet\bullet} \oplus \mathbb{S}_{\bullet\,\vdots}\,,$$

where $\bullet\bullet$ is a set of cardinality 2, ordered by equality, and $\bullet\,\vdots$ is a poset of cardinality 3 with 2 connected components.

In particular, when k is a field of characteristic different from 2, the functor F_D is semisimple. Since D is not distributive, the functor F_D is not projective, by Theorem 9.3. The only non projective direct summand in the above decomposition is $\mathbb{S}_{\bullet\,\vdots}$.

- Let T be the following lattice :

$T =$ 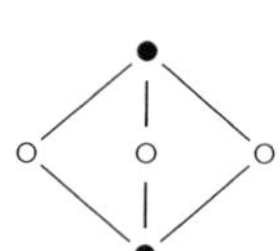

One can show that F_T splits as

$$F_T \cong \mathbb{S}_0 \oplus 4\mathbb{S}_1 \oplus 3\mathbb{S}_2 \oplus 3\mathbb{S}_{\bullet\bullet} \oplus \mathbb{S}_{\bullet\bullet\bullet}\,,$$

where $\bullet\bullet\bullet$ is a set of cardinality 3 ordered by equality. All the summands in this decomposition of F_T, except possibly $\mathbb{S}_{\bullet\bullet\bullet}$, are projective functors.

Since the lattice T is not distributive, the functor F_T is not projective, thus $\mathbb{S}_{\bullet\bullet\bullet}$ is actually not projective either.

REMARK 11.6. The sequence of multiplicities of $\mathbb{S}_n$ as a direct summand of F_D is 1, 4, 4, 1, 0 for $n = 0, 1, 2, 3, 4$, and the number of chains $0_D = x_0 < x_1 < \ldots < x_n$ in D is equal to 1, 4, 4, 1, 0 for $n = 0, 1, 2, 3, 4$. Similarly, the sequence of multiplicities of $\mathbb{S}_n$ as a direct summand of F_T is 1, 4, 3, 0 for $n = 0, 1, 2, 3$, and the number of chains $0_T = x_0 < x_1 < \ldots < x_n$ in T is equal to 1, 4, 3, 0 for $n = 0, 1, 2, 3$. This is *not* a coincidence... (see [**BT19a**] Section 10 and Section 11 for details).

THEOREM 11.7 ([**BT18a**, Theorem 7.1 and Theorem 7.2]). *Let X be a finite set, and $\mathcal{R}_X = k\mathcal{C}(X, X)$, where k is a field.*

(a) *The set of isomorphism classes of simple $\mathcal{R}_X$-modules is parametrized by the set of isomorphism classes of triples (E, R, W), where E is a finite set with $|E| \leq |X|$, R is an order relation on E, and W is a simple $k\operatorname{Aut}(E, R)$-module.*
(b) *The simple $\mathcal{R}_X$-module parametrized by the triple (E, R, W) is $S_{E,R,W}(X)$.*
(c) *In particular, the dimension of this simple module is equal to*

$$\dim_k S_{E,R,W}(X) = \frac{\dim_k W}{|\operatorname{Aut}(E, R)|} \sum_{i=0}^{|E|} (-1)^i \binom{|E|}{i} \big(|G(E, R)| - i\big)^{|X|}.$$

COROLLARY 11.8 ([**BT18a**, Theorem 7.4]). *Let k be a field of characteristic 0. Let X be a finite set, and $J(\mathcal{R}_X)$ denote the Jacobson radical of $\mathcal{R}_X = k\mathcal{C}(X, X)$. Set $n = |X|$. Then*

$$\dim_k J(\mathcal{R}_X) = 2^{n^2} - \sum_{e=0}^{n} \sum_{R} \frac{1}{|\operatorname{Aut}(E, R)|} \Big(\sum_{i=0}^{e} (-1)^i \binom{e}{i} \big(|G(E, R)| - i\big)^n \Big)^2,$$

where R runs through a set of Σ_E-conjugacy classes of order relations on $E = \{1, \ldots, e\}$, and $E = \emptyset$ if $e = 0$.

For small values of n, the dimension of $J(\mathcal{R}_X)$ is given in the following table. For $n \leq 2$, this dimension is 0, so the algebra $k\mathcal{R}_X$ is semisimple. For $n = 3$, the dimension of $J(\mathcal{R}_X)$ is 42, and follows from a computer calculation in [**Bre14**]. This value can be confirmed by hand using the above formula. This value has also been confirmed by a direct computation using the computer software GAP [**GAP15**]. For $n \geq 4$, a direct computation using the algebra $\mathcal{R}_X$ is not possible on current computers, because the dimension 2^{n^2} of this algebra is too big. The table below has been completed via a computer calculation using GAP and the above formula.

n	$\dim_k J(\mathcal{R}_X)$
≤ 2	0
3	42
4	$32,616$
5	$29,446,050$
6	$67,860,904,320$
7	$562,649,705,679,642$
8	$18,446,568,932,288,588,616$

References

[Ben91] D. J. Benson, *Representations and cohomology. I: Basic representation theory of finite groups and associative algebras*, Cambridge Studies in Advanced Mathematics, vol. 30, Cambridge University Press, Cambridge, 1991. MR1110581

[Bir46] Garrett Birkhoff, *On groups of automorphisms* (Spanish), Rev. Un. Mat. Argentina **11** (1946), 155–157. MR15387

[BM09] Jonathan Ariel Barmak and Elias Gabriel Minian, *Automorphism groups of finite posets*, Discrete Math. **309** (2009), no. 10, 3424–3426, DOI 10.1016/j.disc.2008.09.026. MR2528204

[Bou96] Serge Bouc, *Foncteurs d'ensembles munis d'une double action* (French), J. Algebra **183** (1996), no. 3, 664–736, DOI 10.1006/jabr.1996.0238. MR1401673

[Bre14] Murray R. Bremner, *Structure of the rational monoid algebra for Boolean matrices of order 3*, Linear Algebra Appl. **449** (2014), 381–401, DOI 10.1016/j.laa.2014.02.035. MR3191874

[Bro09] Michel Broué, *Higman's criterion revisited*, Michigan Math. J. **58** (2009), no. 1, 125–179, DOI 10.1307/mmj/1242071686. MR2526081

[BST13] Serge Bouc, Radu Stancu, and Jacques Thévenaz, *Simple biset functors and double Burnside ring*, J. Pure Appl. Algebra **217** (2013), no. 3, 546–566, DOI 10.1016/j.jpaa.2012.08.005. MR2974230

[BT16] Serge Bouc and Jacques Thévenaz, *The algebra of essential relations on a finite set*, J. Reine Angew. Math. **712** (2016), 225–250, DOI 10.1515/crelle-2014-0019. MR3466554

[BT18a] Serge Bouc and Jacques Thévenaz. The algebra of boolean matrices, correspondence functors, and simplicity. Preprint, 2018. arXiv:1902.05422. To appear in J. Comb. Algebra.

[BT18b] Serge Bouc and Jacques Thévenaz, *Correspondence functors and finiteness conditions*, J. Algebra **495** (2018), 150–198, DOI 10.1016/j.jalgebra.2017.11.010. MR3726107

[BT19a] Serge Bouc and Jacques Thévenaz, *Correspondence functors and lattices*, J. Algebra **518** (2019), 453–518, DOI 10.1016/j.jalgebra.2018.10.019. MR3873948

[BT19b] Serge Bouc and Jacques Thévenaz. Simple and projective correspondence functors. Preprint, 2019. arXiv:1902.09816.

[CEF15] Thomas Church, Jordan S. Ellenberg, and Benson Farb, *FI-modules and stability for representations of symmetric groups*, Duke Math. J. **164** (2015), no. 9, 1833–1910, DOI 10.1215/00127094-3120274. MR3357185

[GAP15] The GAP Group. *GAP – Groups, Algorithms, and Programming, Version 4.7.8*, 2015. (`http://www.gap-system.org`).

[Git18] Andrew Gitlin, *New examples of dimension zero categories*, J. Algebra **505** (2018), 271–278, DOI 10.1016/j.jalgebra.2018.03.009. MR3789913

[Hal35] Philip Hall. On representatives of subsets. *J. London Math. Soc.*, 10:26–30, 1935.

[Hal86] Marshall Hall Jr., *Combinatorial theory*, 2nd ed., Wiley-Interscience Series in Discrete Mathematics, John Wiley & Sons, Inc., New York, 1986. A Wiley-Interscience Publication. MR840216

[HY14] Akihiko Hida and Nobuaki Yagita, *Representations of the double Burnside algebra and cohomology of the extraspecial p-group*, J. Algebra **409** (2014), 265–319, DOI 10.1016/j.jalgebra.2014.03.021. MR3198843

[Kim82] Ki Hang Kim, *Boolean matrix theory and applications*, Monographs and Textbooks in Pure and Applied Mathematics, vol. 70, Marcel Dekker, Inc., New York, 1982. With a foreword by Gian-Carlo Rota. MR655414

[Mat89] Hideyuki Matsumura, *Commutative ring theory*, 2nd ed., Cambridge Studies in Advanced Mathematics, vol. 8, Cambridge University Press, Cambridge, 1989. Translated from the Japanese by M. Reid. MR1011461

[Pir00] Teimuraz Pirashvili, *Dold-Kan type theorem for Γ-groups*, Math. Ann. **318** (2000), no. 2, 277–298, DOI 10.1007/s002080000120. MR1795563

[PS17] Andrew Putman and Steven V. Sam, *Representation stability and finite linear groups*, Duke Math. J. **166** (2017), no. 13, 2521–2598, DOI 10.1215/00127094-2017-0008. MR3703435

[Rom08] Steven Roman, *Lattices and ordered sets*, Springer, New York, 2008. MR2446182

[Sta12] Richard P. Stanley, *Enumerative combinatorics. Volume 1*, 2nd ed., Cambridge Studies in Advanced Mathematics, vol. 49, Cambridge University Press, Cambridge, 2012. MR2868112

[Tho72] M. C. Thornton, *Spaces with given homeomorphism groups*, Proc. Amer. Math. Soc. **33** (1972), 127–131, DOI 10.2307/2038184. MR292054

[WG19] John D. Wiltshire-Gordon, *Categories of dimension zero*, Proc. Amer. Math. Soc. **147** (2019), no. 1, 35–50, DOI 10.1090/proc/14040. MR3876729

CNRS-LAMFA, Université de Picardie - Jules Verne, 33, rue St Leu, F-80039 Amiens Cedex 1, France

Email address: serge.bouc@u-picardie.fr

Contemporary Mathematics
Volume **758**, 2020
https://doi.org/10.1090/conm/758/15232

Three lectures on quiver Grassmannians

Giovanni Cerulli Irelli

ABSTRACT. This paper contains the material discussed in the series of three lectures that I gave during the workshop of the ICRA 2018 in Prague. I will introduce the reader to some of the techniques used in the study of the geometry of quiver Grassmannians. The notes are quite elementary and thought for phd students or young researchers. I assume that the reader is familiar with the representation theory of quivers.

Introduction

Given a finite quiver Q and a finite dimensional Q–representation M, the quiver Grassmannian $\mathrm{Gr}_{\mathbf{e}}(M)$ is the projective variety of Q–subrepresentations $N \subseteq M$ of dimension vector $\mathbf{dim}\, N = \mathbf{e}$. Quiver Grassmannians were considered in the seminal paper of Schofield [**57**] for the study of general representations of Q. It is shown there that a general representation of dimension vector $\mathbf{d}$ admits a subrepresentation of dimension vector $\mathbf{e}$ if and only if the minimal value of the dimension of the extension space between a representation of dimension vector $\mathbf{e}$ and one of dimension vector $\mathbf{d}-\mathbf{e}$ is zero. This is shown by considering a universal family

$$\pi_{\mathbf{e},\mathbf{d}} : \mathcal{Y}_{\mathbf{e},\mathbf{d}} \to R_{\mathbf{d}}(Q)$$

over the representation space $R_{\mathbf{d}}(Q)$ of Q–representations of dimension vector $\mathbf{d}$ whose fiber over a point $M \in R_{\mathbf{d}}(Q)$ is $\mathrm{Gr}_{\mathbf{e}}(M)$. This is a proper family whose total space is smooth and irreducible. It is nowadays called the universal quiver Grassmannian.

Quiver Grassmannians then appeared in the Fomin and Zelevinsky theory of cluster algebras ([**33**–**35**]), by work of Caldero-Chapoton [**12**], Caldero-Keller [**13**, **14**] and Derksen-Weyman-Zelevinsky [**28**]. In those papers it is shown that for any (non-initial) generator u of the cluster algebra $\mathcal{A}_Q$ associated with Q there exists a Q–representation M such that u has the following form

$$u = \mathbf{x}^{\mathbf{g}_M}\left(\sum_{\mathbf{e}} \chi(\mathrm{Gr}_{\mathbf{e}}(M))\mathbf{x}^{B\mathbf{e}}\right) \tag{0.1}$$

where $\mathbf{g}_M$ is the index of M and χ denotes the Euler characteristic (see section 8). This is a remarkable fact, because the generators of the cluster algebra $\mathcal{A}_Q$ are defined recursively, starting from the initial seed $(B_Q, \mathbf{x})$. Thus, formula (0.1) is a solution of this complicated recurrence relation and it is given in terms of quiver

2010 *Mathematics Subject Classification.* Primary 16G20; Secondary 14N20.

Grassmannians. It is then a natural question to see if a better understanding of the geometry of the projective variety $\mathrm{Gr}_{\mathbf{e}}(M)$ can provide useful information about the cluster algebra $\mathcal{A}_Q$. This turned out to be true in the affine type A: in [**17**] and [**18**] it is shown that by considering only the smooth part of $\mathrm{Gr}_{\mathbf{e}}(M)$ in the formula (0.1) one gets the elements of the atomic basis of $\mathcal{A}_Q$. The atomic basis is a $\mathbb{Z}$–basis such that its positive span coincides with the set of elements of $\mathcal{A}_Q$ which have positive coefficients with respect to any cluster. Nowadays it is known that an atomic basis exists only in very particular cases.

When the quiver is acyclic, in formula (0.1) the Q–representation M is rigid, i.e. $\mathrm{Ext}^1(M, M) = 0$. The positivity conjecture of Fomin and Zelevinky hence implies that the Euler characteristic of the quiver Grassmannians associated with a rigid Q-representation is non-negative. This was proved by Nakajima [**48**]. Caldero-Keller and others conjectured that much more is true, namely that those quiver Grassmannians admit a cellular decomposition. This conjecture is still open. In [**16**] it is proved a little less: namely that those quiver Grassmannians have property (S), i.e. no odd homology, no torsion in even homology, and the cycle map is an isomorphism. This refines the proof of Nakajima. For Dynkin and affine quivers much more is true: for Dynkin quivers every quiver Grassmannian admits a cellular decomposition and for affine quivers, every quiver Grassmannian associated with a representation M whose regular part is rigid, admits a cellular decomposition. See section 7.

Apart from this motivation, the geometry of quiver Grassmannians is an interesting object of study, due to the fact that many geometric properties can be studied via the representation theory of quivers. But one has to be careful here: Reineke showed that every projective variety can be realized as a quiver Grassmannian in an elementary way and Ringel straightened considerably this result by showing that every projective variety arises as a quiver Grassmannian of every wild quiver. See section 4.7 for this, and for some examples.

It is then natural to restrict attention to particular quivers and dimension vectors. The most fruitful restriction is when Q is an equioriented quiver of type A_n, $\mathbf{d} = (n+1, \cdots, n+1)$ and $\mathbf{e} = (1, 2, \cdots, n)$. In this case the generic fiber of the universal quiver Grassmannian is the complete flag variety for SL_{n+1} and the other fibers can be hence considered as "linear" degenerations of the complete flag variety. Among all fibers one is of particular interest: the Feigin degenerate flag variety. In [**20–23**], we have studied degenerate flag varieties (and more general quiver Grassmannians of Dynkin type) from this point of view and get interesting new results and new proofs of known results. In [**19**] we have explored the universal quiver Grassmannian for the special case mentioned above, and find a very interesting variety which is a flat degeneration of the complete flag variety and having the n-th Catalan number of irreducible components. See Section 5 for details concerning quiver Grassmannians of type A, linear degenerations of flag varieties and quiver Grassmannians of Dynkin type.

In the last section 9 a collection of exercises is provided. The exercises are divided according to the different sections of the paper and they are thought to provide a better understanding of the techniques mentioned in the main body of the paper. I encourage the reader to solve the exercises corresponding to a given section during the study of the section.

1. Notations

Let Q be a finite acyclic and connected quiver. We denote by Q_0 the finite set of vertices (whose cardinality is always denoted with the letter n), by Q_1 the finite set of edges, and the two functions $s, t : Q_1 \to Q_0$ provide an orientation of the edges. For an oriented edge α we write $\alpha : s(\alpha) \to t(\alpha)$. The base field is the field of complex numbers, denoted either by K or with the usual symbol $\mathbb{C}$. We denote by $\mathrm{Rep}(Q)$ the category of finite-dimensional complex representations of Q. Recall that the objects of $\mathrm{Rep}(Q)$ are tuples $M = ((M_i)_{i\in Q_0}, (M_\alpha)_{\alpha\in Q_1})$ where M_i is a (finite-dimensional) vector space and $M_\alpha : M_{s(\alpha)} \to M_{t(\alpha)}$ is a linear map. A Q–morphism $\psi : M \to N$ between two Q–representations is a collections $(\psi_i : M_i \to N_i)_{i\in Q_0}$ of linear maps such that the following square

$$\begin{array}{ccc} M_{s(\alpha)} & \xrightarrow{M_\alpha} & M_{t(\alpha)} \\ \Big\downarrow{\scriptstyle \psi_{s(\alpha)}} & & \Big\downarrow{\scriptstyle \psi_{t(\alpha)}} \\ N_{s(\alpha)} & \xrightarrow{N_\alpha} & N_{t(\alpha)} \end{array}$$

commutes for every arrow α of Q. We denote by $\mathrm{Hom}_Q(M,N)$ the vector space of Q–morphisms between the two Q–representations M and N. We denote its dimension by

$$[M,N] := \dim \mathrm{Hom}_Q(M,N).$$

To a quiver Q is associated its (complex) path-algebra $A = KQ$, which is the algebra formed by concatenation of arrows. The category $\mathrm{Rep}_K(Q)$ is equivalent to the category A–mod of KQ–modules. Notice that KQ is finite–dimensional since the quiver Q is acyclic, i.e. it does not have oriented cycles (even if its underlying graph can have a cycle). The category $\mathrm{Rep}_K(Q)$ is abelian and Krull-Schmidt, moreover it is hereditary, i.e. $\mathrm{Ext}_Q^{\geq 2}(-,-) = 0$. We use the standard notation

$$[M,N]^1 := \dim \mathrm{Ext}^1_Q(M,N).$$

The set $\{e_i\}_{i\in Q_0}$ of paths of length zero form a complete set of pairwise orthogonal idempotents of A. Since Q is acyclic, and hence the path algebra $A = KQ$ is finite-dimensional, there are only finitely many simple A–modules parametrized by the vertices of Q. We denote by S_k the simple corresponding to vertex k, by P_k its projective cover and by I_k it injective hull. Recall that as Q–representation, P_k is described as follows: the vector space at vertex i has a basis given by paths from vertex k to vertex i, and the arrows act by "concatenation". Notice that if Q is an orientation of a tree (for example if Q is Dynkin), then every projective P_k is thin, which means that the vector space $(P_k)_i$ at every vertex i is at most one–dimensional. Dually, the injective indecomposable (left) A–modules are the indecomposable direct summands of DA (viewed as left A–module), where D is the standard K–duality. As Q–representation, I_k has at vertex j a vector space with basis consisting of all the paths of Q starting in j and ending in k, and the arrows act by "concatenation".

For a Q–representation M, the collection $(\dim M_i)_{i\in Q_0} \in \mathbb{Z}_{\geq 0}^{Q_0}$ of non–negative integers is called the *dimension vector* of M, and it is denoted in bold by $\mathbf{dim}\, M$. Once the dimension vector is fixed, a Q–representation is determined by linear maps: this leads us to the variety of Q–representations. Let $\mathbf{d} = (d_i)_{i\in Q_0} \in \mathbb{Z}_{\geq 0}^{Q_0}$ be a dimension vector. The vector space

$$R_{\mathbf{d}} := \bigoplus_{\alpha \in Q_1} \mathrm{Hom}_K(K^{d_{s(\alpha)}}, K^{d_{t(\alpha)}})$$

is called the variety of Q–representations of dimension vector $\mathbf{d}$. The group

$$G_{\mathbf{d}} := \prod_{i \in Q_0} \mathrm{GL}_{d_i}(K)$$

acts on $R_{\mathbf{d}}$ by base change: $(g_i)_i \cdot (V_\alpha)_\alpha := (g_{t(\alpha)} V_\alpha g_{s(\alpha)}^{-1})_\alpha$ and $G_{\mathbf{d}}$–orbits are in bijection with isoclasses of Q–representations. The stabilizer of a point $M \in R_{\mathbf{d}}$ is

$$\mathrm{Stab}_{G_{\mathbf{d}}}(M) = \mathrm{Aut}_Q(M)$$

where $\mathrm{Aut}_Q(M)$ denotes the open subvariety of $\mathrm{Hom}_Q(M, M)$ consisting of invertible Q-morphisms. In particular, $\dim \mathrm{Aut}_Q(M) = \dim \mathrm{Hom}_Q(M, M)$. Given another dimension vector $\mathbf{e} \in \mathbb{Z}_{\geq 0}^{Q_0}$ we consider the vector space of ("degree zero") K–morphisms

$$\mathrm{Hom}(\mathbf{e}, \mathbf{d}) = \bigoplus_{i \in Q_0} \mathrm{Hom}_K(K^{e_i}, K^{d_i})$$

and the vector space of ("degree one") K–morphisms

$$\mathrm{Hom}(\mathbf{e}, \mathbf{d}[1]) = \bigoplus_{\alpha \in Q_1} \mathrm{Hom}_K(K^{e_{s(\alpha)}}, K^{d_{t(\alpha)}}).$$

In particular, if $\mathbf{d}$ is a dimension vector we get

$$\dim \mathrm{Hom}(\mathbf{d}, \mathbf{d}[1]) = \dim R_{\mathbf{d}}. \tag{1.1}$$

Given $N \in R_{\mathbf{e}}$ and $M \in R_{\mathbf{d}}$ we consider the map

$$\Phi_N^M : \mathrm{Hom}(\mathbf{e}, \mathbf{d}) \to \mathrm{Hom}(\mathbf{e}, \mathbf{d}[1]) : \ (f_i)_{i \in Q_0} \mapsto (M_\alpha \circ f_{s(\alpha)} - f_{t(\alpha)} \circ N_\alpha)_{\alpha \in Q_1}$$

This is a linear map between finite dimensional vector spaces and one can show quite easily (see e.g. [**53**], [**4**]):

$$\mathrm{Ker}\, \Phi_N^M = \mathrm{Hom}_Q(N, M), \quad \mathrm{CoKer}\, \Phi_N^M \simeq \mathrm{Ext}_Q^1(N, M).$$

From these formulas we immediately get:

$$[N, M] - [N, M]^1 = \dim \mathrm{Hom}(\mathbf{e}, \mathbf{d}) - \dim \mathrm{Hom}(\mathbf{e}, \mathbf{d}[1]). \tag{1.2}$$

We have that $\dim \mathrm{Hom}(\mathbf{e}, \mathbf{d}) = \sum_{i \in Q_0} e_i d_i$ and $\dim \mathrm{Hom}(\mathbf{e}, \mathbf{d}[1]) = \sum_{\alpha \in Q_1} e_{s(\alpha)} d_{t(\alpha)}$. Given two arbitrary integer vectors $\mathbf{e}, \mathbf{d} \in \mathbb{Z}^{Q_0}$ the Euler form of Q is the integral bilinear form $\langle -, - \rangle_Q : \mathbb{Z}^{Q_0} \times \mathbb{Z}^{Q_0} \to \mathbb{Z}$ given by

$$\langle \mathbf{e}, \mathbf{d} \rangle := \sum_{i \in Q_0} e_i d_i - \sum_{\alpha \in Q_1} e_{s(\alpha)} d_{t(\alpha)}.$$

From (1.2) above, we immediately get

$$\dim \mathrm{Hom}_Q(N, M) - \dim \mathrm{Ext}_Q^1(N, M) = \langle \mathbf{dim}\, N, \mathbf{dim}\, M \rangle. \tag{1.3}$$

Formula (1.3) is called the homological interpretation of the Euler form.

In view of (1.1) and (1.3), we have

$$\mathrm{codim}_{R_{\mathbf{d}}}(G_{\mathbf{d}} \cdot M) = \dim R_{\mathbf{d}} - \dim \mathrm{Stab}_{G_{\mathbf{d}}}(M) = \dim \mathrm{Ext}_Q^1(M, M).$$

We conclude that the orbit of M is dense in $R_{\mathbf{d}}$ if and only if $\mathrm{Ext}_Q^1(M, M) = 0$. A representation M such that $\mathrm{Ext}_Q^1(M, M) = 0$ is called *rigid*.

A famous theorem of P. Gabriel [**37**] (see also [**8**] for a different proof and [**4**, Section VII.5] for a survey) states that a quiver Q admits only a finite number of isoclasses of indecomposable representations if and only if Q is a Dynkin quiver i.e. it is an orientation of a simply-laced Dynkin diagram of type A, D, E. The quiver Q is called tame or affine if it is an acyclic orientation of a simply-laced extended Dynkin diagram of type ADE. A quiver which is neither Dynkin nor affine is called wild. The classification of the indecomposable Q–representations is possible if and only if Q is either Dynkin or tame and this explains the terminology. Table 1 shows the Dynkin and the extended Dynkin diagrams.

Type	Dynkin	ExtendedDynkin
A		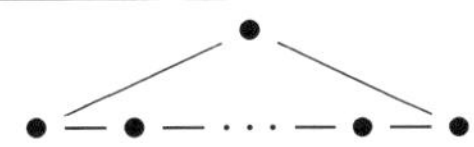
D	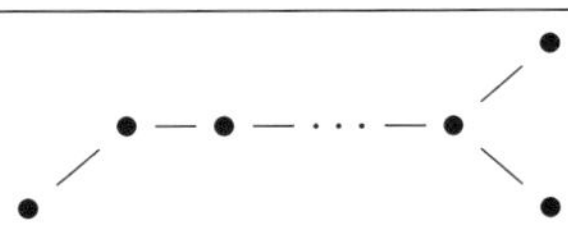	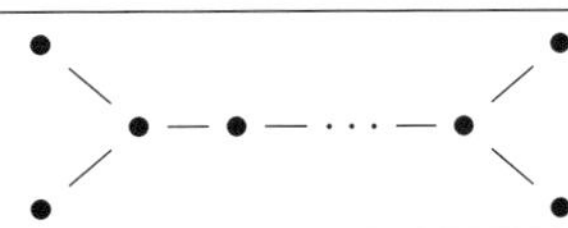
E_6	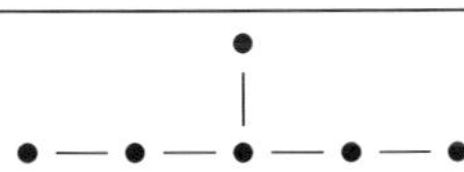	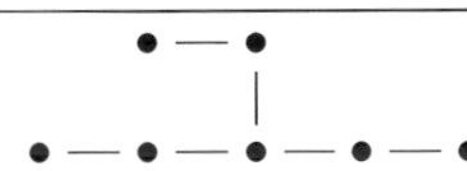
E_7	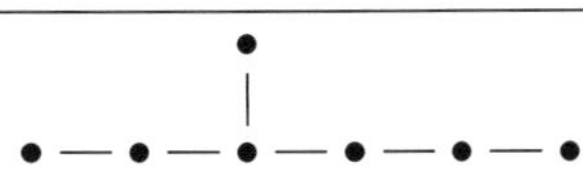	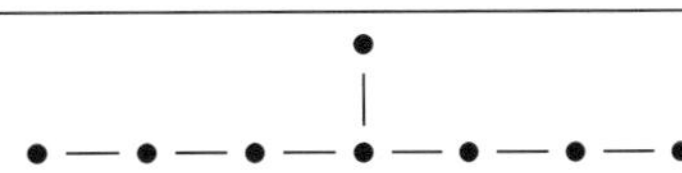
E_8	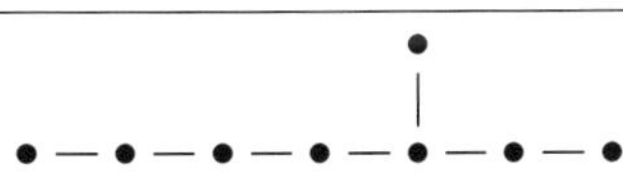	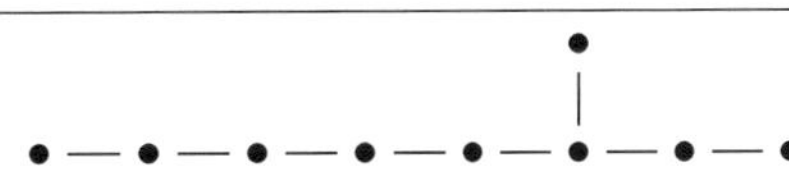

TABLE 1. Dynkin and extendend Dynkin diagrams

If Q is Dynkin, then $R_{\mathbf{d}}$ consists of finitely many $G_{\mathbf{d}}$–orbits, and hence, since such orbits are connected and locally closed, there is a unique orbit which is dense. The corresponding representation is hence a generic representation of dimension vector $\mathbf{d}$ and we denote it by $\tilde{M}_{\mathbf{d}}$. In particular, for Dynkin quivers a representation is generic if and only if it is rigid.

For an arbitrary acyclic quiver Q most dimension vectors do not admit a dense orbit. By Kac's theorem ([**40**, Theorem 1]) there exists an indecompsable representation of dimension vector $\mathbf{d}$ if and only if $\mathbf{d}$ is a positive root for the Kac-Moody algebra associated with the underlying graph of Q; in this case $R_{\mathbf{d}}$ admits a dense orbit if and only if $\mathbf{d}$ is a positive real root.

1.1. Almost split sequences. We conclude this section by recalling the fundamental notions of *almost split sequence*, *irreducible morphism* and of *Auslander–Reiten quiver* of a quiver Q (see e.g. [**11**], [**6**], [**4**]). A short exact sequence

$$\delta : \qquad 0 \longrightarrow N \xrightarrow{f} E \xrightarrow{g} M \longrightarrow 0$$

is called *almost split* if it is non–split, both N and M are indecomposable and for any morphism $h : X \to M$ which is not a split epi (i.e. it does not admit a right inverse), there exists $t : X \to E$ such that $h = g \circ t$. In particular, if δ is an almost split sequence, and M is not a direct summand of X, then $[X,E]=[X,N\oplus M]$. Dually, it can be shown that δ is almost split if and only if it is non–split, both N and M are indecomposable and for any morphism $h:N\to X$ which is not a split mono (i.e. it does not admit a left inverse), there exists $t:E\to X$ such that $h=t\circ f$. A fundamental result of Auslander and Reiten [**6**, Theorem V.1.15] states that for every indecomposable M which is not projective, there exists an almost split sequence δ as above (ending in M), which is unique up to scalar multiples [**6**, Theorem V.1.16]. Dually, for every indecomposable N which is not injective, there exists an almost split sequence δ as above (starting from N).

One can show that almost split sequences are *rigid*, in the sense that they are uniquely determined (up to scalar multiples as elements of $\mathrm{Ext}^1(N,M)$) by the three modules N, E and M [**6**, Proposition V.2.3].

Almost split sequences are closely related to the so-called Auslander-Reiten translate τ and its quasi-inverse τ^-. In general the definition of τ and τ^- is quite involved since they are not functors, but in our situation, which is the case of an hereditary basic and finite dimensional algebra, it reduces to two simple functors:

$$\tau=D\,\mathrm{Ext}^1(-,A)\quad \tau^-=\mathrm{Ext}^1(D(-),A).$$

They are uniquely determined by the Auslander-Reiten formulas:

$$\mathrm{Hom}(M,\tau N)\simeq D\,\mathrm{Ext}^1(N,M)\simeq \mathrm{Hom}(\tau^-M,N). \tag{1.4}$$

If there is an amost split sequence δ as above then

$$N\simeq \tau\mathrm{M},\quad M\simeq\tau^-N.$$

1.2. Auslander-Reiten quiver. A morphism $f:M\to N$ between two indecomposable Q–representations M and N is called *irreducible* if f is neither split mono, nor split epi (i.e. it does not admit neither a left nor a right inverse) and whenever there is a factorization $f=f_2\circ f_1$, then either f_1 is split mono or f_2 is split epi (see [**11**]). The irreducible morphisms from M to N are parametrized by the quotient space $\mathrm{Irr}(M,N)=\mathrm{rad}_Q(M,N)/\mathrm{rad}^2(M,N)$ (see e.g. [**11**, Section 1]). Here $\mathrm{rad}(M,N)=\{f:M\to N \text{ not an isomorphism}\}$ and $\mathrm{rad}^2(M,N)=\{f:M\to N\text{which factor as hg with g not split mono and h not split epi}\}$. The Auslander–Reiten quiver of Q is a quiver denoted by Γ_Q whose vertices are isoclasses of indecomposable Q–representations, and there are k arrows $[M]\to[N]$ if the dimension of the quotient space $\mathrm{Irr}(M,N)$ has dimension k (see e.g. [**11**, Section 1] or [**4**, Section IV.4]).

The Auslander-Reiten quiver Γ_Q consists of several connected components which can be grouped together to form a decomposition

$$\Gamma_Q=\mathcal{P}_Q\amalg\mathcal{R}_Q\amalg\mathcal{I}_Q$$

where $\mathcal{P}_Q$ is the component containg all the indecomposable projectives and dually, $\mathcal{I}_Q$ is the component containg all the indecomposable injectives. The remaining piece $\mathcal{R}_Q$ consists of all connected components which do not contain neither an injective nor a projective module. It is easy to see that both $\mathcal{P}_Q$ and $\mathcal{I}_Q$ are connected components. An indecomposable module M lies in $\mathcal{P}_Q$ (resp. $\mathcal{I}_Q$) if and only if there exists a vertex $k\in Q_0$ and an index $j\geq 0$ such that $M\simeq\tau^{-j}P_k$ (resp. $M\simeq\tau^jI_k$). Such a module is called preprojective (resp. preinjective). With abuse of notation, we denote by $\mathcal{P}_Q$ (resp. $\mathcal{I}_Q$) the full additive subcategory of $\mathrm{Rep}(Q)$ whose indecomposable objects are preprojectives (resp. preinjectives) and its objects are still called preprojectives (resp. preinjectives). The connected components of $\mathcal{R}_Q$ are called regular and their modules are also called regular.

The components $\mathcal{P}_Q$ and $\mathcal{I}_Q$ are called the preprojective and preinjective component of Γ_Q, respectively. These components of the quiver Γ_Q can be described combinatorially via the knitting algorithm. (I recommend the introductory book [**58**] for more details about the construction of AR quivers of Dynkin quivers.)

The main property of the graphs $\mathcal{P}_Q$ and $\mathcal{I}_Q$ is that they encode the information that one needs to understand homomorphisms and extensions between the indecomposable representations corresponding to their vertices. Namely, the dimension of $\mathrm{Hom}(M, N)$ is given by counting paths from $[M]$ to $[N]$ modulo mesh relations. To get the extension spaces one uses the Auslander-Reiten formulas (1.4). Moreover $\mathcal{P}_Q$ is a directed category, in the sense that given two indecomposables $M, N \in \mathcal{P}_Q$ either $[M,N]^1 = 0$ or $[N,M]^1 = 0$. The same holds for $\mathcal{I}_Q$. The regular components encode much less information due to the fact that the infinite radical of the module category contains many maps (if Q is wild). They are described by Ringel [**55**]. They are far from being directed (see [**42**]).

Every module M admits a unique split filtration $M' \subseteq M'' \subseteq M$ where $M' \in \mathcal{I}_Q$, $M''/M' \in \mathcal{R}_Q$ and $M/M'' \in \mathcal{P}_Q$; these are called the preinjective, regular and preprojective parts of M, respectively.

2. Quiver Grassmannians

Let Q be a finite quiver with n vertices and let $A = KQ$ be the associated (complex) path algebra. Given a dimension vector $\mathbf{d}$, an A–module $M \in R_{\mathbf{d}}$ and another dimension vector $\mathbf{e}$ such that $\mathbf{d} - \mathbf{e} \in \mathbb{Z}_{\geq 0}^{Q_0}$, in this section we define the projective variety $Gr_{\mathbf{e}}(M)$ whose points parametrize submodules of M of dimension vector $\mathbf{e}$. We need to ask ourselves "what is a submodule?". This question has two answers: first of all, a submodule is a collection $(N_i)_{i\in Q_0}$ of vector subspaces $N_i \subseteq M_i$ such that $M_\alpha(N_i) \subseteq N_j$ for every arrow $\alpha : i \to j$ of Q. On the other hand, a submodule $N \subset M$ is an A–module N endowed with an injective A–morphism $\iota : N \to M$. The two answers provide two different realizations of $Gr_{\mathbf{e}}(M)$.

2.1. First realization: Universal quiver Grassmannians.

Schofield noticed that quiver Grassmannians come in families: Let $\mathbf{d}$ and $\mathbf{e}$ be two dimension vector for Q such that $e_i \leq d_i$ for all $i \in Q_0$. Let us consider the product of usual Grassmannians of vector spaces over the field K of complex numbers: $Gr_{\mathbf{e}}(\mathbf{d}) := \prod_{i\in Q_0} Gr_{e_i}(K^{d_i})$. Given $M \in R_{\mathbf{d}}(Q)$ and a point $N \in Gr_{\mathbf{e}}(\mathbf{d})$, the condition that N defines a sub-representation of M is $M_\alpha(N_{s(\alpha)}) \subseteq N_{t(\alpha)}$. We hence consider the incidence variety inside $Gr_{\mathbf{e}}(\mathbf{d}) \times R_{\mathbf{d}}$ given by:

$$Gr_{\mathbf{e}}^Q(\mathbf{d}) := \{(N, M) \in Gr_{\mathbf{e}}(\mathbf{d}) \times R_{\mathbf{d}} | \, M_\alpha(N_{s(\alpha)}) \subseteq N_{t(\alpha)}, \, \forall \alpha \in Q_1\}. \tag{2.1}$$

The two projections $p_1 : Gr_{\mathbf{e}}(\mathbf{d}) \times R_{\mathbf{d}} \to Gr_{\mathbf{e}}(\mathbf{d})$ and $p_2 : Gr_{\mathbf{e}}(\mathbf{d}) \times R_{\mathbf{d}} \to R_{\mathbf{d}}$ induce two maps

$$\begin{array}{ccccc} & & Gr_{\mathbf{e}}^Q(\mathbf{d}) & & \\ & \swarrow^{p_{\mathbf{e}}} & & \searrow^{p_{\mathbf{e},\mathbf{d}}} & \\ Gr_{\mathbf{e}}(\mathbf{d}) & & & & R_{\mathbf{d}} \end{array}$$

The group $G_{\mathbf{d}}$ acts diagonally on $Gr_{\mathbf{e}}^Q(\mathbf{d})$ (see exercise 9.5) and the two maps $p_{\mathbf{e}}$ and $p_{\mathbf{e},\mathbf{d}}$ are $G_{\mathbf{d}}$–equivariant. The map p_2 is proper; moreover $Gr_{\mathbf{e}}^Q(\mathbf{d})$ is closed in $Gr_{\mathbf{e}}(\mathbf{d}) \times R_{\mathbf{d}}$ and the closed embedding $Gr_{\mathbf{e}}^{Q}(\mathbf{d}) \to Gr_{\mathbf{e}}(\mathbf{d}) \times R_{\mathbf{d}}$ is proper. It follows that the map $p_{\mathbf{e},\mathbf{d}}$ is proper, being the composition of two proper maps. Its image is the *closed* subset of $R_{\mathbf{d}}$ consisting of those points $M \in R_{\mathbf{d}}$ which admit a sub-representation of dimension vector $\mathbf{e}$. The *quiver Grassmannian* $Gr_{\mathbf{e}}(M)$ associated with a point $M \in R_{\mathbf{d}}$ is defined as the (scheme-theoretic) fiber of $p_{\mathbf{e},\mathbf{d}}$ over M.

Thus quiver Grassmannians come in families: they are fibers of the proper map

$$p_{\mathbf{e},\mathbf{d}} : Gr_{\mathbf{e}}^{Q}(\mathbf{d}) \to R_{\mathbf{d}}$$

which is called the universal quiver Grassmannian. As shown in [**20**, section 2.2], the map $p_{\mathbf{e}}$ realizes $Gr_{\mathbf{e}}^{Q}(\mathbf{d})$ as the total space of an homogeneous vector bundle over $Gr_{\mathbf{e}}(\mathbf{d})$ of rank

$$\sum_{\alpha \in Q_1} d_{s(\alpha)} d_{t(\alpha)} + e_{s(\alpha)} e_{t(\alpha)} - e_{s(\alpha)} d_{t(\alpha)}.$$

In particular, $Gr_{\mathbf{e}}^{Q}(\mathbf{d})$ is smooth and irreducible of dimension

$$\dim Gr_{\mathbf{e}}^{Q}(\mathbf{d}) = \langle \mathbf{e}, \mathbf{d} - \mathbf{e} \rangle + \dim R_{\mathbf{d}}.$$

By upper–semicontinuity of the fiber dimension, we see that for any point M in the image of $p_{\mathbf{e},\mathbf{d}}$ we have

$$\dim Z \geq \dim Gr_{\mathbf{e}}^{Q}(\mathbf{d}) - \dim Im(p_{\mathbf{e},\mathbf{d}}) \geq \langle \mathbf{e}, \mathbf{d} - \mathbf{e} \rangle. \tag{2.2}$$

for every irreducible component Z of $Gr_{\mathbf{e}}(M)$.

Let $D : Rep(Q) \to Rep(Q^{op})$ be the standard duality which associates to a Q–representation M its linear dual DM. There is an isomorphism of projective varieties

$$\zeta : Gr_{\mathbf{e}}(M) \to Gr_{\mathbf{d}-\mathbf{e}}(DM) : L \mapsto \mathrm{Ann}_M(L) := \{\varphi \in DM \,|\, \varphi(L) = 0\}. \tag{2.3}$$

2.2. Second realization: Quiver Grassmannians as geometric quotients and stratification. Following Caldero and Reineke [**15**], one can realize quiver Grassmannians as geometric quotients. Recall the two vector spaces $\mathrm{Hom}(\mathbf{e}, \mathbf{d})$ and $\mathrm{Hom}(\mathbf{e}, \mathbf{d}[1])$ of section 1 and the linear map $\Phi_L^M : \mathrm{Hom}(\mathbf{e}, \mathbf{d}) \to \mathrm{Hom}(\mathbf{e}, \mathbf{d}[1])$ associated with $L \in R_{\mathbf{e}}(Q)$ and $M \in R_{\mathbf{d}}(Q)$. Let us assume that $e_i \leq d_i$ for all $i \in Q_0$. Given $M \in R_{\mathbf{d}}(Q)$ the algebraic map

$$\Phi^M : R_{\mathbf{e}} \times \mathrm{Hom}(\mathbf{e}, \mathbf{d}) \to \mathrm{Hom}(\mathbf{e}, \mathbf{d}[1]) : (L, f) \mapsto \Phi_L^M(f)$$

is used to define the following closed subvariety of $R_{\mathbf{e}} \times \mathrm{Hom}(\mathbf{e}, \mathbf{d})$:

$$\mathrm{Hom}(\mathbf{e}, M) := \{(L, f) \in R_{\mathbf{e}} \times \mathrm{Hom}(\mathbf{e}, \mathbf{d}) |\, \Phi_L^M(f) = 0\}.$$

Inside $\mathrm{Hom}(\mathbf{e}, \mathbf{d})$ there is the open (and dense) subvariety $\mathrm{Hom}^0(\mathbf{e}, \mathbf{d})$ consisting of collections of injective linear maps; the induced open subvariety $\mathrm{Hom}^0(\mathbf{e}, M) := \mathrm{Hom}(\mathbf{e}, M) \cap \big(R_{\mathbf{e}} \times \mathrm{Hom}^0(\mathbf{e}, \mathbf{d})\big)$ is of particular importance for us. Indeed the map

$$\phi : \mathrm{Hom}^0(\mathbf{e}, M) \to Gr_{\mathbf{e}}(M) : (L, f) \mapsto f(L)$$

is surjective and each fiber of ϕ is a free orbit for the algebraic group $G_{\mathbf{e}} = \prod_{i \in Q_0} GL(e_i)$ (see [**15**, Lemma 2]). This implies that the quiver Grassmannian $Gr_{\mathbf{e}}(M)$ is a geometric quotient:

$$Gr_{\mathbf{e}}(M) \simeq \mathrm{Hom}^0(\mathbf{e}, M)/G_{\mathbf{e}}. \tag{2.4}$$

With this formulation, a point p of $Gr_{\mathbf{e}}(M)$ is represented (up to the $G_{\mathbf{e}}$–action) by a pair (L, ι) where $L \in R_{\mathbf{e}}(Q)$ and $\iota : L \to M$ is an injective homomorphism of Q–representations; in this case we use the notation $p = [(L, \iota)]$. As shown by Caldero and Reineke, formula (2.4) implies the following description of the (scheme-theoretic) tangent space $T_p(Gr_{\mathbf{e}}(M))$ at a point p of the quiver Grassmannian.

THEOREM 2.1. *Given $M \in R_{\mathbf{d}}(Q)$ and a point $p = [(L, \iota)] \in Gr_{\mathbf{e}}(M)$, the tangent space $T_p(Gr_{\mathbf{e}}(M))$ at p is isomorphic to* $\mathrm{Hom}_Q(L, M/\iota(L))$.

REMARK 2.2. The tangent space formula only holds at level of schemes. The usual example in this sense is given by considering a regular (indecomposable) representation R_2 of the Kronecker quiver of quasi–length 2 whose dimension vector is $(2, 2)$. The quiver Grassmannian $Gr_{(1,1)}(R_2)$ is a point, but the tangent space has dimension one.

PROPOSITION 2.3. *A non-empty quiver Grassmannian* $\mathrm{Gr}_{\mathbf{e}}(M)$ *associated with a rigid representation M is smooth of dimension $\langle \mathbf{e}, \mathbf{dim}\, M - \mathbf{e}\rangle$.*

PROOF. For every subrepresentation $N \subseteq M$ we have surjective morphisms $\mathrm{Ext}^1(M, M) \twoheadrightarrow \mathrm{Ext}^1(N, M) \twoheadrightarrow \mathrm{Ext}^1(N, M/N)$. This proves that if $[M, M]^1 = 0$ then $[N, M/N]^1 = 0$ and hence, by (1.3), the tangent space $T_N(Gr_{\mathbf{e}}(M))$ has dimension $\langle \mathbf{e}, \mathbf{dim}\, M - \mathbf{e}\rangle$. This shows that $\mathrm{Gr}_{\mathbf{e}}(M)$ is smooth. The dimension is computed by using (2.2). □

As a consequence of the tangent space formula Schofield proved the following.

THEOREM 2.4. [**57**, Theorem 3.3] *The universal quiver Grassmannian $p_{\mathbf{e},\mathbf{d}}$ is surjective if and only if there exist $N \in R_{\mathbf{e}}$ and $R \in R_{\mathbf{d}-\mathbf{e}}$ such that $[N, R]^1 = 0$.*

The realization of a quiver Grassmannian as a geometric quotient (2.4) allows to define a stratification of $Gr_{\mathbf{e}}(M)$ as follows (see [**20**, Section 2.3] for more details): let p be the projection from $\mathrm{Hom}^0_Q(\mathbf{e}, M)$ to $R_{\mathbf{e}}$; its fiber over a point $N \in R_{\mathbf{e}}$ is the space of injective linear maps $\mathrm{Hom}^0_Q(N, M)$. For each isoclass $[N]$ in $R_{\mathbf{e}}$ we can consider the subset $\mathcal{S}_{[N]}$ of $Gr_{\mathbf{e}}(M)$ corresponding under the previous isomorphism to $p^{-1}(G_{\mathbf{e}} \cdot N)/G_{\mathbf{e}}$. The locally closed subset $\mathcal{S}_{[N]}$ is sometimes called an iso-stratum of the quiver Grassmannian. In [**20**, Lemma 2.4] it is shown that $\mathcal{S}_{[N]}$ is a locally closed subset of dimension

$$\dim \mathcal{S}_{[N]} = [N, M] - [N, N].$$

In particular, a quiver Grassmannian $Gr_{\mathbf{e}}(M)$ admits a stratification

$$Gr_{\mathbf{e}}(M) = \coprod_{[N]} \mathcal{S}_{[N]}.$$

In case M is preprojective this stratification is finite. In this case, the irreducible components of $Gr_{\mathbf{e}}(M)$ are hence closures of some strata which we called the *generic sub–representation types* of $Gr_{\mathbf{e}}(M)$ (see [**22**]). See [**39**] for a description of the irreducible components of Grassmannians of submodules of a module over an algebra. In case of rigid modules the following holds

THEOREM 2.5. [**16**, Prop. 37] *A quiver Grassmannian associated with a rigid quiver representation is irreducible.*

3. Degeneration of Q–representations: Bongartz's theorem and applications to quiver Grassmannians

Given $M, N \in R_{\mathbf{d}}$, M is said to *degenerate* to N and in this case it is customary to write $M \leq_{\deg} N$, if the closure of the orbit of M contains N:

$$M \leq_{\deg} N \overset{def}{\iff} \overline{G_{\mathbf{d}} \cdot M} \supseteq G_{\mathbf{d}} \cdot N.$$

For arbitrary finite–dimensional algebras, it is a hard problem to control such a notion. On the other hand, for algebras of finite representation type (i.e. admitting a finite number of indecomposable modules) the following very useful characterization holds:

$$M \leq_{\deg} N \iff \begin{array}{c}[X, M] \leq [X, N] \\ \forall\, X \in \mathrm{Rep}(Q)\end{array} \iff \begin{array}{c}[M, X] \leq [N, X]. \\ \forall\, X \in \mathrm{Rep}(Q)\end{array} \tag{3.1}$$

For Dynkin quivers this result was obtained by Bongartz [**9**] (partial results were obtained by Riedtmann [**51**], Abeasis-Del Fra [**1**–**3**]). The surprising generalization to any algebra of finite representation type was obtained by Zwara [**59**] (the second equivalence follows from Auslander–Reiten theory [**59**, section 2.2], [**5**]). For general quivers, not necessarily Dynkin, the equivalence (3.1) holds true in case both M and N are preprojective or preinjective [**9**].

In the analysis of the geometry of quiver Grassmannians the following result of Bongartz can be useful. In order to formulate it we need to recall the notion of a generic quotient from Bongartz's paper [**9**, Section 2.4]. Suppose that $U \in R_{\mathbf{e}}$ and $M \in R_{\mathbf{d}}$ are given, and also that there exists a monomorphism $\iota : U \to M$; in particular $\mathbf{d} - \mathbf{e} \in \mathbb{Z}_{\geq 0}^{Q_0}$ is a dimension vector. The set of all possible quotients of M by U is an irreducible constructible subset of $R_{\mathbf{d}-\mathbf{e}}$ which is $G_{\mathbf{d}-\mathbf{e}}$–invariant. If this set is the closure of the orbit of a point S then S is called the *generic quotient* of M by U. In general, generic quotients may not exist. They exist for Dynkin quivers, or if M is preinjective or if M is regular over an affine quiver, since there are only finitely many isoclasses of quotients.

THEOREM 3.1. *([**9**, Theorem 2.4]) Let $M, N \in R_{\mathbf{d}}$ such that $M \leq_{\deg} N$. Let U be a representation such that $[U, M] = [U, N]$ then the following holds:*

(1) *if U embeds into N, it embeds into M too;*
(2) *in this case every quotient of N by U is a degeneration of the generic quotient of M by U, in case it exists.*

Bongartz's theorem 3.1 can be used to prove that a certain quiver Grassmannian is non-empty. For example it can be used to prove Schofield's theorem 2.4. For Dynkin quivers, an interesting homological criterion that guarantees the non–emptiness of a quiver Grassmannian associated with an arbitrary representation can be found in [**47**].

4. Examples of quiver Grassmannians

In this section we collect examples of quiver Grassmannians.

4.1. Example 1: The Grassmannian. Let $Q = \cdot$ be the quiver with one vertex and no arrows. A Q–representation of dimension vector $\mathbf{d} = (d) \in \mathbb{Z}_{\geq 0}$ is a vector space $M = K^d$. Let $\mathbf{e} = (e)$ be a subdimension vector, i.e. $0 \leq e \leq d$. The quiver Grassmannian $\mathrm{Gr}_{\mathbf{e}}(M)$ is an ordinary Grassmannian of vector subspaces of M:

$$\mathrm{Gr}_{\mathbf{e}}(M) = \{W \subseteq \mathbb{C}^d | \dim W = e\}.$$

Choosing a basis $\{w_1, \cdots, w_e\}$ of a point $W \in \mathrm{Gr}_{\mathbf{e}}(M)$ defines a $e \times d$-matrix

$$A_W = \begin{pmatrix} w_1 \\ \hline w_2 \\ \hline \vdots \\ \hline w_e \end{pmatrix}$$

of maximal rank e whose rows are the elements of the chosen basis. On the other hand every $e \times d$-matrix A of maximal rank e defines a point $W_A \in \mathrm{Gr}_{\mathbf{e}}(M)$ which is the span of the rows of A. This defines a surjective map

$$\mathrm{Mat}_{e\times d}^{\max} \twoheadrightarrow \mathrm{Gr}_{\mathbf{e}}(M) : A \longmapsto W_A$$

where $\mathrm{Mat}_{e\times d}^{\max}$ is the open subset of maximal rank $e \times d$–matrices. The group GL_e acts freely on $\mathrm{Mat}_{e\times d}$ by left multiplication and does not change the span of the rows. The Grassmannian is hence a geometric quotient

$$\mathrm{Gr}_{\mathbf{e}}(\mathbb{C}^d) \simeq \mathrm{Mat}_{e\times d}^{\max}/\mathrm{GL}_e.$$

The quotient map $\mathrm{Mat}_{e\times d}^{\max} \twoheadrightarrow \mathrm{Gr}_{\mathbf{e}}(M)$ can be (locally) trivialized as follows: given $A \in \mathrm{Mat}_{e\times d}^{\max}$ there exists column indices $1 \le j_1 < j_2 < \cdots < j_e \le d$ such that the $e \times e$-submatrix A^J supported on the columns $J = (j_1, \cdots, j_e)$ of A is invertible. Define

$$\Delta_J(A) := \det(A^J) \neq 0.$$

For every $J = (j_1 < \cdots < j_e)$ define

$$\tilde{\mathcal{U}}_J = \left\{A \in \mathrm{Mat}_{e\times d}^{\max} | \Delta_J(A) \neq 0\right\} \subset \mathrm{Mat}_{e\times d}.$$

This is an open subset. Given $A \in U_J$ we can multiply on the left by the inverse of A^J and we get a matrix $\overline{A}$ such that $\overline{A}^J = \mathbf{1}_e$. We hence see that the restriction of the quotient map to $\tilde{\mathcal{U}}_J$ provides a trivial quotient

$$\tilde{\mathcal{U}}_J \twoheadrightarrow \mathcal{U}_J = \left\{W_A | A^J = \mathbf{1}_e\right\} \simeq \mathbb{A}^{e(d-e)} \ : A \mapsto W_A$$

so that $\tilde{\mathcal{U}}_J \simeq \mathcal{U}_J \times \mathrm{GL}_e$. This provides $\mathrm{Gr}_{\mathbf{e}}(M)$ with the structure of an $e(d-e)$–manifold. The Grasssmannian is covered by the affine spaces $\mathcal{U}_J$:

$$\mathrm{Gr}_{\mathbf{e}}(M) = \bigcup_J \mathcal{U}_J$$

an hence $\{\mathcal{U}_J\}$ is an affine covering of the Grassmannian, called the standard affine covering.

EXAMPLE 4.1. The standard affine covering for $\mathrm{Gr}_2(\mathbb{C}^4)$ is formed by

$$\mathcal{U}_{(1,2)} = \left\{\left\langle \begin{matrix} 1 & 0 & * & * \\ 0 & 1 & * & * \end{matrix} \right\rangle\right\}, \quad \mathcal{U}_{(1,3)} = \left\{\left\langle \begin{matrix} 1 & * & 0 & * \\ 0 & * & 1 & * \end{matrix} \right\rangle\right\},$$

$$\mathcal{U}_{(1,4)} = \left\{\left\langle \begin{matrix} 1 & * & * & 0 \\ 0 & * & * & 1 \end{matrix} \right\rangle\right\}, \quad \mathcal{U}_{(2,3)} = \left\{\left\langle \begin{matrix} * & 1 & 0 & * \\ * & 0 & 1 & * \end{matrix} \right\rangle\right\},$$

$$\mathcal{U}_{(2,4)} = \left\{\left\langle \begin{matrix} * & 1 & * & 0 \\ * & 0 & * & 1 \end{matrix} \right\rangle\right\}, \quad \mathcal{U}_{(3,4)} = \left\{\left\langle \begin{matrix} * & * & 1 & 0 \\ * & * & 0 & 1 \end{matrix} \right\rangle\right\},$$

where $*$ denotes an arbitrary complex number and $\langle A\rangle$ denotes the span of the rows of A.

It is well-known (see e.g. [**38**, Example 6.6]) that the Grassmannian $\mathrm{Gr}_e(\mathbb{C}^d)$ is a smooth, projective and irreducible algebraic variety of dimension

$$\dim \mathrm{Gr}_e(\mathbb{C}^d) = e(d-e). \tag{4.1}$$

The tangent space at $W \in \mathrm{Gr}_e(\mathbb{C}^d)$ is

$$T_W(\mathrm{Gr}_e(\mathbb{C}^d)) \simeq \mathrm{Hom}_{\mathbb{C}}(W, \mathbb{C}^d/W).$$

We notice that $M = \mathbb{C}^d$ is a rigid Q–representation (cf. Proposition 2.3).

We now highlight another fundamental property of $\mathrm{Gr}_e(\mathbb{C}^d)$: it admits a *cellular decomposition.* We say that two matrices A and B of the same size are row-equivalent if there exists an invertible matrix C (of the appropriate size) such that $B = CA$. Since invertible matrices are products of elementary matrices, we see that this happens if and only if A can be transformed into B via Gaussian elimination. It is not hard to prove that every matrix $A \in \mathrm{Mat}_{e\times d}$ is row equivalent to a matrix, denoted rref(A), with the following properties: 1) The zero rows are at the bottom; 2) The pivot of every (non-zero) row is one; 3) the columns containg the pivots have only the pivots as non-zero elements; 4) the pivot of the i-th row is on the left of the pivot of the $(i+1)$-th row. The matrix rref(A) is called the row reduced echelon form of A (and this explains the notation). The columns of rref(A) containing the pivots are called dominants. We hence get a partition (i.e. a disjoint union)

$$\mathrm{Gr}_e(\mathbb{C}^d) = \amalg_J \mathcal{C}_J \tag{4.2}$$

where $J = (1 \leq j_1 < \cdots < j_e \leq d)$ is an e-set of column indices and $\mathcal{C}_J$ has the following equivalent definitions

$$\begin{aligned} \mathcal{C}_J &= \{W_A |\, \text{rref(A) has dominant columns J}\} \\ &= \left\{ W \subseteq_{\mathbb{C}} \mathbb{C}^d |\, \dim W_k = \left\{ \begin{array}{ll} \dim W_{k-1} & \text{if } k \notin I \\ \dim W_{k-1} + 1 & \text{if } k \in J \end{array} \right. \right\} \end{aligned}$$

where $e_i = (0, \cdots, 0, 1, 0, \cdots, 0)$ is the i–th row of the identity matrix $\mathbf{1}_d$ and $W_k := W \cap \langle e_k, \cdots, e_n\rangle$ (notice that we identify $\mathbb{C}^d$ with the row matrices). It is clear that $\mathcal{C}_J$ is an affine space (i.e. a *cell*). To make this more explicit we consider the subgroup $\mathrm{U} \subset \mathrm{GL}_d$ of unipotent upper triangular $d \times d$ matrices (i.e. its elements are upper triangular matrices with 1 on the diagonal). This group acts on $\mathrm{Mat}^{\max}_{e\times d}$ to the right, i.e. it acts on the columns. We immediately get

$$\mathcal{C}_J = \langle e_j |\, j \in J\rangle \mathrm{U}.$$

This has the following interesting consequence: the closure of a cell is a union of cells of smaller dimension

$$\overline{\mathcal{C}_J} = \amalg_I \mathcal{C}_I \tag{4.3}$$

where I varies on a subset of index-sets that can be explicitely described. Conditions (4.2) and (4.3) imply at once the following properties of $X = \mathrm{Gr}_e(M)$:

(1) $H_i(X) = 0$ if i is odd, and it is torsion-free if i is even.
(2) The cycle map $A_\bullet(X) \to H_\bullet(X)$ is an isomorphism.

Varieties having those two properties are said to have property (S). The cellular decomposition (4.2) depends on the choice of an ordering of the standard basis vectors of $\mathbb{C}^d$. It is sometimes called the standard cellular decomposition of the Grassmannian.

EXAMPLE 4.2. The standard cells of $\mathrm{Gr}_2(\mathbb{C}^4)$ are

$$\mathcal{C}_{(1,2)} = \left\{ \left\langle \begin{matrix} 1 & 0 & * & * \\ 0 & 1 & * & * \end{matrix} \right\rangle \right\}, \quad \mathcal{C}_{(1,3)} = \left\{ \left\langle \begin{matrix} 1 & * & 0 & * \\ 0 & 0 & 1 & * \end{matrix} \right\rangle \right\},$$

$$\mathcal{C}_{(1,4)} = \left\{ \left\langle \begin{matrix} 1 & * & * & 0 \\ 0 & 0 & 0 & 1 \end{matrix} \right\rangle \right\}, \quad \mathcal{C}_{(2,3)} = \left\{ \left\langle \begin{matrix} 0 & 1 & 0 & * \\ 0 & 0 & 1 & * \end{matrix} \right\rangle \right\},$$

$$\mathcal{C}_{(2,4)} = \left\{ \left\langle \begin{matrix} 0 & 1 & * & 0 \\ 0 & 0 & 0 & 1 \end{matrix} \right\rangle \right\}, \quad \mathcal{C}_{(3,4)} = \left\{ \left\langle \begin{matrix} 0 & 0 & 1 & 0 \\ 0 & 0 & 0 & 1 \end{matrix} \right\rangle \right\},$$

The Hasse diagram of the closure relation is the following:

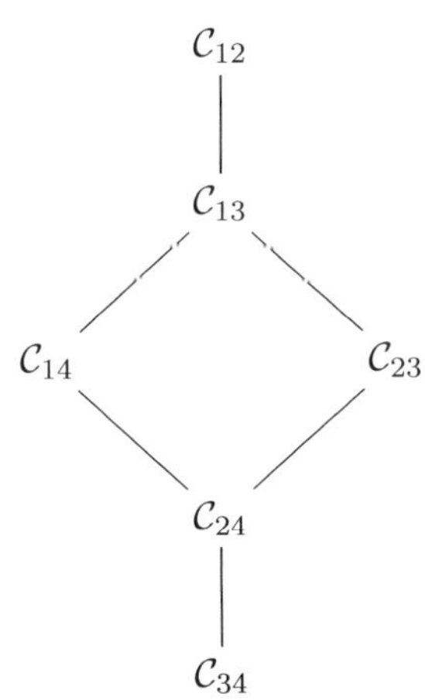

Another useful way to describe the cells is via torus action. Let us consider the following action of the one-dimensional torus $T = \mathbb{C}^*$ on $\mathrm{Gr}_e(M)$: for every $\lambda \in T$ we rescal the standard basis vectors as $\lambda \cdot e_i = \lambda^{i-1} e_i$. This defines a linear automorphism of the vector space K^d and hence descends to an action on $\mathrm{Gr}_e(M)$. Notice that this action depends on the ordering of the standard basis. It is hence immediate to see that

$$\mathcal{C}_J = \{ W \in \mathrm{Gr}_e(\mathbb{C}^d) | \lim_{\lambda \to 0} \lambda \cdot W = \langle e_j | j \in J \rangle \}.$$

For example, $\lim_{\lambda \to 0} \lambda \cdot \langle e_1 + e_2 \rangle = \lim_{\lambda \to 0} \langle e_1 + \lambda e_2 \rangle = \langle e_1 \rangle$.

4.2. Example 2: The complete flag variety. The complete flag variety is

$$\mathcal{F}l_{n+1} = \left\{U_1 \subset U_2 \subset \cdots \subset U_n \subset \mathbb{C}^{n+1} | \dim U_i = i\right\}.$$

This is naturally a quiver Grassmannian: Let

$$Q : 1 \longrightarrow 2 \longrightarrow \cdots \longrightarrow n-1 \longrightarrow n \tag{4.4}$$

be the equioriented quiver of type A_n; let

$$M = P_1^{n+1} : \mathbb{C}^{n+1} \xrightarrow{\mathbf{1}_{n+1}} \mathbb{C}^{n+1} \xrightarrow{\mathbf{1}_{n+1}} \cdots \longrightarrow \mathbb{C}^{n+1} \xrightarrow{\mathbf{1}_{n+1}} \mathbb{C}^{n+1}$$

and let $\mathbf{e} = (1, 2, 3, \cdots, n)$. Then

$$\mathcal{F}l_{n+1} \simeq \mathrm{Gr}_{\mathbf{e}}(M).$$

The complete flag variety is smooth, irreducibile of minimal dimension

$$\dim \mathcal{F}l_{n+1} = \frac{n(n+1)}{2} = \langle \mathbf{e}, \mathbf{dim}\, M - \mathbf{e} \rangle.$$

We notice that M is indeed a rigid Q–representation (cf. Proposition 2.3)

4.3. Example 3: The complete degenerate flag variety. Let $V = \mathbb{C}^{n+1}$ with standard basis $\{e_1, \cdots, e_{n+1}\}$. For every $k = 1, \cdots, n+1$ consider the projection along e_k:

$$pr_k : V \to V : \sum_i x_i e_i \mapsto \sum_{i \neq k} x_i e_i.$$

Motivated by the study of abelian degenerations of simple Lie algebras, E. Feigin introduced the projective variety $\mathcal{F}l^a_{n+1}$ called the (complete) sl_n-degenerate flag variety [**30–32**]. He showed that it has a realization in terms of linear algebra as follows

$$\mathcal{F}l^a_{n+1} \simeq \{(U_1, \cdots, U_n) \in \prod_{k=1}^{n} \mathrm{Gr}_k(V) | \, pr_{k+1}(U_k) \subseteq U_{k+1}\}$$

and he proved that this projective variety has marvellous properties: it is a (typically) singular, irreducible projective variety of dimension

$$\dim \mathcal{F}l^a_{n+1} = \dim \mathcal{F}l_{n+1} = \frac{n(n+1)}{2}$$

which is a flat degeneration of the complete flag variety $\mathcal{F}l_{n+1}$; moreover, it is a normal, locally complete intersection variety which admits a cellular decomposition. Let Q be the equioriented quiver (4.4) of type A_n. Let $A = KQ$ be its path algebra. As a Q–representation $A = \oplus_{i=1}^n P_i$ is the direct sum of all the indecomposable projectives and hence $\mathbf{dim}\, A = (1, 2, \cdots, n)$. Dually, $DA = \oplus_{k=1}^n I_k$ is the sum of all the indecomposable injectives and $\mathbf{dim}\, DA = (n, n-1, n-2, \cdots, 1)$. We notice that

$$\mathbf{dim}\, (A \oplus DA) = (n+1, n+1, \cdots, n+1) = \mathbf{dim}\, P_1^{n+1}.$$

From the definition it is immediate to check that

$$\mathcal{F}l^a_{n+1} = \mathrm{Gr}_{\mathbf{dim}\, A}(A \oplus DA).$$

4.4. Example 4: Singular quiver Grassmannians. We give two easy examples of singular quiver Grassmannians. Let $Q: 1 \to 2$ be an A_2 quiver.

The easiest example of a non-smooth quiver Grassmannian is the following: let

$$M = \mathbb{C}^2 \xrightarrow{\begin{pmatrix} 1 & 0 \\ 0 & 0 \end{pmatrix}} \mathbb{C}^2 \simeq S_1 \oplus P_1 \oplus S_2$$

and let $\mathbf{e} = (1,1)$. Then $\mathrm{Gr}_{\mathbf{e}}(M)$ is the union of two $\mathbf{P}^1$'s crossing in one point. Thus $\mathrm{Gr}_{\mathbf{e}}(M)$ is a connected, equidimensional curve of dimension one with two irreducible components and one singular point.

An easy example of a singular non-equidimensional quiver Grassmannian is the following: let

$$M = \mathbb{C}^2 \xrightarrow{\begin{pmatrix} 1 & 0 \\ 0 & 0 \\ 0 & 0 \end{pmatrix}} \mathbb{C}^3 \simeq S_1 \oplus P_1 \oplus S_2^2$$

and let $\mathbf{e} = (1,1)$. Then $\mathrm{Gr}_{\mathbf{e}}(M)$ is the union of a $\mathbf{P}^2$ and a $\mathbf{P}^1$ crossing in one point. Thus $\mathrm{Gr}_{\mathbf{e}}(M)$ is a connected projective variety of dimension two with two irreducible components (one of dimension 1 and one of dimension 2) and one singular point.

4.5. Example 5: A non–connected quiver Grassmannian. Let us give an easy example of a non–connected quiver Grassmannian. Let $Q: 1 \rightrightarrows 2$ be the Kronecker quiver. Let A be a 2×2 complex matrix with distinct eigenvalues. Let us consider the Q-representation

$$M = \mathbb{C}^2 \underset{A}{\overset{\mathbf{1}_2}{\rightrightarrows}} \mathbb{C}^2$$

and let $\mathbf{e} = (1,1)$. Then $\mathrm{Gr}_{\mathbf{e}}(M)$ consists of two distinct points (the two eigenspaces). The tangent space at those two points is zero dimensional and hence $\mathrm{Gr}_{\mathbf{e}}(M)$ is a reduced projective variety of dimension 0 with two connected components.

4.6. Example 6: a smooth quiver Grassmannian with negative Euler characteristic. We borrow this example from Derksen-Weyman-Zelevinsky's paper [**28**, Example 3.6]. Let $Q: 1 \overset{\longrightarrow}{\underset{\longrightarrow}{\rightrightarrows}} 2$ be the 4-Kronecker quiver. Let $\mathbf{d} = (3,4)$ and let $\mathbf{e} = (1,3)$. We notice that $\langle \mathbf{e}, \mathbf{d} - \mathbf{e} \rangle = 1$. It is easy to construct representations $N \in R_{\mathbf{e}}$ and $R \in R_{\mathbf{d}-\mathbf{e}}$ such that $[N,R] = 1 = \langle \mathbf{e}, \mathbf{d} - \mathbf{e} \rangle$ and hence $[N,R]^1 = 0$. By Theorem 2.4, the universal quiver Grassmannian $\mathrm{Gr}_{\mathbf{e}}^Q(\mathbf{d}) \to R_{\mathbf{d}}(Q)$ is surjective. Thus, there exists an open and dense subset U of $R_{\mathbf{d}}$ such that the fiber $\mathrm{Gr}_{\mathbf{e}}(M)$ over a point $M \in U$ is smooth of minimal dimension $\langle \mathbf{e}, \mathbf{d} - \mathbf{e} \rangle = 1 > 0$. Let $X = \mathrm{Gr}_{\mathbf{e}}(M)$ be a generic fiber (i.e. $M \in U$). Then X is a smooth curve of degree 4 and genus 3. It hence follows that its Euler characteristic is negative: $\chi(\mathrm{Gr}_{\mathbf{e}}(M)) = -4 < 0$. Notice that since $\langle \mathbf{d}, \mathbf{d} \rangle = -11 < 0$, U does not contain an open orbit, i.e. there is no rigid representation of dimension vector $\mathbf{d}$.

4.7. Every projective variety is a quiver Grassmannian. It is well known that every projective variety can be realized as the intersection of a Veronese variety with a linear space (see [**38**, Example 2.9]). Markus Reineke noticed that this construction can be used in a straightforward way to realize every projective variety as a quiver Grassmannian [**50**]. In the next section 4.8 it is shown how this works in an example. In this costruction, the quiver and the quiver representation depends on the chosen projective variety. Surprisingly, Ringel proved that the choice of the quiver does not depend on the variety, as long as the quiver is *wild.*

THEOREM 4.3. [**52**] *Let X be a projective variety and let Q be a* wild *quiver. Then there exists a Q–representation M and a dimension vector* $\mathbf{e}$ *such that*

$$X \simeq \mathrm{Gr}_{\mathbf{e}}(M).$$

4.8. Realization of an elliptic curve as a quiver Grassmannian. Following Reineke [**50**], in this section we realize an elliptic curve as a quiver Grassmannian. This example appeared in the blog of Le Bruyn [**44**]. Let us consider the elliptic curve

$$\mathcal{E} = \{[x:y:z] \in \mathbf{P}^2 |\, y^2 z = x^3 + z^3\}.$$

We fix the complex vector space $V = \mathbb{C}^3$ with standard basis $\{e_1, e_2, e_3\}$ and its linear dual V^* with dual basis $\{e_1^*, e_2^*, e_3^*\}$. The first thing we need to do is to linearize the equation defining the elliptic curve: we do this using the Veronese embedding $j : V \to \mathrm{Sym}^3(V) : \; v \mapsto v \otimes v \otimes v$. Consider the linear form $\varphi = (e_2^*)^2 e_3^* - (e_1^*)^3 - (e_3^*)^3 : \mathrm{Sym}^3(V) \to \mathbb{C}$ then

$$\mathcal{E} = \{[v] \in \mathrm{Gr}_1(V) |\, \varphi(j(v)) = 0\} \simeq \{[w] \in \mathrm{Gr}_1(\mathrm{Sym}^3 V) |\, \varphi(w) = 0\} \tag{4.5}$$

where $[v]$ denotes the line generated by a non-zero vector v and the isomorphism is induced by the embedding j.

Next, we need to describe the image of the Veronese embedding. Recall that a tensor $\omega \in V^{\otimes n}$ of the form $\omega = v^{\otimes n}$ is called decomposable. Thus, the image of j consists of the decomposable tensors of $V^{\otimes 3}$. For $1 \leq i \leq j \leq k \leq 3$ we define

$$e_i e_j := e_i \otimes e_j + e_j \otimes e_i, \quad e_i e_j e_k := \textstyle\sum_{\sigma \in S_3} e_{\sigma(i)} \otimes e_{\sigma(j)} \otimes e_{\sigma(k)}.$$

The sets $\{e_i e_j\}$ and $\{e_i e_j e_k\}$ form a basis of $\mathrm{Sym}^2(V)$ and of $\mathrm{Sym}^3(V)$, respectively. It is easy to describe the decomposable vectors of $V \otimes V$: consider the canonical isomorphism

$$\zeta_? : V \otimes V \simeq \mathrm{Hom}(V^*, V) : \omega = v \otimes w \mapsto \zeta_\omega = (u \mapsto u(v) w)$$

defined on the decomposable tensors and extended by linearity. It is immediate to verify that ω is decomposable if and only if $\mathrm{rk}(\zeta_\omega) \leq 1$. We use this criterion to detect the decomposable tensors of $V \otimes V \otimes V$: We consider the linear map

$$\iota : \mathrm{Sym}^3(V) \to V \otimes \mathrm{Sym}^2(V)$$

defined on the basis elements by $e_i e_j e_k \mapsto e_i \otimes e_j e_k + e_j \otimes e_i e_k + e_k \otimes e_i e_j$. This is the injective linear map compatible with the inclusions $\mathrm{Sym}^3(V) \subset V^{\otimes 3}$ and $V \otimes \mathrm{Sym}^2(V) \subset V^{\otimes 3}$. In particular, it sends a decomposable vector to a decomposable vector: Indeed

$$\iota(\sum_{i \leq j \leq k} \alpha_i \alpha_j \alpha_k \, e_i e_j e_k) = (\sum_i \alpha_i e_i) \otimes (\sum_{j \leq k} \alpha_j \alpha_k e_j e_k).$$

The standard basis $\{e_1, e_2, e_3\}$ of V determines an isomorphism of vector spaces $V \otimes \mathrm{Sym}^2(V) \simeq \mathrm{Sym}^2(V) \oplus \mathrm{Sym}^2(V) \oplus \mathrm{Sym}^2(V)$. Let

$$\pi_1, \pi_2, \pi_3 : V \otimes \mathrm{Sym}^2(V) \to \mathrm{Sym}^2(V)$$

be the projections onto the three factors, respectively. Thus, by definition, a vector ω of $V \otimes \mathrm{Sym}^2(V)$ is written as $\omega = e_1 \otimes \pi_1(\omega) + e_2 \otimes \pi_2(\omega) + e_3 \otimes \pi_3(\omega)$. Let

$$\psi_1, \psi_2, \psi_3 : \mathrm{Sym}^3(V) \xrightarrow{\ \iota\ } V \otimes \mathrm{Sym}^2(V) \xrightarrow{\pi_1,\pi_2,\pi_3} \mathrm{Sym}^2(V)$$

be the composite maps $\psi_i = \pi_i \circ \iota$. From the criterion above we see that ω is decomposable if and only if $\pi_1(\omega)$, $\pi_2(\omega)$ and $\pi_3(\omega)$ are all contained in a same line. We hence have:

$$t \in \mathrm{Im}\, j \Leftrightarrow \iota(t) \text{ is decomposable} \Leftrightarrow \mathrm{dimSpan}\{\psi_1(t), \psi_2(t), \psi_3(t)\} \leq 1. \tag{4.6}$$

We consider the quiver $\bullet \longleftarrow \bullet \rightrightarrows\!\!\!\rightarrow \bullet$ and its representation

$$M = \mathbb{C} \xleftarrow{\ \varphi\ } \mathrm{Sym}^3(V) \xrightarrow{\psi_1,\psi_2,\psi_3} \mathrm{Sym}^2(V)$$

Then, putting together (4.5) and (4.6), we have $\mathcal{E} \simeq \mathrm{Gr}_{\mathbf{e}}(M)$ where $\mathbf{e} = (0, 1, 1)$.

5. Quiver Grassmannians of type A

In sections (4.2) and (4.3) we saw that the complete flag variety and the corresponding degenerate flag variety are quiver Grassmannians attached to representations of the equioriented quiver of type A_n. In this section we study some general properties of quiver Grassmannians of this sort, that we briefly call "of type A".

Let Q be the equioriented quiver (4.4) of type A_n. A Q–representation $M = ((M_i)_{i=1}^n, (f_i)_{i=1}^{n-1})$ is

$$M : M_1 \xrightarrow{f_1} M_2 \xrightarrow{f_2} \cdots \xrightarrow{f_{n-2}} M_{n-1} \xrightarrow{f_{n-1}} n\ .$$

It is an interesting problem of linear algebra to find the normal form of a collection of linear maps $(f_1, \cdots, f_{n-1})$ by base change. It turns out that the indecomposable Q–representations are thin and they are supported on connected subgraphs of Q. Given $1 \leq i \leq j \leq n$ we denote by $U_{i,j}$ the indecomposable supported on the interval $[i, j]$. The projectives are $P_i = U_{i,n}$ and the injectives are $I_k = U_{1,k}$. The AR-translate $\tau U_{i,j}$ of $U_{i,j}$ is $\tau U_{i,j} = U_{i+1,j+1}$ and the AR-quiver is the following (for $n = 4$):

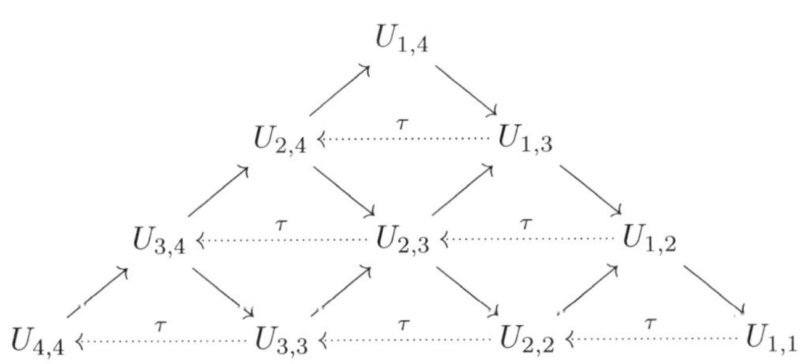

The space of homomorphisms (and of extensions) between them are at most one-dimensional and are given as follows

$$[U_{ij}, U_{k\ell}] = \begin{cases} 1 & \text{if } k \leq i \leq \ell \leq j \\ 0 & \text{otherwise} \end{cases} \tag{5.1}$$

$$[U_{k\ell}, U_{ij}]^1 = [U_{ij}, U_{k+1\ell+1}] = \begin{cases} 1 & \text{if } k+1 \leq i \leq \ell+1 \leq j \\ 0 & \text{otherwise} \end{cases} \tag{5.2}$$

We can order the indecomposable Q–representations as $M(1) < M(2) < \cdots < M(N)$ (where $N = \frac{n(n+1)}{2}$) so that

$$M(i) < M(\ell) \Longrightarrow \mathrm{Ext}^1(M(i), M(\ell)) = 0. \tag{5.3}$$

As shown in [**19**, Remark 7], a natural choice is the following (for $n = 4$)

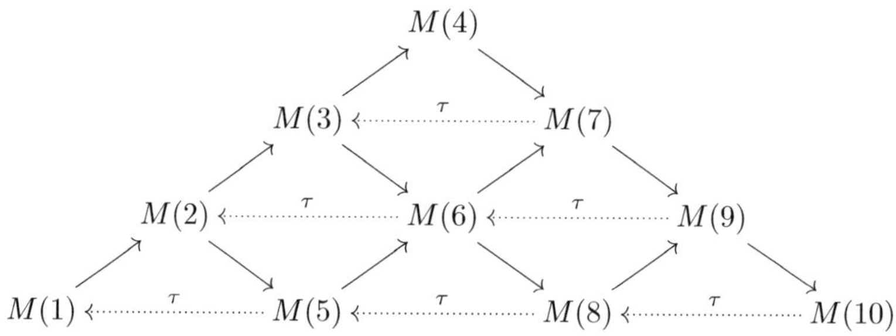

To a Q-representation $M = ((M_i), (f_i))$ we associate the sequence of non-negative integers $\mathbf{r}^M = (r^M_{i,j} | 1 \leq j \leq n)$ given by the ranks of the composite linear maps f_i's, i.e. $r^M_{i,j} = rk(f_{j-1} \circ \cdots \circ f_i)$ for $i < j$ and $r^M_{i,i} = \dim M_i$. If M decomposes as direct sum of indecomposables as $M = \oplus_{i,j} U_{i,j}^{m_{i,j}}$, the relation between the rank tuple $\mathbf{r}^M$ and the tuple of multiplicities $(m_{i,j})$ is given by

$$r^M_{i,j} = \textstyle\sum_{k \leq i \leq j \leq \ell} m_{k,\ell}, \quad m^M_{ij} = r^M_{i,j} - r^M_{i-1,j} - r^M_{i,j+1} + r^M_{i-1,j+1}.$$

In particular, the sequence $(r^M_{i,j})$ satisfies the inequalities

$$r^M_{i,j} + r^M_{i-1,j+1} \geq r^M_{i,j+1} + r^M_{i-1,j}.$$

Conversely, let $\mathbf{r} = (r_{i,j} | 1 \leq i \leq j \leq n)$ be a sequence of non-negative integers which fulfill the inequalities

$$r_{i,j} + r_{i-1,j+1} \geq r_{i,j+1} + r_{i-1,j} \tag{5.4}$$

for all $1 \leq i \leq j \leq n$, with the convention $r_{i,j} = 0$ if $i = 0$ or $j = n + 1$. An easy induction shows that (5.4) are equivalent to

$$r_{i,k} + r_{j,\ell} \geq r_{j,\ell} + r_{i,k} \tag{5.5}$$

for every $1 \leq i \leq j \leq k \leq \ell \leq n$. Then one can easily contruct a Q-representation M such that $\mathbf{r}^M = \mathbf{r}$. A sequence of non-negative integers $\mathbf{r} = (r_{i,j})$ satisfying the inequalities (5.5) is called a rank sequence. By the above, $M \simeq N$ if and only if $\mathbf{r}^M = \mathbf{r}^N$, i.e. the isoclasses of Q–representations are parametrized by rank sequences. Moreover it is proved in [**1**] that $M \leq_{deg} N$ if and only if $r^M_{ii} = r^N_{ii}$ and $r^M_{ij} \geq r^N_{ij}$ for every i, j; in this case we briefly write $\mathbf{r}^M \geq \mathbf{r}^N$.

5.1. Cellular decomposition. Let $M = (M_i, f_i)$ be a Q–representation. Decompose $M = \oplus_{k=1}^s M(k)$ as a direct sum of its indecomposable direct summands, so that $[M(i), M(j)]^1 = 0$ for $i < j$.

For every $\lambda \in \mathbb{C}^*$ consider the automorphism $f_\lambda : M \to M$ of M which rescales the $M(k)$'s as follows

$$f_\lambda(m) = \lambda^{k-1} m \quad \forall m \in M(k).$$

This gives an action

$$T \times \mathrm{Gr}_{\mathbf{e}}(M) \to \mathrm{Gr}_{\mathbf{e}}(M) : \; (\lambda, N) \mapsto \lambda \cdot N$$

of the one-dimensional torus $T = \mathbb{C}^*$ on every quiver Grassmannian associated with M. This action has finitely many T–fixed points which are coordinate subrepresentations:

$$\mathrm{Gr}_{\mathbf{e}}(M)^T = \coprod_{\mathbf{f}_1+\cdots+\mathbf{f}_s=\mathbf{e}} \mathrm{Gr}_{\mathbf{f}_1}(M(1)) \times \cdots \times \mathrm{Gr}_{\mathbf{f}_s}(M(s)) \tag{5.6}$$

Notice that since the $M(k)$'s are thin, every non-empty quiver Grassmannian associated to them is a point and $\mathrm{Gr}_{\mathbf{e}}(M)^T$ is a finite collection of points. For a T-fixed point L consider the attracting space

$$\mathcal{C}_L = \{N \in \mathrm{Gr}_{\mathbf{e}}(M) | \lim_{\lambda \to 0} \lambda.N = L\}.$$

We can represent the representation M as a collection of strings, ordered from top to bottom, according to (5.3). This collection of strings is called the coefficient quiver of M. Given a T-fixed point $L \in \mathrm{Gr}_{\mathbf{e}}(M)^T$ we colour black the vertices corresponding to L and to compute the dimension of $\mathcal{C}_L$ we only need to count how many white vertices there are below each black *source* of L. For example, if $n = 3$ and $M = A \oplus DA$, $\mathbf{e} = \mathbf{dim}\, A = (1, 2, 3)$ the following is coefficient quiver of M together with a T-fixed point L highlighted by black vertices

$$\begin{array}{ccccccc}
 & & & & \bullet & & 1\\
 & & \circ & \longrightarrow & \bullet & & 2\\
\circ & \longrightarrow & \bullet & \longrightarrow & \bullet & & 3\\
\circ & \longrightarrow & \circ & \longrightarrow & \circ & & 4\\
\bullet & \longrightarrow & \bullet & & & & 5\\
\circ & & & & & & 6
\end{array}$$

Then the dimension of $\mathcal{C}_L$ is 4 (see [**19**, Sec. 6.4]). By elementary linear algebra techniques one can prove the following result.

THEOREM 5.1. [**19**, Thm 12] *For every T–fixed point L, the attracting set $\mathcal{C}_L$ is an affine space and the quiver Grassmannian $\mathrm{Gr}_{\mathbf{e}}(M)$ admits a cellular decomposition*

$$\mathrm{Gr}_{\mathbf{e}}(M) = \coprod_{L \in \mathrm{Gr}_{\mathbf{e}}(M)^T} \mathcal{C}_L.$$

Moreover the points of every cell $\mathcal{C}_L$ are isomorphic to L and hence every iso-stratum $\mathcal{S}_L$ decomposes as union of cells.

The choice of the ordering of the indecomposable direct summands $M(i)$'s of M is necessary to make theorem 5.1 working. Indeed, consider the case $n = 2$ and the representation $M = S_1 \oplus P_1 \oplus S_2$ of example 4 with the ordering given by

$$\begin{array}{ccccc}
\bullet & & & & 1\\
\circ & \longrightarrow & \bullet & & 2\\
 & & \circ & & 3
\end{array}$$

and let L be the highlighted point of $\mathrm{Gr}_{(1,1)}(M)$. One can easily verify that $\mathcal{C}(L) \simeq \{([1 : \lambda], [1 : \mu]) \in \mathbf{P}^1 \times \mathbf{P}^1 | \lambda\mu = 0\}$ which is not a cell.

5.2. Schubert quiver Grassmannians. Let Q be the equioriented quiver of type A_n as in the previous section. A Q–representation $R = ((R_i), (f_i))$ is projective if and only if every linear map $f_i : R_i \to R_{i+1}$ is injective. In other words, R is projective if and only if $R_\bullet$ is a flag in R_n. The automorphism group $Aut(R)$ of R consists of those $g \in GL(R_n)$ which fix the flag $R_\bullet$; thus, $Aut(R)$ is a parabolic subgroup of $GL(R_n)$. This implies that the quiver Grassmannians associated with R are Schubert varieties inside the partial flag variety $GL(R_n)/Aut(R)$. Given a Q-representation M, let us consider its minimal projective resolution

$$0 \longrightarrow P \xrightarrow{\iota} R \xrightarrow{\pi} M \longrightarrow 0\,.$$

For example if $n = 3$ and $M = A \oplus DA$ then the diagram

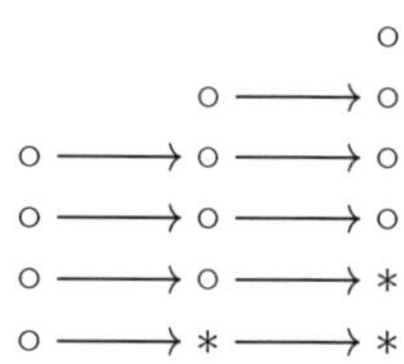

describes the minimal projective resolution of M: the $*$ form (the coefficient-quiver of) P, the whole diagram is (the coefficient-quiver of) R and the diagram without $*$ is (the coefficient-quiver of) M. We can use the surjective morphism $\pi : R \to M$ to embed a quiver Grassmannian $\mathrm{Gr}_{\mathbf{e}}(M)$ associated with M into the partial flag variety $GL(R_n)/Aut(R)$ (see [**23**, Prop. 2.1]): we define the map

$$\zeta : \mathrm{Gr}_{\mathbf{e}}(M) \to \mathrm{Gr}_{\mathbf{e}+\mathbf{dim}\,P}(R) : \; N \mapsto \pi^{-1}(N).$$

This is a closed embedding and we refer to it as the standard embedding of $\mathrm{Gr}_{\mathbf{e}}(M)$ inside a partial flag variety. Let $B \subseteq Aut(R)$ be the Borel subgroup of $GL(R_n)$ contained in the parabolic subgroup $Aut(R)$. It is a natural problem to find conditions on the Q–representation M and to the dimension vector $\mathbf{e}$ so that the standard embedding of $\mathrm{Gr}_{\mathbf{e}}(M)$ into a partial flag manifold is stable by B. Indeed, if this happens, then the irreducible components of $\mathrm{Gr}_{\mathbf{e}}(M)$ are Schubert varieties. For this reason, we called such varieties Schubert quiver Grassmannians.

In [**23**] we studied this problem and found a purely combinatorial solution. To state the result we introduced the following terminology:

Definition 5.2. We say that a Q–representation M is a *catenoid* if the indecomposable direct summands of M lie in an oriented (connected) path of the AR-quiver of Q.

To illustrate definition 5.2 consider the case $n = 4$; given a Q-representation M we highlight with $\bullet$ the vertices of the AR-quiver corresponding to the indecomposable direct summands of M. Consider the following two configurations:

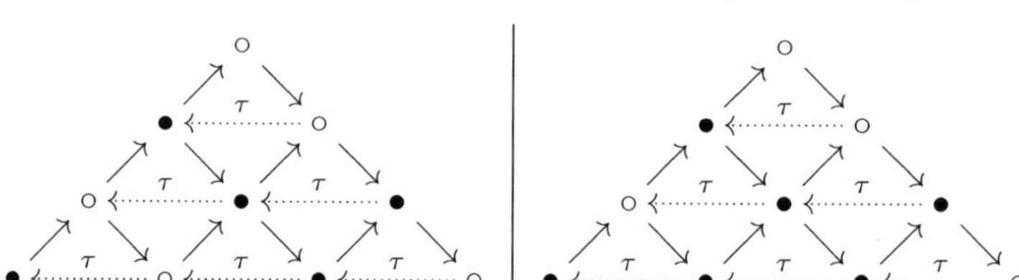

Catenoid | NotaCatenoid

Then a representation whose configuration of its indecomposables is shown on the left is a catenoid, while the one on the right is not a catenoid. Notice that the multiplicities do not play any rôle.

THEOREM 5.3. *A quiver Grassmannian* $\mathrm{Gr}_{\mathbf{e}}(M)$ *is a Schubert quiver Grassmannian if and only if* M *is a catenoid.*

Theorem 5.3 has the following interesting corollary.

COROLLARY 5.4. *Degenerate flag varieties are Schubert varieties.*

PROOF. The complete degenerate flag variety is $\mathrm{Gr}_{\mathbf{dim}\,A}(A \oplus DA)$ and the representation $M = A \oplus DA$ is a catenoid. By Theorem 5.3 its irreducible components are hence Schubert varieties. Since the complete degenerate flag variety is irreducible the result follows. □

Corollary 5.4 was first proved in collaboration with Martina Lanini [**24**] and it was the starting point for the study of Schubert quiver Grassmannians. It holds for partial degenerate flag varieties, and for symplectic degenerate flag varieties. For a characteristic-free approach see [**25**].

5.3. Linear degeneration of the complete flag variety. Let Q be the equioriented quiver of type A_n and let $A = KQ$ be its path algebra. The dimension vector $\mathbf{d} = (n+1, n+1, \cdots, n+1)$ is of special intereset. Indeed, both P_1^{n+1} and $A \oplus DA$ have dimension vector equal to $\mathbf{d}$. The complete flag variety $\mathcal{F}l_{n+1}$ and the degenerate flag variety $\mathcal{F}l_{n+1}^a$ are hence fibers of the universal quiver Grassmannian

$$p_{\mathbf{e},\mathbf{d}} : \mathrm{Gr}_{\mathbf{e}}^Q(\mathbf{d}) \to R_{\mathbf{d}}$$

where $\mathbf{e} = \mathbf{dim}\,A$. Notice that $p_{\mathbf{e},\mathbf{d}}$ is surjective because the rigid representation P_1^{n+1} belongs to its image. The family $p_{\mathbf{e},\mathbf{d}}$ is hence of special interest and it was studied in [**19**]. The main result of [**19**] describes the locus where $p_{\mathbf{e},\mathbf{d}}$ is flat with irreducible fibers, denoted with $\mathcal{U}_{flat,irr}$ and the locus where $p_{\mathbf{e},\mathbf{d}}$ is flat, denoted with $\mathcal{U}_{flat}$. Since $\mathcal{F}l_{n+1} = p_{\mathbf{e},\mathbf{d}}^{-1}(P_1^{n+1})$ and $\mathcal{F}l_{n+1}^a = p_{\mathbf{e},\mathbf{d}}^{-1}(A \oplus DA)$ are irreducible, both P_1^{n+1} and $A \oplus DA$ lie in $\mathcal{U}_{flat,irr}$. Let us consider the rank sequences $\mathbf{r}^0$, $\mathbf{r}^1$ and $\mathbf{r}^2$ given by

$$r_{i,j}^0 = n+1, \quad r_{i,j}^1 = n+1-(j-i), \quad r_{i,j}^2 = n-(j-i).$$

We put $M^0 = P_1^{n+1}$, $M^1 = A \oplus DA$ and notice that $\mathbf{r}^0 = \mathbf{r}^{M^0}$ and $\mathbf{r}^{M^1}$. The representation M^2 such that $\mathbf{r}^2 = \mathbf{r}^{M^2}$ is

$$M^2 = \bigoplus_{i=1}^{n} P_i \oplus \bigoplus_{j=1}^{n-1} I_j \oplus \bigoplus_{k=1}^{n} S_k = A \oplus DA/(\mathrm{soc}\,DA) \oplus \mathrm{soc}\,DA$$

The following are the coefficient quivers of M^0, M^1 and M^2 respectively, for $n = 3$

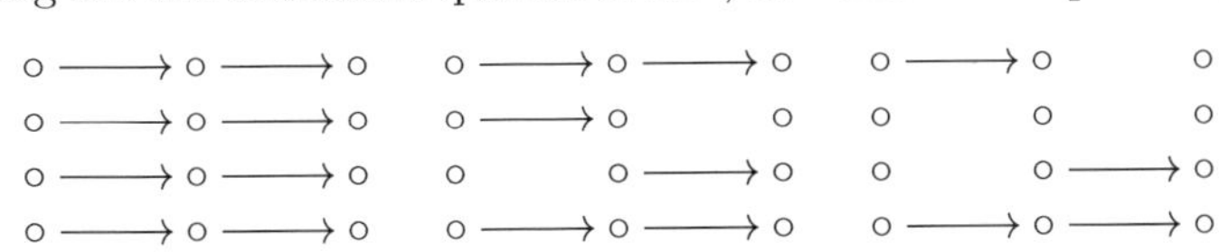

M^0 M^1 M^2

The following theorem describes the loci $\mathcal{U}_{\mathrm{flat,Irr}}$ and $\mathcal{U}_{\mathrm{flat}}$ defined above.

THEOREM 5.5. [**19**]

(1) $\mathcal{U}_{\mathrm{flat,Irr}} = \{M \in R_{\mathbf{d}} | \, M \leq_{deg} M^1\} = \{M \in R_{\mathbf{d}} | \, \mathbf{r}^M \geq \mathbf{r}^1\}$.
(2) $\mathcal{U}_{\mathrm{flat}} = \{M \in R_{\mathbf{d}} | \, M \leq_{deg} M^2\} = \{M \in R_{\mathbf{d}} | \, \mathbf{r}^M \geq \mathbf{r}^2\}$.

It is worth noting that the irreducible flat locus $\mathcal{U}_{\mathrm{flat,Irr}}$ coincides with the normal flat locus, i.e. the locus consisting of points M whose fiber $p_{\mathbf{e},\mathbf{d}}^{-1}(M)$ is normal and of the same dimension as the complete flag variety $\mathcal{F}l_{n+1}$.

In view of Theorem 5.5 we call $\mathrm{Gr}_{\mathbf{dim}\, A}(M^2)$ the most-flat linear degeneration of the complete flag variety, or, in short, mf-linear degeneration of the flag variety. It turns out that the mf-linear degeneration has an interesting geometric structure: it is an equi-dimensional, locally complete intersection variety whose irreducible components are naturally parametrized by non-crossing arc diagrams on n vertices and hence they are in number of $C_n = \frac{1}{n}\binom{2n}{n}$, the n-th Catalan number.

The flat irreducible locus contains two Schubert varieties, namely the complete flag variety $\mathcal{F}l_{n+1}$ and the degenerate flag variety $\mathcal{F}l^a_{n+1}$. We call $\mathcal{U}_{PBW} \subset \mathcal{U}_{irr,flat}$ the locus of points whose fiber is a Schubert quiver Grassmannian. In [**19**, Section 5] those Schubert quiver Grassmannian are shown to be PBW-degenerations of the complete flag variety, and hence the name. Moreover a complete description of the realization as a Schubert variety is provided.

6. Quiver Grassmannians of Dynkin type

In section 5 we saw that even quiver Grassmannians of type A can have a complicated geometric structure. We hence cannot expect to find general results concerning quiver Grassmannians associated with an arbitrary Dynkin quiver. What we can do is to restrict our attention to particular quiver Grassmannians. We have in mind the complete degenerate flag variety $\mathcal{F}l_{n+1} \simeq \mathrm{Gr}_{\mathbf{dim}\, A}(P_1^{n+1})$, and the degenerate flag variety $\mathcal{F}l^a_{n+1} \simeq \mathrm{Gr}_{\mathbf{dim}\, A}(A \oplus DA)$, where A is the path algebra of the equioriented quiver of type A_n. Those varieties share many nice properties. It is hence natural to consider quiver Grassmannians of the form $\mathrm{Gr}_{\mathbf{dim}\, P}(P \oplus I)$ where P is a projective and I is an injective representation of an arbitrary Dynkin quiver. We call such varieties well-behaved quiver Grassmannians [**20**]. We prove the following result.

THEOREM 6.1. [**20**] *Let Q be a Dynkin quiver. Let P and I be a projective and an injective Q-representation, respectively. Let $Z = \mathrm{Gr}_{\mathbf{dim}\, P}(P \oplus I)$. Then Z is a reduced, irreducible and rational locally complete intersection scheme of minimal dimension $\langle \mathbf{dim}\, P, \mathbf{dim}\, I\rangle$. Moreover Z has normal singularities and it is acted upon by an algebraic group $G \subset \mathrm{Aut}(P \oplus I)$ with finitely many orbits.*

We can extend a little bit the class of well-behaved quiver Grassmannians by considering varieties of the form $Z = \mathrm{Gr}_{\mathbf{dim}\, X}(X \oplus Y)$ where X and Y are rigid and such that $[X, Y]^1 = 0$. In this case the variety Z has the same properties stated in theorem 6.1 but not the group action and the normality. Normality does not hold for such quiver Grassmannians, in general, since they can have singularities in codimension 1 (see example 4 in section 4.4).

7. Cell decomposition and property (S)

A finite partition (X_i) of a complex algebraic variety X is said to be an α–*partition* if

$$X_1 \amalg \cdots \amalg X_i \text{ is closed in } X \text{ for every } i. \tag{7.1}$$

Clearly, every piece of an α–partition is locally closed. Property (7.1) can be reformulated by

$$\overline{X_i} \subseteq X_1 \amalg \cdots \amalg X_{i-1} \tag{7.2}$$

for every i. A *cellular decomposition* of X is an α–partition whose parts X_i are (complex) affine spaces. By (4.2) and (4.3) we see that the Grassmannian admits an α-partition into affine spaces; more precisely the standard cells of the Grassmannian form an α-partition with the stronger property that in (7.2) the equality holds. The following is an example of a variety admitting a partition into affine spaces which do not admit a cellular decomposition.

EXAMPLE 7.1. Let $X = \{[x : y : z] \in \mathbf{P}^2 | \, xyz = 0\}$ be the union of three $\mathbf{P}^1$'s crossing in the three distinct points $[1 : 0 : 0]$, $[0 : 1 : 0]$ and $[0 : 0 : 1]$. We can represent X as a triangle:

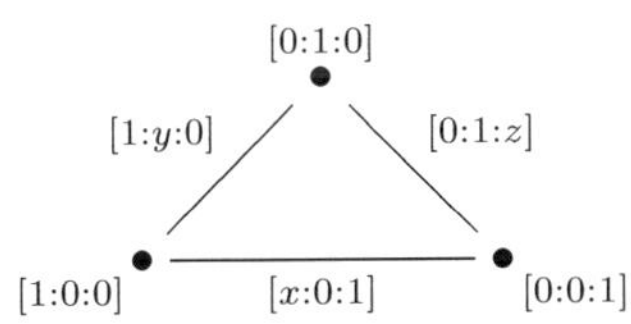

Then $X = \mathcal{C}_1 \cup \mathcal{C}_2 \cup \mathcal{C}_3$ is the disjoint union of three affine lines $\mathbb{A}^1$ given by $\mathcal{C}_1 = \{[1 : y : 0]\}$, $\mathcal{C}_2 = \{[1 : 0 : z]\}$, $\mathcal{C}_3 = \{[0 : 1 : z]\}$ which do not form an α-partition.

The existence of a cellular decomposition for X is rare but when happens it implies wonderful homological properties: we denote by $H_i(X)$ the i th space of the Borel–Moore homology of X (see [**26**]). Following [**27**, Sec. 1.7] we say that an algebraic variety X has *property (S)* if:

(**S1**) $H_i(X)$ is zero if i is odd and it has no torsion if i is even;
(**S2**) the cycle map $\varphi_i : A_i(X) \to H_{2i}(X)$ is an isomorphism for all i.

(Here $A_k(X)$ denotes the Chow group generated by K–dimensional irreducible subvarieties modulo rational equivalences (see [**36**, Sec. 1.3])). It is easy to prove (see e.g. [**27**, Section 1.10]) that

$$\text{Cell decompositon} \Longrightarrow \text{Property (S)}$$

but the opposite is not true. A counter-example for the reverse implication was communicated to me by Antonio Rapagnetta: it is the first example of an irrational surface with trivial H^1 [**7**]. Those surfaces have property (S) but they cannot admit a cellular decomposition, since cellular decomposition implies rationality. The variety X of example 7.1 has non-trivial H_1 and hence it cannot admit a cellular decomposition.

If X is a smooth complex algebraic variety, we can formulate property (S) equivalently in terms of singular cohomology as: 1) the odd cohomology groups $H^{2i+1}(X)$ vanish and 2) the cycle map $A^i(X) \to H^{2i}(X)$ is an isomorphism.

By [**27**, Lemma 1.9], if $f : X \to Y$ is a locally-trivial affine bundle, and Y has property (S) (resp. admits a cellular decomposition) then X has property (S) (resp. admits a cellular decomposition) too. If X admits an α-partition into pieces having property (S) (resp. admitting a cellular decomposition) then X has property (S) (resp. admits a cellular decomposition).

In [**27**] it is shown that Springer fibers for classical groups admit a cellular decomposition and the Springer fibers for exceptional groups have property (S). It is conjectured that cell decomposition exists also for the exceptional groups.

We say that a quiver representation M has property (C) (resp. (S)) if every quiver Grassmannian $\mathrm{Gr}_{\mathbf{e}}(M)$ associated with M admits a cellular decomposition (resp. has property (S)). In [**16**] the following result is proved.

THEOREM 7.2. [**16**] *Let Q be a connected quiver and M a Q–representation.*

(1) *If Q is Dynkin, then M has property (C).*
(2) *If Q is affine and M has indecomposable or rigid regular part, then M has property (C).*
(3) *If Q is arbitrary and M is rigid, then M has property (S).*

It is conjectured that every rigid quiver representation has property (C). For quivers with two vertices this has been proved in [**56**]. For quivers of type $\tilde{D}$, theorem 7.2 was partially proved in [**45**] and [**46**]. It is not clear to us if the restrictions assumed in part (2) of Theorem 7.2 on the regular part of M for an affine quiver are necessary. At the moment, our proof only works for those cases. The proof of part (3) is based on a general theorem of Ellingsrund and Strømme [**29**] concerning the decomposition of the diagonal in the Chow group.

7.1. Decomposition induced by short exact sequences. In this section we illustrate the idea of the proof of part (1) and part (2) of theorem 7.2. Let Q be an acyclic quiver and let

$$\eta:\ 0 \longrightarrow M' \stackrel{\iota}{\longrightarrow} M \stackrel{\pi}{\longrightarrow} M'' \longrightarrow 0$$

be a short exact sequence in $\mathrm{Rep}(Q)$. This induces the map

$$\Psi^\eta : \mathrm{Gr}_{\mathbf{e}}(M) \to \coprod_{\mathbf{f}+\mathbf{g}=\mathbf{e}} \mathrm{Gr}_{\mathbf{f}}(M') \times \mathrm{Gr}_{\mathbf{g}}(M'') :\ N \mapsto (\iota^{-1}N, \pi(N))$$

between quiver Grassmannians. By taking the preimage $\mathcal{S}^\eta_{\mathbf{f},\mathbf{g}} = (\Psi^\eta)^{-1}(\mathrm{Gr}_{\mathbf{f}}(M') \times \mathrm{Gr}_{\mathbf{g}}(M''))$ of each piece, we get the algebraic map

$$\Psi^\eta_{\mathbf{f},\mathbf{g}} : \mathcal{S}^\eta_{\mathbf{f},\mathbf{g}} \to \mathrm{Gr}_{\mathbf{f}}(M') \times \mathrm{Gr}_{\mathbf{g}}(M'') :\ N \mapsto (\iota^{-1}N, \pi(N)).$$

The finite partition $\mathrm{Gr}_{\mathbf{e}}(M) = \coprod_{\mathbf{f}+\mathbf{g}=\mathbf{e}} \mathcal{S}^\eta_{\mathbf{f},\mathbf{g}}$ is an α-partition (see [**16**, Lemma 20]). It is hence natural to investigate the map $\Psi^\eta_{\mathbf{f},\mathbf{g}}$ in order to deduce nice properties of each piece $\mathcal{S}^\eta_{\mathbf{f},\mathbf{g}}$. The first thing to study is its image. It is basically by definition that the image of $\Psi^\eta_{\mathbf{f},\mathbf{g}}$ consists of those pairs $(N', N'') \in \mathrm{Gr}_{\mathbf{f}}(M') \times \mathrm{Gr}_{\mathbf{g}}(M'')$ such that in the commutative diagram

$$\begin{array}{ccc} \mathrm{Ext}^1(M'', M') & \twoheadrightarrow & \mathrm{Ext}^1(M'', M'/N') \\ \downarrow & & \downarrow \\ \mathrm{Ext}^1(N'', M') & \twoheadrightarrow & \mathrm{Ext}^1(N'', M'/N') \end{array} \tag{7.3}$$

the element $\eta \in \mathrm{Ext}^1(M'', M')$ is mapped to zero in $\mathrm{Ext}^1(N'', M'/N')$ (see [**16**, Lemma 21]). For a general η the image of $\Psi^{\eta}_{\mathbf{f},\mathbf{g}}$ is hence difficult to control. Nevertheless, there are some short exact sequences for which this image is under control. They are called generating.

7.2. Generating short exact sequences.

DEFINITION 7.3. An element $\xi \in \mathrm{Ext}^1(S, X)$ is *generating* if $\mathrm{Ext}^1(S, X) = \mathbb{C}\xi$.

In other words $\xi \in \mathrm{Ext}^1(S, X)$ is generating if either $[S, X]^1 = 0$ and $\xi = 0$ or $[S, X]^1 = 1$ and $\xi \neq 0$.

If a generating sequence $\xi \in \mathrm{Ext}^1(S, X)$ is split then $\mathrm{Ext}^1(S, X) = 0$ and hence the map $\Psi^{\xi}_{\mathbf{f},\mathbf{g}}$ is surjective by the above description of its image.

If a generating sequence $\xi \in \mathrm{Ext}^1(S, X)$ is not split then $\mathrm{Ext}^1(S, X) \simeq \mathbb{C}$ and hence for every $(N_1, N_2) \in \mathrm{Gr}_{\mathbf{f}}(X) \times \mathrm{Gr}_{\mathbf{g}}(S)$ the diagram (7.3) becomes

$$\begin{array}{ccc} \mathbb{C} & \twoheadrightarrow & \mathrm{Ext}^1(S, X/N_1) \\ \downarrow\!\!\!\downarrow & & \downarrow\!\!\!\downarrow \\ \mathrm{Ext}^1(N_2, X) & \twoheadrightarrow & \mathrm{Ext}^1(N_2, X/N_1) \end{array} \tag{7.4}$$

forcing $[N_2, X]^1 \leq 1$, $[S, X/N_1]^1 \leq 1$ and $[N_2, X/N_1]^1 \leq 1$. A pair (N_1, N_2) is *not* in the image of $\Psi^{\xi}_{\mathbf{f},\mathbf{g}}$ if and only if $[N_2, X/N_1]^1 = 1$. By diagram chasing one shows that a pair (N_1, N_2) is *not* in the image of $\Psi^{\xi}_{\mathbf{f},\mathbf{g}}$ if and only if $[N_2, X]^1 = [S, X/N_1]^1 = 1$. It turns out [**16**, Lemma 27] that the following subrepresentations are well–defined

$$X_S := \max\{N \subset X \,|\, [S, X/N]^1 = 1\}, \quad S^X := \min\{N \subset S \,|\, [N, X]^1 = 1\}.$$

Let us give a better description of those subrepresentations. The subrepresentation X_S is the maximal subrepresentation of X such that the pushout sequence

$$\begin{array}{lccccccccc} \xi: & 0 & \longrightarrow & X & \longrightarrow & Y & \longrightarrow & S & \longrightarrow & 0 \\ & & & \downarrow\!\!\!\downarrow{\scriptstyle p} & & \downarrow\!\!\!\downarrow & & \downarrow{\scriptstyle =} & & \\ p_*\xi: & 0 & \longrightarrow & X/X_S & \longrightarrow & \overline{Y} & \longrightarrow & S & \longrightarrow & 0 \end{array}$$

does not split. Dually, the subrepresentation $S^X \subseteq S$ is the minimal subrepresentation such that the pull-back sequence

$$\begin{array}{lccccccccc} i^*\xi: & 0 & \longrightarrow & X & \longrightarrow & \tilde{Y} & \longrightarrow & S^X & \longrightarrow & 0 \\ & & & \downarrow{\scriptstyle =} & & \downarrow & & \downarrow{\scriptstyle i} & & \\ \xi: & 0 & \longrightarrow & X & \longrightarrow & Y & \longrightarrow & S & \longrightarrow & 0 \end{array}$$

does not split. If ξ is almost split then this description implies that $S^X = S$ and $X_S = 0$. We hence say that a generating extension is a *generalized almost split* sequence if $S^X = S$ and $X_S = 0$. It is easy to find examples of generalized almost split sequences which are not almost split (see exercise 9.22). If $\xi \in \mathrm{Ext}^1(S, X)$ is generating and not split, then, by the Auslander-Reiten formula, $[X, \tau S] = [\tau^- X, S] = 1$; Let $f : X \to \tau S$ and $g : \tau^- X \to S$ be two non-zero maps then

$$X_S = \mathrm{Ker}(f) \quad \text{and} \quad S^X = \mathrm{Im}\,(g). \tag{7.5}$$

EXAMPLE 7.4. Let Q be the equioriented quiver (4.4) of type A_n. Given indices $1 \leq i < k < j < \ell \leq n$ let us consider the indecomposable Q–representations $X = U_{k,\ell}$ and $S = U_{i,j}$. In view of (5.2), $[S, X]^1 = 1$ and there is a generating extension $\xi : 0 \to X \to Y \to S \to 0$ where $Y = U_{i,\ell} \oplus U_{k,j}$. A non-zero map between X and τS is injective, and hence, by (7.5), $X_S = 0$. The image of a non-zero map $\tau^- X = U_{k-1,\ell-1} \to U_{i,j}$ is $S^X = U_{k-1,\ell-1}$. For quivers of this type the only generalized almost split sequences are the almost split sequences (see exercise 9.23). In general, one can easily find examples of generilized almost split sequences which are not almost split (see exercise 9.22).

Turning back to $\mathrm{Im}(\Psi^\xi_{\mathbf{f},\mathbf{g}})$, it follows from the discussion above that (N_1, N_2) is *not* in the image of $\Psi^\xi_{\mathbf{f},\mathbf{g}}$ if and only if $N_1 \subseteq X_S$ and $N_2 \supseteq S^X$. We hence see that for a non-split generating extension ξ the image of $\Psi^\xi_{\mathbf{f},\mathbf{g}}$ is given by:

$$\mathrm{Im}\,\Psi^\xi_{\mathbf{f},\mathbf{g}} = (\mathrm{Gr}_{\mathbf{f}}(X) \times \mathrm{Gr}_{\mathbf{g}}(S)) \setminus \left(\mathrm{Gr}_{\mathbf{f}}(X_S) \times \mathrm{Gr}_{\mathbf{g}-\mathbf{dim}\,S^X}(S/S^X)\right). \tag{7.6}$$

The following technical result, that we call the "reduction theorem", provides a way to reduce the problem of checking property (C) or (S) for a Q-representation Y which is the center of a generating extension.

THEOREM 7.5. *Let $\xi \in \mathrm{Ext}^1(S, X)$ be a generating extension. Then $\Psi^\xi_{\mathbf{f},\mathbf{g}}$ is a locally trivial affine bundle over its image of rank $\langle \mathbf{g}, \mathbf{dim}\, X - \mathbf{f} \rangle$.*

Theorem 7.5 is called "reduction theorem" because of the following corollary.

COROLLARY 7.6. *Let $\xi : 0 \longrightarrow X \longrightarrow Y \longrightarrow S \longrightarrow 0$ be a generating extension. If $\mathrm{Im}\,\Psi^\xi_{\mathbf{f},\mathbf{g}}$ admits a cellular decomposition for all $\mathbf{f}$ and $\mathbf{g}$, then Y has property (C). In particular, if $[S, X]^1 = 0$ and both X and S have property (C) (or (S)) then Y has property (C) (or (S)).*

Corollary 7.6 has the following immediate consequence: if M is a preprojective Q–representation and all its indecomposable direct summands have property (S) or (C) then M itself has property (S) or (C). This is because the preprojective component is directed.

In order to prove that every representation of a Dynkin quiver has property (C) it is hence enough to prove it for the indecomposables. This is done by induction using an elementary technique (see [**16**, Theorem 45]).

If Q is an affine quiver then one can use its well-known representation theory to deduce that every indecomposable preprojective Y fits as the middle term of a generating extension ξ. Moreover the subrepresentations X_S and S^X are under control and hence is the image of $\Psi^\xi_{\mathbf{f},\mathbf{g}}$. By corollary 7.6 one gets the proof of part (2) of theorem 7.2 by induction. Proving property (C) for preprojective representations of an arbitrary quiver Q seems to be much harder, due to the fact that the reduction theorem only allow us to use generating extensions. This is done in [**56**] for quivers with two vertices by combining the reduction theorem 7.5 with covering theory.

The proof of theorem 7.2 has the following easy corollaries.

COROLLARY 7.7. *Given an acyclic quiver Q, every Q-representation M whose regular part is rigid has property (S).*

PROOF. Decompose $M = M_{\mathcal{P}} \oplus M_{\mathcal{R}} \oplus M_{\mathcal{I}}$ as the sum of a preprojective, a regular and a preinjective Q–representation. By theorem 7.2, every indecomposable direct summand of both $M_{\mathcal{P}}$, $M_{\mathcal{R}}$ and $M_{\mathcal{I}}$ has property (S). By the reduction theorem 7.5 it follows that $M_{\mathcal{P}}$, $M_{\mathcal{R}}$ and $M_{\mathcal{I}}$ have property (S). Since $[M_{\mathcal{P}}, M_{\mathcal{R}} \oplus M_{\mathcal{I}}]^1 = [M_{\mathcal{R}}, M_{\mathcal{I}}]^1 = 0$ again by the reduction theorem 7.5 we get that M has property (S). □

COROLLARY 7.8. [**16**, Corollary 42] *Let M be a rigid representation of a quiver Q and let $\iota : \mathrm{Gr}_{\mathbf{e}}(M) \to \prod_{i\in Q_0} \mathrm{Gr}_{\mathbf{e}_i}(M_i)$ be the closed embedding. Then the induced map in cohomology $\iota^* : H^\bullet(\prod_{i\in Q_0} \mathrm{Gr}_{\mathbf{e}_i}(M_i)) \to H^\bullet(\mathrm{Gr}_{\mathbf{e}}(M))$ is surjective.*

COROLLARY 7.9. [**16**, Corollary 2] *Let M be a rigid representation of a quiver Q and let $X = \mathrm{Gr}_{\mathbf{e}}(M)$ be a quiver Grassmannian attached to it. Then X is defined over $\mathbb{Z}$ and it has polynomial point count, i.e.*

$$\#\mathrm{Gr}_{\mathbf{e}}(M)|_{\mathbf{F}_q} = \sum_i \dim_{\mathbb{Q}} H^{2i}(\mathrm{Gr}_{\mathbf{e}}(M), \mathbb{Q}) q^i.$$

8. The cluster multiplication formula

In this section we provide an application of the reduction theorem 7.5 to cluster algebras. Let Q be an acyclic quiver with n vertices and let M be a Q-representation. The *F-polynomial* of M is

$$F_M(\mathbf{y}) = \sum_{\mathbf{e}} \chi(\mathrm{Gr}_{\mathbf{e}}(M))\mathbf{y}^{\mathbf{e}} \in \mathbb{Z}[y_1, \cdots, y_n]$$

where χ denotes the Euler-Poincarè characteristic. The $\mathbf{g}$*–vector* or *index* of M is

$$\mathbf{g}_M = [I_1^M] - [I_0^M] \in K_0(\mathrm{Rep}(Q)) \simeq \mathbb{Z}^{Q_0}$$

where $0 \to M \to I_0^M \to I_1^M \to 0$ is the minimial injective resolution of M. Notice that $(\mathbf{g}_M)_i = -\langle S_i, M\rangle$. The *exhange matrix* $B = (b_{i,j})_{i,j\in Q_0} \in \mathrm{Mat}_{n\times n}(\mathbb{Z})$ of Q is the integer matrix given by

$$b_{i,j} = \#\{j \to i \in Q_1\} - \#\{i \to j \in Q_1\}$$

The cluster character of M is the Laurent polynomial

$$CC_M(\mathbf{x}, \mathbf{y}) = \sum_{\mathbf{e}} \chi(\mathrm{Gr}_{\mathbf{e}}(M))\mathbf{x}^{B\mathbf{e}+\mathbf{g}_M}\mathbf{y}^{\mathbf{e}} =: CC(M) \in \mathbb{Z}[y_1, \cdots, y_n][x_1^{\pm 1}, \cdots, x_n^{\pm 1}].$$

The reduction theorem 7.5 implies the following multiplication formula. To state the precise result we need to recall that given a generating extension $\xi \in \mathrm{Ext}^1(S, X)$ there exists an exact sequence $0 \to X/X_S \to \tau S^X \to I \to 0$ where I is injective.

COROLLARY 8.1. [**16**, Theorem 66] *Let $\xi : 0 \to X \to Y \to S \to 0$ be a generating extension. Then*

$$CC(X)CC(S) = CC(Y) + \mathbf{y}^{\mathbf{dim}\, S^X} CC(X_S)CC(S/S^X)\,\mathbf{x}^{\mathbf{f}} \tag{8.1}$$

where $I = \oplus_{j\in Q_0} I_j^{f_j}$.

The multiplication formula (8.1) is a slight generalization of the multiplication formula of Caldero and Keller [**14**] and it can be interpreted as a categorification of the exchange relations in the cluster algebra associated with Q. The apperence of $\mathbf{dim}\, S^X$ in the formula seems to be new and provides a representation theoretic description of the $\mathbf{c}$–vectors.

9. Exercises

We conclude the notes with a list of exercises divided by arguments.

9.1. Generalities on quiver Grassmannians.

EXERCISE 9.1. Recall that the complete flag variety $\mathcal{F}l_{n+1} = \{U_1 \subset U_2 \subset \cdots \subset U_n \subset \mathbb{C}^{n+1} | \dim U_i = i\}$ can be realized as the quiver Grassmannian $\mathrm{Gr}_{\mathbf{e}}(P_1^{n+1})$, where P_1 is the projective cover of the simple S_1 for the equioriented type A quiver $Q : 1 \to \cdots \to n$ and $\mathbf{e} = (1, 2, \cdots, n)$. Show that the dimension of $\mathcal{F}l_n$ is $\langle \mathbf{e}, \mathbf{d} - \mathbf{e} \rangle$ where $\mathbf{d} = \mathbf{dim}\, P_1^{n+1}$.

EXERCISE 9.2. Consider the quiver $Q : 1 \to 2$ and let $M = P_1 \oplus S_1 \oplus S_2$. Show that $\mathrm{Gr}_{(1,1)}(M)$ is isomorphic to two $\mathbf{P}^1$ crossing in one point.

EXERCISE 9.3. Given an e-subset $I = (1 \leq i_1 < i_2 < \cdots < i_e \leq n)$ of $[1, n]$, compute the dimension of the affine space $\mathcal{C}_I \subset \mathrm{Gr}_e(\mathbb{C}^d)$.

EXERCISE 9.4. Show that the universal quiver Grassmannian $\mathrm{Gr}_{\mathbf{e}}^Q(\mathbf{d})$ has dimension $\langle \mathbf{e}, \mathbf{d} - \mathbf{e} \rangle + \dim R_{\mathbf{d}}(Q)$.

EXERCISE 9.5. Given $g = (g_i) \in G_{\mathbf{d}}$, $N = (N_i) \in \mathrm{Gr}_{\mathbf{e}}(\mathbf{d})$ and $M \in R_{\mathbf{d}}(Q)$ consider the action: $g \cdot N = (g_i(N_i))_{i \in Q_0}$ and $g \cdot M = (g_j M_\alpha g_i^{-1})_{\alpha: i \to j \in Q_1}$. Show that the universal quiver Grassmannian $\mathrm{Gr}_{\mathbf{e}}^Q(\mathbf{d}) \subset \mathrm{Gr}_{\mathbf{e}}(\mathbf{d}) \times R_{\mathbf{d}}(Q)$ is invariant under the diagonal $G_{\mathbf{d}}$ action and that the map $p_{\mathbf{d}} : \mathrm{Gr}_{\mathbf{e}}^Q(\mathbf{d}) \to R_{\mathbf{d}}(Q)$ is $G_{\mathbf{d}}$-equivariant.

EXERCISE 9.6. The group $G_{\mathbf{e}}$ acts on $\mathrm{Hom}(\mathbf{e}, \mathbf{d})$ by: $g = (g_i) \in G_{\mathbf{e}}$, $(N, f) \in \mathrm{Hom}(\mathbf{e}, \mathbf{d})$, $\mathbf{g} \cdot (N, f) := ((g_{t(\alpha)} N_\alpha g_{s(\alpha)}^{-1}), (f_i g_i^{-1}))$. Verify that given $M \in R_{\mathbf{d}}(Q)$, $\mathrm{Hom}(\mathbf{e}, M) \subset \mathrm{Hom}(\mathbf{e}, \mathbf{d})$ is $G_{\mathbf{e}}$-stable. Prove that $G_{\mathbf{e}}$ acts freely on $\mathrm{Hom}^0(\mathbf{e}, M)$.

EXERCISE 9.7. Let M be a rigid quiver representation. Using the tangent space formula, show that every non-empty quiver Grassmannian $\mathrm{Gr}_{\mathbf{e}}(M)$ attached to M is smooth of minimal dimension $\langle \mathbf{e}, \mathbf{dim}\, M - \mathbf{e} \rangle$.

EXERCISE 9.8. Let M be a rigid quiver representation of dimension vector $\mathbf{d}$. Show that $\mathrm{Gr}_{\mathbf{e}}(M)$ is non–empty if and only if $\mathrm{Gr}_{\mathbf{e}}(M')$ is non–empty for every $M' \in R_{\mathbf{d}}(Q)$.

EXERCISE 9.9. Consider the quiver $Q : 1 \to 2$ and let $M^1 = P_1 \oplus P_2 \oplus I_1 \oplus I_2$. Find the generic subrepresentation type of the degenerate flag variety $\mathrm{Gr}_{(1,2)}(M^1)$.

EXERCISE 9.10. Let X and Y be rigid representations of an acyclic quiver Q such that $[X, Y]^1 = 0$ and let M be a Q–representation such that $M \leq_{deg} X \oplus Y$. Use Bongartz's theorem 3.1 to show that there exists a short exact sequence $0 \to X \to M \to Y \to 0$.

EXERCISE 9.11. Consider the quiver $Q : 1 \to 2$ and let $M^2 = P_1 \oplus S_1^2 \oplus S_2^2$. Find all iso-strata of the mf-liner degeneration of the flag variety $\mathrm{Gr}_{(1,2)}(M)$ and show that there are two generic subpresentation types. [Hint: compute the dimension of the iso-strata]

EXERCISE 9.12. Let Q be an acyclic quiver and let $M = P \oplus I$ where P is projective and I is injective. Consider the quiver Grassmannian $\mathrm{Gr}_{\mathbf{e}}(M)$ where $\mathbf{e} = \mathbf{dim}\, P$. Let $N \in \mathrm{Gr}_{\mathbf{dim}\, P}(M)$. Show that the iso-stratum $\mathcal{S}_{[N]}$ has dimension less or equal than $\langle \mathbf{e}, \mathbf{dim}\, M - \mathbf{e} \rangle$. Using Bongartz's theorem, show that the only generic iso-stratum is $\mathcal{S}_{[P]}$. Conclude that $\mathrm{Gr}_{\mathbf{e}}(M)$ is irreducible of minimal dimension.

9.2. Quiver Grassmannians of type A.

EXERCISE 9.13. Find a normal form for pairs of matrices $(A, B) \in \mathrm{Mat}_{m\times k} \times \mathrm{Mat}_{k\times n}$ by base change. In other words, find the decomposition of a representation V of $Q : 1 \to 2 \to 3$ as direct sum of indecomposable Q–representations.

EXERCISE 9.14. Let $Q = 1 \to 2$ and let $M = S_2 \oplus P_1 \oplus S_1 \in \mathrm{Rep}(Q)$. Order the indecomposable direct summands of M as $M(1) = S_1$, $M(2) = P_1$ and $M(3) = S_2$. Consider the quiver Grassmannian $\mathrm{Gr}_{(1,1)}(M)$ and the $\mathbb{C}^*$-action given by $\lambda \cdot m = \lambda^{k-1}m$ for every $m \in M(k)$. Prove that there is a torus fixed point L such that its attracting set $\mathcal{C}_L$ is not an affine space.

EXERCISE 9.15. Let Q be a Dynkin quiver, and let X and Y be two rigid Q–representations such that $[X, Y]^1 = 0$. Show that the dimension of each isostratum $\mathcal{S}_{[N]}$ of a quiver Grassmannian $\mathrm{Gr}_{\mathbf{dim}\,X}(X \oplus Y)$ satisfies:

$$\dim \mathcal{S}_{[N]} \leq \langle \mathbf{dim}\, X, \mathbf{dim}\, Y\rangle.$$

and equality holds if and only if $N \simeq X$. ([HINT: use the fact that the degeneration order for Dynkin quivers is equivalent to the Hom-order: $M \leq_{deg} M'$ if and only if $[M, L] \leq [M', L]$, for every L]). Conclude that $\mathrm{Gr}_{\mathbf{dim}\,X}(X \oplus Y)$ is irreducible of minimal dimension.

EXERCISE 9.16. Realize the degenerate flag variety $\mathcal{F}l_4^a$ as a Schubert variety.

EXERCISE 9.17. Let $Q : 1 \to 2 \to 3$ and $M = P_3 \oplus P_2^2 \oplus S_2 \oplus I_2 \oplus I_1$. Verify that M is catenoid and describe the natural embedding of $\mathrm{Gr}_{(1,2,1)}(M)$ inside a partial flag manifold. Show that there are two irreducible components.

EXERCISE 9.18. Let Q be the equioriented quiver of type A_n, and let $A = KQ$ be its path algebra. Put $\mathbf{e} = \mathbf{dim}\, A$ and $\mathbf{d} = \mathbf{dim}\,(A \oplus DA)$ as in section 5.3. Use theorem 5.5 and exercise 9.10 to show that a point $M \in R_{\mathbf{d}}(Q)$ belongs to $\mathcal{U}_{\mathrm{flat,Irr}}$ if and only if there exists a short exact sequence $0 \to A \to M \to DA \to 0$.

9.3. Cellular decomposition of quiver Grassmannians.

EXERCISE 9.19. Let $\eta : 0 \longrightarrow \tau S \xrightarrow{\iota} Y \xrightarrow{\pi} S \longrightarrow 0$ be an almost split sequence. Describe the image of the map

$$\begin{aligned} \Psi^{\eta}_{\mathbf{f},\mathbf{g}} : \qquad \mathrm{Gr}_{\mathbf{e}}(E) &\longrightarrow \mathrm{Gr}_{\mathbf{f}}(\tau S) \times \mathrm{Gr}_{\mathbf{g}}(S) \\ N &\longmapsto (N \cap \iota(\tau S), \pi(N)) \end{aligned}$$

EXERCISE 9.20. Show that if $\xi \in \mathrm{Ext}^1(S, X)$ is generating then

$$\mathrm{Im}(\Psi^{\xi}_{\mathbf{f},\mathbf{g}}) = \{(N_1, N_2) \in \mathrm{Gr}_{\mathbf{f}}(X) \times \mathrm{Gr}_{\mathbf{g}}(S) |\, [N_2, X/N_1]^1 = 0\}.$$

EXERCISE 9.21. Let $\xi \in \mathrm{Ext}^1(S, X)$ be a non–split generating extension. Prove that X_S and S^X are well-defined as follows. (For X_S) for every $N, N' \subset X$ such that $[S, X/N]^1 = [S, X/N']^1 = 1$, one has $[S, X/(N + N')]^1 = 1$. Dually (for S^X) for every $N, N' \subset S$ such that $[N, X]^1 = [N', X]^1 = 1$, one has $[N \cap N', X]^1 = 1$.

EXERCISE 9.22. Prove that an almost split sequence $\xi \in \mathrm{Ext}^1(X, \tau X)$ ending in a *brick* X (i.e. a Q–representation such that $[X, X] = 1$) is generalized almost split. Find an example of a generalized almost split which is not almost split. (Hint: look among the representations of a quiver of type D_4.)

EXERCISE 9.23. Let Q be the equioriented quiver of type A_n. Prove that a non-split generating extension (between two indecomposable Q-representations) is generalized almost split if and only if it is almost split.

EXERCISE 9.24. Prove the equalities in (7.5).

EXERCISE 9.25. Let X and Y be indecomposable preprojectives, such that $\mathrm{Hom}(X,Y) = \mathbb{C}\iota$ with $\iota : X \to Y$ an irreducible monomorphism. Let $S = Coker(\iota)$. Show that the short exact sequence

$$\xi : \ 0 \longrightarrow X \xrightarrow{\ \iota\ } Y \xrightarrow{\ \pi\ } S \longrightarrow 0$$

induced by ι is generating and $S^X = S$.

EXERCISE 9.26. A short exact sequence

$$0 \longrightarrow A \xrightarrow{f=(f_1,f_2)^t} B_1 \oplus B_2 \xrightarrow{g=(g_2,g_1)} C \longrightarrow 0$$

gives rise to a commutative diagram

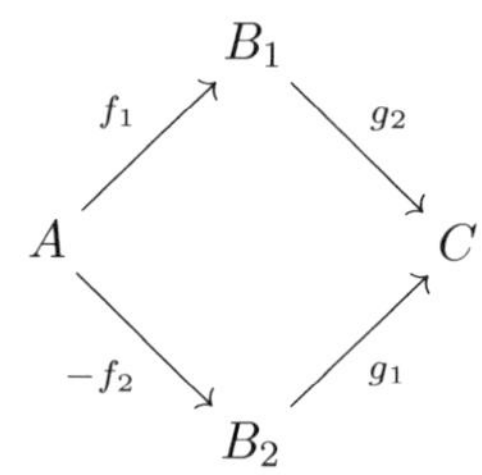

which is both a push–out and a pull-back square. Prove that

$$Ker(g_i) \simeq Ker(f_i), \quad Coker(g_i) \simeq Coker(f_i) \quad (i = 1, 2).$$

EXERCISE 9.27. Let Q be the following quiver of type $\tilde{A}_2$:

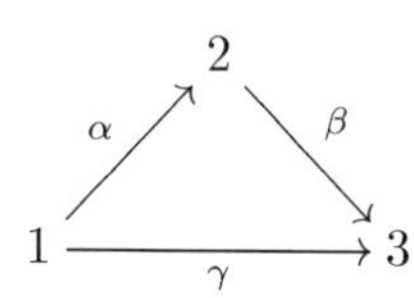

Let M be the indecomposable Q representation of dimension vector $(3,3,4)$. Using the construction seen during the lecture, find a cellular decomposition of the quiver Grassmannian $\mathrm{Gr}_{\mathbf{e}}(M)$ for $\mathbf{e} = (1,2,3)$.

EXERCISE 9.28. Let Q be the following quiver of type $\tilde{D}_4$:

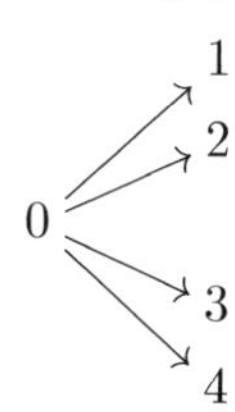

and let M be the indecomposable preprojective Q–representation of dimension vector $(3,2,2,2,2)$. Let $\mathbf{e} = (1,1,1,1,1)$. Find a cellular decomposition of the quiver Grassmannian $\mathrm{Gr}_{\mathbf{e}}(M)$, using the techniques seen at the lecture. Find a geometric interpretation.

EXERCISE 9.29. Let X be a rigid brick (i.e. $[X, X] = 1$ and $[X, X]^1 = 0$) which is not projective and let $\xi : 0 \to \tau X \to E \to X \to 0$ be the almost split sequence ending in X. Show that E is rigid and $[X \oplus \tau X, E]^1 = [E, X \oplus \tau X]^1 = 0$. Show that ξ is generalized almost split.

EXERCISE 9.30. Let $\xi : 0 \to X \to Y \to S \to 0$ be a generating extension. Prove that the reduction theorem 7.5 implies the following multiplication formula of F-polynomials $F_X F_S = F_Y + \mathbf{y}^{\mathbf{dim} S^X} F_{X_S} F_{S/S^X}$.

Acknowledgments

The author would like to thank the organizers of the ICRA 2018 for inviting me to give a series of lectures on this topic. The author also wants to thank the Ph.D. students who asked several questions during the lectures; they hope that this paper can serve to them as a handy guide into this subject. The author is indebted to all my coauthors, in particular Markus Reineke and Evgeny Feigin, on whose work most of this paper is based on. Finally, they sincerely thank the anonymous referee and Alex Puntz for a careful reading of a previous version of this paper and for many helpful suggestions.

References

[1] S. Abeasis and A. Del Fra, *Degenerations for the representations of an equioriented quiver of type D_m*, Adv. in Math. **52** (1984), no. 2, 81–172, DOI 10.1016/0001-8708(84)90047-1. MR742658

[2] S. Abeasis and A. Del Fra, *Degenerations for the representations of a quiver of type $\mathcal{A}_m$*, J. Algebra **93** (1985), no. 2, 376–412, DOI 10.1016/0021-8693(85)90166-8. MR786760

[3] S. Abeasis and A. Del Fra, *Degenerations for the representations of an equioriented quiver of type D_m*, Adv. in Math. **52** (1984), no. 2, 81–172, DOI 10.1016/0001-8708(84)90047-1. MR742658

[4] Ibrahim Assem, Daniel Simson, and Andrzej Skowroński, *Elements of the representation theory of associative algebras. Vol. 1*, London Mathematical Society Student Texts, vol. 65, Cambridge University Press, Cambridge, 2006. Techniques of representation theory. MR2197389

[5] Maurice Auslander and Idun Reiten, *Modules determined by their composition factors*, Illinois J. Math. **29** (1985), no. 2, 280–301. MR784524

[6] Maurice Auslander, Idun Reiten, and Sverre O. Smalø, *Representation theory of Artin algebras*, Cambridge Studies in Advanced Mathematics, vol. 36, Cambridge University Press, Cambridge, 1997. Corrected reprint of the 1995 original. MR1476671

[7] Rebecca Barlow, *A simply connected surface of general type with $p_g = 0$*, Invent. Math. **79** (1985), no. 2, 293–301, DOI 10.1007/BF01388974. MR778128

[8] I. N. Bernšteĭn, I. M. Gel′fand, and V. A. Ponomarev, *Coxeter functors, and Gabriel's theorem* (Russian), Uspehi Mat. Nauk **28** (1973), no. 2(170), 19–33. MR0393065

[9] Klaus Bongartz, *On degenerations and extensions of finite-dimensional modules*, Adv. Math. **121** (1996), no. 2, 245–287, DOI 10.1006/aima.1996.0053. MR1402728

[10] W. Crawley-Boevey, *Lectures on representations of quivers*. Preprint 1992. Available at `https://www.math.uni-bielefeld.de/~wcrawley/quivlecs.pdf`.

[11] W. Crawley-Boevey, *More lectures on representations of quivers*. Preprint 1992. Available at `https://www.math.uni-bielefeld.de/~wcrawley/morequivlecs.pdf`.

[12] Philippe Caldero and Frédéric Chapoton, *Cluster algebras as Hall algebras of quiver representations*, Comment. Math. Helv. **81** (2006), no. 3, 595–616, DOI 10.4171/CMH/65. MR2250855

[13] Philippe Caldero and Bernhard Keller, *From triangulated categories to cluster algebras*, Invent. Math. **172** (2008), no. 1, 169–211, DOI 10.1007/s00222-008-0111-4. MR2385670

[14] Philippe Caldero and Bernhard Keller, *From triangulated categories to cluster algebras. II* (English, with English and French summaries), Ann. Sci. École Norm. Sup. (4) **39** (2006), no. 6, 983–1009, DOI 10.1016/j.ansens.2006.09.003. MR2316979

[15] Philippe Caldero and Markus Reineke, *On the quiver Grassmannian in the acyclic case*, J. Pure Appl. Algebra **212** (2008), no. 11, 2369–2380, DOI 10.1016/j.jpaa.2008.03.025. MR2440252

[16] G Cerulli Irelli, F. Esposito, H. Franzen, M. Reineke, *Cellular decomposition and algebraicity of cohomology for quiver Grassmannians*, arXiv: 1804.07736.

[17] Giovanni Cerulli Irelli and Francesco Esposito, *Geometry of quiver Grassmannians of Kronecker type and applications to cluster algebras*, Algebra Number Theory **5** (2011), no. 6, 777–801, DOI 10.2140/ant.2011.5.777. MR2923727

[18] Giovanni Cerulli Irelli, Grégoire Dupont, and Francesco Esposito, *A homological interpretation of the transverse quiver Grassmannians*, Algebr. Represent. Theory **16** (2013), no. 2, 437–444, DOI 10.1007/s10468-011-9314-2. MR3036000

[19] G. Cerulli Irelli, X. Fang, E. Feigin, G. Fourier, and M. Reineke, *Linear degenerations of flag varieties*, Math. Z. **287** (2017), no. 1-2, 615–654, DOI 10.1007/s00209-016-1839-y. MR3694690

[20] Giovanni Cerulli Irelli, Evgeny Feigin, and Markus Reineke, *Quiver Grassmannians and degenerate flag varieties*, Algebra Number Theory **6** (2012), no. 1, 165–194, DOI 10.2140/ant.2012.6.165. MR2950163

[21] Giovanni Cerulli Irelli, Evgeny Feigin, and Markus Reineke, *Degenerate flag varieties: moment graphs and Schröder numbers*, J. Algebraic Combin. **38** (2013), no. 1, 159–189, DOI 10.1007/s10801-012-0397-6. MR3070124

[22] G. Cerulli Irelli, E. Feigin, and M. Reineke, *Desingularization of quiver Grassmannians for Dynkin quivers*, Adv. Math. **245** (2013), 182–207, DOI 10.1016/j.aim.2013.05.024. MR3084427

[23] Giovanni Cerulli Irelli, Evgeny Feigin, and Markus Reineke, *Schubert quiver Grassmannians*, Algebr. Represent. Theory **20** (2017), no. 1, 147–161, DOI 10.1007/s10468-016-9634-3. MR3606484

[24] Giovanni Cerulli Irelli and Martina Lanini, *Degenerate flag varieties of type A and C are Schubert varieties*, Int. Math. Res. Not. IMRN **15** (2015), 6353–6374, DOI 10.1093/imrn/rnu128. MR3384481

[25] Giovanni Cerulli Irelli, Martina Lanini, and Peter Littelmann, *Degenerate flag varieties and Schubert varieties: a characteristic free approach*, Pacific J. Math. **284** (2016), no. 2, 283–308, DOI 10.2140/pjm.2016.284.283. MR3544301

[26] Neil Chriss and Victor Ginzburg, *Representation theory and complex geometry*, Birkhäuser Boston, Inc., Boston, MA, 1997. MR1433132

[27] C. De Concini, G. Lusztig, and C. Procesi, *Homology of the zero-set of a nilpotent vector field on a flag manifold*, J. Amer. Math. Soc. **1** (1988), no. 1, 15–34, DOI 10.2307/1990965. MR924700

[28] Harm Derksen, Jerzy Weyman, and Andrei Zelevinsky, *Quivers with potentials and their representations II: applications to cluster algebras*, J. Amer. Math. Soc. **23** (2010), no. 3, 749–790, DOI 10.1090/S0894-0347-10-00662-4. MR2629987

[29] Geir Ellingsrud and Stein Arild Strømme, *Towards the Chow ring of the Hilbert scheme of* $\mathbf{P}^2$, J. Reine Angew. Math. **441** (1993), 33–44. MR1228610

[30] Evgeny Feigin, *$\mathbb{G}_a^M$ degeneration of flag varieties*, Selecta Math. (N.S.) **18** (2012), no. 3, 513–537, DOI 10.1007/s00029-011-0084-9. MR2960025

[31] Evgeny Feigin, *Degenerate flag varieties and the median Genocchi numbers*, Math. Res. Lett. **18** (2011), no. 6, 1163–1178, DOI 10.4310/MRL.2011.v18.n6.a8. MR2915473

[32] Evgeny Feigin and Michael Finkelberg, *Degenerate flag varieties of type A: Frobenius splitting and BW theorem*, Math. Z. **275** (2013), no. 1-2, 55–77, DOI 10.1007/s00209-012-1122-9. MR3101796

[33] Sergey Fomin and Andrei Zelevinsky, *Cluster algebras. I. Foundations*, J. Amer. Math. Soc. **15** (2002), no. 2, 497–529, DOI 10.1090/S0894-0347-01-00385-X. MR1887642

[34] Sergey Fomin and Andrei Zelevinsky, *Cluster algebras. II. Finite type classification*, Invent. Math. **154** (2003), no. 1, 63–121, DOI 10.1007/s00222-003-0302-y. MR2004457

[35] Sergey Fomin and Andrei Zelevinsky, *Cluster algebras. IV. Coefficients*, Compos. Math. **143** (2007), no. 1, 112–164, DOI 10.1112/S0010437X06002521. MR2295199

[36] William Fulton, *Intersection theory*, Ergebnisse der Mathematik und ihrer Grenzgebiete (3) [Results in Mathematics and Related Areas (3)], vol. 2, Springer-Verlag, Berlin, 1984. MR732620

[37] Peter Gabriel, *Unzerlegbare Darstellungen. I* (German, with English summary), Manuscripta Math. **6** (1972), 71–103; correction, ibid. 6 (1972), 309, DOI 10.1007/BF01298413. MR332887

[38] Joe Harris, *Algebraic geometry: A first course*, Graduate Texts in Mathematics, vol. 133, Springer-Verlag, New York, 1992. MR1182558

[39] Andrew Hubery, *Irreducible components of quiver Grassmannians*, Trans. Amer. Math. Soc. **369** (2017), no. 2, 1395–1458, DOI 10.1090/tran/6693. MR3572278

[40] V. G. Kac, *Infinite root systems, representations of graphs and invariant theory*, Invent. Math. **56** (1980), no. 1, 57–92, DOI 10.1007/BF01403155. MR557581

[41] V. G. Kac, *Infinite root systems, representations of graphs and invariant theory. II*, J. Algebra **78** (1982), no. 1, 141–162, DOI 10.1016/0021-8693(82)90105-3. MR677715

[42] Otto Kerner, *Exceptional components of wild hereditary algebras*, J. Algebra **152** (1992), no. 1, 184–206, DOI 10.1016/0021-8693(92)90095-4. MR1190411

[43] Birger Iversen, *A fixed point formula for action of tori on algebraic varieties*, Invent. Math. **16** (1972), 229–236, DOI 10.1007/BF01425495. MR299608

[44] L. Le Bruyn, *Neverending books*. `http://web.archive.org/web/20120918072747/http://www.neverendingbooks.org/index.php/quiver-grassmannians-can-be-anything.html`

[45] Oliver Lorscheid and Thorsten Weist, *Quiver Grassmannians of extended Dynkin type D: Part 1: Schubert systems and decompositions into affine spaces*, Mem. Amer. Math. Soc. **261** (2019), no. 1258, v+78, DOI 10.1090/memo/1258. MR4031099

[46] O. Lorscheid, T. Weist, *Quiver Grassmannians of extended Dynkin type D- Part 2: Schubert decompositions and F-polynomials*. Preprint (2015). arXiv: 1507.00395.

[47] Kathrin Möllenhoff and Markus Reineke, *Embeddings of representations*, Algebr. Represent. Theory **18** (2015), no. 4, 977–987, DOI 10.1007/s10468-015-9528-9. MR3372128

[48] Hiraku Nakajima, *Quiver varieties and cluster algebras*, Kyoto J. Math. **51** (2011), no. 1, 71–126, DOI 10.1215/0023608X-2010-021. MR2784748

[49] Fan Qin, *Quantum cluster variables via Serre polynomials*, J. Reine Angew. Math. **668** (2012), 149–190. With an appendix by Bernhard Keller. MR2948875

[50] Markus Reineke, *Every projective variety is a quiver Grassmannian*, Algebr. Represent. Theory **16** (2013), no. 5, 1313–1314, DOI 10.1007/s10468-012-9357-z. MR3102955

[51] Christine Riedtmann, *Degenerations for representations of quivers with relations*, Ann. Sci. École Norm. Sup. (4) **19** (1986), no. 2, 275–301. MR868301

[52] Claus Michael Ringel, *Quiver Grassmannians for wild acyclic quivers*, Proc. Amer. Math. Soc. **146** (2018), no. 5, 1873–1877, DOI 10.1090/proc/13882. MR3767342

[53] Claus Michael Ringel, *Tame algebras and integral quadratic forms*, Lecture Notes in Mathematics, vol. 1099, Springer-Verlag, Berlin, 1984. MR774589

[54] Claus Michael Ringel, *Distinguished bases of exceptional modules*, Algebras, quivers and representations, Abel Symp., vol. 8, Springer, Heidelberg, 2013, pp. 253–274, DOI 10.1007/978-3-642-39485-0_11. MR3183888

[55] Claus Michael Ringel, *Finite dimensional hereditary algebras of wild representation type*, Math. Z. **161** (1978), no. 3, 235–255, DOI 10.1007/BF01214506. MR501169

[56] Dylan Rupel and Thorsten Weist, *Cell decompositions for rank two quiver Grassmannians*, Math. Z. **295** (2020), no. 3-4, 993–1038, DOI 10.1007/s00209-019-02379-6. MR4125679

[57] Aidan Schofield, *General representations of quivers*, Proc. London Math. Soc. (3) **65** (1992), no. 1, 46–64, DOI 10.1112/plms/s3-65.1.46. MR1162487

[58] Ralf Schiffler, *Quiver representations*, CMS Books in Mathematics/Ouvrages de Mathématiques de la SMC, Springer, Cham, 2014. MR3308668

[59] Grzegorz Zwara, *Degenerations for modules over representation-finite algebras*, Proc. Amer. Math. Soc. **127** (1999), no. 5, 1313–1322, DOI 10.1090/S0002-9939-99-04714-0. MR1476404

Sapienza-Università di Roma, Dipartimento S.B.A.I., Via Antonio Scarpa 10, 00164, Roma, Italy

Email address: `giovanni.cerulliirelli@uniroma1.it`

Contemporary Mathematics
Volume **758**, 2020
https://doi.org/10.1090/conm/758/15233

Higher symmetries in abstract stable homotopy theories

Moritz Rahn

ABSTRACT. This survey offers an overview of an on-going project on uniform symmetries in abstract stable homotopy theories. This project has calculational, foundational, and representation-theoretic aspects, and key features of this emerging field on abstract representation theory include the following. First, generalizing the classical focus on representations over fields, it is concerned with the study of representations over rings, differential-graded algebras, ring spectra, and in more general abstract stable homotopy theories. Second, restricting attention to specific shapes, it offers an explanation of the axioms of triangulated categories, higher triangulations, and monoidal triangulations. This has led to fairly general results concerning additivity of traces. Third, along similar lines of thought it suggests the development of abstract cubical homotopy theory as an additional calculational toolkit. An interesting symmetry in this case is given by a global form of Serre duality. Fourth, abstract tilting equivalences give rise to non-trivial elements in spectral Picard groupoids and hence contribute to their calculation. And, finally, it stimulates a deeper digression of the notion of stability itself, leading to various characterizations and relative versions of stability.

CONTENTS

1. Introduction

A good part of modern mathematics is centered around the study of symmetrics.[1] Symmetries arise in various parts of mathematics, and in many specific situations good control over the available symmetries is helpful in (if not crucial to) applications. Here we are interested in global symmetries which are common to all abstract stable homotopy theories. In one form or another (see further below), abstract stable homotopy theories are in the background of many areas of pure

2010 *Mathematics Subject Classification.* Primary 55U35, 55U40, 18E30; Secondary 16E35, 55P42.

[1]Always begin an introduction by drawing a big picture.

mathematics. This certainly includes algebra and representation theory, homotopy theory and algebraic topology as well as algebraic geometry. The overall goal of this project is a detailed study of uniform symmetries which arise in these fields. The credo is that some of these symmetries provide a convenient calculational toolkit that is common to all these situations. Other symmetries are more interesting from an abstract representation-theoretic perspective.

In order to fill this first paragraph with more life, for the time being we focus on *triangulated categories* as one of the classical approaches to an "abstract stable homotopy theory". Triangulated categories originated in algebraic geometry in the study of derived categories $D(X)$ of schemes [**Ver67**, **Ver96**] and in homotopy theory in the study of *the* stable homotopy category of spectra $\mathcal{SHC}$ [**Boa64**, **Pup67**, **Vog70**]. In both of these situations, the classical triangulations on $D(X)$ and $\mathcal{SHC}$ encode aspects of the calculus of cones or, equivalently, derived cokernels – as it is visible to the respective category alone. And symmetries show up very prominently in the definition of a triangulated category [**Nee01**, **HJR10**]. First, a triangulated category $\mathcal{T}$ is by definition endowed with a suspension functor $\Sigma\colon \mathcal{T} \xrightarrow{\sim} \mathcal{T}$ which is an equivalence of categories. Second, the rotation axiom allows us to rotate distinguished triangles forward and backwards. Triangulated categories also encode information about pairs of composable morphisms, but here the classical axioms miss some of the existing symmetries.

Correspondingly, our first goal is to give a different explanation of the axioms of a triangulated category. In order to achieve this in a fairly elementary way, in §2 we stick to the specific situation of chain complexes of modules over a ring R. The main goal in that section is to give a somewhat unorthodox construction of the classical Verdier triangulation on the derived category $D(R)$. In fact, we take a representation-theoretic perspective and study representations of the A_n-quivers

$$\vec{A}_1 = 1, \quad \vec{A}_2 = (1 \to 2), \quad \text{and} \quad \vec{A}_3 = (1 \to 2 \to 3)$$

with values in chain complexes. Suitable combinations of derived cokernels and more general derived pushouts reveal interesting symmetries in the representation theories of these A_n-quivers. For instance, a representation

$$x \xrightarrow{f} y \xrightarrow{g} z \tag{1.1}$$

of $\vec{A}_3$ in chain complexes can be equivalently encoded by a diagram of chain complexes as in Figure 1. In that diagram all squares are derived pushouts (or, equivalently, derived pullbacks) and the diagram vanishes on the boundary stripes.

A suitable restriction of Figure 1 gives rise to the octahedral diagram of the representation (1.1). For instance, the key distinguished triangle from the octahedral axiom ("the cone of cones is a cone") is induced by the square $\boxed{1}$ which exhibits the cone $C(g)$ as the cone of $C(f) \to C(gf)$ (see Remark 2.52 for more details). But the diagrams in Figure 1 have the advantage that there are two obvious symmetry operations. For instance, there is the symmetry which shifts such diagrams along the diagonal. And this symmetry results in a relation between the octahedral diagram of (1.1) and the one of the induced representation

$$C(f) \to C(gf) \to \Sigma x.$$

Also representations of $\vec{A}_2$ and $\vec{A}_1$ can be encoded by similar symmetric diagrams (see Figure 3 and Figure 7). And in that case the symmetries induce the familiar rotation symmetry and suspension at the level of $D(R)$.

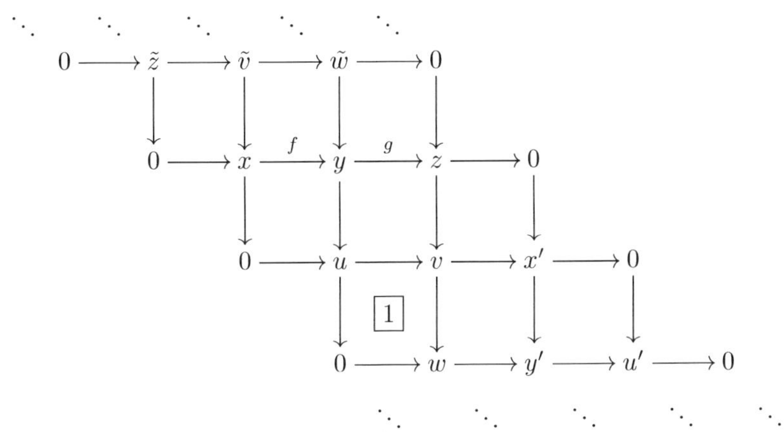

FIGURE 1. A symmetric presentation of representations of $\vec{A}_3$.

Following this line of thought, we are led to a construction of the classical Verdier triangulation on $D(R)$. The way we present this construction suggests that these arguments do not rely in an essential way on the specifics of chain complexes in R-modules. Instead, only formal aspects of the calculus of chain complexes were invoked, namely the existence of a well-behaved calculus of derived limits and colimits with the following additional key properties.

(i) There is a zero object.
(ii) A square is a derived pushout square if and only if it is a derived pullback square.

These two properties are the typical defining features of an "abstract stable homotopy theory". There is an entire zoo of models all of which make such a notion precise. This zoo includes stable cofibration categories [**Sch13**, **Len17**], stable model categories [**Hov99**], stable ∞-categories [**Lur09**, **Lur14**], and stable derivators [**Hel97**, **Fra96**, **Mal01a**] (see Disclaimer 3.2 for more references). For many of our purposes the precise choice of the model is not too relevant. In fact, many of the proofs are quite formal (and that is one of the points of the project) so that they are essentially model independent. Let us hasten to emphasize that we believe in diversity of technology.

These abstract stable homotopy theories are enhancements of triangulated categories. In fact, in the above cases one can use the defining exactness properties of *stability* in order to construct canonical triangulations. More interesting than this result itself are the techniques leading to its proof. These are developed in §3 in the framework of stable derivators (but see Disclaimer 3.2), and these techniques are crucial to our later studies of symmetries. One convenient feature of stable derivators (which is, for instance, also enjoyed by stable ∞-categories [**Lur14**, Prop. 1.1.3.1] but not by triangulated categories) is the following. Given a stable derivator $\mathscr{D}$ and a small category A there is the stable derivator $\mathscr{D}^A$ of representations of shape A with values in $\mathscr{D}$ (exponentials exist in stable derivators).

And this opens the door for the study of symmetries of stable homotopy theories of representations of more complicated shapes than the A_n-quivers for $n = 1, 2, 3$ which show up implicitly in the axioms of triangulated categories.

In §4 we finally get to the heart of this project, and this is where we take the above exponentials serious. Given a small category A we associate to it the 2-functor

$$(-)^A\colon \mathcal{DER}_{\mathrm{St,ex}} \to \mathcal{DER}\colon \mathscr{D} \mapsto \mathscr{D}^A \tag{1.2}$$

which sends a stable derivator $\mathscr{D}$to the stable derivator of A-shaped representations in $\mathscr{D}$. By specializing to specific $\mathscr{D}$, this yields the homotopy theory of representations over fields, over rings, in quasi-coherent modules over schemes, of differential-graded or of spectral representations (see Examples 4.2 for details). We think of $(-)^A$ as a gadget that neatly encodes the abstract representation theory of the shape A. Correspondingly, given two small categories A and B (which could be the same), a *strong stable equivalence* $\Phi\colon A \overset{s}{\sim} B$ is a family of equivalences

$$\Phi_{\mathscr{D}}\colon \mathscr{D}^A \simeq \mathscr{D}^B, \quad \mathscr{D} \text{ a stable derivator,}$$

which is pseudo-natural with respect to exact morphisms in $\mathscr{D}$. As a slogan, such a strong stable equivalence shows that A and B have the same abstract representation theory (see §4.1 for a more detailed discussion).

A priori, any strong stable equivalence $\Phi\colon A \overset{s}{\sim} B$ reveals a potentially interesting symmetry in abstract stable homotopy theories. However, we also believe in the following slogan.

"Some shapes are more important than others."

In fact, some shapes are more foundational than others, and strong stable equivalences between such shapes buy us more. We illustrate this by the following three shapes which are closely related to triangulated categories, higher triangulated categories and the Waldhausen $S_\bullet$-construction, and monoidal triangulated categories and Goodwillie calculus.

(i) The abstract representation theories of the Dynkin quivers $\vec{A}_1, \vec{A}_2$, and $\vec{A}_3$ are intimately related with the axioms of triangulated categories. This relation is provided through the calculus of derived (co)kernels of morphisms and pairs of composable morphisms [**Fra96, Mal01a, Gro13, Gro16**]. Similarly, longer chains of composable morphisms are the same as abstract representations of Dynkin quivers

$$\vec{A}_n = (1 \to 2 \to \ldots \to n).$$

And their abstract representation theory is directly connected with higher octahedral diagrams and ∞-triangulated categories [**BBD82, Mal05a, Bal11, GŠ16a**]. Interesting symmetries are made visible by diagrams as in Walhausen's $S_\bullet$-construction in algebraic K-theory [**Wal85, GŠ16a**]. Higher-dimensional versions of these diagrams have recently been studied in [**Pog17, Bec18, DJW18**].

(ii) Similarly, the abstract representation theory of the commutative square

$$\square = \vec{A}_2 \times \vec{A}_2$$

is closely related to refined axioms for monoidal, triangulated or tensor triangulated categories. Of course, representations of the square arise in monoidal contexts as pointwise products of pairs of morphisms. A prominent operation in this context is the pushout product operation (Example 4.62). Shadows of this operation together with various compatibilities with rotations of the two morphisms, symmetry, and dualization have been axiomatized by May [**May01**]. The square is strongly stably equivalent to the trivalent source and this offers a more representation-theoretic perspective on May's axioms [**KN02, GŠ14**].

This calculus leads to applications in the study of dualizability phenomena. More specifically, part of Grothendieck's original motivation to look for enhancements of triangulated categories was the following observation: traces of endomorphisms do not satisfy additivity at the level of triangulated categories (see Remark 4.72 for a precise formulation). In [**GPS14a**] a result concerning additivity of traces is established for stable, monoidal derivators (following May's proof [**May01**] for model categories). And, again, possibly more interesting than the additivity result are the techniques developed in that paper. (For further developments of this additivity result see [**PS16, PS14, GAdS14, JY18**].)

(iii) Finally, also the abstract representation theory of the n-cube

$$\square^n = \vec{A}_2 \times \ldots \times \vec{A}_2$$

and some of its subposets is fairly foundational. It leads to a detailed understanding of the calculus of iterated partial cofiber constructions, total cofibers, and of an interpolation between cocartesian and strongly cocartesian n-cubes [**BG18**]. Such shapes show up naturally in monoidal contexts and also in Goodwillie calculus [**Goo92, MV15**]. Moreover, there is a global form of an abstract Serre duality on representations of these posets [**BG18**, §12], which, conjecturally, can be extended to arbitrary homotopy finite shapes. Jointly with Falk Beckert, we intend to come back to this elsewhere.

Besides these more foundational cases, there are various shapes that have been studied very systematically in representation theory. In representation theory the focus is often on representations of quivers, posets, groups, or more general small categories with values in vector spaces over a field. (Of course a good deal of representation theory is concerned with more general algebras, but here we focus on those aspects that can be captured by (1.2).) Tilting theory is a derived version of Morita theory, and it offers many tools to study the corresponding derived categories and the calculus of derived equivalences (see for example [**APR79, BB80, HR82, Ric91, Kel94**] or the survey articles in [**AHHK07**]). It turns out that some aspects of tilting theory have interesting counterparts in the context of abstract stable homotopy theories.

First, this is the case for the classical BGP reflection functors. Let us recall that Bernšteĭn, Gel′fand, and Ponomarev [**BGP73**] introduced reflection functors and used them to give an elegant proof of Gabriel's classification result of connected hereditary representation-finite algebras over an algebraically closed field [**Gab72**]. For every quiver Q and every source or sink $q \in Q$, the reflected quiver $Q' = \sigma_q Q$ of Q is obtained by reversing the orientations of all edges adjacent to q. Associated with these reflections at the level of quivers are reflection functors

$$\mathrm{Mod}(kQ) \to \mathrm{Mod}(kQ')$$

between the module categories of the corresponding path algebras. While these reflection functors fail to be equivalences, Happel [**Hap87**] showed that for acyclic quivers derived reflection functors are exact equivalences

$$D(kQ) \overset{\Delta}{\simeq} D(kQ').$$

Later Ladkani [**Lad07a**] established a similar equivalence $D(\mathcal{A}^Q) \simeq D(\mathcal{A}^{Q'})$ for arbitrary abelian categories, thereby yielding a universal derived equivalence in his terminology. It was shown in [**GŠ16b, GŠ15**] that Happel's theorem can be strengthened further by showing that reflection functors yield a strong stable equivalence

$$Q \overset{\mathrm{s}}{\sim} Q'.$$

And this also holds for not necessarily finite or acyclic quivers. By purely combinatorial arguments one concludes from this that if T is a finite oriented tree and if T' is an arbitrary reorientation of T, then there is a strong stable equivalence

$$T \overset{\mathrm{s}}{\sim} T'.$$

Using the more flexible framework of stable ∞-categories further generalizations were recently constructed by Dyckerhoff, Jasso, and Walde [**DJW19**]. In *loc. cit.* certain procedures to glue stable ∞-categories are developed, and this gluing works neither for triangulated categories nor for stable derivators. These techniques are then applied to offer a fairly general construction of reflection functors for stable ∞-categories. Their result specializes both to the reflection functors in [**GŠ16b, GŠ15**] and also to additional examples considered by Ladkani [**Lad07a, Lad07c, Lad07b**].

A second aspect from tilting theory that generalizes nicely is the calculus of tilting complexes. Recall that these are chain complexes of bimodules such that the corresponding derived tensor and hom functors are derived equivalences. To obtain a counterpart of this for abstract stable homotopy theories, one invokes the universality of the homotopy theory of spectra: it is the free stable homotopy theory generated by the sphere spectrum. In the framework of derivators references related to this result include [**Hel88, Hel97, Fra96, Cis08, Tab08, CT11, CT12**] (but see also [**Dug01b, Len10**] and [**Lur09, Lur14, GGN15**] for closely related results in model categories and ∞-categories, respectively). As a consequence of this universality and Brown representability, every stable derivator is enriched over spectra. And this allows us to define tensor and hom functors associated to spectral bimodules (aka. weighted colimits and weighted limits [**Kel05b**]).

It turns out that the strong stable equivalences mentioned above are induced by *universal tilting modules*. These are certain spectral bimodules and the terminology is justified by the following two facts. First, these bimodules realize strong stable equivalences in arbitrary stable derivators and, second, they are spectral refinements of the classical tilting complexes. In fact, the classical tilting complexes are recovered from the universal tilting modules by smashing with the Eilenberg–MacLane spectrum. One convenient feature of these spectral bimodules is that they can be calculated quite explicitly. For instance, there are explicit spectral bimodules that govern the calculus of homotopy finite limits and colimits. We illustrate this in §4.5 by some examples related to the calculus of cofibers such as the universal constructor for cofiber sequences.

Universal tilting modules realize strong stable *equivalences* and they are consequently *invertible* spectral bimodules. Considered from this perspective, the construction of strong stable equivalences contributes to the calculation of the *spectral Picard groupoid.* This bicategory has small categories as objects and invertible spectral bimodules as morphisms, and it is the natural home for all the higher symmetries and relations between them. To the best of the knowledge of the author, the only shape for which the spectral Picard group is known is the trivial shape $A = *$. In that case there is an isomorphism $\mathbb{Z} \cong \mathrm{Pic}_{\mathcal{S}p}(*)$ and a generator is given by $\mathbb{S}^1 = \Sigma\mathbb{S}$ ([**HMS94**] or [**Str92**, Thm. 2.2]). One of the current goals of this project is to obtain a calculation of spectral Picard groups at least for some shapes, but for the time being only first steps have been achieved (Remark 4.109).

We refrain from giving a more detailed description of the content. Instead we refer the reader to the respective introductions of the individual sections §2, §3, and §4.

Philosophy of this writing. It was our main goal to write an account which is readable to representation theorists, to triangulated category theorists, and also to (abstract) homotopy theorists. This explains the high level of detail which is offered at various places, and we want to apologize for this to these different communities. Still our feeling is that this is a reasonable compromise. As additional guiding principles, we tried to stress the main ideas in the project, we refer to the original literature for most of the proofs, we advertise a few loose ends which are to be pursued further, and we include many references to indicate the various connections to other parts of mathematics.

The level of sophistication increases from section to section. In §2 we deliberately only invoke elementary techniques from homological algebra and some category theory. In §3 we give a survey on derivators which offers one of the many approaches to abstract homotopy theory (but see Disclaimer 3.2). We just develop the basics which are needed for the applications in the final section (and here we gloss over the more technical aspects of the theory despite the fact that the author actually likes them). Finally, in §4 we freely use the language of derivators (but see Disclaimer 3.2) in order to summarize and explain the main results of this project on higher symmetries.

Introduction of new coauthor. It was very tempting to also add Moritz Groth as an author to this paper and to make various jokes such as "the first author thanks the second author for having joined the project while the second author in turn thanks the first author for fruitful discussions". But to cut this short, the author recently got married and changed his last name from Groth to Rahn. All future publications will appear under this new last name.

2. A representation-theoretic perspective on triangulated categories

Triangulated categories were introduced in the 1960's, originally motivated by situations in algebraic geometry [**Ver67, Ver96**] and topology [**Boa64, Pup67, Vog70**]. Since their invention triangulated categories have become a valuable tool in many areas of pure mathematics and this ubiquity is one of the features of triangulated categories. In algebraic geometry triangulated categories arise as derived categories of schemes ([**Ver67, Ver96**] or [**Huy06**]), in representation theory they come up as derived categories of algebras (see [**Hap88**] or [**AHHK07**]), in modular representation theory as stable module categories [**BCR97**], and in homotopy theory as homotopy categories of spectra or related (stable model) categories ([**Vog70**] or [**Hov99**]). Nice surveys on this ubiquity of triangulated categories can be found in [**HJR10**], in [**Bal10**], and from a different perspective in [**SS03**].

In this first section we want to contribute to the understanding of the axioms of triangulated categories. The main idea we intend to transport is that the axioms of triangulated categories are closely related to (abstract) representation theory of A_1-quivers (!), A_2-quivers, and A_3-quivers. To keep things simple and very explicit, in most of this section we specialize to chain complexes of modules over a ring R. Assuming only elementary techniques from homological algebra, we revisit the classical Verdier triangulation on the derived category $D(R)$ of a ring and offer a somewhat unorthodox construction of it. In §3 we will see that the arguments invoked here easily extend to more general situations.

In §2.1 we revisit the classical derived cokernel and derived pushout construction in abelian categories. In §2.2 we specialize to the case of a module category and collect key aspects of the calculus of more general derived limits and Kan extensions. As an illustration we express derived cokernels in terms of derived Kan extensions. In §2.3 we iterate the construction of cofibers. This shows that morphisms can be equivalently encoded by cofiber sequences and also by Barratt–Puppe sequences. This latter reformulation showcases the existing derived symmetries of $\vec{A}_2$-representations. In §2.4 we consider derived representations of $\vec{A}_3$ and encode these in terms of refined ocatahedral diagrams. We discuss at some length in which sense these diagrams are more symmetric versions of the diagrams in the classical octahedral axiom.

2.1. Functorial cones. To set the stage, let us begin by establishing some basic notation.

NOTATION 2.1. Throughout this section, $\mathcal{A}$ denotes an abelian category. From §2.2 on we focus on the case $\mathcal{A} = \mathrm{Mod}(R)$, the category of (not necessarily finitely generated) left modules over a (not necessarily commutative) ring R.

We recall that abelian categories essentially axiomatize the usual calculus of finite direct sums of modules and kernels and cokernels of homomorphisms of modules. In classical homological algebra, the focus is on additive functors between abelian categories and on an investigation of their behavior with respect to short exact sequences. In particular, various techniques are developed in order to approximate a given functor by exact functors in a universal way. To make such a statement more specific, we collect the following lemma.

NOTATION 2.2. For every abelian category $\mathcal{A}$, we denote by $\mathrm{Ch}(\mathcal{A})$ the category of unbounded chain complexes in $\mathcal{A}$.

LEMMA 2.3. *Let $\mathcal{A}$ and $\mathcal{B}$ be abelian categories and let $F\colon \mathcal{A} \to \mathcal{B}$ be an additive functor. The following are equivalent.*

(i) *The functor $F\colon \mathcal{A} \to \mathcal{B}$ is exact.*
(ii) *The induced functor $F\colon \mathrm{Ch}(\mathcal{A}) \to \mathrm{Ch}(\mathcal{B})$ preserves quasi-isomorphisms.*

PROOF. The passage to homology objects is obtained by the formation of subquotients, and the proof is hence straightforward. $\square$

This suggests that the above-mentioned 'exact approximations' of more general additive functors are defined at the level of derived categories.

CONSTRUCTION 2.4. For every abelian category $\mathcal{A}$, we denote by $W_{\mathcal{A}}$ the class of quasi-isomorphisms in $\mathrm{Ch}(\mathcal{A})$. Up to set-theoretic considerations, there is the localization functor

$$\mathrm{Ch}(\mathcal{A}) \xrightarrow{\qquad\qquad\gamma\qquad\qquad} D(\mathcal{A}) = \mathrm{Ch}(\mathcal{A})[(W_{\mathcal{A}})^{-1}], \tag{2.5}$$

which universally inverts the quasi-isomorphisms $W_{\mathcal{A}}$ [**GZ67**]. In the case of a module category $\mathcal{A} = \mathrm{Mod}(R)$ we use the standard notation $\mathrm{Ch}(R) \to D(R)$.

It is common to think of the derived category $D(\mathcal{A})$ as a rather refined invariant of the abelian category $\mathcal{A}$. A good deal of information about homological invariants is encoded by derived categories, and many such invariants are invariant under derived equivalences (see, for instance, [**Kel07b**] and the many references therein).

Here we want to take a different, but related perspective and focus on formal properties of derived categories. The reason that we drew the morphism in (2.5) in such a long way is that we want to stress that the categories $\mathrm{Ch}(\mathcal{A})$ and $D(\mathcal{A})$ live in rather different worlds. The category $\mathrm{Ch}(\mathcal{A})$ is an abelian category and hence enjoys rather nice properties. In particular, it has all finite limits and finite colimits. In contrast to this, the category $D(\mathcal{A})$ is rather ill-behaved (the 1-categorical localization procedure is an act of violence, and it destroys these nice properties). There are two relevant ways to make this more precise.

(i) First, derived categories $D(\mathcal{A})$ tend to not have many limits or colimits. In fact, the only typical (co)limits which exist are finite biproducts and in some cases also infinite (co)products.
(ii) More importantly, the calculus of *derived (co)limits* is not visible to the category $D(\mathcal{A})$ *alone.*

There are various ways to deal with these observations (see Disclaimer 3.2 and references there), and here we begin by following Verdier and Grothendieck. The following classical theorem offers one way to react upon the above issues, and the remaining goal of this first section is to explain that result from the perspective of symmetries.

THEOREM 2.6 (Verdier [**Ver67, Ver96**]). *The derived category $D(\mathcal{A})$ "is" a triangulated category.*

In the formulation of this theorem, we were rather picky and put the verb into quotation marks. This is in order to stress that a triangulation amounts to the specification of additional *structure* on $D(\mathcal{A})$ (in contrast to asking for mere *properties* of $D(\mathcal{A})$). In particular, this means that there are

(i) an auto-equivalence $\Sigma\colon D(\mathcal{A}) \xrightarrow{\simeq} D(\mathcal{A})$ and
(ii) a class of distinguished triangles $x \xrightarrow{f} y \to z \to \Sigma x$ in $D(\mathcal{A})$,

and these are subject to certain axioms [**Nee01, HJR10**]. Given such a distinguished triangle, the object z is referred to as "the" cone or cofiber of f. This time, the quotation marks are meant to allude to the following well-known fact. Approached this way, cones are only unique up to non-canonical isomorphism and only weakly functorial.

The following proposition describes an alternative approach to the cone construction. It exhibits the cone as the total left derived cokernel functor, thereby guaranteeing that the cone is a canonical and functorial construction.

PROPOSITION 2.7. *Let $\mathcal{A}$ be an abelian category, let $[1]$ be the poset $(0 < 1)$, let $\mathcal{A}^{[1]}$ be the category of morphisms in $\mathcal{A}$, and let* $\mathrm{cok}\colon \mathcal{A}^{[1]} \to \mathcal{A}$ *be the cokernel functor.*

(i) *The cokernel functor is right exact.*
(ii) *The cokernel functor is exact on monomorphisms.*
(iii) *The cokernel functor has a total left derived functor given by the cone construction,*

$$C \cong L\mathrm{cok}\colon D(\mathcal{A}^{[1]}) \to D(\mathcal{A}).$$

PROOF. A short exact sequence in $\mathcal{A}^{[1]}$ is simply a morphism of short exact sequences in $\mathcal{A}$,

$$\begin{array}{ccccccccc} 0 & \longrightarrow & x' & \longrightarrow & x & \longrightarrow & x'' & \longrightarrow & 0 \\ & & \downarrow{\scriptstyle f'} & & \downarrow{\scriptstyle f} & & \downarrow{\scriptstyle f''} & & \\ 0 & \longrightarrow & y' & \longrightarrow & y & \longrightarrow & y'' & \longrightarrow & 0. \end{array}$$

(i) This is an immediate consequence of the snake lemma, which yields a 6-term exact sequence

$$0 \to \ker(f') \to \ker(f) \to \ker(f'') \to \mathrm{cok}(f') \to \mathrm{cok}(f) \to \mathrm{cok}(f'') \to 0.$$

(ii) In the case of monomorphisms, the above sequence reduces to a short exact sequence, so also this statement is immediate from the snake lemma.

(iii) While the third statement does not follow as directly from the snake lemma, the snake lemma suggests what we are supposed to do. Namely, we should approximate an arbitrary chain map up to quasi-isomorphism by a monomorphic one and apply the cokernel to it instead. The details of this are as follows.

Given a chain map $f\colon x \to y$, we note that $\mathrm{cok}(f)$ is the pushout in the quadrilateral on the right in

(2.8)

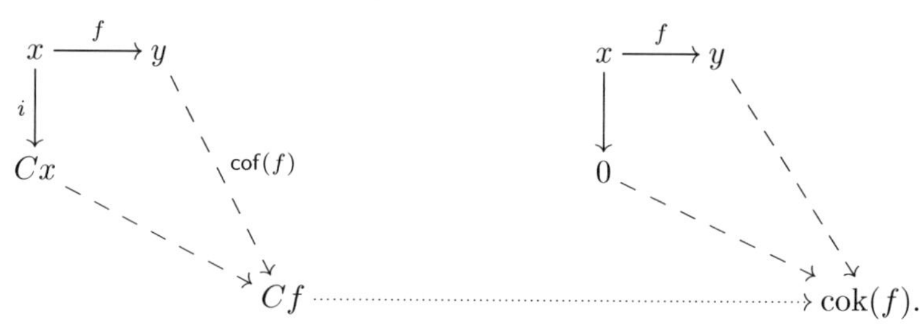

The intended monomorphic approximation up to quasi-isomorphism is displayed in the remaining quadrilateral. Therein, $x \mapsto Cx$ denotes the classical functor which sends a *chain complex* to its cone (see, for instance, [**GM03**, **Wei94**]). The relevant key facts about this construction are that, first, the cone Cx is an acyclic complex and, second, it comes with a natural monomorphic chain map $i\colon x \to Cx$. Now, the cone Cf of the *chain map* f is defined by the pushout square on the left in (2.8). Note that, since the induced map $x \to Cx \oplus y$ is a monomorphism, by the first two parts the resulting functor $C\colon \mathrm{Ch}(\mathcal{A}^{[1]}) \to \mathrm{Ch}(\mathcal{A})\colon f \mapsto Cf$ preserves quasi-isomorphisms. Consequently, it descends to a cone functor at the level of derived categories

$$C\colon D(\mathcal{A}^{[1]}) \to D(\mathcal{A}).$$

In order to exhibit this as the total left derived functor of the cokernel, it only remains to construct a corresponding natural transformation. Since the cone Cx is acyclic, the unique chain map $Cx \to 0$ together with the identities id_x and id_y induces a levelwise quasi-isomorphism

$$(Cx \leftarrow x \to y) \to (0 \leftarrow x \to y),$$

i.e., the intended monomorphic resolution. At the level of pushouts this yields a natural chain map $Cf \to \mathrm{cok}(f)$ (the dotted arrow in (2.8)), which in turn gives rise to the intended natural transformation in

$$\begin{array}{ccc} \mathrm{Ch}(\mathcal{A}^{[1]}) & \xrightarrow{C} & \mathrm{Ch}(\mathcal{A}) \\ {\scriptstyle\gamma}\downarrow & {\scriptstyle\varepsilon}\nearrow & \downarrow{\scriptstyle\gamma} \\ D(\mathcal{A}^{[1]}) & \xrightarrow[C]{} & D(\mathcal{A}). \end{array}$$

Standard techniques imply that (C, ε) is a model for $L\mathrm{cok}$.

(For more details and a more systematic explanation we refer to [**Gro18**, §3].) □

We want to briefly discuss this result.

REMARK 2.9. For every abelian category $\mathcal{A}$ the category of morphisms $\mathcal{A}^{[1]}$ is again abelian. It is important to distinguish the two categories

$$D(\mathcal{A}^{[1]}) \qquad \text{and} \qquad D(\mathcal{A})^{[1]},$$

the *derived category of the morphism category* and the *morphism category of the derived category*. A relation between these categories is provided by a forgetful functor

$$\mathrm{dia}_{[1]}\colon D(\mathcal{A}^{[1]}) \to D(\mathcal{A})^{[1]}$$

which is constructed as follows. An object $X \in D(\mathcal{A}^{[1]})$ is a chain complex of morphisms which we can hence rewrite as a functor $X\colon [1] \to \mathrm{Ch}(\mathcal{A})$. Written this way, we can compose X with the localization functor $\gamma\colon \mathrm{Ch}(\mathcal{A}) \to D(\mathcal{A})$, and we make the definition

$$\mathrm{dia}_{[1]}(X) = \gamma \circ X\colon [1] \to D(\mathcal{A}).$$

The crucial observation now is that this functor discards relevant information. In particular, the functorial cone from Proposition 2.7 does not factor through this underlying diagram functor,

$$\begin{array}{ccc} D(\mathcal{A}^{[1]}) & \xrightarrow{C} & D(\mathcal{A}) \\ {\scriptstyle\mathrm{dia}_{[1]}}\downarrow & \nearrow{\scriptstyle\nexists C} & \\ D(\mathcal{A})^{[1]}. & & \end{array}$$

(This fails already for vector spaces over a field as is detailed in [**Gro18**, §5].)

Thus, to put this into plain English, there is a functorial cone construction on the *derived category of the morphism category* but not on the *morphism category of the derived category*. In order to develop more intuition for these underlying diagram functors, we collect the following straightforward result [**BG18**, Cor. 3.9].

EXAMPLE 2.10. Let R be a semi-simple ring, let B be a category with finitely many objects, and let RB be the R-linear category algebra. There is a similar underlying diagram functor

$$\mathrm{dia}_B\colon D(\mathrm{Mod}(R)^B) \to D(R)^B\colon X \mapsto \gamma \circ X,$$

which is equivalent to the $\mathbb{Z}$-graded homology functor $H_*\colon D(RB) \to \mathrm{Mod}(RB)^{\mathbb{Z}}$.

This example illustrates very nicely that underlying diagram functors discard relevant information. It turns out that not only these underlying diagram functors tend to fail to be equivalences, but also that the categories

$$D(\mathcal{A}^B) \qquad \text{and} \qquad D(\mathcal{A})^B$$

are not equivalent (see [**Gro18**, §5] for an explicit example). Moreover, there is no simple recipe which allows us to reconstruct derived categories of diagram categories (categories of the form $D(\mathcal{A}^B)$) from the derived category $D(\mathcal{A})$. If we want to study derived (co)limits at the level of derived categories only (in contrast to the approaches in Disclaimer 3.2), then we have to keep track of the categories $D(\mathcal{A}^B)$ themselves.

Before we state a general theorem about the existence of derived (co)limits (Theorem 2.19), we give an elementary proof in the following central example. We denote by $\ulcorner$ the full subposet of the square looking like:

(2.11)

$$\begin{array}{ccc} (0,0) & \longrightarrow & (0,1) \\ \downarrow & & \\ (1,0) & & \end{array}$$

Correspondingly, the diagram category $\mathcal{A}^{\ulcorner}$ is the category of spans in $\mathcal{A}$ and the colimit functor

$$\mathrm{colim}_{\ulcorner}\colon \mathcal{A}^{\ulcorner} \to \mathcal{A} \tag{2.12}$$

simply forms pushouts.

PROPOSITION 2.13. *Let $\mathcal{A}$ be an abelian category. The pushout functor* (2.12) *has a total left derived functor*

$$L\,\mathrm{colim}_{\ulcorner}\colon D(\mathcal{A}^{\ulcorner}) \to D(\mathcal{A}).$$

PROOF. The strategy of the proof is very similar to the one in Proposition 2.7, and we are more sketchy. As a left adjoint the pushout functor is right exact, and it turns out to be exact on those spans

$$\begin{array}{ccc} x & \longrightarrow & y \\ \downarrow & & \\ z, & & \end{array}$$

such that the induced map $x \to y \oplus z$ is a monomorphism. Functorial resolutions adapted to the pushout are hence given by any of the spans

$$\begin{array}{ccc} x & \longrightarrow & y \\ \downarrow & & \\ Cx \oplus z, & & \end{array} \qquad \begin{array}{ccc} x & \longrightarrow & Cx \oplus y \\ \downarrow & & \\ z, & & \end{array} \qquad \begin{array}{ccc} x & \longrightarrow & Cx \oplus y \\ \downarrow & & \\ Cx \oplus z, & & \end{array}$$

together with the obvious maps collapsing the cones. Choosing for instance the resolution of the left, the derived pushout functor

$$L\,\mathrm{colim}_{\ulcorner}\colon D(\mathcal{A}^{\ulcorner}) \to D(\mathcal{A})$$

is induced by applying $\mathrm{colim}_{\ulcorner}$ to these resolution. The canonical natural transformation

$$\begin{array}{ccc} \mathrm{Ch}(\mathcal{A}^{\ulcorner}) & \xrightarrow{\mathrm{colim}_{\ulcorner}} & \mathrm{Ch}(\mathcal{A}) \\ {\scriptstyle\gamma}\downarrow & {\scriptstyle\varepsilon}\nearrow & \downarrow{\scriptstyle\gamma} \\ D(\mathcal{A}^{\ulcorner}) & \xrightarrow[L\,\mathrm{colim}_{\ulcorner}]{} & D(\mathcal{A}) \end{array}$$

again comes from the chain maps which collapse the cones. (For a more systematic approach see [**Gro18**, §7].) □

2.2. Derived limits and derived Kan extensions. In this short subsection we collect key formal properties of the calculus of derived limits in categories of modules. More generally, restriction functors between derived diagram categories have adjoints on both sides. These derived Kan extension functors can be calculated in terms of derived (co)limits. It was the philosophy of Grothendieck [**Gro83**, pp. 196-200] that the "entire triangulated information" of a derived category is encoded by this system of derived diagram categories and derived limit functors.

NOTATION 2.14. For every small category B, we denote the derived category of the diagram category $\mathrm{Mod}(R)^B$ by

$$\mathscr{D}_R(B) = D(\mathrm{Mod}(R)^B).$$

As a special case $\mathscr{D}_R(*)$ is simply the derived category $D(R)$. The diagonal functor $\Delta_B\colon \mathrm{Mod}(R) \to \mathrm{Mod}(R)^B$ is exact and hence induces a similar diagonal functor

$$\Delta_B\colon \mathscr{D}_R(*) \to \mathscr{D}_R(B).$$

Recall that (co)limits are simply adjoints to diagonal functors [**ML98**, § IV.2]. Generalizing Proposition 2.13 there is the following derived version of this.

THEOREM 2.15. *For every small category B there are adjunctions*

$$(L\,\mathrm{colim}_B, \Delta_B)\colon \mathscr{D}_R(B) \rightleftarrows \mathscr{D}_R(*) \quad \textit{and} \quad (\Delta_B, R\lim_B)\colon \mathscr{D}_R(*) \rightleftarrows \mathscr{D}_R(B).$$

PROOF. This is part of the statement of Theorem 2.19. □

In order to build towards a further generalization, we recall the following definition. Given a functor $u\colon A \to B$ between small categories, there is the *restriction* or *precomposition functor*

$$u^*\colon \mathrm{Mod}(R)^B \to \mathrm{Mod}(R)^A\colon X \mapsto X \circ u. \tag{2.16}$$

In particular, for an object $b \in B$, we obtain the *evaluation functor*

$$b^*: \mathrm{Mod}(R)^B \to \mathrm{Mod}(R).$$

In this case we simplify notation by writing $f_b\colon X_b \to Y_b$ for the value of $f\colon X \to Y$ under the evaluation functor.

Kan extensions are some kind of relative versions of (co)limits. To explain what we mean by this it suffices to identify diagonal functors as restriction functors. In more detail, for every small category B let $\pi_B\colon B \to *$ be the unique functor to the terminal category. Note that under the isomorphism $\mathrm{Mod}(R)^* \cong \mathrm{Mod}(R)$ the functors Δ_B and $(\pi_B)^*$ are identified. Thus, to put this differently, the (co)limit functors are adjoints to restriction along these particular functors. It turns out that the existence of limits and colimits in $\mathrm{Mod}(R)$ also guarantees the existence of adjoints to arbitrary restriction functors (2.16). Given a functor $u\colon A \to B$ between small categories, there are adjunctions

$$(u_!, u^*)\colon \mathrm{Mod}(R)^A \rightleftarrows \mathrm{Mod}(R)^B \quad \text{and} \quad (u^*, u_*)\colon \mathrm{Mod}(R)^B \rightleftarrows \mathrm{Mod}(R)^A.$$

The functor $u_!$ is the **left Kan extension** functor along u, and u_* is referred to as the **right Kan extension** along u. For the basic theory of these Kan extension functors see [**ML98**, §X] or [**Bor94b**, §3.7]. For a discussed geared towards our current situation we also refer to [**Gro18**, §6].

The restriction functors (2.16) are again exact. Hence, for trivial reasons we obtain derived restriction functors

$$u^*: \mathscr{D}_R(B) \to \mathscr{D}_R(A).$$

As a relative version of Theorem 2.15, also the Kan extension adjunctions have derived versions.

THEOREM 2.17. *For every functor $u\colon A \to B$ between small categories there are adjunctions*

$$(u_!, u^*)\colon \mathscr{D}_R(A) \rightleftarrows \mathscr{D}_R(B) \quad \textit{and} \quad (u^*, u_*)\colon \mathscr{D}_R(B) \rightleftarrows \mathscr{D}_R(A).$$

PROOF. This is part of the statement of Theorem 2.19. □

For every functor $u\colon A \to B$ we hence have a derived left Kan extension functors and a derived right Kan extension functor

$$u_!\colon \mathscr{D}_R(A) \to \mathscr{D}_R(B) \quad \text{and} \quad u_*\colon \mathscr{D}_R(A) \to \mathscr{D}_R(B).$$

To be of any use, it is important to be able to calculate these derived Kan extension functors. More specifically, given $X \in \mathscr{D}_R(A)$ and $b \in B$ we would like to be able to express the values

$$u_!(X)_b, u_*(X)_b \in \mathscr{D}_R(*) \cong D(R)$$

of the derived Kan extensions $u_!(X), u_*(X) \in \mathscr{D}_R(B)$ in terms of X, u, and b only. Before dealing with this derived situation, let us recall the corresponding fact for module categories.

CONSTRUCTION 2.18. Let $u\colon A \to B$ be a functor between small categories and let $X\colon A \to \mathrm{Mod}(R)$ be a diagram of modules. In order to construct the left Kan extension $u_!(X)\colon B \to \mathrm{Mod}(R)$ we recall the definition of **slice categories**. For every object $b \in B$ the slice category (u/b) of objects u-over b has the following description. Objects in (u/b) are pairs $(a, f\colon ua \to b)$ consisting of an object $a \in A$ and a morphism $f\colon ua \to b$ in B. A morphism $g\colon (a, f) \to (a', f')$ in (u/b) is a morphism $g\colon a \to a'$ in A such that $f' \circ u(g) = f$. Note that there is forgetful functor $p\colon (u/b) \to A\colon (a,f) \mapsto a$, and to construct $u_!(X)_b$ it suffices to form a colimit over this slice category. More precisely, for every $b \in B$ we define

$$u_!(X)_b = \operatorname{colim}_{(u/b)} X \circ p \in \operatorname{Mod}(R).$$

One checks that this defines a diagram $u_!(X)\colon B \to \operatorname{Mod}(R)$, the left Kan extension of X along u.

Dually, for every $b \in B$ there is the slice category (b/u) of objects u-under b. And also this category comes with a forgetful functor $q\colon (b/u) \to A\colon (a, f\colon b \to ua) \mapsto a$. It turns out that also the right Kan extension $u_*(X)$ can be constructed pointwise by the formula

$$u_*(X)_b = \lim_{(b/u)} X \circ q \in \operatorname{Mod}(R).$$

This concludes the recap of the pointwise formulas for Kan extensions of diagrams of modules (which, of course, generalizes to arbitrary complete and cocomplete categories). The following theorem shows that similar formulas extend to the derived Kan extensions in Theorem 2.17. For convenience, we also summarize some additional key properties of the calculus of derived limits, derived colimits, and derived Kan extensions over a ring.

THEOREM 2.19 (Grothendieck). *Let R be a ring. The formation of derived categories of diagram categories defines a 2-functor*

$$\mathscr{D}_R\colon B \mapsto \mathscr{D}_R(B) = D(\operatorname{Mod}(R)^B)$$

from small categories to not necessarily small categories. The 2-functor $\mathscr{D}_R$ enjoys the following properties.

(Der1) *The canonical inclusion functors $B_j \to \coprod_{i\in I} B_i, j \in I$, induce an equivalence of categories*

$$\mathscr{D}_R(\coprod_{i\in I} B_i) \xrightarrow{\sim} \prod_{i\in I} \mathscr{D}_R(B_i).$$

(Der2) *A morphism $f\colon X \to Y$ in $\mathscr{D}_R(B)$ is an isomorphism if and only if the morphisms $f_b\colon X_b \to Y_b, b \in B$, are isomorphisms in $\mathscr{D}_R(*) = D(R)$.*

(Der3) *For every functor $u\colon A \to B$, the restriction functor $u^*\colon \mathscr{D}_R(B) \to \mathscr{D}_R(A)$ has a left adjoint $u_!$ and a right adjoint u_*,*

$$(u_!, u^*)\colon \mathscr{D}_R(A) \rightleftarrows \mathscr{D}_R(B), \qquad (u^*, u_*)\colon \mathscr{D}_R(B) \rightleftarrows \mathscr{D}_R(A).$$

(Der4) *For every functor $u\colon A \to B$, the functors $u_!, u_*\colon \mathscr{D}_R(A) \to \mathscr{D}_R(B)$ can be calculated pointwise. More specifically, for $X \in \mathscr{D}_R(A)$ and $b \in B$ there are canonical isomorphisms*

$$L\operatorname{colim}_{(u/b)} p^*(X) \xrightarrow{\sim} u_!(X)_b \quad \text{and} \quad u_*(X)_b \xrightarrow{\sim} R\lim_{(b/u)} q^*(X).$$

PROOF. While an elementary proof building on classical homological algebra only does not seem to exist in the literature, we think that it would be nice to expand on the techniques behind Proposition 2.7 and Proposition 2.13 in order to obtain such a proof. Currently, this result is a consequence of more general theorems. The category $\operatorname{Mod}(R)$ is Grothendieck abelian and $\operatorname{Ch}(R)$ can hence be endowed with a combinatorial Quillen model structure (see [**Bek00**, Prop. 3.13], [**Hov01**, §2], or [**CD09**, §§2-3]). For combinatorial Quillen model categories a reasonably simple proof of a variant of this theorem can be found in [**Gro13**, §1.3]. □

REMARK 2.20. We want to stress that, by definition, $\mathscr{D}_R$ sends a small category B to the derived category $D(\mathrm{Mod}(R)^B)$. In this assignment the derived category is considered as a plain category only and hence not as a triangulated category. In fact, the main goal of §2.3 and §2.4 now consists of showing how to construct the Verdier triangulation on $D(R)$ (Theorem 2.6) using certain nice properties of this 2-functor $\mathscr{D}_R$ only. Those arguments will indicate the above-mentioned intimate relation between the triangulation and symmetries.

Formally, in the remainder of this section we focus on modules over a ring. We want to stress, however, that variants of Theorem 2.19 arise in many other situations as well. Moreover, many of the arguments which are sketched further below apply almost verbatim to those more general situations (see §3).

The following key property will be used over and over again.

PROPOSITION 2.21. *Let R be a ring and let $u\colon A \to B$ a fully faithful functor. The Kan extension functors $u_!, u_*\colon \mathscr{D}_R(A) \to \mathscr{D}_R(B)$ are fully faithful.*

PROOF. This is a special case of Proposition 3.23. □

This result is of central importance in many constructions, and to explain its relevance we collect the following reformulation (based on standard results on adjunctions): For every fully faithful $u\colon A \to B$ and every $X \in \mathscr{D}_R(A)$ the adjunction units and counits

$$\eta\colon X \xrightarrow{\sim} u^*u_!X \qquad \text{and} \qquad \varepsilon\colon u^*u_*X \xrightarrow{\sim} X$$

are invertible. Thus, this result allows us to extend diagrams defined on smaller categories to larger ones without affecting the diagram on the smaller category, and in this sense *Kan extensions* are proper "extensions".

As a first example of the derived Kan extension functors from Theorem 2.19, we reformulate the proof of Proposition 2.7 in terms of these functors. This needs some preparation.

NOTATION 2.22. We denote by $\square$ the category $[1] \times [1]$,

$$\begin{array}{ccc} (0,0) & \longrightarrow & (0,1) \\ \downarrow & & \downarrow \\ (1,0) & \longrightarrow & (1,1). \end{array}$$

For every ring R, a **square** in $\mathscr{D}_R$ is an object $X \in \mathscr{D}_R(\square)$. Such a square is hence a commutative square of chain complexes

$$\begin{array}{ccc} x & \longrightarrow & y \\ \downarrow & & \downarrow \\ z & \longrightarrow & w \end{array}$$

considered as an object in the derived category $D(\mathrm{Mod}(R)^\square)$. Associated to every square, there are the following canonical maps.

(i) Ignoring the chain complex w for a moment, there is the corresponding left derived pushout $(L\operatorname{colim}_{\ulcorner})(X|_{\ulcorner})$ (Proposition 2.13). This chain complex comes with a canonical comparison map

$$(L\operatorname{colim}_{\ulcorner})(X|_{\ulcorner}) \to w. \tag{2.23}$$

(ii) Dually, ignoring the chain complex x, there is the right derived pullback $R\mathrm{lim}_\lrcorner(X|_\lrcorner)$, and this chain complex comes with a canonical comparison map

(2.24)
$$x \to (R\mathrm{lim}_\lrcorner)(X|_\lrcorner).$$

DEFINITION 2.25. Let R be a ring and let $X \in \mathscr{D}_R(\square)$.

(i) The square X is **cocartesian** if (2.23) is an isomorphism in $\mathscr{D}_R(*) = D(R)$.
(ii) The square X is **cartesian** if (2.24) is an isomorphism in $\mathscr{D}_R(*) = D(R)$.

The following two properties of $\mathscr{D}_R$ are crucial to all later constructions.

PROPOSITION 2.26 (Stability of $\mathscr{D}_R$). *Let R be a ring.*

(i) *The category $\mathscr{D}_R(*) = D(R)$ has a zero object.*
(ii) *A square $X \in \mathscr{D}_R(\square)$ is cocartesian if and only if it is cartesian. We refer to these squares as **bicartesian squares**.*

PROOF. The first statement is obvious since $D(R)$ is an additive category. We invite the reader to come up with a direct proof of the second statement. That result also follows more indirectly from the fact that $\Sigma\colon D(R) \to D(R)$ is an equivalence combined with Theorem 2.19 and Theorem 3.31. $\square$

Whenever we want to stress that a given square is bicartesian, we decorate it by a square in the middle such as in

$$\begin{array}{ccc} x & \longrightarrow & y \\ \downarrow & \square & \downarrow \\ z & \longrightarrow & w. \end{array}$$

REMARK 2.27. On a first view, it seems to be a bit strange, that in Definition 2.25 we introduce the notions of cocartesian and cartesian squares in $\mathscr{D}_R$ for every ring R, in order to then observe in Proposition 2.26 that these two classes coincide. The point is, of course, that these notions make sense in many other situations as well (see §3), and that often these classes do not agree. For instance, in a complete and cocomplete category $\mathcal{C}$ with a zero object, the classes of pushout and pullback squares agree if and only if $\mathcal{C}$ is equivalent to the terminal category (which is equivalent to $\mathcal{C}$ being a contractible groupoid). In this precise sense the conjunction of the two properties in Proposition 2.26 are invisible to ordinary category theory and they require more refined techniques (see also Remark 3.33).

With these basic notions in place, we can now revisit Proposition 2.7.

DEFINITION 2.28. Let R be a ring. A **cofiber square** in $\mathscr{D}_R$ is a cocartesian square $X \in \mathscr{D}_R(\square)$ which vanishes at the lower left corner,

(2.29)
$$\begin{array}{ccc} x & \xrightarrow{f} & y \\ \downarrow & \square & \downarrow \\ 0 & \longrightarrow & z. \end{array}$$

We denote by $\mathscr{D}_R(\square)^{\mathsf{cof}} \subseteq \mathscr{D}_R(\square)$ the full subcategory spanned by the cofiber squares.

We want to stress that the lower left corner is a zero object in $D(R)$, so it really could be any acyclic chain complex. The intuition is that a cofiber square is essentially determined by the upper horizontal morphism $(f\colon x \to y)$. This morphism is obtained formally from the cofiber square by restriction along

$$k\colon [1] \to \square\colon i \mapsto (0,i). \tag{2.30}$$

PROPOSITION 2.31. *For every ring R restriction along* (2.30) *induces an equivalence of categories*

$$k^*\colon \mathscr{D}_R(\square)^{\mathrm{cof}} \xrightarrow{\sim} \mathscr{D}_R([1]).$$

PROOF. We sketch the proof and begin by noting that the functor (2.30) factors as

$$k = j \circ i\colon [1] \to \ulcorner \to \square.$$

Here, $\ulcorner$ is again suggestive notation for the poset (2.11) which corepresents spans. By Theorem 2.19 there are derived Kan extension adjunctions

$$(i^*, i_*)\colon \mathscr{D}_R(\ulcorner) \rightleftarrows \mathscr{D}_R([1]) \qquad \text{and} \qquad (j_!, j^*)\colon \mathscr{D}_R(\ulcorner) \rightleftarrows \mathscr{D}_R(\square).$$

And in this case the construction of i_* and $j_!$ can be given very explicitly.

(i) Before passing to derived categories, it is straightforward to check that the functor i_* which sends a chain map $(x \to y)$ to the span $(0 \leftarrow x \to y)$ is right adjoint to i^* (alternatively, this also follows from the pointwise formulas in Construction 2.18). Since both functors are exact, this immediately yields the first adjunction.

(ii) In the case of the second adjunction a direct argument can be sketched as follows. While pushouts fail to be exact on all spans, they are exact on spans such that at least one of the maps is a monomorphism. Hence, as in the proof of Proposition 2.13, we can resolve an arbitrary span by adding inclusions into cones on at least one of the edges, and the corresponding pushout squares are models for the left derived pushout square functor

$$j_!\colon \mathscr{D}_R(\ulcorner) \to \mathscr{D}_R(\square).$$

With these adjunctions at our disposal we can now consider the composition

$$j_! \circ i_*\colon \mathscr{D}_R([1]) \to \mathscr{D}_R(\ulcorner) \to \mathscr{D}_R(\square),$$

and this yields the desired equivalence. In more detail, the functors i_* and $j_!$ are both fully faithful (Proposition 2.21), and one checks that the essential image of their composition consists precisely of the cofiber squares. □

The above sketch has shown that the left quadrilateral in (2.8) can be constructed in terms of derived Kan extension functors. This justifies the terminology *cofiber square* since for every such square (2.29) there is a canonical isomorphism between z and the cone or cofiber of f,

$$Cf \xrightarrow{\sim} z.$$

There is the following important special case of cofiber squares.

DEFINITION 2.32. Let R be a ring. A **suspension square** in $\mathscr{D}_R$ is a cocartesian square which vanishes at the lower left and the upper right corner,

$$\begin{array}{ccc} x & \longrightarrow & 0 \\ \downarrow & \square & \downarrow \\ 0 & \longrightarrow & x'. \end{array} \tag{2.33}$$

REMARK 2.34. For every ring R we denote by $\mathscr{D}_R(\square)^\Sigma \subseteq \mathscr{D}_R(\square)$ the full subcategory spanned by the suspension squares. Similar to the case of Proposition 2.31, evaluation at $(0,0) \in \square$ induces an equivalence of categories

$$(0,0)^* \colon \mathscr{D}_R(\square)^\Sigma \xrightarrow{\sim} \mathscr{D}_R(*) = D(R).$$

We invite the reader to adapt the arguments of the previous case in order to sketch a proof of this result. An inverse of this equivalence is given by forming the cofiber square associated to $(x \to 0)$. Hence, the proof of Proposition 2.7 shows that we simply form the cokernel of the inclusion $x \to Cx$, which is clearly isomorphic to the usual suspension Σx of the chain complex x. The terminology *suspension squares* is hence justified by the fact that every such square (2.33) induces a canonical isomorphism

$$\Sigma x \xrightarrow{\sim} x'.$$

2.3. A_2-quivers and Barratt–Puppe sequences. The point of the constructions at the end of the previous subsection was the following. A morphism in $\mathscr{D}_R$ can be equivalently encoded by a cofiber square (Proposition 2.31) and, similarly, an object in $\mathscr{D}_R$ is as good as a suspension square (Remark 2.34). In this subsection we want to expand a bit on this observation for the case of morphisms. Representations of $\vec{A}_2$ in $\mathscr{D}_R$ turn out to be equivalent to *Barratt–Puppe sequences*. Moreover, the symmetry of these Barratt–Puppe sequence corresponds to the rotation at the level of triangulated categories.

As a warmup we consider *cofiber sequences* which are essentially obtained by forming two cofiber squares. To carry this out in detail, we introduce the following notation.

NOTATION 2.35. For every natural number n, we denote by $[n]$ the finite linear order $(0 < 1 < \ldots < n)$. The product $\boxminus = [1] \times [2]$ hence agrees with the shape of two adjacent commuting squares

$$\begin{array}{ccccc} (0,0) & \longrightarrow & (0,1) & \longrightarrow & (0,2) \\ \downarrow & & \downarrow & & \downarrow \\ (1,0) & \longrightarrow & (1,1) & \longrightarrow & (1,2). \end{array}$$

In representation theoretic parlance, the poset $[n]$ is the linearly oriented A_{n+1}-quiver

$$\vec{A}_{n+1} = (1 \to \ldots \to n+1).$$

Note that the labeling conventions only match up to a shift by one. There is a unique isomorphism between the two shapes which will always be used implicitly, hopefully not arising in a confusion concerning the various labels.

DEFINITION 2.36. Let R be a ring. A **cofiber sequence** in $\mathscr{D}_R$ is an object $X \in \mathscr{D}_R(\boxbar)$ such that the square on the left and the square on the right are cocartesian and such that X vanishes at $(1,0)$ and $(0,2)$,

$$
\begin{array}{ccccc}
x & \xrightarrow{f} & y & \longrightarrow & 0 \\
\big\downarrow & \square & \big\downarrow{\scriptstyle g} & \square & \big\downarrow \\
0 & \longrightarrow & z & \longrightarrow & x'.
\end{array}
\tag{2.37}
$$

We denote by $\mathscr{D}_R(\boxbar)^{\mathsf{cof}} \subseteq \mathscr{D}_R(\square)$ the full subcategory spanned by the cofiber squares.

Morally, a cofiber sequence is as good as a morphism only. In fact, given a cofiber sequence (2.37), the morphism $z \to x'$ can be reconstructed from $g\colon y \to z$ by Proposition 2.31. Similarly, by that same result also the morphism $g\colon y \to z$ is determined by $f\colon x \to y$. This makes the following result plausible in which we denote the inclusion of the upper left morphism by

$$
k\colon [1] \to \boxbar\colon i \mapsto (0,i). \tag{2.38}
$$

PROPOSITION 2.39. *For every ring R restriction along* (2.38) *induces an equivalence of categories*

$$
k^*\colon \mathscr{D}_R(\boxbar)^{\mathsf{cof}} \xrightarrow{\sim} \mathscr{D}_R([1]).
$$

PROOF. The strategy is very similar to the one in the earlier cases. The functor k in (2.38) factors as a composition

$$
j \circ i\colon [1] \to B \to [1] \times [2]
$$

of fully faithful inclusions where B is given by

$$
\begin{array}{ccccc}
(0,0) & \longrightarrow & (0,1) & \longrightarrow & (0,2) \\
\big\downarrow & & & & \\
(1,0). & & & &
\end{array}
$$

At the level of chain complexes it is straightforward to check that the restriction functor i^* has a right adjoint i_* which simply adds two zero chain complexes and the unique chain maps. Alternatively, this follows immediately from Construction 2.18 since the corresponding slice categories are empty. Thus, both i^* and i_* are exact, and the induced functor $i_*\colon \mathscr{D}_R([1]) \to \mathscr{D}_R(B)$ is hence under control. It is fully faithful (Proposition 2.21) and the essential image consists of all $X \in \mathscr{D}_R(B)$ which vanish at $(1,0)$ and $(0,2)$. Similarly, at the level of chain complexes a left adjoint $j_!$ to j^* is given by the functor which adds two pushout squares. The corresponding left derived functor $j_!$ hence adds two derived pushout squares, and one checks that the essential image of this fully faithful functor can be characterized by this. As an upshot, both functors in

$$
j_! \circ i_*\colon \mathscr{D}_R([1]) \to \mathscr{D}_R(B) \to \mathscr{D}_R(\boxbar)
$$

are fully faithful and the essential image of their composition consists precisely of the cofiber sequences. □

$$
\begin{array}{ccccccccccccc}
\ddots & & \ddots & & \ddots & & \ddots & & & & & & \\
(-1,-2) & \to & (-1,-1) & \to & (-1,0) & \to & (-1,1) & & & & & & \\
 & & \downarrow & & \downarrow & & \downarrow & & & & & & \\
 & & (0,-1) & \longrightarrow & (0,0) & \longrightarrow & (0,1) & \longrightarrow & (0,2) & & & & \\
 & & & & \downarrow & & \downarrow & & \downarrow & & & & \\
 & & & & (1,0) & \longrightarrow & (1,1) & \longrightarrow & (1,2) & \to & (1,3) & & \\
 & & & & & & \downarrow & & \downarrow & & \downarrow & & \\
 & & & & & & (2,1) & \longrightarrow & (2,2) & \to & (2,3) & \to & (2,4) \\
 & & & & & & \ddots & & \ddots & & \ddots & & \ddots
\end{array}
$$

FIGURE 2. The full subposet $M_2 \subseteq \mathbb{Z} \times \mathbb{Z}$

In order to analyze cofiber sequences in more detail, we recall the following pasting and cancellation property of bicartesian squares in $\mathscr{D}_R$. There are three obvious inclusions of the square $\square$ in $\boxbar$. Let us denote by

$$\iota_{01}\colon \square \to \boxbar, \quad \iota_{12}\colon \square \to \boxbar, \quad \text{and} \quad \iota_{02}\colon \square \to \boxbar$$

the inclusion of the left square, the right square, and the composed square, respectively.

PROPOSITION 2.40. *Let R be a ring and let $X \in \mathscr{D}_R(\boxbar)$. If two of the squares $\iota_{01}^*(X)$, $\iota_{12}^*(X)$, and $\iota_{02}^*(X)$ are bicartesian, then so is the third square.*

PROOF. This is a special case of [**Gro13**, Prop. 4.6]. $\square$

CONSTRUCTION 2.41. Let R be a ring and let $X \in \mathscr{D}_R(\boxbar)^{\mathsf{cof}}$ be a cofiber sequence looking like (2.37). By Proposition 2.40 the composed square $\iota_{02}^*(X)$ is again bicartesian and it takes the form

$$
\begin{array}{ccc}
x & \longrightarrow & 0 \\
\downarrow & \square & \downarrow \\
0 & \longrightarrow & x'.
\end{array}
$$

Thus, $\iota_{02}^*(X)$ is a suspension square (Definition 2.32), and by Remark 2.34 we obtain a canonical isomorphism $\Sigma x \xrightarrow{\sim} x'$. It follows that associated to every cofiber sequence there is a triangle

$$x \xrightarrow{f} y \xrightarrow{g} z \xrightarrow{h} \Sigma x.$$

This is an ordinary diagram in $\mathscr{D}_R(*) = D(R)$, and the morphism h is obtained from the unlabeled morphism $z \to x'$ in (2.37) thanks to the isomorphism $\Sigma x \xrightarrow{\sim} x'$.

This already suggests that cofiber sequences are closely related to distinguished triangles, and we will come back to this at the end of §2.4. Here, instead, we summarize some of the constructions carried out so far. For every ring R, at the level of derived categories of diagram categories there are equivalences of categories

$$\mathscr{D}_R(\boxbar)^{\mathsf{cof}} \xrightarrow{\sim} \mathscr{D}_R(\square)^{\mathsf{cof}} \xrightarrow{\sim} \mathscr{D}_R([1]).$$

These equivalences are induced by suitable restriction functors, and inverse equivalences are obtained by extending a morphism to a cofiber square or by iterating this construction twice. We now simply iterate this construction countably many times in both directions. The combinatorial details are as follows.

NOTATION 2.42. Let $M_2 \subseteq \mathbb{Z}\times\mathbb{Z}$ be the full subposet given by all $(i,j)\in\mathbb{Z}\times\mathbb{Z}$ such that $i-1\leq j\leq i+2$.

A part of this poset is displayed in Figure 2. The reason that we consider this poset is that, for every ring R, this shape allows us to simultaneously encode a morphism $(f\colon x\to y)\in\mathscr{D}_R([1])$, all its iterated derived cokernels, and also all its iterated derived kernels. This is achieved by forming *Barratt–Puppe sequences* in the following sense.

DEFINITION 2.43. Let R be a ring. A diagram $X\in\mathscr{D}_R(M_2)$ is a **Barratt–Puppe sequence** in $\mathscr{D}_R$ if the diagram X vanishes on the two boundary stripes and all squares in X are bicartesian. We denote by $\mathscr{D}_R(M_2)^{\mathrm{ex}}\subseteq\mathscr{D}_R(M_2)$ the full subcategory spanned by all Barratt–Puppe sequences.

The terminology is motivated by similar constructions with pointed topological spaces. In that case this notion yields the classical Barratt–Puppe sequences as in [**Pup58**], which provide the space-level origin of many long exact sequences in algebraic topology. The following result makes precise that Barratt–Puppe sequences are determined by their restrictions along the inclusion

$$i\colon [1]\to M_2\colon k\mapsto (0,k). \tag{2.44}$$

THEOREM 2.45. *For every ring R restriction along* (2.44) *induces an equivalence of categories*

$$i^*\colon \mathscr{D}_R(M_2)^{\mathrm{ex}}\xrightarrow{\sim}\mathscr{D}_R([1]).$$

PROOF. We only sketch the arguments and refer the reader to [**GŠ16a**, §4] for a detailed proof of a more general result. The idea is to construct an inverse equivalence

$$F_{[1]}\colon \mathscr{D}_R([1])\xrightarrow{\sim}\mathscr{D}_R(M_2)^{\mathrm{ex}}$$

as a composition of suitable derived Kan extensions. To this end we observe that the inclusion (2.44) factors as

$$i\colon [1]\xrightarrow{i_1}K_1\xrightarrow{i_2}K_2\xrightarrow{i_3}K_3\xrightarrow{i_4}M_2,$$

where the K_j are the following full subposets of M_2 and the i_j are the obvious inclusions.

(i) K_1 is obtained from the image of i by adding those objects on the boundary stripe which sit under the image of i, i.e., the objects $(n,n+2), n\geq 0$, and the objects $(n,n-1), n>0$.
(ii) K_2 contains besides the objects in K_1 also all remaining objects under the image of i, i.e., the objects (n,n) and $(n,n+1)$ for $n>0$.
(iii) K_3 is obtained from K_2 by also adding the remaining objects on the boundary stripes, i.e., the objects $(n,n+2), n<0$, and the objects $(n,n-1), n\leq 0$.

We know from Proposition 2.21 that the following four functors

$$\mathscr{D}_R([1])\xrightarrow{(i_1)_*}\mathscr{D}_R(K_1)\xrightarrow{(i_2)_!}\mathscr{D}_R(K_2)\xrightarrow{(i_3)_!}\mathscr{D}_R(K_3)\xrightarrow{(i_4)_*}\mathscr{D}_R(M_2)$$

are fully faithful. Invoking arguments similar to the previous cases in combination

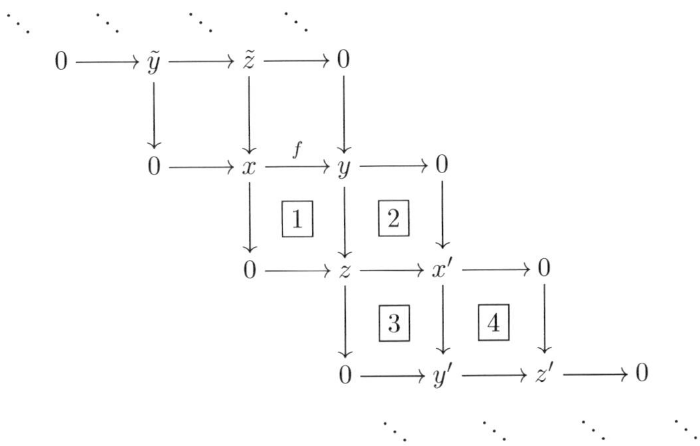

FIGURE 3. Iterated derived (co)kernels of $(f\colon x \to y) \in \mathscr{D}_R([1])$

with homotopy (co)finality arguments, one checks that

(i) $(i_1)_*\colon \mathscr{D}_R([1]) \to \mathscr{D}_R(K_1)$ simply adds zero objects at the new components,
(ii) $(i_2)_!\colon \mathscr{D}_R(K_1) \to \mathscr{D}_R(K_2)$ inductively adds bicartesian squares,
(iii) $(i_3)_!\colon \mathscr{D}_R(K_2) \to \mathscr{D}_R(K_3)$ again adds zero objects at the new components,
(iv) and that $(i_4)_*\colon \mathscr{D}_R(K_3) \to \mathscr{D}_R(M_2)$ constructs new bicartesian squares.

Thus, the composition of these four functors indeed extends $(f\colon x \to y) \in \mathscr{D}_R([1])$ to a Barratt–Puppe sequence. Moreover, in each of these four steps the above describes precisely the corresponding essential image, and we hence conclude that

$$F_{[1]} = (i_4)_* \circ (i_3)_! \circ (i_2)_! \circ (i_1)_*$$

is the desired inverse equivalence of $i^*\colon \mathscr{D}_R(M_2)^{\mathrm{ex}} \to \mathscr{D}_R([1])$. □

Let us discuss these Barratt–Puppe sequences to some extent.

REMARK 2.46. Let R be a ring and let $X \in \mathscr{D}_R(M_2)^{\mathrm{ex}}$ be a Barratt–Puppe sequence looking like Figure 3. In that Barratt–Puppe sequence any two adjacent squares determine a cofiber sequence and hence give rise to a triangle in $\mathscr{D}_R(*)$ (Construction 2.41). In particular, the cofiber sequence determined by the squares $\boxed{1}, \boxed{2}$ induces a triangle associated to $f\colon x \to y$. Similarly, the squares $\boxed{2}, \boxed{3}$ determine a triangle associated to the morphism $g\colon y \to z$, and this is a rotated triangle of the previous one. As an upshot, the Barratt–Puppe sequence in Figure 3 encodes all iterated rotations of triangles associated to $f\colon x \to y$.

The defining exactness properties of Barratt–Puppe sequences (vanishing on boundary stripes and all squares are bicartesian) are invariant under the obvious symmetries of the shape M_2. Hence, as alluded to in the introduction, the rotation at the level of distinguished triangles is a shadow of certain symmetries on representations of $\vec{A}_2 = (1 \to 2) \cong [1]$ with values in chain complexes. As a

$$
\begin{array}{ccccccccccccccc}
\ddots & & \ddots & & \ddots & & \ddots & & \ddots & & & & & & \\
(-1,-2) & \to & (-1,-1) & \to & (-1,0) & \to & (-1,1) & \to & (-1,2) & & & & & & \\
 & & \downarrow & & \downarrow & & \downarrow & & \downarrow & & & & & & \\
 & & (0,-1) & \longrightarrow & (0,0) & \longrightarrow & (0,1) & \longrightarrow & (0,2) & \to & (0,3) & & & & \\
 & & & & \downarrow & & \downarrow & & \downarrow & & \downarrow & & & & \\
 & & & & (1,0) & \longrightarrow & (1,1) & \longrightarrow & (1,2) & \to & (1,3) & \to & (1,4) & & \\
 & & & & & & \downarrow & & \downarrow & & \downarrow & & \downarrow & & \\
 & & & & & & (2,1) & \longrightarrow & (2,2) & \to & (2,3) & \to & (2,4) & \to & (2,5) \\
 & & & & & & \ddots & & \ddots & & \ddots & & \ddots & & \ddots
\end{array}
$$

FIGURE 4. The full subposet $M_3 \subseteq \mathbb{Z} \times \mathbb{Z}$

variant of this, let us consider the square $\boxed{1}$ in Figure 3 as a "triangular fundamental domain" of the Barratt–Puppe sequence. The flip symmetry of the shape M_2 identifies the squares $\boxed{1}$ and $\boxed{4}$. At the level of diagrams this flip symmetry corresponds to the suspension, and Figure 3 encodes the fundamental domain and all its iterated (de)suspensions. Of course, the suspension is simply the cube of the cofiber ([**GPS14b**, Lem. 5.13]), and in that generality this can be interpreted as an example of the abstract fractionally Calabi–Yau property of $\vec{A}_n$-quivers ([**Kel08**],[**GŠ16a**, Cor. 5.20]).

REMARK 2.47. The strategy behind the constructions discussed so far is fairly typical and it relies on two ingredients. First, for every fully faithful functor $u\colon A \to B$ the derived Kan extension functors $u_!, u_*\colon \mathscr{D}_R(A) \to \mathscr{D}_R(B)$ are again fully faithful. In particular, they induce equivalences onto their essential images. Second, for suitable classes of fully faithful functors, the functors $u_!, u_*$ simply add zero chain complexes at the new objects. We refer the reader to [**Gro13**] for a more systematic discussion of this.

2.4. A_3-quivers and refined octahedral diagrams. In this subsection, we conclude the unorthodox proof of Theorem 2.6 and there are only two main steps left. First, we repeat to some extent the discussion in §2.3 but starting with representations of $\vec{A}_3$ instead. Second, for suitable shapes the underlying diagram functors from Example 2.10 are full and essentially surjective, and this can be invoked to establish the existence of the Verdier triangulation.

The octahedral axiom of triangulated categories asks for a version of the third Noether isomorphism for cones in triangulated categories. Hence, it is essentially a statement about the situation of two composable morphisms, which is to say of a representation of the linearly oriented A_3-quiver

$$\vec{A}_3 \cong [2] = (0 < 1 < 2).$$

The following is a variant for $\vec{A}_3$ of the above discussion of Barratt–Puppe sequences.

NOTATION 2.48. Let $M_3 \subseteq \mathbb{Z}\times\mathbb{Z}$ be the full subposet given by all $(i,j) \in \mathbb{Z}\times\mathbb{Z}$ such that $i-1 \leq j \leq i+3$.

A part of this poset is displayed in Figure 4, and this poset is the shape of *refined octahedral diagrams* in the following precise sense.

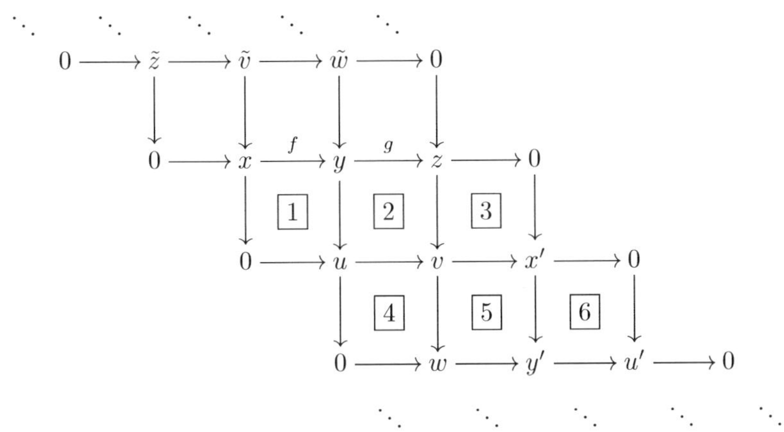

FIGURE 5. The refined octahedral diagram of $(x \xrightarrow{f} y \xrightarrow{g} z) \in \mathscr{D}_R([2])$

DEFINITION 2.49. Let R be a ring. A diagram $X \in \mathscr{D}_R(M_3)$ is a **refined octahedral diagram** in $\mathscr{D}_R$ if the diagram X vanishes on the two boundary stripes and all squares in X are bicartesian. We denote by $\mathscr{D}_R(M_3)^{\mathrm{ex}} \subseteq \mathscr{D}_R(M_3)$ the full subcategory spanned by all refined octahedral diagrams.

Similar to the case of Barratt–Puppe sequences, a refined octahedral diagram is determined by restriction along various embeddings of $\vec{A}_3 \cong [2]$ to M_3. To be specific, we consider the standard embedding

$$i\colon [2] \to M_3\colon k \mapsto (0, k). \tag{2.50}$$

THEOREM 2.51. *For every ring R restriction along* (2.50) *induces an equivalence of categories*

$$i^*\colon \mathscr{D}_R(M_3)^{\mathrm{ex}} \xrightarrow{\sim} \mathscr{D}_R([2]).$$

PROOF. It is straightforward to adapt the strategy of the proof of Theorem 2.45 in order to also cover this situation. For a detailed proof we refer to [**GŠ16a**, §4]. For later reference we denote the inverse equivalence by $F_{[2]}\colon \mathscr{D}_R([2]) \xrightarrow{\sim} \mathscr{D}_R(M_3)^{\mathrm{ex}}$. □

The connection to the octahedral axiom is as follows, and this justifyies the above terminology.

REMARK 2.52. Let R be a ring and let $X \in \mathscr{D}_R(M_3)^{\mathrm{ex}}$ be a refined octahedral diagram looking like Figure 5. This diagram is determined by the restriction $i^*(X) \in \mathscr{D}_R([2])$, and this restriction in turn gives rise to the three morphisms

$$f\colon x \to y, \quad g\colon y \to z, \quad \text{and} \quad g \circ f\colon x \to z.$$

Given three such morphisms in $\mathscr{D}_R(*)$, let us recall that the octahedral axiom relates triangles of f, g, and $g \circ f$ by an additional triangle incorporating the various cones. These four triangles can be deduced from Figure 5 as follows.

(i) The square $\boxed{1}$ is a cofiber square, and there is hence a canonical isomorphism $Cf \xrightarrow{\sim} u$. Similarly, if we consider the square $\boxed{1}$ and the horizontal composition $\boxed{2}+\boxed{3}$, then we obtain by Proposition 2.40 a cofiber sequence

$$\begin{array}{ccccc} x & \xrightarrow{f} & y & \longrightarrow & 0 \\ \downarrow & \square & \downarrow{\scriptstyle g} & \square & \downarrow \\ 0 & \longrightarrow & u & \longrightarrow & x'. \end{array}$$

This cofiber sequence of f induces by Construction 2.41 the first of the four triangles.

(ii) Similarly, the vertical composition $\boxed{2}+\boxed{4}$ is by Proposition 2.40 a cofiber square and we obtain the identification $Cg \xrightarrow{\sim} w$. Jointly with the vertical composition $\boxed{3}+\boxed{5}$ this yields the desired cofiber sequence of g.

(iii) As of the cofiber sequence of $g\circ f$, we note that the horizontal composition $\boxed{1}+\boxed{2}$ is a cofiber square for $g \circ f$, leading to $C(g \circ f) \xrightarrow{\sim} v$. In combination with the square $\boxed{3}$ we obtain the cofiber sequence of $g \circ f$.

(iv) Finally, the square $\boxed{4}$ and the horizontal composition $\boxed{5}+\boxed{6}$ define a cofiber sequence

$$\begin{array}{ccccc} u & \longrightarrow & v & \longrightarrow & 0 \\ \downarrow & \square & \downarrow & \square & \downarrow \\ 0 & \longrightarrow & w & \longrightarrow & u'. \end{array}$$

Invoking the above identifications of u, v, and w as cones, this cofiber sequence induces the fourth triangle

$$Cf \to C(g \circ f) \to Cg \to \Sigma Cf.$$

One checks directly that all the necessary compatibilities of the octahedral axiom are satisfied (see [**Gro13**, §4]).

For completeness, let us include a few comments on a variant of Theorem 2.45 and Theorem 2.51 for $\vec{A}_1 \cong *$.

REMARK 2.53. Let R be a ring. We denote by $M_1 \subseteq \mathbb{Z} \times \mathbb{Z}$ the full subposet given by all $(i,j) \in \mathbb{Z} \times \mathbb{Z}$ such that $i-1 \leq j \leq i+1$. A part of this poset is displayed in Figure 6. Moreover, let us write

$$\mathscr{D}_R(M_1)^{\mathrm{ex}} \subseteq \mathscr{D}_R(M_1)$$

for the full subcategory spanned by all diagrams which vanish on the boundary stripes and which make all squares bicartesian. Evaluation at $(0,0) \in M_1$ induces an equivalence of categories

$$(0,0)^* \colon \mathscr{D}_R(M_1)^{\mathrm{ex}} \xrightarrow{\sim} D(R), \tag{2.54}$$

as one checks by copying the proof of Theorem 2.45 (see [**GŠ16a**, §4]). A typical diagram $X \in \mathscr{D}_R(M_1)^{\mathrm{ex}}$ is displayed in Figure 7, and therein all squares are suspension squares (Remark 2.34).

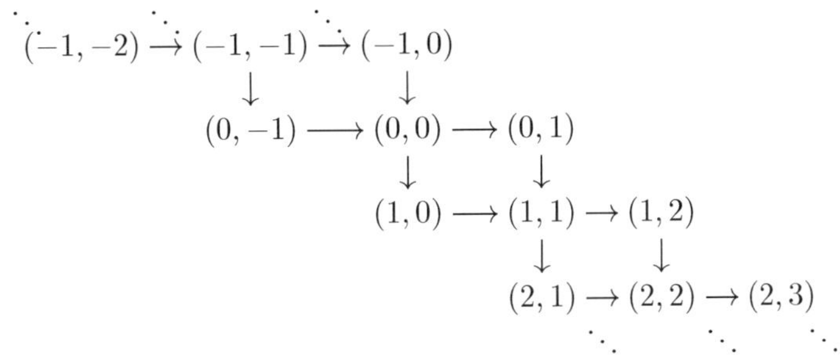

FIGURE 6. The full subposet $M_1 \subseteq \mathbb{Z} \times \mathbb{Z}$

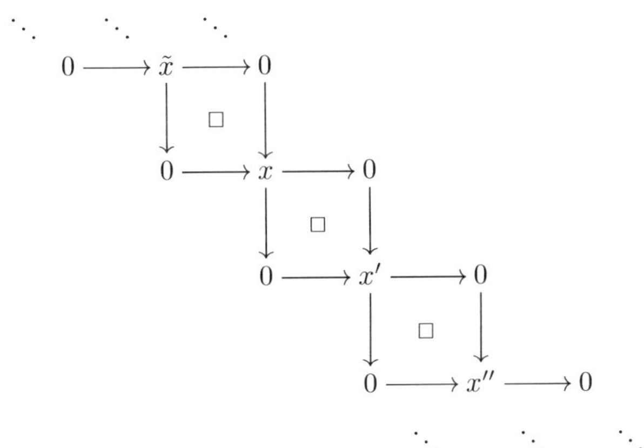

FIGURE 7. A diagram $X \in \mathscr{D}_R(M_1)^{\mathrm{ex}}$

The shape M_1 has two obvious generating symmetries, namely translation along the diagonal direction and the swap symmetry which keeps the diagonal fixed while it interchanges the objects $(n, n+1)$ and $(n+1, n)$ for all n. These symmetries of M_1 induce symmetries on $\mathscr{D}_R(M_1)^{\mathrm{ex}}$ since the defining exactness properties are clearly invariant under these symmetries. Interpreted at the level of $D(R)$ (by a conjugation with (2.54)), we obtain induced self-equivalences on $D(R)$. The first symmetry gives rise to the suspension equivalence $\Sigma\colon D(R) \xrightarrow{\sim} D(R)$, and the second symmetry can be identified with $-\mathrm{id}\colon D(R) \xrightarrow{\sim} D(R)$. One can make precise in which sense this latter symmetry is responsible for the sign in the rotation axiom and we will come back to this in §3.2 (see Remark 3.34 and Remark 3.36).

In order to turn towards the alternative proof of Theorem 2.6, there is only one ingredient missing. Note that both in Theorem 2.45 and in Theorem 2.51 we gave alternative presentations of the derived categories of diagram categories

$$\mathscr{D}_R([1]) = D(\mathrm{Mod}(R)^{[1]}) \qquad \text{and} \qquad \mathscr{D}_R([2]) = D(\mathrm{Mod}(R)^{[2]}).$$

In contrast to this, in the classical Verdier triangulation we consider morphisms or pairs of composable morphisms in $D(R)$, which is to say objects in the categories

$$D(R)^{[1]} \qquad \text{and} \qquad D(R)^{[2]}.$$

In order to connect these two perspectives, the following result proves crucial.

PROPOSITION 2.55. *For every ring R and every $n \geq 0$ the underlying diagram functor*

$$\mathrm{dia}_{[n]} \colon \mathscr{D}_R([n])) = D(\mathrm{Mod}(R)^{[n]}) \to D(R)^{[n]}$$

is essentially surjective and full.

PROOF. We invite the reader to check this directly. Alternatively, see for instance [**Cis10**, Prop. 2.15] or [**RB06**, Thm. 10.3.3] for much more general statements. □

REMARK 2.56. Let R be a ring.

(i) The functor $\mathrm{dia}_{[0]}$ is, of course, an equivalence of categories, but already in the case of $n = 1$ the functor $\mathrm{dia}_{[1]}$ is not faithful. For instance, in the case of a field $R = k$, one can invoke Auslander–Reiten theory to make precise that the extension group $\mathrm{Ext}^1_{k\vec{A}_2}((k \to 0), (0 \to k)) \cong k$ is precisely the reason for this failure (see also [**Gro18**, §5]).
(ii) One can show, that, more generally for every free category F the underlying diagram functors dia_F is essentially surjective and full ([**Cis10**, Prop. 2.15] or [**RB06**, Thm. 10.3.3]). (Let us recall that a free category is the path category of an oriented graph.)
(iii) In contrast to this, already in the case of a field $R = k$ and the non-free category $\square$, the underlying diagram functor $\mathrm{dia}_\square$ does not preserve isomorphism types (see [**BG18**, Example 3.17]) and is hence not full and essentially surjective.

We now sketch the slightly unorthodox proof of Theorem 2.6.

PROOF. *(of Theorem 2.6)* We carry out the sketch proof in the case of $\mathrm{Mod}(R)$, but the arguments also apply to arbitrary abelian categories (see Remark 2.59). Moreover, we focus on those aspects which allow for nice "representation theoretic explanations". Let us take for granted that the derived category $D(R)$ is additive (see also Remark 3.36). The suspension functor $\Sigma \colon D(R) \to D(R)$ is simply the shift functor (Remark 2.34), and it clearly is an equivalence of categories.

We now define the class of distinguished triangles in $D(R)$. To this end, for every $(f \colon x \to y) \in \mathscr{D}_R([1])$ we can consider by Proposition 2.39 the corresponding cofiber sequence $X \in \mathscr{D}_R([1] \times [2])^{\mathsf{cof}}$ as in (2.37). As detailed in Construction 2.41 we obtain an associated triangle

$$x \xrightarrow{f} y \to z \to \Sigma x. \tag{2.57}$$

This is defined to be the *standard triangle* of $f \in \mathscr{D}_R([1])$, and a triangle in $D(R)$ is distinguished when it is isomorphic to a standard triangle. It is worthwhile to summarize the construction by the following diagram

$$\begin{array}{ccc} \mathscr{D}_R([1]) & \xrightarrow[\sim]{F_{[1]}} & \mathscr{D}_R(M_2)^{\mathrm{ex}} \\ {\scriptstyle \mathrm{dia}_{[1]}}\downarrow & \searrow^{\mathrm{tria}} & \downarrow \\ D(R)^{[1]} & & D(R)^{[3]}. \end{array}$$

The horizontal functor is the equivalence constructed in the proof of Theorem 2.45. The unlabeled vertical functor takes a Barratt–Puppe sequence, restricts it to the appropriate cofiber sequence and then passes to the underlying distinguished triangle. Note that the vertical functors and hence also the diagonal functor amount to a loss of information since in all three cases underlying diagram functors are involved.

In order to show that every morphism $f\colon x \to y$ in $D(R)$ extends to a distinguished triangle we invoke that $\mathrm{dia}_{[1]}$ is essentially surjective (Proposition 2.55). Thus, we can find $Y \in \mathscr{D}_R([1])$ and an isomorphism $\alpha\colon \mathrm{dia}_{[1]}(Y) \xrightarrow{\sim} f$, and a combination of the standard triangle $\mathrm{tria}(Y)$ of Y and the isomorphism α yields the intended distinguished triangle. Similarly, the weak functoriality of distinguished triangles is a consequence of $\mathrm{dia}_{[1]}$ being essentially surjective and full. Up to the sign issue, the rotation axiom is essentially a consequence of the symmetries of Barratt–Puppe sequences (see also Remark 2.46). The sign issue will be taken up again §3 (see the discussion of additivity in Remark 3.36).

Instead, we content ourselves by taking care of the octahedral axiom. Let B be the shape of the diagrams showing up in the octahedral axiom. For every

$$X = (x \to y \to z) \in \mathscr{D}_R([2])$$

we can consider the corresponding refined octahedral diagram $F_{[2]}(X) \in \mathscr{D}_R(M_3)^{\mathrm{ex}}$ (Theorem 2.51). The above construction of standard triangles and Remark 2.52 imply that from $F_{[2]}(X)$ we can construct the intended octahedral diagram

$$\mathrm{octa}(X)\colon B \to D(R).$$

Similarly to the previous case, the situation can be summarized by the diagram

$$\begin{array}{ccc} \mathscr{D}_R([2]) & \xrightarrow[\sim]{F_{[2]}} & \mathscr{D}_R(M_3)^{\mathrm{ex}} \\ {\scriptstyle \mathrm{dia}_{[2]}}\downarrow & \searrow^{\mathrm{octa}} & \downarrow \\ D(R)^{[2]} & & D(R)^{B}. \end{array} \tag{2.58}$$

Here, $F_{[2]}$ is the equivalence from Theorem 2.51 and the remaining functors amount to a loss of information. Since also the functor $\mathrm{dia}_{[2]}$ is essentially surjective (Proposition 2.55), this establishes the octahedral axiom. □

Let us collect some of the benefits which we obtain from this alternative construction of the classical Verdier triangulation on $D(R)$.

REMARK 2.59. Let R be a ring, let $(x \xrightarrow{f} y \xrightarrow{g} z) \in \mathscr{D}_R([2])$, and let $F_{[2]}(X)$ be the associated refined actahedral diagram.

(i) The shape M_3 (Figure 4) has two obvious generating symmetries (translation and flip symmetry). By definition refined octahedral diagrams are preserved by restriction along these symmetries. This implies, for instance, that X also encodes the refined octahedral diagram associated to the canonical representation of $\vec{A}_3$ looking like

$$(Cf \to C(g \circ f) \to \Sigma x).$$

And this leads to a relation between the respective octahedral diagrams.

(ii) The refined octahedral diagram $F_{[2]}(X)$ encodes additional distinguished triangles teaching us something about the cone $C(g \circ f)$ of the composition. In fact, let us recall that every bicartesian square

$$\begin{array}{ccc} x & \longrightarrow & y \\ \downarrow & \square & \downarrow \\ z & \longrightarrow & w \end{array}$$

in $\mathscr{D}_R$ gives rise to a cofiber square

$$\begin{array}{ccc} x & \longrightarrow & y \oplus z \\ \downarrow & \square & \downarrow \\ 0 & \longrightarrow & w, \end{array}$$

and hence to a distinguished triangle

$$x \to y \oplus z \to w \to \Sigma x.$$

See, for instance, [**GPS14b**] for a construction of these **Mayer–Vietoris sequences.** Let us specialize this to bicartesian squares occurring in $F_{[2]}(X)$ (Figure 5). In this situation, the square $\boxed{2}$ gives rise to an additional distinguished triangle

$$y \to z \oplus Cf \to C(g \circ f) \to \Sigma y.$$

Similarly, the Mayer–Vietoris sequence associated to the square $\boxed{5}$ looks like

$$C(g \circ f) \to \Sigma x \oplus Cg \to \Sigma y \to \Sigma C(g \circ f).$$

(iii) The proof of the octahedral axiom (as summarized by (2.58)) used only that $\mathrm{dia}_{[2]}$ is essentially surjective. Since this functor is also full, we conclude that octahedral diagrams depend weakly functorially on pairs of composable morphisms in derived categories.

(iv) For simplicity, we presented the proof in the case of $\mathrm{Mod}(R)$ only. However, there are variants of the main ingredients (such as Theorem 2.45 and Theorem 2.51) for arbitrary abelian categories $\mathcal{A}$. In fact, by (co)finality arguments the constructions only rely on the existence of finite limits and colimits, and these exist in all abelian categories. Also the above remarks generalize to this more general situation.

This sketch proof exhibits some of the axioms of triangulated categories as certain shadows of symmetries of Barratt–Puppe sequences and of refined octahedral diagrams. Put differently, this strategy makes precise the connection of these axioms to abstract representations of the quivers $\vec{A}_2$ and $\vec{A}_3$. The quivers $\vec{A}_n$ for $n \geq 4$ play a similar role for higher triangulations and we will study their abstract representations in more detail in §4.2 (for the connection to higher triangulations see Remark 4.32).

3. A crash course on derivators

In this section we give a short introduction to derivators, which offer one of the many approaches to higher category theory, abstract homotopy theory, or homotopical algebra. The precise choice of this model is not too relevant for our later purposes (see Disclaimer 3.2), and here we essentially only present what is necessary for our later discussion of strong stable equivalences of quivers (or small categories). In fact, the main goal of this section is to provide enough background in order to turn the following pseudo-definition into an actual definition.

PSEUDO-DEFINITION 3.1. Two quivers Q and Q' are **strongly stably equivalent** if for every stable homotopy theory $\mathscr{D}$there is an equivalence between the homotopy theory $\mathscr{D}^Q$ of representations of shape Q and the homotopy theory $\mathscr{D}^{Q'}$ of representations of shape Q',

$$\Phi_{\mathscr{D}}\colon \mathscr{D}^Q \xrightarrow{\sim} \mathscr{D}^{Q'},$$

which is pseudo-natural with respect to exact morphisms $F\colon \mathscr{D} \to \mathscr{E}$ of stable homotopy theories.

This goal pretty much dictates what we are supposed to do. In particular, as a first step we have to make precise what we mean by an "abstract homotopy theory". Here we choose to work with *derivators* but see the following disclaimer.

DISCLAIMER 3.2. By now there are various approaches to formalize an "abstract homotopy theory". All of these notions enhance certain defects of the more classical homotopy categories (which arise as 1-categorical localizations).

(i) One of the most classical approaches is given by Quillen model categories [**Qui67**, **Hov99**, **Hir03**]. The homotopy theory is in this case encoded by a class of weak equivalences accompanied by classes of fibrations and cofibrations which are subject to certain axioms.
(ii) There is the zoo of different models for a theory of $(\infty, 1)$-categories. One way to think of an individual $(\infty, 1)$-category is as the result of a "higher categorical localization" of a category with weak equivalences. A very prominent model is given by ∞-categories, and this model was developed to an impressive extent most notably in [**Joy08**] and [**Lur09**, **Lur14**]. With all these techniques at hand, this is currently the most flexible notion. Among the alternative approaches to such a theory are simplicial categories [**Ber07**], Segal categories [**HS01**], and complete Segal spaces [**Rez01**]. Additional references and more details can for example be found in [**Ber10**, **Sim12**, **Cam13**, **Gro10**]. Moreover, a model-independent approach to higher category theory can be found in [**RV**] and the sequels.
(iii) Derivators [**Gro**, **Hel88**] axiomatize key properties of the calculus of homotopy limits and homotopy Kan extensions. In this language, these are characterized by ordinary universal properties making them accessible to elementary techniques.

The notion of a derivator is a compromise: a derivator encodes more information than a mere homotopy category but somewhat less information than a full homotopy theory. Some of the defects of classical homotopy categories and triangulated categories are addressed successfully by the theory of derivators – and this is done in a reasonably elementary way by means of (at most 2-) categorical techniques only. For instance, derivators allow for a construction of exponentials and for a formal study of stability, thereby offering a first framework for this project.

At the same time there are obvious limitations to the theory of derivators (such as the absence of gluing of derivators or the incoherence of morphisms and natural transformations), and for various purposes it is more convenient to use the more flexible theory of ∞-categories. Going even beyond this, for instance for the calculus of universal tilting modules (as in §4.5), it would be desirable to have a more well-developed *theory* of $(\infty, 2)$-categories and monoidal such.

In any case, the author believes in diversity of technology. Most of the results of this project were established in the language of derivators and this is hence also the model we choose in this account. However, it is very likely that sooner or later ∞-categories will enter the picture more prominently (see already [**DJW18**, **DJW19**]).

The structure of this section is dictated by the above-mentioned main goal. In §3.1 we introduce derivators and collect some examples. In §3.2 we turn to stable derivators and discuss their relation to triangulated categories. In §3.3 we construct exponentials for derivators which encode the calculus of parametrized limits and Kan extensions. Finally, in §3.4 we discuss morphisms of derivators, adjunctions, equivalences, and related notions.

3.1. Derivators. In this section we discuss the philosophy of derivators which offer one of the many approaches to abstract homotopy theory. Derivators were introduced independently by Grothendieck [**Gro**] and Heller [**Hel88**], and closely related notions were also considered by Franke [**Fra96**] and Keller [**Kel91**] (see Remark 3.16 for some additional references). The main focus of derivators is on diagram categories and the related calculus of (derived or homotopy) limits, colimits, and Kan extensions.

NOTATION 3.3. We denote by $\mathcal{C}at$ the 2-category of small categories, functors, and natural transformation, and similarly, by $\mathcal{CAT}$ the 2-category of not necessarily small categories.

In this section we use very basic terminology related to 2-categories. The discussions in [**ML98**, §XII] and [**Bor94a**, §7] suffice for our purposes (but see also [**KS74**, **Lac10**]).

DEFINITION 3.4. A **prederivator** is a 2-functor $\mathscr{D}\colon \mathcal{C}at^{\mathrm{op}} \to \mathcal{CAT}$.

Thus, a prederivator $\mathscr{D}$ associates to every small category $A \in \mathcal{C}at$ a category $\mathscr{D}(A)$, the category of **coherent diagrams of shape** A **in** $\mathscr{D}$. For the trivial shape $*$, we refer to $\mathscr{D}(*)$ as the **underlying category** of $\mathscr{D}$. Moreover, every functor $u\colon A \to B$ induces a **restriction functor** or **precomposition functor** $u^*\colon \mathscr{D}(B) \to \mathscr{D}(A)$. Similarly, every natural transformation $\alpha\colon u \to v$ of functors $u, v\colon A \to B$ induces a natural transformation $\alpha^*\colon u^* \to v^*$, but in these notes we on purpose decide to not be too explicit about these 2-dimensional aspects.

WARNING 3.5. The above is only terminology. Given a prederivator $\mathscr{D}$ and $A \in \mathcal{C}at$, an object $X \in \mathscr{D}(A)$ is just an abstract object, and not a diagram in any precise sense. However, as we will see, every such X gives rise to a usual diagram $A \to \mathscr{D}(*)$, and it is crucial to distinguish between these two.

EXAMPLE 3.6. Let $\mathcal{C}$ be a category and $A \in \mathcal{C}at$. We denote by $\mathcal{C}^A$ the category of functors $X\colon A \to \mathcal{C}$ and natural transformation between them. The formation of these diagram categories defines a 2-functor

$$y_{\mathcal{C}}\colon \mathcal{C}at^{\mathrm{op}} \to \mathcal{CAT}\colon A \mapsto \mathcal{C}^A,$$

the **prederivator represented by** $\mathcal{C}$. The underlying category is $\mathcal{C}$ itself.

More interesting examples arise from this one by *localization*. The following example collects two key examples of a construction which applies to arbitrary relative categories (categories with a class of weak equivalences), and this relies on the fact that localizations are 2-localizations.

EXAMPLES 3.7. (i) Let $\mathcal{A}$ be an abelian category. The prederivator of $\mathcal{A}$ is the 2-functor

$$\mathscr{D}_{\mathcal{A}}\colon \mathcal{C}at^{\mathrm{op}} \to \mathcal{CAT}\colon B \mapsto D(\mathcal{A}^B).$$

The underlying category of $\mathscr{D}_{\mathcal{A}}$ is the derived category $D(\mathcal{A})$.

(ii) Let $\mathcal{M}$ be a Quillen model category with class of weak equivalences W. For every $B \in \mathcal{C}at$ we denote by W^B those natural transformations $\alpha\colon X \to Y$ between diagrams $X, Y\colon B \to \mathcal{M}$ such that the components $\alpha_b\colon X_b \to Y_b$ are weak equivalences for all $b \in B$, i.e., W^B is the class of levelwise weak equivalences. The homotopy prederivator of $\mathcal{M}$ is the 2-functor

$$\mathscr{H}o_{\mathcal{M}}\colon \mathcal{C}at^{\mathrm{op}} \to \mathcal{CAT}\colon B \mapsto \mathrm{Ho}(\mathcal{M}^B) = \mathcal{M}^B[(W^B)^{-1}].$$

The underlying category of $\mathscr{H}o_{\mathcal{M}}$ is the homotopy category $\mathrm{Ho}(\mathcal{M})$.

CONSTRUCTION 3.8. Let $\mathscr{D}$be a prederivator, $B \in \mathcal{C}at$, and $b \in B$. We can consider the functor $b\colon * \to B$ which sends the unique object to $b \in B$, and the corresponding restriction functor

$$b^*\colon \mathscr{D}(B) \to \mathscr{D}(*)$$

is referred to as an **evaluation functor** (at b). For every $f\colon X \to Y$ in $\mathscr{D}(B)$ we write $f_b\colon X_b \to Y_b$ for its image under the evaluation functor. As an exercise in 2-functoriality, we invite the reader to check that every $X \in \mathscr{D}(B)$ gives rise to an underlying diagram

$$\mathrm{dia}_B(X)\colon B \to \mathscr{D}(*)\colon b \mapsto X_b,$$

and that this defines an **underlying diagram functor**

$$\mathrm{dia}_B\colon \mathscr{D}(B) \to \mathscr{D}(*)^B.$$

WARNING 3.9. Let $\mathscr{D}$be a derivator and $B \in \mathcal{C}at$. In general, the underlying diagram functor dia_B fails to be an equivalence. In fact, in many cases the categories $\mathscr{D}(B)$ and $\mathscr{D}(*)^B$ are not equivalent. We refer to the category $\mathscr{D}(*)^B$ as the category of **incoherent diagrams of shape** B in $\mathscr{D}$.

However, the underlying diagram functors are useful in order to visualize abstract coherent diagrams, i.e., we often draw $\mathrm{dia}_B(X)$ and say that $X \in \mathscr{D}(B)$ looks like $\mathrm{dia}_B(X)$. To reissue this warning, it is important to distinguish between these two.

In specific situations the underlying diagram functors take the following form.

EXAMPLE 3.10. Let $B \in \mathcal{C}at$.

(i) In the case of a represented prederivator $y_{\mathcal{C}}$, the underlying diagram dia_B is the isomorphism of categories

$$\mathrm{dia}_B \colon \mathcal{C}^B \xrightarrow{\sim} (\mathcal{C}^*)^B.$$

(ii) For every abelian category $\mathcal{A}$ and the corresponding prederivator $\mathscr{D}_{\mathcal{A}}$, the underlying diagram functors

$$\mathrm{dia}_B \colon D(\mathcal{A}^B) \to D(\mathcal{A})^B$$

were already considered in §2 (see, in particular, Remark 2.9, Example 2.10, and their discussion). From that discussion we know that these functors often fail to be equivalences.

(iii) Similarly, for every Quillen model category $\mathcal{M}$, the corresponding underlying diagram functor of $\mathscr{H}o_{\mathcal{M}}$ takes the form

$$\mathrm{dia}_B \colon \mathrm{Ho}(\mathcal{M}^B) \to \mathrm{Ho}(\mathcal{M})^B.$$

In general, homotopy categories of diagram categories and diagram categories of homotopy categories are not equivalent, and these underlying diagram functors hence again fail to be equivalences.

The point is that coherent diagrams (such as objects in $D(\mathcal{A}^B)$ and $\mathscr{H}o(\mathcal{M}^B)$) carry more information than incoherent ones. This information is crucial when one wants to calculate their derived or homotopy (co)limits. In fact, as we have seen in §2, morphisms in derived categories do not suffice to canonically determine their derived (co)kernels, and similarly for derived pushouts. The key properties listed in Theorem 2.19 are now turned into an abstract definition, thereby capturing a formal calculus of abstract limits and colimits. Axiom (Der4) will be made more precise in the discussion that follows the definition.

DEFINITION 3.11. A prederivator $\mathscr{D} \colon \mathcal{C}at^{\mathrm{op}} \to \mathcal{CAT}$ is a **derivator** if it enjoys the following properties.

(Der1) The canonical inclusion functors $B_j \to \coprod_{i \in I} B_i, j \in I$, induce an equivalence of categories

$$\mathscr{D}(\coprod_{i\in I} B_i) \xrightarrow{\sim} \prod_{i\in I} \mathscr{D}(B_i).$$

(Der2) A morphism $f \colon X \to Y$ in $\mathscr{D}(B)$ is an isomorphism if and only if the morphisms $f_b \colon X_b \to Y_b, b \in B$, are isomorphisms in $\mathscr{D}(*)$.

(Der3) For every functor $u \colon A \to B$, the restriction functor $u^* \colon \mathscr{D}(B) \to \mathscr{D}(A)$ has a left adjoint $u_!$ and a right adjoint u_*,

$$(u_!, u^*) \colon \mathscr{D}(A) \rightleftarrows \mathscr{D}(B), \qquad (u^*, u_*) \colon \mathscr{D}(B) \rightleftarrows \mathscr{D}(A). \tag{3.12}$$

(Der4) For every functor $u \colon A \to B$, the functors $u_!, u_* \colon \mathscr{D}(A) \to \mathscr{D}(B)$ can be calculated pointwise.

REMARK 3.13. Some discussion of this definition is in order.

(i) Axiom (Der1) makes precise that that coherent diagrams of shape given by a disjoint union are determined by the canonical restrictions to the respective summands. Axiom (Der2) is motivated by the following two examples.

(a) A natural transformation $\alpha\colon X \to Y$ in a diagram category $\mathcal{C}^B$ is an isomorphism if and only if all components $\alpha_b\colon X_b \to Y_b$ are isomorphisms.

(b) For every abelian category $\mathcal{A}$ and $B \in \mathcal{C}at$ there is an obvious isomorphism $\mathrm{Ch}(\mathcal{A}^B) \xrightarrow{\simeq} \mathrm{Ch}(\mathcal{A})^B$. Under this isomorphism the quasi-isomorphisms $W_{\mathcal{A}^B}$ in $\mathcal{A}^B$ correspond precisely to the levelwise quasi-isomorphisms $W_{\mathcal{A}}^B$.

(ii) Axioms (Der3) and (Der4) jointly encode, first, a "*homotopical* completeness and cocompleteness property", thereby guaranteeing that "homotopical versions" of limits, colimits, and Kan extensions exist, and, second, formulas which allow us to calculate the (homotopy) Kan extensions. Let us expand a bit on this.

As a special case, for every $A \in \mathcal{C}at$, there is the unique functor $\pi_A\colon A \to *$. Correspondingly, by (Der3) the functor $\pi_A^*\colon \mathscr{D}(*) \to \mathscr{D}(A)$ has a left adjoint $(\pi_A)_!$ and a right adjoint $(\pi_A)_*$, which we respectively also denote by

$$\operatorname{colim}_A = (\pi_A)_!\colon \mathscr{D}(A) \to \mathscr{D}(*) \quad \text{and} \quad \lim_A = (\pi_A)_*\colon \mathscr{D}(A) \to \mathscr{D}(*).$$

These abstract **colimit** and **limit** functors generalize ordinary categorical (co)limits, derived (co)limits, and homotopy (co)limits (see Examples 3.15).

More generally, the adjoints $u_!$ and u_* are referred to as **left** and **right Kan extensions**, respectively. To be able to work with these adjoints, it is crucial that they can be calculated as in the classical case (Construction 2.18). To formulate this abstractly let us assume that $\mathscr{D}$is a prederivator which satisfies (Der3). For every functor $u\colon A \to B$, for every $b \in B$, and every coherent diagram $X \in \mathscr{D}(A)$ there are canonical maps

$$\operatorname{colim}_{(u/b)} p^*(X) \to u_!(X)_b \quad \text{and} \quad u_*(X)_b \to \lim_{(b/u)} q^*(X) \tag{3.14}$$

(Here, p and q are the projection functors associated to slice categories as in Construction 2.18.) Axiom (Der4) asks that these canonical maps are isomorphisms, and in this precise sense Kan extensions in derivators are pointwise.

(iii) The construction of the canonical maps in (3.14) is an instance of the calculus of canonical mates [**KS74**]. While working with derivators one often runs into the situation that outputs of certain universal constructions "obviously are isomorphic". In many cases, the formalism of mates and related notion of *homotopy exact squares* allows one to actually conclude this by, first, providing canonical maps between such gadgets and, second, guaranteeing that these maps are isomorphisms [**Gro13, Mal12**]. For a more detailed discussion of the formalism behind it we refer to [**Gro18**, §8 and §10]. In many cases, establishing a rule to manipulate Kan extensions amounts to showing that a certain square is homotopy exact (see [**Gro83, Gro**] and [**Cis04, Cis06, Mal05b**] for many more advanced examples).

(iv) All axioms of a derivator ask for certain *properties* of the underlying prederivators, which is the only actual *structure*. In contrast to this, in other approaches such as triangulated categories some *non-canonical structure* is put on an underlying category.

EXAMPLES 3.15. Let us take up again the above examples of prederivators.

(i) Let $\mathcal{C}$ be an ordinary category. The category $\mathcal{C}$ is complete and cocomplete if and only if $y_{\mathcal{C}}\colon B \mapsto \mathcal{C}^B$ is a derivator, the **derivator represented by $\mathcal{C}$**. In this case the abstract (co)limits are the usual ones from ordinary category theory, and the more general adjoints $u_!, u_*$ are the classical Kan extensions. For an introduction to these Kan extensions we refer to [**Gro18**, §6] and for a detailed proof that $y_{\mathcal{C}}$ indeed is a derivator to [**Gro18**, §9.2].

(ii) For every Grothendieck abelian category $\mathcal{A}$, $\mathscr{D}_{\mathcal{A}}\colon B \mapsto D(\mathcal{A}^B)$ is a derivator, the **derivator of (chain complexes in) $\mathcal{A}$**. In this case abstract (co)limits are derived (co)limits, and similarly for Kan extensions. The fact that $\mathscr{D}_{\mathcal{A}}$ is a derivator follows from the next example.

As special cases, there are hence the derivator $\mathscr{D}_k$ of a field k and the derivator $\mathscr{D}_R$ of a ring R (see Theorem 2.19). In contrast to the derived category $D(k)$ of a field, the derivator $\mathscr{D}_k$ of a field already is quite interesting as it encodes, for instance, derived categories of path algebras, incidence algebras, and group algebras. As an additional class of examples, associated to every scheme X there is the derivator $\mathscr{D}_X$ of (chain complexes of) quasi-coherent $\mathcal{O}_X$-modules. In fact, by [**EE05**, §3] the category of quasi-coherent $\mathcal{O}_X$-modules is Grothendieck abelian for arbitrary schemes X.

(iii) For every Quillen model category $\mathcal{M}$, the prederivator $\mathscr{H}o_{\mathcal{M}}\colon B \mapsto \mathrm{Ho}(\mathcal{M}^B)$ is a derivator, the **homotopy derivator of $\mathcal{M}$**. In full generality this was established by Cisinski as [**Cis03**, Thm. 6.11], and related to this see [**CS02**]. For an even more general version we also refer to [**Cis10**, Thm. 2.21, Cor. 2.24, and Cor. 2.28]. For combinatorial Quillen model categories an alternative proof can be found in [**Gro13**, §1.3]. For homotopy derivators abstract (co)limits specialize to homotopy (co)limits and similarly for Kan extensions. Let us specialize this important class to some examples of central interest.

   (a) For every Grothendieck abelian category $\mathcal{A}$, the category $\mathrm{Ch}(\mathcal{A})$ admits a combinatorial model structure with quasi-isomorphisms as weak equivalences (see [**Bek00**, Prop. 3.13], [**Hov01**, §2], or [**CD09**, §§2-3]). Hence, the derivator $\mathscr{D}_{\mathcal{A}}$ of $\mathcal{A}$ can be described as homotopy derivator $\mathscr{H}o_{\mathrm{Ch}(\mathcal{A})}$.

   (b) As an universal example there is the **derivator of spaces**

$$\mathcal{S} = \mathscr{H}o_{\mathrm{Top}},$$

   which arises homotopy derivator of the classical Serre model structure [**Qui67**]. For alternative approaches to $\mathcal{S}$ we also refer to [**Tho80, Gro83, Mal05b, Cis06**]. In the case of $\mathcal{S}$ abstract limits are the classical homotopy limits of topological spaces. Systematic classical accounts are in [**BK72, BV73**] and special instances can, for example, be found in [**Mat76**]. Similarly, the **derivator of pointed spaces** is $\mathcal{S}_* = \mathscr{H}o_{\mathrm{Top}_*}$.

(c) An important role in this project is played by the derivator of spectra. This derivator is defined as

$$\mathcal{S}p = \mathscr{H}o_{\mathrm{Sp}},$$

the homotopy derivator with respect to the classical model structure on sequential spectra [**BF78**]. The underlying category of $\mathcal{S}p$ is *the* stable homotopy category $\mathcal{SHC}$ [**Boa64, Pup67, Vog70**]. Alternative (and monoidal) approaches to this derivator are based on [**HSS00, EKMM97, MMSS01**].

There is also a variant of homotopy derivators of complete and cocomplete ∞-categories [**RV17b, Len17**]. Morally, the same should be true for all other approaches to a theory of complete and cocomplete $(\infty, 1)$-categories, This moral is, for example, based on the model-independent approach to $(\infty, 1)$-category theory of Riehl–Verity [**RV, RV17a, RV17b**].

REMARK 3.16. A few more comments related to Definition 3.11 are in order.

(i) There is some flexibility with respect to the choice of allowable shapes in the definition of a derivator. In this paper we always allow for arbitrary small shapes, but in some situations more restrictive assumptions are useful. For instance, if one uses bounded derived categories of diagram categories of suitably finite shapes only, then by a result of Keller [**Kel07a**] for every exact category in the sense of Quillen [**Qui73**, §2] there is a corresponding derivator.

(ii) Following the convention of Anderson [**And79**], Heller [**Hel88**], and Franke [**Fra96**], the axioms in Definition 3.11 encode aspects of the calculus of *diagrams* (covariant functors) in a fixed abstract homotopy theory. Consequently, the domain of definition is $\mathcal{C}at^{\mathrm{op}}$. Grothendieck [**Gro**], Cisinski [**Cis03, Cis08**], Maltsiniotis [**Mal01b, Mal07**] consider *presheaves* (contravariant functors) instead, and consequently their domain of definition is $\mathcal{C}at^{\mathrm{coop}}$, which is obtained from $\mathcal{C}at$ by changing the orientation of functors and natural transformations. The resulting theories are equivalent.

(iii) Up to the fact that we give preference to diagrams and not presheaves, the precise form of Definition 3.11 is due to Grothendieck [**Gro**, pp. 43-46]. Similar axioms were also proposed by Anderson [**And79**, §2], Heller [**Hel88**], and Franke [**Fra96**]. Moreover, related notions were consider by Keller [**Kel91**] and Garkusha [**Gar06, Gar05**]. For an additional introductory account we also refer to work of Maltsiniotis [**Mal01b, Mal01a**].

We conclude this section by one trivial example.

EXAMPLE 3.17. Let $\mathscr{D}$ be a derivator. The **opposite derivator** $\mathscr{D}^{\mathrm{op}}$ of $\mathscr{D}$ is given by

$$\mathscr{D}^{\mathrm{op}}\colon \mathcal{C}at^{\mathrm{op}} \to \mathcal{CAT}\colon B \mapsto \mathscr{D}(B^{\mathrm{op}})^{\mathrm{op}}.$$

As in ordinary category theory, this example is important because of the resulting *duality principle*. In many later statements we allow ourselves to focus on results on colimits and left Kan extensions. An application to opposite derivators yields the corresponding dual statement. The formation of opposites of derivators is compatible with the formation of opposites of complete and cocomplete categories, abelian categories and model categories.

3.2. Stable derivators. Having introduced derivators as a compromise model for abstract homotopy theories in §3.1, in this subsection we briefly discuss stable derivators. The main goal is to develop some intuition for the notion and to discuss the relation to triangulated categories. It will turn out that many of the arguments in §2 apply verbatim in this more general context.

To build towards stable derivators, we begin by the following definition.

DEFINITION 3.18. A derivator $\mathscr{D}$ is **pointed** if the underlying category $\mathscr{D}(*)$ has a zero object.

REMARK 3.19. Let $\mathscr{D}$ be a derivator.

(i) As a consequence of (Der1), the underlying category $\mathscr{D}(*)$ has an initial object and a final object, and these objects are hence asked to be isomorphic. As usual, any such object is denoted by $0 \in \mathscr{D}(*)$.
(ii) It follows from (Der3) that for every $B \in \mathcal{C}at$ also $\mathscr{D}(B)$ has a zero object, and that all restriction and Kan extension functors preserve these zero objects.

EXAMPLES 3.20. Let us take up again Examples 3.15.

(i) Let $\mathcal{C}$ be a complete and cocomplete category. The underlying category of the represented derivator $y_{\mathcal{C}}$ is $\mathcal{C}$, and $y_{\mathcal{C}}$ is hence pointed if and only if $\mathcal{C}$ is pointed (has a zero object).
(ii) Let $\mathcal{A}$ be a Grothendieck abelian category. The underlying category of the derivator $\mathscr{D}_{\mathcal{A}}$ is the derived category $D(\mathcal{A})$. As an additive category, $D(\mathcal{A})$ clearly has a zero object, and $\mathscr{D}_{\mathcal{A}}$ is hence pointed.
(iii) Similarly, for a pointed Quillen model category $\mathcal{M}$, the homotopy derivator $\mathscr{H}o_{\mathcal{M}}$ is pointed. Among the specific examples $\mathcal{S}$, $\mathcal{S}_*$, and $\mathcal{S}p$ only the last two are pointed.

DEFINITION 3.21. Let $\mathscr{D}$ be a derivator and let $X \in \mathscr{D}(\square)$.

(i) The square X is **cocartesian** if it lies in the essential image of the left Kan extension functor
$$(i_{\ulcorner})_! \colon \mathscr{D}(\ulcorner) \to \mathscr{D}(\square).$$
(ii) The square X is **cartesian** if it lies in the essential image of the right Kan extension functor
$$(i_{\lrcorner})_* \colon \mathscr{D}(\lrcorner) \to \mathscr{D}(\square).$$

REMARK 3.22. There are the following remarks related to Definition 3.21.

(i) Let $\mathscr{D}$ be a derivator and let $X \in \mathscr{D}(\square)$ be a square in $\mathscr{D}$. It turns out that X is cocartesian if and only if a certain canonical map
$$\operatorname{colim}_{\ulcorner}(i_{\ulcorner})^* X \to X_{1,1}$$
is an isomorphism. The formalism behind these canonical maps (based on the calculus of canonical mates) will not be made explicit in this paper (but see [**KS74**] or [**Gro13**]). In this paper, we will occasionally refer to certain maps in derivators as "canonical", and in such situations they always arise by means of this calculus.
(ii) Let us take up again our standard examples (Examples 3.15). A square in a represented derivator is cocartesian if and only if it is a pushout square. For every Grothendieck abelian category $\mathcal{A}$, a square in $\mathscr{D}_{\mathcal{A}}$ is

cocartesian if and only if is a derived pushout square. In particular, in the derivator $\mathscr{D}_R$ of a ring we recover Definition 2.25, which played a key role in §2. Finally, a square in the homotopy derivator of a Quillen model category is cocartesian if and only if it is a homotopy pushout square.

PROPOSITION 3.23. *Let $\mathscr{D}$ be a derivator and $u\colon A \to B$ a fully faithful functor. The Kan extension functors $u_!, u_*\colon \mathscr{D}(A) \to \mathscr{D}(B)$ are fully faithful.*

PROOF. For a proof based on the calculus of canonical mates and the corresponding formalism of homotopy exact squares we refer to [**Gro13**, Prop. 1.20]. □

This property applied to $\mathscr{D}_R$ (Proposition 2.21) was central to many constructions in §2. In order to develop some intuition for derivators we generalize some of these constructions to pointed derivators.

CONSTRUCTION 3.24. Let $\mathscr{D}$ be a pointed derivator.

(i) We begin with a derivator version of Proposition 2.31. Let $f \in \mathscr{D}([1])$ be a morphism looking like $(f\colon x \to y)$, which is to say that $\mathrm{dia}_{[1]}(f)\colon [1] \to \mathscr{D}(*)$ takes the form $(f\colon x \to y)$ (see Warning 3.9). In order to extend f to a cofiber square, we again consider the fully faithful inclusion functors

$$i_\ulcorner \circ i\colon [1] \to \ulcorner \to \square$$

and their corresponding Kan extension functors

$$(i_\ulcorner)_! \circ i_*\colon \mathscr{D}([1]) \to \mathscr{D}(\ulcorner) \to \mathscr{D}(\square).$$

The image $(i_\ulcorner)_! \circ i_*(f) \in \mathscr{D}(\square)$ can be restricted along the inclusion $k\colon [1] \to \square$ pointing at the vertical morphism on the right. We define the **cone functor**

$$\mathsf{cof} = k^* \circ (i_\ulcorner)_! \circ i_*\colon \mathscr{D}([1]) \to \mathscr{D}([1]).$$

The object obtained from $\mathsf{cof}(f)$ by an evaluation at $1 \in [1]$ is also referred to as the cone of f, but the corresponding construction is distinguished notationally by

$$C = 1^* \circ \mathsf{cof}\colon \mathscr{D}([1]) \to \mathscr{D}(*).$$

With this notation, the **cofiber square** $(i_\ulcorner)_! \circ i_*(f)$ associated to f looks like

$$\begin{array}{ccc} x & \xrightarrow{f} & y \\ \downarrow & & \downarrow{\scriptstyle \mathsf{cof}(f)} \\ 0 & \longrightarrow & C(f). \end{array} \tag{3.25}$$

We invite the reader to dualize these construction in order to define **fiber functors**

$$\mathsf{fib}\colon \mathscr{D}([1]) \to \mathscr{D}([1]) \qquad \text{and} \qquad F\colon \mathscr{D}([1]) \to \mathscr{D}(*).$$

(ii) As a minor variant of the previous case, we now construct suspension and loop functors in pointed derivators, thereby generalizing Remark 2.34. There are at least two ways of defining the suspension. First, for $x \in \mathscr{D}(*)$ we would like to make the definition

$$\Sigma x = C(x \to 0).$$

To explain the right hand side in terms of Kan extensions, it suffices to consider the functor $0\colon * \to [1]$ pointing at zero and the corresponding right Kan extension functor $0_*\colon \mathscr{D}(*) \to \mathscr{D}([1])$. In fact, 0_* extends x to a morphism with 0 as target as follows from (Der4). Correspondingly, we define the **suspension functor** as

$$\Sigma = C \circ 0_*\colon \mathscr{D}(*) \to \mathscr{D}(*).$$

Second, a canonically isomorphic way of defining the suspension is by means of

$$\Sigma = (1,1)^* \circ (i_{\ulcorner})_! \circ (0,0)_*\colon \mathscr{D}(*) \to \mathscr{D}(\ulcorner) \to \mathscr{D}(\square) \to \mathscr{D}(*).$$

In both cases the suspension Σx sits as lower right corner in the corresponding **suspension square**

(3.26)
$$\begin{array}{ccc} x & \longrightarrow & 0 \\ \downarrow & & \downarrow \\ 0 & \longrightarrow & \Sigma x. \end{array}$$

It is straightforward to dualize these constructions in order to obtain a **loop functor**

$$\Omega\colon \mathscr{D}(*) \to \mathscr{D}(*).$$

Examples 3.27. Let us specialize these constructions in our examples of pointed derivators (Examples 3.20).

(i) Let $\mathcal{C}$ be a pointed, complete and cocomplete category and $y_{\mathcal{C}}$ its represented derivator. The cofiber squares (3.25) and the suspension squares (3.26) in $y_{\mathcal{C}}$ are ordinary pushout squares. Hence, (3.25) implies that cof and C in $y_{\mathcal{C}}$ are the usual cokernel functors (once considered with the structure map and once without). As a special case of this, the suspension squares (3.26) imply that Σx is isomorphic to a zero object, and Σ is hence naturally isomorphic to the constant functor on the zero object,

(3.28)
$$\Sigma \cong 0\colon \mathcal{C} \to \mathcal{C}.$$

(ii) Let $\mathcal{A}$ be a Grothendieck abelian category and $\mathscr{D}_{\mathcal{A}}$ its derivator. In this case the cone functor $C\colon D(\mathcal{A}^{[1]}) \to D(\mathcal{A})$ reproduces the functorial cone construction from Proposition 2.7 (see also the proof of Proposition 2.13). As noted in §2, the functor

$$\mathsf{cof}\colon D(\mathcal{A}^{[1]}) \to D(\mathcal{A}^{[1]})$$

is in the background of the rotation of distinguished triangles (as a shadow of the symmetries of Barratt–Puppe sequences in Figure 3). The suspension functor in $\mathscr{D}_{\mathcal{A}}$ specializes to the usual shift functor $\Sigma\colon D(\mathcal{A}) \to D(\mathcal{A})$ (Remark 2.34).

(iii) As a final example, we consider the homotopy derivator $\mathcal{S}_*$ of pointed spaces which also provides the motivation for some of the terminology employed here. Given a morphism $(f\colon X \to Y) \in \mathrm{Ho}(\mathrm{Top}_*^{[1]})$ of pointed topological spaces, the cone $C(f)$ is the usual mapping cone construction. Moreover, the abstract suspension specializes to the reduced suspension functor

$$\Sigma\colon \mathrm{Ho}(\mathrm{Top}_*) \to \mathrm{Ho}(\mathrm{Top}_*),$$

and

$$\Omega\colon \mathrm{Ho}(\mathrm{Top}_*) \to \mathrm{Ho}(\mathrm{Top}_*)$$

is the usual loop space functor (see also Remark 3.34).

PROPOSITION 3.29. *Let $\mathscr{D}$ be a pointed derivator.*

(i) *There is an adjunction* $(\mathsf{cof}, \mathsf{fib})\colon \mathscr{D}([1]) \rightleftarrows \mathscr{D}([1])$.
(ii) *There is an adjunction* $(\Sigma, \Omega)\colon \mathscr{D}(*) \rightleftarrows \mathscr{D}(*)$.

PROOF. For a proof we refer to [**Gro13**, §3]. □

In triangulated categories we are used to be able to rotate distinguished triangles without a loss of information and similarly that the suspension functor is an equivalence. With the few exceptions of exotic triangulated categories [**MSS07**], for most triangulated categories arising in nature these features are consequences of properties of *stable* homotopy theories in the background.

DEFINITION 3.30. A pointed derivator $\mathscr{D}$ is **stable** if every square in $\mathscr{D}$ is cocartesian if and only if it is cartesian, and these squares are then referred to as **bicartesian**.

In order to find examples of stable derivators, it is useful to have simpler characterizations of stability.

THEOREM 3.31. *The following are equivalent for a pointed derivator $\mathscr{D}$.*

(i) *The derivator $\mathscr{D}$ is stable.*
(ii) *The adjunction* $(\mathsf{cof}, \mathsf{fib})\colon \mathscr{D}([1]) \rightleftarrows \mathscr{D}([1]$ *is an equivalence.*
(iii) *The adjunction* $(\Sigma, \Omega)\colon \mathscr{D}(*) \rightleftarrows \mathscr{D}(*)$ *is an equivalence.*

PROOF. If $\mathscr{D}$ is stable, then cofiber squares (3.25) and fibre squares are essentially the same, and similarly for suspension squares (3.26) and loop squares. From this it is fairly straightforward to deduce that (i) implies (ii) and (iii). The converse implications are more involved and we refer the reader to [**GPS14b**, Thm. 7.1]. □

EXAMPLES 3.32. With this preparation we can revisit our examples of pointed derivators (Examples 3.20).

(i) Let $\mathcal{C}$ be a pointed, complete and cocomplete category and $y_{\mathcal{C}}$ its represented derivator. In this case the abstract suspension functor is naturally isomorphic to the constant functor on the zero object (3.28). Hence, by Theorem 3.31 we conclude that $y_{\mathcal{C}}$ is stable if and only if $\Sigma \cong 0$ is an equivalence of categories if and only if $\mathcal{C}$ is equivalent to the terminal category $*$.
(ii) Let $\mathcal{A}$ be a Grothendieck abelian category and $\mathscr{D}_{\mathcal{A}}$ its derivator. In $\mathscr{D}_{\mathcal{A}}$ the abstract suspension functor agrees with the shift functor $\Sigma\colon D(\mathcal{A}) \to D(\mathcal{A})$, which clearly is an equivalence of categories. Hence, derivators of the form $\mathscr{D}_{\mathcal{A}}$ are stable.
(iii) A stable model category is a pointed model category such that the abstract suspension functor is an equivalence in its homotopy derivator ([**Hov99**, Def. 7.1.1]). By Theorem 3.31 we deduce that homotopy derivators of stable model categories are stable (as it also follows from [**Hov99**, Rmk. 7.1.12]). As a special case, the derivator $\mathcal{S}p$ of spectra is stable, and there are many additional interesting examples (see Examples 4.2 or [**SS03**]).

Additional classes of examples arise as homotopy derivators of exact categories [**Gil11, Što14**], stable cofibration categories [**Sch13, Len17**], or stable ∞-categories [**Lur09, Lur14, Len17**].

REMARK 3.33. We want to stress the observation that a represented derivator $y_{\mathcal{C}}$ is stable if and only if $\mathcal{C} \simeq *$. This essentially says that stability is *invisible to ordinary category theory.* In order to capture the phenomenon of stability as it arises in interesting examples in homological algebra, stable homotopy theory or homotopical algebra we have to use more refined techniques. Among these are triangulated categories, stable derivators, stable model categories, and stable ∞-categories (see again Disclaimer 3.2).

We conclude this subsection by sketching the relation to triangulated categories. Many arguments are completely parallel to the discussion in §2, and here we hence focus on the remaining arguments. We begin by sketching that stability implies additivity. In order to motivate the general approach, we revisit the classical situation in topology.

REMARK 3.34. Let (X, x_0) be a pointed topological space. The *loop space* ΩX of X at x_0 is the space of paths $[0,1] \to X$ which send both boundary points $0, 1$ to x_0. The loop space is the homotopy pullback (in the usual Serre model structure) of the cospan on the left in

$$\begin{array}{ccc} & & * \\ & & \big\downarrow{\scriptstyle x_0} \\ * & \xrightarrow[x_0]{} & X, \end{array} \qquad\qquad \begin{array}{ccc} \Omega X & \longrightarrow & PX \\ \big\downarrow & \lrcorner & \big\downarrow{\scriptstyle \mathrm{ev}_1} \\ * & \xrightarrow[x_0]{} & X. \end{array}$$

In fact, if one wants to calculate this homotopy pullback, then it suffices to replace one of the maps $x_0 \colon * \to X$ by a weakly homotopy equivalent Serre fibration and then to calculate the categorical pullback. The standard example of such an approximation is the path space PX of paths $[0,1] \to X$ starting at x_0 endowed with the evaluation map $\mathrm{ev}_1 \colon PX \to X$ at the end point 1. This space is weakly contractible since all such paths are homotopic to the constant path at x_0, and the pullback of the cospan on the right gives the above description of the loop space. The concatenation of loops can be used to show that the loop space ΩX is a group object in the homotopy category $\mathrm{Ho}(\mathrm{Top}_*)$.

In order to motivate the general approach in pointed derivators, we explain a purely categorical way of modelling the inversion of loops in topology. And for this purpose it is convenient to use a different model for the above homotopy pullback, which is obtained by replacing *both* maps by the above Serre fibration and then calculating the pullback

$$\begin{array}{ccc} \Omega X & \xrightarrow{p_2} & PX \\ {\scriptstyle p_1}\big\downarrow & \lrcorner & \big\downarrow{\scriptstyle \mathrm{ev}_1} \\ PX & \xrightarrow[\mathrm{ev}_1]{} & X. \end{array} \tag{3.35}$$

Up to an obvious homeomorphism, this model of the loop space has as points paths $[-1,1] \to X$ which send $-1, 1$ to x_0, and in this model the inversion of loops $\iota \colon \Omega X \to \Omega X$ is given by a reparametrization via the reflection at $0 \in [-1,1]$. The point is that this reparametrization simply amounts to interchanging the two copies of PX in the pullback square (3.35). More formally, the outer commutative square in the diagram

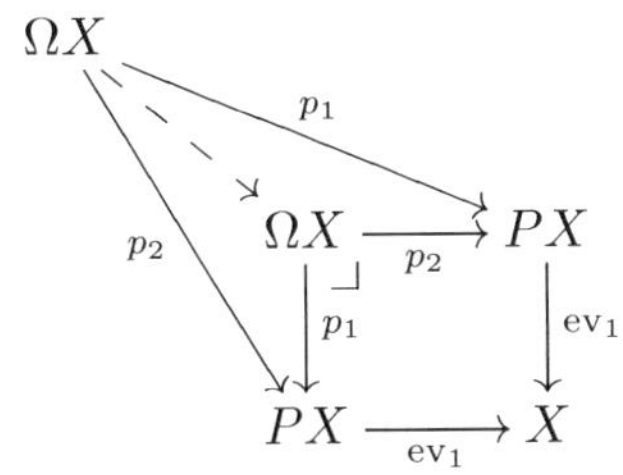

induces by the universal property of the pullback square a canonical dashed morphism making both triangles commute. And this morphism of course is the above reparametrization map $\iota\colon \Omega X \to \Omega X$.

This can be abstracted to pointed derivators as follows.

REMARK 3.36. Let $\mathscr{D}$be a pointed derivator. For every $x \in \mathscr{D}(*)$, the loop object Ωx can be naturally turned into a group object. The vague idea is to model the concatenation of loops from topology purely categorically. The details are a bit more involved (see [**Gro13**, §4.1]) and they rely on the Segalian approach to A_∞-multiplications [**Seg74**]. (In fact, using fibrational techniques the arguments in [**Gro13**, §4.1] can be adapted to construct coherently associative multiplication maps.) One can make precise that the inversion map $(-)^{-1}\colon \Omega x \to \Omega x$ comes from the swap symmetry of loop squares (the duals of (3.26)) which interchanges $(1,0)$ and $(0,1)$. A variant of the *Eckmann–Hilton argument* [**EH62**, Thm. 4.17] implies that two-fold loop objects $\Omega^2 x$ are abelian group objects.

COROLLARY 3.37. *The underlying category of a stable derivator is additive.*

PROOF. In a stable derivator $\mathscr{D}$, the suspension $\Sigma\colon \mathscr{D}(*) \to \mathscr{D}(*)$ is an equivalence (Theorem 3.31), and for every object $x \in \mathscr{D}(*)$ there is a natural isomorphism $x \cong \Omega^2\Sigma^2 x$. By Remark 3.36 this implies that every object in $\mathscr{D}(*)$ is an abelian group object, showing that $\mathscr{D}(*)$ is additive. □

With this preparation the construction of triangulations on stable derivators (Theorem 3.42) is fairly parallel to §2. As a variant of the underlying diagram functor (Construction 3.8), for every prederivator $\mathscr{D}$ there are partial underlying diagram functors

$$\mathrm{dia}_{A,B}\colon \mathscr{D}(A \times B) \to \mathscr{D}(A)^B, \qquad A, B \in \mathcal{C}at,$$

making a diagram incoherent in the B-direction only. More formally, for a coherent diagram $X \in \mathscr{D}(A \times B)$ and $b \in B$ we set

$$\mathrm{dia}_{A,B}(X)_b = (\mathrm{id}_A \times b)^*(X) \in \mathscr{D}(A),$$

the restriction along $\mathrm{id}_A \times b\colon A \cong A \times * \to A \times B$.

DEFINITION 3.38. A derivator $\mathscr{D}$ is **strong** if the partial underlying diagram functor

$$\mathrm{dia}_{A,F}\colon \mathscr{D}(A \times F) \to \mathscr{D}(A)^F$$

is essentially surjective and full for all free categories $F \in \mathcal{C}at$ and all $A \in \mathcal{C}at$.

CONSTRUCTION 3.39. Let $\mathscr{D}$be a stable derivator. By Construction 3.24 the underlying category $\mathscr{D}(*)$ can be endowed with a suspension functor

$$\Sigma\colon \mathscr{D}(*) \to \mathscr{D}(*) \tag{3.40}$$

which is an equivalence by Theorem 3.31. For every morphism $(f\colon x \to y) \in \mathscr{D}([1])$, completely parallel to the sketch proof of Theorem 2.6 (see the discussion around (2.57)) one constructs the corresponding standard triangle of f. A triangle

$$x \xrightarrow{f} y \xrightarrow{g} z \xrightarrow{h} \Sigma x \tag{3.41}$$

in $\mathscr{D}(*)$ is distinguished if it is isomorphic to a standard triangle.

The following theorem has some history, and we refer to work of Franke [**Fra96**], Maltsiniotis [**Mal01a**], and the author [**Gro13**].

THEOREM 3.42. *For every strong and stable derivator $\mathscr{D}$, the underlying category $\mathscr{D}(*)$ admits a triangulation given by* (3.40) *and* (3.41).

PROOF. With this preparation, the proof is essentially the same as the proof of Theorem 2.6. We expand a bit on the rotation axiom in order to explain where the sign comes from, and refer the reader to *loc. cit.* for the remaining aspects. For every morphism $(f\colon x \to y) \in \mathscr{D}([1])$ there is an associated Barratt–Puppe sequence (Figure 3). Let us redraw the relevant part of it and add decorations to the various zero objects in order to distinguish them notationally. So let B be the following poset and let $Y \in \mathscr{D}(B)$ look like

$$\begin{array}{ccccc} x & \xrightarrow{f} & y & \longrightarrow & 0_2 \\ \downarrow & \square & \downarrow g & \square & \downarrow \\ 0_1 & \longrightarrow & z & \xrightarrow{h} & x' \\ & & \downarrow & \square & \downarrow f' \\ & & 0_3 & \longrightarrow & y'. \end{array} \tag{3.43}$$

In order to identify the underlying morphism of f' as Σf, we make the identifications

$$\varphi_1\colon \Sigma x \xrightarrow{\sim} x' \qquad \text{and} \qquad \psi_1\colon \Sigma y \xrightarrow{\sim} y',$$

and these lead to the commutative square

$$\begin{array}{ccc} \Sigma x & \xrightarrow[\sim]{\varphi_1} & x' \\ {\scriptstyle \Sigma f}\downarrow & & \downarrow {\scriptstyle f'} \\ \Sigma y & \xrightarrow[\psi_1]{\sim} & y'. \end{array} \tag{3.44}$$

Strictly speaking these identification φ_1 and ψ_1 respectively are the canonical maps associated to the suspension squares

$$\begin{array}{ccc} x & \longrightarrow & 0_2 \\ \downarrow & \square & \downarrow \\ 0_1 & \longrightarrow & x', \end{array} \qquad\qquad \begin{array}{ccc} y & \longrightarrow & 0_2 \\ \downarrow & \square & \downarrow \\ 0_3 & \longrightarrow & y'. \end{array}$$

Similarly, in order to identify the final object in the distinguished triangles of f and g as suitable suspensions, we restrict (3.43) in order to obtain the cofiber sequences

$$
\begin{array}{ccccc}
x & \xrightarrow{f} & y & \longrightarrow & 0_2 \\
\downarrow & \square & \downarrow{\scriptstyle g} & \square & \downarrow \\
0_1 & \longrightarrow & z & \xrightarrow[h]{} & x'
\end{array}
\qquad\qquad
\begin{array}{ccccc}
y & \xrightarrow{g} & z & \longrightarrow & 0_3 \\
\downarrow & \square & \downarrow{\scriptstyle h} & \square & \downarrow \\
0_2 & \longrightarrow & x' & \xrightarrow[f']{} & y',
\end{array}
$$

paste the respective bicartesian squares in order to obtain the suspension squares

$$
\begin{array}{ccc}
x & \longrightarrow & 0_2 \\
\downarrow & \square & \downarrow \\
0_1 & \longrightarrow & x',
\end{array}
\qquad\qquad
\begin{array}{ccc}
y & \longrightarrow & 0_3 \\
\downarrow & \square & \downarrow \\
0_2 & \longrightarrow & y',
\end{array}
$$

and deduce from these the resulting identifications

$$\varphi_2 \colon \Sigma x \xrightarrow{\sim} x' \qquad \text{and} \qquad \psi_2 \colon \Sigma y \xrightarrow{\sim} y'.$$

Now, the suspension squares giving rise to the identifications φ_1 and φ_2 are literally the same and we deduce $\varphi_1 = \varphi_2$. In contrast to this, the suspension squares inducing ψ_1 and ψ_2 differ by a restriction along the symmetry exchanging $(1,0)$ and $(0,1)$, and by Remark 3.36 it follows that ψ_1 agrees with the *negative* of ψ_2. Thus, in order to write the third morphism in the distinguished triangle associated to g as a morphism $\Sigma x \to \Sigma y$, invoking (3.44) we are led to

$$
\begin{array}{ccc}
\Sigma x & \xrightarrow[\sim]{\varphi_1} & x' \\
{\scriptstyle -\Sigma f}\downarrow & & \downarrow{\scriptstyle f'} \\
\Sigma y & \xrightarrow[\psi_2]{\sim} & y'.
\end{array}
$$

Consequently, the distinguished triangle associated to g takes the form

$$y \xrightarrow{g} z \xrightarrow{h} \Sigma x \xrightarrow{-\Sigma f} \Sigma y,$$

as asked for by the rotation axiom of triangulated categories. □

3.3. Exponentials of derivators. In this short subsection we discuss the construction of derivators of representations or exponentials in more categorical terminology. This is a derivatorish version of diagram categories and these exponentials govern the calculus of parametrized limits and parametrized Kan extensions. The formation of represented derivators, of derivators of abelian categories and of homotopy derivators are compatible with the passage to exponentials.

Let $\mathscr{D}$ be a derivator and let $A \in \mathcal{C}at$. Taking the philosophy of derivators serious, we should not be happy with a *category* $\mathscr{D}(A)$ of coherent diagrams of shape A in $\mathscr{D}$, but we should rather ask for a *derivator* $\mathscr{D}^A$ of such diagrams. The heuristics are as follows: diagrams of shape B in $\mathscr{D}^A$ are diagrams of shape A in diagrams of shape B in $\mathscr{D}$ which by the exponential law should be the same as diagrams of shape $A \times B$ in $\mathscr{D}$. This suggests the following construction.

CONSTRUCTION 3.45. Let $\mathscr{D}$ be a prederivator and let $A \in \mathcal{C}at$. We denote by $\mathscr{D}^A$ the 2-functor

$$\mathscr{D}^A \colon \mathcal{C}at^{\mathrm{op}} \to \mathcal{CAT} \colon B \mapsto \mathscr{D}(A \times B).$$

This is the prederivator of diagrams of shape A in $\mathscr{D}$. The underlying category is $\mathscr{D}^A(*) = \mathscr{D}(A \times *) \cong \mathscr{D}(A)$.

The following example illustrates the relation to representation theory. In that example we already invoke the notion of a morphism of (pre-)derivators which we introduce formally in §3.4.

EXAMPLE 3.46. Let k be a field, let Q be a quiver with finitely many vertices only, and let kQ be the k-linear path algebra. Identifying the quiver Q with the free category Q generated by it, we want to describe the prederivator

$$\mathscr{D}_k^Q : \mathcal{C}at^{\mathrm{op}} \to \mathcal{CAT} : B \mapsto \mathscr{D}_k(Q \times B)$$

differently. For this purpose, given an abelian category $\mathcal{A}$, let us write $\mathrm{Ho}(\mathrm{Ch}(\mathcal{A}))$ for the localization of $\mathrm{Ch}(\mathcal{A})$ at the class of quasi-isomorphisms, so that we have $D(\mathcal{A}) = \mathrm{Ho}(\mathrm{Ch}(\mathcal{A}))$. With this notation, by means of the exponential law and the equivalence $\mathrm{Mod}(k)^Q \simeq \mathrm{Mod}(kQ)$ there is the following chain of equivalences of categories

$$\begin{aligned}
\mathscr{D}_k^Q(B) &= \mathscr{D}_k(Q \times B) \\
&= \mathrm{Ho}(\mathrm{Ch}(k)^{Q \times B}) \\
&\cong \mathrm{Ho}\left((\mathrm{Ch}(k)^Q)^B\right) \\
&\simeq \mathrm{Ho}\left(\mathrm{Ch}(kQ)^B\right) \\
&= \mathscr{D}_{kQ}(B).
\end{aligned}$$

This holds for every $B \in \mathcal{C}at$ in a compatible way, and we hence conclude that there is an equivalence of derivators

$$\mathscr{D}_k^Q \simeq \mathscr{D}_{kQ}.$$

Thus the formation of shifted derivators models the passage to the path algebra, and there are of course variants for incidence algebras, group algebras, and more general category algebras.

In the above example, the prederivator $\mathscr{D}_k^Q$ of Q-shaped diagrams in $\mathscr{D}_k$ is again a derivator. Let us recall from classical category theory that functor categories $\mathcal{C}^A$ inherit "exactness properties" from $\mathcal{C}$ by means of pointwise constructions. The following is a derivatorish version of this.

THEOREM 3.47. *Let $\mathscr{D}$ be a derivator and let $A \in \mathcal{C}at$. The prederivator $\mathscr{D}^A$ is a derivator, the* ***derivator of A-shaped diagrams in $\mathscr{D}$****. Moreover, if $\mathscr{D}$ is pointed, stable, or strong, then $\mathscr{D}^A$ is pointed, stable, or strong, respectively.*

PROOF. Most of the axioms of a derivator are straightforward. In order to show that $\mathscr{D}^A$ also satisfies (Der4), some basics on the calculus of homotopy exact squares are needed [**Gro13**, Thm. 1.25]. The fact that $\mathscr{D}^A$ is strong if this is the case for $\mathscr{D}$ is immediate from Definition 3.38. Moreover, by Remark 3.19 the property of being pointed is inherited as well, and for the remaining case of stable derivators we refer to [**Gro13**, Prop. 4.3]. □

We also refer to these derivators $\mathscr{D}^A$ as **exponentials** or **shifted derivators**.

REMARK 3.48. Let us expand a bit on the calculus of Kan extensions in shifted derivators. Let $\mathscr{D}$be a derivator, let $A \in \mathcal{C}at$, and let $u\colon B \to B'$ be a functor. The proof of Theorem 3.47 shows that left Kan extension along u in $\mathscr{D}^A$ is given by left Kan extension along $\mathrm{id}_A \times u$ in $\mathscr{D}$,

$$u_!^{\mathscr{D}^A} = (\mathrm{id}_A \times u)_!^{\mathscr{D}} \colon \mathscr{D}(A \times B) \to \mathscr{D}(A \times B').$$

Moreover, Kan extensions and restrictions in unrelated variables commute ([**Gro13**, Prop. 2.5]). This specializes to the fact that for every $a \in A$ the following square

$$\begin{array}{ccc} \mathscr{D}(A \times B) & \xrightarrow{(\mathrm{id}\times u)_!} & \mathscr{D}(A \times B') \\ {\scriptstyle (a\times\mathrm{id})^*}\downarrow & \cong & \downarrow{\scriptstyle (a\times\mathrm{id})^*} \\ \mathscr{D}(B) & \xrightarrow[u_!]{} & \mathscr{D}(B') \end{array}$$

commutes up to a canonical isomorphism. In particular, for every $X \in \mathscr{D}(A \times B)$ there is a canonical isomorphism

$$\mathrm{colim}_B^{\mathscr{D}}(a^*X) \xrightarrow{\sim} a^* \, \mathrm{colim}_B^{\mathscr{D}^A} X$$

in $\mathscr{D}(*)$. In this precise sense (co)limits and Kan extensions in $\mathscr{D}^A$ treat the A-variable as parameters.

Remark 3.49. This closure under the formation of exponentials is one of the technical advantages of derivators over triangulated categories. Recall that, in general, for a triangulated category $\mathcal{T}$ and a small category A there is no natural triangulation on the functor category $\mathcal{T}^A$ (which, for instance, is compatible with all evaluations). And even if there would be such a triangulation, in most examples, the category $\mathcal{T}^A$ is not the one we would like to study (for instance, for a field k and a quiver Q, we are interested in $D(kQ)$ which is different from $D(k)^Q$, the category of $\mathbb{Z}$-graded representations of Q).

Similarly to derivators, ∞-categories also enjoy this closure property [**Lur09**, Prop. 1.2.7.3], while for arbitrary Quillen model category the situation is somewhat more subtle (related to this see [**BK72, Hel88, Jar87**] for early special cases, and [**Hir03**, Thm. 11.6.1] or [**Lur09**, §A2.8] for published general results).

Examples 3.50. Let B be a small category.

(i) For every complete and cocomplete category $\mathcal{C}$ the derivators $y_{\mathcal{C}}^B$ and $y_{\mathcal{C}^B}$ are equivalent. In these derivators the abstract calculus of limits and Kan extensions agrees with the classical parametrized versions of limits and Kan extensions ([**ML98**, §V.3]).

(ii) For every Grothendieck abelian category $\mathcal{A}$ there is an equivalence of derivators

$$\mathscr{D}_{\mathcal{A}}^B \simeq \mathscr{D}_{\mathcal{A}^B}.$$

(iii) For every model category $\mathcal{M}$ there is an equivalene of derivators

$$\mathscr{H}o_{\mathcal{M}^B} \simeq \mathscr{H}o_{\mathcal{M}}^B.$$

There is a similar variant for ∞-categories.

Remark 3.51. Let $\mathscr{D}$ be a derivator and $A \in \mathcal{C}at$.

(i) The underlying category of $\mathscr{D}^A$ is $\mathscr{D}(A)$, which is to say that every stage of a derivator is the underlying category of a derivator. In particular, this often allows us to focus on the underlying category when we want to establish certain properties for arbitrary stages. For instance, the class of strong and stable derivators is closed under exponentials (Theorem 3.47), and we conclude by Theorem 3.42 that all stages of such derivators admit canonical triangulations. These triangulations are compatible with each other as it follows from the discussion of morphisms in §3.4.

(ii) The calculus of limits and Kan extensions in $\mathscr{D}^A$ is the calculus of *parametrized* limits and Kan extensions in $\mathscr{D}$. Implicitly, this parametrized calculus already appeared in proof of the rotation axiom. In fact, the vertical coherent morphism f' in (3.43) really is the value of the parametrized suspension functor $\Sigma\colon \mathscr{D}([1]) \to \mathscr{D}([1])$ at $f \in \mathscr{D}([1])$. In fact, by [**GPS14b**, Lem. 5.13] there is a natural isomorphism

$$\Sigma \cong \mathsf{cof}^3\colon \mathscr{D}([1]) \to \mathscr{D}([1]). \tag{3.52}$$

3.4. Morphisms of derivators. In this subsection we briefly discuss morphisms of derivators, the interaction of morphisms with limits and the relation to exact functors of triangulated categories. We also introduce natural transformations between morphisms of derivators and the related notions of adjunctions and equivalences of derivators. To begin with, there is the following definition.

DEFINITION 3.53. A **morphism** of derivators is a pseudo-natural transformation.

Let us unravel this definition. For derivators $\mathscr{D}$and $\mathscr{E}$, a morphism $F\colon \mathscr{D} \to \mathscr{E}$ consists of

(i) functors $F_A\colon \mathscr{D}(A) \to \mathscr{E}(A)$ for $A \in \mathcal{C}at$ and
(ii) natural isomorphisms $\gamma_u\colon u^*F_B \xrightarrow{\sim} F_Au^*$ for $u\colon A \to B$,

$$\begin{array}{ccc} \mathscr{D}(A) & \xrightarrow{F_A} & \mathscr{E}(A) \\ {\scriptstyle u^*}\uparrow & \cong\Nwarrow & \uparrow{\scriptstyle u^*} \\ \mathscr{D}(B) & \xrightarrow[F_B]{} & \mathscr{E}(B), \end{array}$$

satisfiying certain coherence axioms. For instance, for every pair of composable functors $u\colon A \to B$ and $v\colon B \to C$ the natural isomorphism γ_{vu} agrees with the following pasting of γ_u and γ_v,

$$\begin{array}{ccc} \mathscr{D}(A) & \xrightarrow{F_A} & \mathscr{E}(A) \\ {\scriptstyle u^*}\uparrow & \cong\Nwarrow & \uparrow{\scriptstyle u^*} \\ \mathscr{D}(B) & \xrightarrow[F_B]{} & \mathscr{E}(B), \\ {\scriptstyle v^*}\uparrow & \cong\Nwarrow & \uparrow{\scriptstyle v^*} \\ \mathscr{D}(C) & \xrightarrow[F_C]{} & \mathscr{E}(C). \end{array}$$

And there is the axiom $\gamma_{\mathrm{id}_A} = \mathrm{id}$ and one additional axiom concerning the interaction of γ_u and γ_v with natural transformations $\alpha\colon u \to v$. A morphism is **strict** if all components γ_u are identities, which is to say that the functors F_A commute with restrictions on the nose (and not only up specified, coherent isomorphisms).

EXAMPLES 3.54. Let us take up again some of Examples 3.15.

(i) Every functor $F\colon \mathcal{C} \to \mathcal{D}$ between complete and cocomplete categories induces a strict morphism of represented derivators

$$F = y_F\colon y_{\mathcal{C}} \to y_{\mathcal{D}}.$$

In fact, for every $A \in \mathcal{C}at$, we define F_A as the postcomposition functor induced by F,

$$F_A\colon \mathcal{C}^A \to \mathcal{D}^A\colon X \mapsto F \circ X,$$

and it is straightforward to verify the axioms of a strict morphism.

(ii) Every left Quillen functor $F\colon \mathcal{M} \to \mathcal{N}$ of model categories induces a total left derived morphism of homotopy derivators

$$LF\colon \mathscr{H}o_{\mathcal{M}} \to \mathscr{H}o_{\mathcal{N}};$$

see [**KM08**] and [**Cis10**, §3]. The explicit construction of the components $(LF)_A$ is given by

$$(LF)_A = L(F_A)\colon \mathrm{Ho}(\mathcal{M}^A) \to \mathrm{Ho}(\mathcal{N}^A).$$

The passage to total left derived functors is pseudo-natural [**Hov99**, §1.4], and this yields the desired pseudo-naturality constraints γ_u. There is, of course, a similar result for right Quillen functors, and these results yield extrinsic constructions of many interesting examples of morphisms of derivators. This includes derived tensor products or derived hom functors, and we refer the reader to the literature for plenty of examples.

(iii) A relevant intrinsic example is given by restriction morphisms. In fact, for every derivator $\mathscr{D}$and every $u\colon A \to B$ there is an induced **restriction morphism**

$$u^*\colon \mathscr{D}^B \to \mathscr{D}^A.$$

For every $C \in \mathcal{C}at$ the corresponding component is defined as

$$u^*_C = (u \times \mathrm{id}_C)^*\colon \mathscr{D}(B \times C) \to \mathscr{D}(A \times C)$$

These functors define a strict morphism u^* as it follows immediately from 2-functoriality of $\mathscr{D}$. (There are also Kan extension morphisms as we will see in a bit.)

Definition 3.55. A morphism of derivators $F\colon \mathscr{D} \to \mathscr{E}$ preserves colimits of shape A if for every $X \in \mathscr{D}(A)$ the canonical map

$$\mathrm{colim}_A FX \xrightarrow{\sim} F\,\mathrm{colim}_A X$$

is an isomorphism in $\mathscr{E}(*)$.

These canonical maps are further instances of the canonical mates mentioned in Remark 3.22. In all detail, these are morphisms of the form

$$\mathrm{colim}^{\mathscr{E}}_A F_A(X) \xrightarrow{\sim} F_*\,\mathrm{colim}^{\mathscr{D}}_A(X),$$

where $F_*\colon \mathscr{D}(*) \to \mathscr{E}(*)$ is the underlying functor of F. In represented derivators this notion reduces to the usual notion of preservation of colimits, while in homotopy derivators of abelian categories or model categories this captures the preservation of derived colimits or homotopy colimits, respectively. Of course, there are refined notions such as the preservation of the A-shaped colimit of a fixed diagram only.

EXAMPLES 3.56. The following classes of morphisms are important.

(i) A morphism of pointed derivators is **pointed** if and only if it preserves zero object (in the usual sense).
(ii) A morphism of derivators is **right exact** if it preserves initial objects and pushouts, i.e., colimits of shape $\ulcorner$. There is the dual notion of a **left exact morphism** and the combined notion of an **exact morphism**.
(iii) Finally, a morphism is **cocontinuous** if it preserves all colimits and **continuous** if it preserves all limits.

LEMMA 3.57. *A morphism of stable derivators is left exact if and only if it is right exact if and only if it is exact.*

PROOF. The key is the following elementary observation: a morphism of derivators preserves colimits of shape A if and only if it is preserves colimiting cocones of shape A [**Gro16**, Prop. 3.9]. As a special case this implies that a morphism of derivators preserves pushouts if and only if it is preserves cocartesian squares, and the claims follow by definition of stability. □

Exact morphisms are the good notion of morphisms of stable derivators. To provide some first evidence for this claim, we include the following discussion. Recall that in ordinary category theory all colimits can be constructed from coproducts and coequalizers [**ML98**, §V.2]. More relevant to our current discussion is a variant of this for finite colimits. To begin with, let us recall that a category is finite if it has finitely many objects and morphisms only. *Finite colimits* can be constructed from finite coproducts and coequalizers or, alternatively, from initial objects and pushouts. Similarly, a functor preserves finite colimits if and only if it is right exact, i.e., it preserves initial objects and pushouts.

A variant of these results also holds for derivators, but in this case we have to invoke a more restrictive notion of finiteness. Suggestively, this is due to the fact that in derivator land colimits also include derived colimits and homotopy colimits.

DEFINITION 3.58. A category A is **strictly homotopy finite** if A is finite, if every endomorphism $f\colon x \to x$ in A is equal to id_x, and if isomorphic objects in A are equal (A is skeletal). A category is **homotopy finite** if it is equivalent to a strictly homotopy finite category.

The following result is due to Ponto–Shulman. There is also a variant for the construction of colimits, but here we only formulate the result for morphisms.

THEOREM 3.59. *A morphism of derivators is right exact if and only if it preserves homotopy finite colimits.*

PROOF. This is [**PS16**, Thm. 7.1]. □

COROLLARY 3.60. *A morphism of stable derivators is exact if and only if it preserves homotopy finite limits and homotopy finite colimits.*

PROOF. This is immediate from Theorem 3.59 and Lemma 3.57. □

REMARK 3.61. Generalizing Definition 3.55, a morphism of derivators $F\colon \mathscr{D} \to \mathscr{E}$ preserves left Kan extensions along $u\colon A \to B$ in $\mathcal{C}at$ if for every $X \in \mathscr{D}(A)$ the canonical morphism

$$u_! F(X) \xrightarrow{\sim} F u_!(X)$$

is an isomorphism in $\mathscr{E}(B)$. Since Kan extensions are pointwise, one checks that a morphism is cocontinuous if and only if it preserves all left Kan extensions [**Gro13**, Prop. 2.3]. As a variant, a morphism of stable derivators is exact if and only if it preserves all sufficiently finite left and right Kan extensions (see [**Gro16**, Thm. 9.14] for more details).

In order to offer a second justification that exact morphisms are the good morphisms for stable derivators we include the following discussion. Since our main focus is on stable derivators, in the following construction we immediately specialize to this context.

CONSTRUCTION 3.62. Let $G\colon \mathscr{D} \to \mathscr{E}$ be a pointed morphism of stable derivators.

(i) For every $x \in \mathscr{D}(*)$ there is the defining suspension square $X \in \mathscr{D}(\square)$ on the left in

$$\begin{array}{ccc} x & \longrightarrow & 0 \\ \downarrow & & \downarrow \\ 0 & \longrightarrow & \ulcorner\,\Sigma x, \end{array} \qquad\qquad \begin{array}{ccc} Gx & \longrightarrow & 0 \\ \downarrow & & \downarrow \\ 0 & \longrightarrow & G\Sigma x. \end{array}$$

Since G preserves zero objects, the image $GX \in \mathscr{E}(\square)$ looks like the above diagram on the right (here, we invoke that morphisms of derivators commute with evaluation functors as special cases of restriction functors). In general, GX will not be cocartesian. Hence, a comparison of this square against the suspension square of Gx yields a canonical comparison map

$$\psi\colon \Sigma Gx \to G\Sigma x.$$

This comparison map is an isomorphism if and only if GX is cocartesian, which is certainly the case if G is not only pointed but even exact. In that case, we can consider its inverse

(3.63) $$\varphi = \psi^{-1}\colon G\Sigma x \xrightarrow{\cong} \Sigma Gx$$

which allows us to 'pull out the suspension'.

(ii) As a variant of the previous case, let $(f\colon x \to y) \in \mathscr{D}([1])$ be a morphism and let $X \in \mathscr{D}(\square)$ be the cofiber square associated to it,

$$\begin{array}{ccc} x & \xrightarrow{f} & y \\ \downarrow & & \downarrow \\ 0 & \longrightarrow & \ulcorner\,Cf, \end{array} \qquad\qquad \begin{array}{ccc} Gx & \xrightarrow{Gf} & Gy \\ \downarrow & & \downarrow \\ 0 & \longrightarrow & GCf. \end{array}$$

The image square $GX \in \mathscr{E}(\square)$ again vanishes at the lower left corner, but, in general, it fails to be cocartesian. This time this leads to a canonical comparison map

$$\psi\colon CGf \to GCx,$$

which is invertible if G is exact.

The details of these constructions are carried out in [**Gro16**, §8], and the connection to exact functors of triangulated categories is provided by the following results. Therein, the underlying categories of strong, stable derivators are of course endowed with the canonical triangulations established in Theorem 3.42.

PROPOSITION 3.64. *Let $F\colon \mathscr{D} \to \mathscr{E}$ be an exact morphism of strong stable derivators. The natural isomorphism $\varphi\colon F\Sigma \xrightarrow{\sim} \Sigma F$ defined by (3.63) turns the underlying functor $F\colon \mathscr{D}(*) \to \mathscr{E}(*)$ into an exact functor.*

PROOF. This is [**Gro16**, Thm. 10.6]. □

REMARK 3.65. The direct proof of Proposition 3.64 is a lengthy direct calculation. There is a more systematic approach based on the notion of *exact formulas in stable derivators* [**BG18**, §13]. An instance of such an exact formula is the canonical isomorphism $\mathsf{cof}^3 \cong \Sigma\colon \mathscr{D}([1]) \to \mathscr{D}([1])$ from (3.52), and Proposition 3.64 follows from the compatibility of exact morphisms with this formula. However, as detailed in *loc. cit.* there are many additional such exact formulas (see for instance [**BG18**, Ex. 13.14]).

Having discussed the notion of exact morphisms, in order to fulfill our duties in this section it only remains to talk about equivalences of derivators. This notion is, of course, defined internally in the 2-category of derivators, and correspondingly we begin by defining natural transformations.

DEFINITION 3.66. A natural transformation of morphisms of derivators is a modification.

Again, for basics on this notion, we refer to [**Gro13**, §2] and [**Gro16**, Lem. 3.11], but for convenience we unravel this definition. Let $F, G\colon \mathscr{D} \to \mathscr{E}$ be morphisms of derivators. A natural transformation $\alpha\colon F \to G$ consists of natural transformations

$$\alpha_A\colon F_A \to G_A\colon \mathscr{D}(A) \to \mathscr{E}(A), \qquad A \in \mathcal{C}at.$$

These components α_A are supposed to satisfy the obvious compatibility with the pseudo-naturality constraints of F and G.

REMARK 3.67. There is a 2-category $\mathcal{DER}$ of derivators, morphisms of derivators, and natural transformations. We denote by $\mathcal{DER}_{\mathrm{St,ex}}$ the sub-2-category given by stable derivators, exact morphisms, and all natural transformations. (The basic calculus of the interaction of morphisms with colimits implies, for instance, that exact morphisms are closed under composition.) This 2-category plays a key role in abstract representation theory, as we discuss in §4.

DEFINITION 3.68. An **adjunction** of derivators is an adjunction internally to $\mathcal{DER}$. An **equivalence** of derivators is an equivalence internally to $\mathcal{DER}$.

Thus, an adjunction consists of morphisms $L\colon \mathscr{D} \to \mathscr{E}$ and $R\colon \mathscr{E} \to \mathscr{D}$ together with natural transformations $\eta\colon \mathrm{id} \to RL$ and $\varepsilon\colon LR \to \mathrm{id}$ which are subject to the triangular identities. We will denote adjunctions by

$$(L, R)\colon \mathscr{D} \rightleftarrows \mathscr{E}.$$

There is the following result which often allows us to exhibit a given morphism as part of an adjunction.

PROPOSITION 3.69. *A morphism $F\colon \mathscr{D} \to \mathscr{E}$ of derivators is a left adjoint if and only if all components $F_A\colon \mathscr{D}(A) \to \mathscr{E}(A), A \in \mathcal{C}at$, are left adjoints and the morphism F is cocontinuous.*

PROOF. This is [**Gro13**, Prop. 2.9], but we want to sketch one direction of the proof. Given a morphism $F\colon \mathscr{D} \to \mathscr{E}$ such that all components F_A are left adjoints, we begin by choosing right adjoints G_A and corresponding levelwise adjunctions

$$(F_A, G_A)\colon \mathscr{D}(A) \rightleftarrows \mathscr{E}(A)$$

for all $A \in \mathcal{C}at$ independently. It turns out that the cocontinuity of F allows us to define pseudo-naturality constraints $u^*G_B \xrightarrow{\sim} G_A u^*$, thereby obtaining the intended adjunction $(F, G)\colon \mathscr{D} \rightleftarrows \mathscr{E}$. □

In order to provide some additional feeling for the notion we mention the following compatibility of the various adjunctions (F_A, G_A) constituting an adjunction $(F, G)\colon \mathscr{D} \rightleftarrows \mathscr{E}$ of derivators. For every $u\colon A \to B$ in $\mathcal{C}at$, $X \in \mathscr{D}(B)$, and $Y \in \mathscr{E}(B)$ the diagram

$$\begin{array}{ccc}
\hom_{\mathscr{E}(B)}(FX, Y) & \xrightarrow{\sim} & \hom_{\mathscr{D}(B)}(X, GY) \\
{\scriptstyle u^*}\big\downarrow & & \big\downarrow{\scriptstyle u^*} \\
\hom_{\mathscr{E}(A)}(u^*FX, u^*Y) & & \hom_{\mathscr{D}(A)}(u^*X, u^*GY) \\
{\scriptstyle \sim}\big\downarrow & & \big\downarrow{\scriptstyle \sim} \\
\hom_{\mathscr{E}(A)}(Fu^*X, u^*Y) & \xrightarrow[\sim]{} & \hom_{\mathscr{D}(A)}(u^*X, Gu^*Y)
\end{array}$$

commutes.

EXAMPLES 3.70. In order to obtain key examples of adjunctions we revisit the situations considered in Examples 3.54.

(i) A functor $F\colon \mathcal{C} \to \mathcal{D}$ between complete and cocomplete categories is a left adjoint if and only if the morphism $y_F\colon y_{\mathcal{C}} \to y_{\mathcal{D}}$ is a left adjoint. This follows from the obvious fact that the passage to represented derivators is 2-functorial.

(ii) For every Quillen adjunction $(F, G)\colon \mathcal{M} \rightleftarrows \mathcal{N}$ there is an induced adjunction

$$(LF, RG)\colon \mathscr{H}o_{\mathcal{M}} \rightleftarrows \mathscr{H}o_{\mathcal{N}}.$$

In particular, if the model categories $\mathcal{M}$ and $\mathcal{N}$ are stable, then LF and RG are exact morphisms (by Proposition 3.69 and Corollary 3.60), and this leads to a rich supply of exact morphisms of stable derivators.

(iii) Let $\mathscr{D}$ be a derivator, let $u\colon A \to B$ in $\mathcal{C}at$, and let $u^*\colon \mathscr{D}^B \to \mathscr{D}^A$ be the restriction morphism. Clearly, all components $(u \times \mathrm{id}_C)^*, C \in \mathcal{C}at$, of the restriction morphism have adjoints on both sides, and the morphism is also continuous and cocontinuous. In fact, as a further application of the calculus of canonical mates, Kan extensions and restrictions in unrelated variables commute up to canonical isomorphisms [**Gro13**, Prop. 2.5]. As an upshot, by Proposition 3.69 there are adjunctions of derivators

$$(u_!, u^*)\colon \mathscr{D}^A \rightleftarrows \mathscr{D}^B \qquad \text{and} \qquad (u^*, u_*)\colon \mathscr{D}^B \rightleftarrows \mathscr{D}^A.$$

The components of these **Kan extension morphisms** are the Kan extension functors $(u \times \mathrm{id}_C)_!$ and $(u \times \mathrm{id}_C)_*$ of $\mathscr{D}$.

COROLLARY 3.71. *Let $\mathscr{D}$ be a strong, stable derivator and let $u\colon A \to B$ be in $\mathcal{C}at$. The functors*

$$u^*\colon \mathscr{D}(B) \to \mathscr{D}(A), \quad u_!\colon \mathscr{D}(A) \to \mathscr{D}(B), \quad \text{and} \quad u_*\colon \mathscr{D}(A) \to \mathscr{D}(B)$$

are exact functors with respect to the canonical triangulations.

PROOF. As adjoint morphisms, $u_!$ and u^* are cocontinuous, while u^* and u_* are continuous (Proposition 3.69). Since stability is inherited by the shifted derivators $\mathscr{D}^A$ and $\mathscr{D}^B$ (Theorem 3.47), all three morphisms $u^*, u_!$, and u_* are exact (Lemma 3.57), and the claim hence follows from Proposition 3.64. □

There is the following more refined version of this result.

REMARK 3.72. Let $\mathcal{T}ria\mathcal{C}at$ be the 2-category of triangulated categories, exact functors, and exact natural transformations. For every strong and stable derivator $\mathscr{D}$, the formation of canonical triangulations and canonical exact structures yields a lift of $\mathscr{D}$ against the forgetful functor $\mathcal{T}ria\mathcal{C}at \to \mathcal{CAT}$,

$$\begin{array}{ccc} & & \mathcal{T}ria\mathcal{C}at \\ & \overset{\mathscr{D}}{\nearrow} & \downarrow \\ \mathcal{C}at^{\mathrm{op}} & \xrightarrow[\mathscr{D}]{} & \mathcal{CAT}; \end{array} \tag{3.73}$$

see [**Gro16**, Thm. 10.14]. We want to use this result to stress once more the distinction between properties and structures. As we discussed, the property of being strong and stable implies the existence of canonical triangulations (Theorem 3.42). Similarly, the property of preserving certain basic (co)limits, implies the existence of canonical exact structures (Proposition 3.64). In particular, equivalences of derivators always are exact.

Along these lines, there is the following remark. Let us recall that an exact natural transformation $\alpha\colon F \to G$ of exact functors between triangulated categories is a natural transformation α which commutes with the natural isomorphisms $F\Sigma \xrightarrow{\sim} \Sigma F$ and $G\Sigma \xrightarrow{\sim} \Sigma G$. Since in derivator land these isomorphisms arise canonically, there is no counterpart for the notion of an exact natural transformation for derivators. In fact, *every* natural transformation between exact morphisms is compatible with the canonical morphisms (3.63) (see [**Gro16**, Cor. 10.12]). This result is used implicitly in the construction of the lifts in (3.73).

4. Higher symmetries

In this section we give an overview over some main results of this project on higher symmetries. In §4.1 we make precise the notion of strong stable equivalences as certain uniform versions of derived equivalences for abstract representations. In §4.2 we briefly illustrate the concept by a discussion of the abstract representation theory of Dynkin quivers of type A. In §4.3 we pass to more general abstract reflection functors. In §4.4 we include a discussion of monoidal derivators, enriched derivators, and the universality of the derivator of spectra. Finally, in §4.5 we apply this to the construction of universal tilting modules. These are invertible, spectral bimodules that realize strong stable derivators in arbitrary stable derivators.

4.1. Strong stable equivalences. Motivated by the compatibility of the formation of derivators of abelian categories and exponentials (Example 3.46), in this subsection we define strong stable equivalences as a variant of the classical derived equivalences of quivers.

CONSTRUCTION 4.1. Let A be a small category and let $\mathscr{D}$ be a stable derivator. We again denote by $\mathscr{D}^A$ the 2-functor constructed in Construction 3.45, which by Theorem 3.47 is a derivator, the derivator of representations of shape A with values in $\mathscr{D}$. The motivation for this terminology stems from Example 3.46. It is straightforward to check that the formation $\mathscr{D} \mapsto \mathscr{D}^A$ extends to a 2-functor

$$(-)^A \colon \mathcal{DER}_{\mathrm{St,ex}} \to \mathcal{DER} \colon \mathscr{D} \mapsto \mathscr{D}^A,$$

where $\mathcal{DER}_{\mathrm{St,ex}} \subseteq \mathcal{DER}$ again denotes the 2-category of stable derivators, exact morphisms, and arbitrary transformations (Remark 3.67).

The slogan is that this 2-functor encodes the abstract representation theory of the small category A. To fill this slogan with more life, we collect the following examples of stable derivators and specialize the shape A to the case of (the path-category of) a quiver Q.

EXAMPLES 4.2. As part of the structure encoded by the 2-functor $(-)^Q$ of abstract representations of a quiver Q there are the following stable derivators of more specific representations.

(i) For every ordinary, not necessarily commutative ring R there is the stable derivator $\mathscr{D}_R$ of the ring (Examples 3.15). In fact, this derivator arises for instance from the projective model structure [**Hov99**, §2.3] on the category $\mathrm{Ch}(R)$ of unbounded chain complexes over R. From this we obtain by shifting the derivator $\mathscr{D}_R^Q$ of representations of Q in $\mathscr{D}_R$, and there is an equivalence of stable derivators

$$\mathscr{D}_R^Q \simeq \mathscr{D}_{RQ}.$$

One way to see this is by observing that $\mathrm{Mod}(RQ) \sim \mathrm{Mod}(R)^Q$ induces a Quillen equivalence at the level of unbounded chain complexes.

(ii) This generalizes immediately to arbitrary Grothendieck abelian categories $\mathcal{A}$. In fact, the injective model structure on the category $\mathrm{Ch}(\mathcal{A})$ (see e.g. [**Hov01**] or [**Lur14**, Chapter 1]) induces a stable derivator $\mathscr{D}_{\mathcal{A}}$. For example, for every quasi-compact, quasi-separated scheme X there is the stable derivator $\mathscr{D}_X$ of unbounded chain complexes of quasi-coherent $\mathcal{O}_X$-modules [**Hov01**]. More generally, associated to every such $\mathcal{A}$ there is the derivators $\mathscr{D}_{\mathcal{A}}^Q$ of representations of Q with values in $\mathscr{D}_{\mathcal{A}}$.

(iii) Still sticking to the framework of classical homological algebra, this can be generalized further to exact categories in the sense of Quillen [**Qui73**, §2]. At least if we are willing to restrict to suitably finite shapes, for every exact category $\mathcal{E}$ there is by [**Kel07a**], [**Gil11**] or [**Što14**] the bounded derivator $\mathscr{D}_{\mathcal{E}}$ of $\mathcal{E}$ enhancing the bounded derived category. Correspondingly, for finite quivers Q there is the derivator $\mathscr{D}_{\mathcal{E}}^Q$ of representations.

(iv) As additional interesting variants, given a differential-graded algebra A over an arbitrary ground ring we can consider *differential-graded representations* of Q over A which is to say functors from Q to dg-modules over A. In order to import this to derivators, we recall that the category of dg-modules over A admits suitable stable model structures (see for instance [**Hin97, SS00, Fre09**]), and consequently we obtain the stable derivator $\mathscr{D}_A$ of dg-modules over A. For our quiver Q there is an equivalence $\mathscr{D}_A^Q \simeq \mathscr{D}_{AQ}$, where AQ is a differential-graded version of the usual path-algebra.

(v) Another algebraic context giving rise to stable derivators is *stable module theory* and *representation theory of groups.* For every quasi-Frobenius ring or, more generally, Iwanaga–Gorenstein ring [**EJ00**, §9.1] the corresponding category Mod(R) of modules can be endowed with the Gorenstein projective and the Gorenstein injective model structure [**Hov02**, Theorem 8.6]. These Quillen equivalent model structures are stable (see for instance [**Bec14**, Corollary 1.1.16]), and hence induce up to equivalence the same stable derivator $\mathscr{D}_R^{\mathrm{Gor}}$. Correspondingly, associated to Q there is the stable derivator $(\mathscr{D}_R^{\mathrm{Gor}})^Q$ of representations. A special case occurs when R is the group algebra kG of a finite group G over a field k , and this important special case was for instance studied in [**BCR97, Ric97, BIK11**].

(vi) Going beyond the algebraic context, we can pass to spectra in the sense of topology (see for example [**HSS00, EKMM97, MMSS01**]). For concreteness, let us stick to one of these monoidal models and assume that E is a symmetric ring spectrum. Then the category of E-module spectra can be endowed with a stable model structure [**HSS00**], and we obtain the associated stable derivator $\mathscr{D}_E$ of E-module spectra. Correspondingly, we obtain the derivator $\mathscr{D}_E^Q$ of *spectral representations* of Q over E.

(vii) Finally, as mentioned in §3.1, typically derivators arise as shadows of the various approaches to axiomatic homotopy theory. Hence, an entire zoo of additional examples of stable derivators is induced by the many examples of stable model categories, stable ∞-categories [**Lur14**], or stable cofibration categories [**Sch10**]. Among others many examples of interest arise in equivariant stable homotopy theory [**MM02, LMSM86**], motivic stable homotopy theory [**Voe98, MV99, Jar00**] or parametrized stable homotopy theory [**MS06, ABG+09, ABG10**]. For many more examples of stable model categories arising in various areas of algebra, geometry, and topology see [**SS03**].

Before defining strong stable equivalences we recall the following classical definition.

DEFINITION 4.3. Two quivers Q and Q' are **derived equivalent** over a field k if the path-algebras kQ and kQ' are derived equivalent, i.e., if there is an exact equivalence of derived categories

$$D(kQ) \overset{\Delta}{\simeq} D(kQ').$$

Such derived equivalences are usually obtained by means of tilting theory (see the handbook [**AHHK07**] and the many references therein) and have been studied systematically (also for more general finite dimensional algebras over a field).

REMARK 4.4. In Definition 4.3 we were very careful and stressed that the existence of derived equivalences potentially depends on the choice of the field or ring of coefficients. It turns out that certain derived equivalences are "more combinatorial in nature" and they even extend to representations with values in arbitrary abelian categories (hence they are *universal derived equivalences* in the sense of Ladkani [**Lad07a, Lad08**]).

Following the line of thought of Ladkani's universal derived equivalences one step further, we are led to the following definition [**GŠ14**, Def. 5.1].

DEFINITION 4.5. Two small categories A and A' are **strongly stably equivalent**, in notation $A \overset{s}{\sim} A'$, if there is a pseudo-natural equivalence

$$\Phi\colon (-)^A \simeq (-)^{A'}\colon \mathcal{DER}_{\mathrm{St,ex}} \to \mathcal{DER}.$$

We call such a pseudo-natural equivalence a **strong stable equivalence**.

In more down-to-earth terms this means the following. Given two small categories A and A', a strong stable equivalence $\Phi\colon A \overset{s}{\sim} A'$ consists of

(i) for *every* stable derivator $\mathscr{D}$an equivalence of derivators

$$\Phi_{\mathscr{D}}\colon \mathscr{D}^A \simeq \mathscr{D}^{A'}$$

(ii) and for every exact morphism of stable derivators $F\colon \mathscr{D} \to \mathscr{E}$ a natural isomorphism $\gamma_F\colon F \circ \Phi_{\mathscr{D}} \to \Phi_E \circ F$,

$$\begin{array}{ccc} \mathscr{D}^A & \xrightarrow[\simeq]{\Phi_{\mathscr{D}}} & \mathscr{D}^{A'} \\ {\scriptstyle F}\downarrow & \swArrow\cong & \downarrow{\scriptstyle F} \\ \mathscr{E}^A & \xrightarrow[\Phi_{\mathscr{E}}]{\simeq} & \mathscr{E}^{A'}. \end{array}$$

And this datum is supposed to satisfy some obvious coherence axioms. Let us comment a bit on this definition.

REMARK 4.6. The notion of strong stable equivalences is in various respects more restrictive than the notion of a derived equivalence or a universal derived equivalence.

(i) Strongly stably equivalent small categories have equivalent homotopy theories of representations in Grothendieck abelian categories, of differential-graded representations, of spectral representations and of more general abstract representations (by choosing specific examples in Examples 4.2).

(ii) The components $\Phi_{\mathscr{D}}$ of a strong stable equivalence are equivalences of *homotopy theories* and not merely of *homotopy categories* (together with the classical triangulation). This means that also the higher-order homotopy theoretic information is supposed to be preserved.

(iii) The various equivalences $\Phi_{\mathscr{D}}$ are suitably compatible with exact morphisms. For instance, Quillen adjunctions between stable model categories induce exact morphisms of homotopy derivators and, similarly, exact functors between stable ∞-categories induce exact morphisms of homotopy derivators. For strongly stably equivalent quivers or categories this implies that the equivalences commute with various kinds of restriction of scalar functors, induction and coinduction functors as well as (Bousfield) localizations and colocalizations.

To put it as a slogan, strongly stably equivalent categories have the same abstract stable representation theory.

We also want to point the following. Definition 4.5 is formulated in the language of derivators. However, for every stable equivalence $A \overset{s}{\sim} B$ we can conclude the following (related to this see [**Ren09**], [**Dug01a**], and [**Lur09**]).

(i) For every combinatorial model category $\mathcal{M}$ the diagram categories $\mathcal{M}^A$ and $\mathcal{M}^B$ are Quillen equivalent.
(ii) For every presentable ∞-category $\mathcal{C}$ the diagram categories $\mathcal{C}^A$ and $\mathcal{C}^B$ are equivalent.

In §§4.2-4.3 we illustrate this notion by some interesting examples of strong stable equivalences. We conclude this section by some obvious closure properties and also obstructions.

LEMMA 4.7. *Let A, A', B, B', and $A_i, A'_i, i \in I$, be small categories.*

(i) *The relation of 'being strongly stably equivalent' $\overset{s}{\sim}$ defines an equivalence relation.*
(ii) *Equivalent categories are strongly stably equivalent.*
(iii) *If $A \overset{s}{\sim} A'$ and $B \overset{s}{\sim} B'$, then $A \times B \overset{s}{\sim} A' \times B'$.*
(iv) *If $A_i \overset{s}{\sim} A'_i$ for $i \in I$, then $\bigsqcup A_i \overset{s}{\sim} \bigsqcup A'_i$.*

PROOF. The straightforward proof is left as an exercise. □

On the other hand, classical results from representation theory provide us with a non-trivial necessary condition for quivers to be strongly stably equivalent.

PROPOSITION 4.8. *If two finite quivers without oriented cycles are strongly stably equivalent, then the underlying non-oriented graphs are isomorphic.*

PROOF. This is [**GŠ14**, Prop. 5.3]. □

4.2. Abstract representation theory of A_n-quivers. In this subsection we illustrate the notion of a strong stable equivalence by a few examples related to Dynkin quivers of type A. We construct reflection functors and briefly study the related Coxeter and Serre functors in this case. This subsection is largely based on the paper [**GŠ16a**] which is joint with Jan Šťovíček.

Let us begin by a few toy examples which make the connection to stability very obvious. The first examples are discussed in quite some detail, but later we allow ourselves to be a bit more concise.

EXAMPLE 4.9. The source $(\bullet \leftarrow \bullet \rightarrow \bullet)$ and the sink $(\bullet \rightarrow \bullet \leftarrow \bullet)$ of valence two are strongly stably equivalent. In fact, let $\mathscr{D}$ be a stable derivator and let X be an abstract representation of the source of valence two with values in $\mathscr{D}$ as displayed on the left in

$$\begin{array}{ccc} x & \longrightarrow & y \\ \downarrow & & \\ z, & & \end{array} \qquad \begin{array}{ccc} x & \longrightarrow & y \\ \downarrow & & \downarrow \\ z & \longrightarrow & w, \end{array} \qquad \begin{array}{ccc} & & y \\ & & \downarrow \\ z & \longrightarrow & w. \end{array}$$

The idea is that the strong stable equivalence is obtained by first forming the cocartesian square in the middle and then restricting it to the sink of valence two as displayed on the very right, thereby obtaining $\Phi(X)$. To formalize this idea, we recall the following two facts.

(i) Every functor $u\colon A \to B$ between small categories induces by Examples 3.70 Kan extension and restriction morphisms of derivators

$$u_!\colon \mathscr{D}^A \to \mathscr{D}^B, \quad u^*\colon \mathscr{D}^B \to \mathscr{D}^A, \quad \text{and} \quad u_*\colon \mathscr{D}^A \to \mathscr{D}^B.$$

(ii) Moreover, Kan extensions along fully faithful functors (Proposition 3.23) are fully faithful and hence induce equivalences on their images.

We are interested in the special case of the fully faithful inclusions $i_\ulcorner\colon \ulcorner \to \square$ and $i_\lrcorner\colon \lrcorner \to \square$. The left Kan extension morphisms $(i_\ulcorner)_!\colon \mathscr{D}^\ulcorner \to \mathscr{D}^\square$ sends an abstract representation of the source to the corresponding cocartesian square. Denoting by $\mathscr{D}^{\square,\mathrm{cocart}}$ the full subderivator of $\mathscr{D}^\square$ consisting of the cocartesian squares, we obtain the equivalence of derivators on the left in

$$(i_\ulcorner)_!\colon \mathscr{D}^\ulcorner \xrightarrow{\sim} \mathscr{D}^{\square,\mathrm{cocart}}, \qquad (i_\lrcorner)_*\colon \mathscr{D}^\lrcorner \xrightarrow{\sim} \mathscr{D}^{\square,\mathrm{cart}}.$$

Similarly, the formation of cartesian squares yields the equivalence on the right, and in both cases inverse equivalences are given by the corresponding restriction morphisms. Now, by definition of stability (Definition 3.30), the derivators $\mathscr{D}^{\square,\mathrm{cocart}}$ and $\mathscr{D}^{\square,\mathrm{cart}}$ agree, and we obtain the desired equivalence $\Phi_\mathscr{D}\colon \mathscr{D}^\ulcorner \xrightarrow{\sim} \mathscr{D}^\lrcorner$ as a composition of equivalences of derivators

$$\Phi_\mathscr{D} = (i_\lrcorner)^* \circ (i_\ulcorner)_!\colon \mathscr{D}^\ulcorner \xrightarrow{\sim} \mathscr{D}^{\square,\mathrm{cocart}} = \mathscr{D}^{\square,\mathrm{cart}} \xrightarrow{\sim} \mathscr{D}^\lrcorner.$$

Since only restrictions and sufficiently finite Kan extensions are involved in this construction, it is straightforward to verify (invoking Remark 3.61) that these equivalences are pseudo-natural with respect to exact morphisms. Consequently, we obtain the desired strong stable equivalence

$$\Phi\colon \ulcorner = (\bullet \leftarrow \bullet \to \bullet) \overset{s}{\sim} \lrcorner = (\bullet \to \bullet \leftarrow \bullet).$$

The following example is similar, but it involves an additional homotopy finality argument.

EXAMPLE 4.10. The source of valence two $\ulcorner = (\bullet \leftarrow \bullet \to \bullet)$ and the linearly oriented quiver $[2] = (\bullet \to \bullet \to \bullet)$ are strongly stably equivalent. In fact, let $\mathscr{D}$ be a stable derivator and let X be an abstract representation of the source of valence two with values in $\mathscr{D}$ which is displayed on the left in:

$$\begin{array}{ccc} x & \xrightarrow{g} & y \\ {\scriptstyle f}\downarrow & & \\ z, & & \end{array} \qquad \begin{array}{ccccc} Ff & \xrightarrow{\mathsf{fib}(f)} & x & \xrightarrow{g} & y \\ \downarrow & \square & \downarrow{\scriptstyle f} & & \\ 0 & \longrightarrow & z, & & \end{array} \qquad Ff \xrightarrow{\mathsf{fib}(f)} x \xrightarrow{g} y$$

The idea is to simply replace the morphism $f\colon x \to z$ by its fiber $\mathsf{fib} f\colon Ff \to x$, thereby obtaining the above representation $\Phi(X)$ on the right. In more detail, starting from either side, by means of fully faithful Kan extension morphisms we can pass to a representation as displayed in the middle. In fact, starting with our representation $X = (z \leftarrow x \to y)$ we first add a zero object and then the cartesian square as in the diagram

$$\begin{array}{ccc} x & \xrightarrow{g} & y \\ {\scriptstyle f}\downarrow & & \\ z, & & \end{array} \qquad \begin{array}{ccc} & & x & \xrightarrow{g} & y \\ & & {\scriptstyle f}\downarrow & & \\ 0 & \longrightarrow & z, & & \end{array} \qquad \begin{array}{ccccc} Ff & \xrightarrow{\mathsf{fib}(f)} & x & \xrightarrow{g} & y \\ \downarrow & \square & \downarrow & & \\ 0 & \longrightarrow & z. & & \end{array}$$

To re-express this in terms of Kan extensions, let $C \subseteq [1] \times [2]$ be the full subposet obtained by removing the lower right corner $(1,2)$, and let $B_1 \subseteq C$ be the result of also removing the upper left corner $(0,0)$. There are the obvious full inclusions $i_1 \colon \ulcorner \to B_1$ and $i_2 \colon B_1 \to C$ with corresponding Kan extension morphisms

$$(i_1)_* \colon \mathscr{D}^{\ulcorner} \to \mathscr{D}^{B_1} \qquad \text{and} \qquad (i_2)_! \colon \mathscr{D}^{B_1} \to \mathscr{D}^{C}.$$

By Proposition 3.23 both Kan extension morphisms are fully faithful and they induce equivalences onto their respective images. The morphism $(i_1)_!$ precisely amounts to adding a zero object (as a consequence of (Der4)) while $(i_2)_*$ adds a cartesian square (and this step invokes a simple homotopy finality argument). As an upshot we obtain an equivalence of derivators

$$(i_2)_* \circ (i_1)_! \colon \mathscr{D}^{\ulcorner} \xrightarrow{\sim} \mathscr{D}^{C,\mathrm{ex}_1}$$

where $\mathscr{D}^{C,\mathrm{ex}_1} \subseteq \mathscr{D}^{C}$ is the full subderivator spanned by all diagrams which vanish on the lower left corner and which make the square cartesian (the square is hence essentially a fiber square).

If we instead begin with a representation of the linearly oriented A_3-quiver, we simply add the cofiber square to the first of the two morphisms. Thus, in suggestive notation we carry out the constructions:

$$w \xrightarrow{f} x \xrightarrow{g} y, \qquad \begin{array}{ccccc} w & \xrightarrow{f} & x & \xrightarrow{g} & y, \\ \downarrow & & & & \\ 0 & & & & \end{array} \qquad \begin{array}{ccccc} w & \xrightarrow{f} & x & \xrightarrow{g} & y \\ \downarrow & \square & \downarrow{\scriptstyle \mathsf{cof}(f)} & & \\ 0 & \longrightarrow & Cf & & \end{array}$$

Denoting by $B_2 \subseteq C$ the full subposet obtained by removing $(1,1)$, there are full inclusion functors $j_1 \colon [2] = (\bullet \to \bullet \to \bullet) \to B_2$ and $j_2 \colon B_2 \to C$. Arguments similar to the previous case imply that we obtain an equivalence of derivators

$$(j_2)_! \circ (j_1)_* \colon \mathscr{D}^{[2]} \xrightarrow{\sim} \mathscr{D}^{C,\mathrm{ex}_2}$$

where $\mathscr{D}^{C,\mathrm{ex}_2} \subseteq \mathscr{D}^{C}$ is the full subderivator spanned by all diagrams which vanish on the lower left corner and which make the square cocartesian (the square is hence a cofiber square). Now, since $\mathscr{D}$is a stable derivator, the two subderivators $\mathscr{D}^{C,\mathrm{ex}_1}$ and $\mathscr{D}^{C,\mathrm{ex}_2}$ agree, and we obtain the desired equivalence

$$\Phi_{\mathscr{D}} = (j_1)^* \circ (j_2)^* \circ (i_2)_* \circ (i_1)_! \colon \mathscr{D}^{\ulcorner} \xrightarrow{\sim} \mathscr{D}^{[2]}.$$

Since only restrictions and sufficiently finite Kan extensions are involved, these equivalences are pseudo-natural with respect to exact morphisms (Remark 3.61), and we obtain the desired strong stable equivalence

$$\Phi \colon \ulcorner = (\bullet \leftarrow \bullet \to \bullet) \overset{\mathrm{s}}{\sim} [2] = (\bullet \to \bullet \to \bullet).$$

COROLLARY 4.11. *All A_3-quivers are strongly stably equivalent.*

PROOF. We have to show that the quivers $Q_1 = (1 \to 2 \to 3)$, $Q_2 = (1 \leftarrow 2 \to 3)$, $Q_3 = (1 \to 2 \leftarrow 3)$, and $Q_4 = (1 \leftarrow 2 \leftarrow 3)$ are strongly stably equivalent. There are strong stable equivalences $Q_1 \overset{\mathrm{s}}{\sim} Q_2$ (Example 4.10) and $Q_2 \overset{\mathrm{s}}{\sim} Q_3$ (Example 4.9), and since Q_1 and Q_4 are equivalent we also deduce $Q_1 \overset{\mathrm{s}}{\sim} Q_4$ (Lemma 4.7). Since $\overset{\mathrm{s}}{\sim}$ is an equivalence relation (Lemma 4.7), we are done. $\square$

There is a version of Corollary 4.11 for longer A_n-quivers as well, and a proof of this essentially follows the above pattern. Given two differently oriented A_n-quivers Q_1 and Q_2 and a stable derivator $\mathscr{D}$, we construct a certain poset $P = P_{Q_1,Q_2}$ together with suitable combinations of fully faithful Kan extensions morphisms

$$\mathscr{D}^{Q_1} \to \mathscr{D}^{P} \qquad \text{and} \qquad \mathscr{D}^{Q_2} \to \mathscr{D}^{P}.$$

The stability of $\mathscr{D}$will then imply that in both cases the essential image consist of precisely the same representations of P (which are determined by certain exactness conditions such as the vanishing on certain objects or the fact that certain squares are *bicartesian*). In fact, in Example 4.9 the poset $P = \square$ was enough, while in Example 4.10 we considered the subposet $P = C \subseteq [1] \times [2]$ obtained by removing the final vertex $(1, 2)$. For a direct proof of the A_n-version of Corollary 4.11 we refer the reader to [**GŠ14**, §6]. Here we instead prefer to present the more systematic approach as considered in [**GŠ16a**] which also makes more precise the connection to (higher) triangulations (and hence the content of §2 and the construction of canonical triangulations in §3). Luckily, it turns out that for every fixed n there is poset which conveniently encodes all A_n-quivers with arbitrary orientations, and this poset will be described in detail in the following construction.

CONSTRUCTION 4.12. We recall that every quiver Q has associated to it a repetitive quiver $\widehat{Q}$ with the following description: vertices in it are pairs (k, q) with $k \in \mathbb{Z}$ and $q \in Q$ while associated to every edge $\alpha\colon q_1 \to q_2$ in Q there are the two edges $\alpha\colon (k, q_1) \to (k, q_2)$ and $\alpha^*\colon (k, q_2) \to (k+1, q_1)$ in $\widehat{Q}$. Here we are only interested in the special case of the linearly oriented A_n-quiver $\vec{A}_n = (1 < \ldots < n)$. In the special case of $n = 3$ the repetitive quiver of $\vec{A}_3$ takes the form:

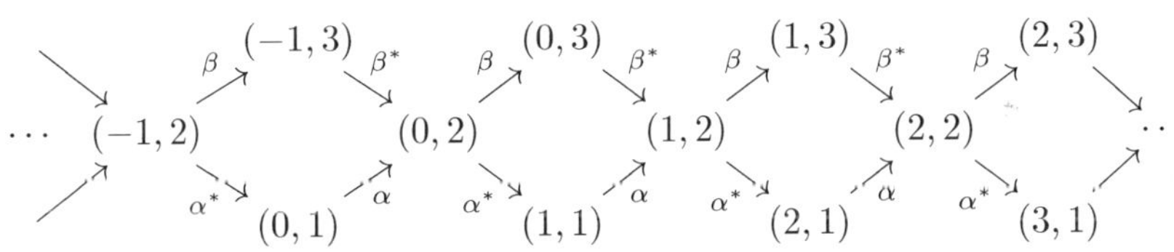

Let $M_{\vec{A}_n}$ be the category which is obtained from the repetitive quiver of $\vec{A}_n$ by making all squares commutative. By abuse of terminology, we call this poset $M_{\vec{A}_n}$ the **mesh category**.

We want to show that, for stable derivators, representations of $\vec{A}_n$ can be equivalently encoded by suitable representations of mesh categories. It is worth to compare the following to the discussion of the rotation axiom in §2.3 and of octahedral axiom in §2.4.

CONSTRUCTION 4.13. For every $n \geq 0$ we introduce the following short-hand notation for mesh categories:

$$M_n = M_{[n+1]} \quad \text{for} \quad [n+1] = (0 < \ldots < n+1).$$

Given a stable derivator $\mathscr{D}$, we denote by $\mathscr{D}^{M_n,\mathrm{ex}} \subseteq \mathscr{D}^{M_n}$ the full subderivator spanned by all representations which

(i) vanish on the boundary stripes (i.e., at $(k, 0), (k, n+1)$ for all $k \in \mathbb{Z}$)
(ii) and which make all squares bicartesian.

The fact that this is a derivator is a consequence of Theorem 4.15 since derivators are closed under equivalences of prederivators. In order to relate $\mathscr{D}^{M_n,\mathrm{ex}}$ to $\mathscr{D}^{\vec{A}_n}$, we note that there is the fully faithful functor

$$i\colon \vec{A}_n \to M_n\colon l \mapsto (0,l) \tag{4.14}$$

The following is a derivatorish version of Theorem 2.45 and Theorem 2.51. In fact, the proofs of these two theorems were modeled after the proof of the following result.

THEOREM 4.15. *For every stable derivator $\mathscr{D}$ and $n \geq 0$ restriction along i* (4.14) *induces an equivalence of derivators*

$$i^*\colon \mathscr{D}^{M_n,\mathrm{ex}} \xrightarrow{\sim} \mathscr{D}^{\vec{A}_n}.$$

This equivalence is pseudo-natural with respect to exact morphisms, and the inclusion $\mathscr{D}^{M_n,\mathrm{ex}} \to \mathscr{D}^{M_n}$ is exact.

PROOF. We only sketch the proof and refer the reader to [**GŠ14**, Thm. 4.6] for details. For every stable derivator $\mathscr{D}$ we want to construct an inverse equivalence of i^*. Thus, given a representation X of $\vec{A}_n$ with values in $\mathscr{D}$ we want to obtain a coherent diagram of shape M_n satisfying the defining exactness conditions of $\mathscr{D}^{M_n,\mathrm{ex}}$. To this end, similarly to the sketch proofs of Theorem 2.45 and Theorem 2.51, we note that the inclusion (4.14) factors as a composition of inclusions of full subcategories

$$i\colon \vec{A}_n \xrightarrow{i_1} K_1 \xrightarrow{i_2} K_2 \xrightarrow{i_3} K_3 \xrightarrow{i_4} M_n$$

where

(i) K_1 contains all objects from $\vec{A}_n$ and the objects $(k,n+1)$ for $k \geq 0$ and $(k,0)$ for $k > 0$,
(ii) K_2 is obtained from K_1 by adding the objects $(k,l), k > 0$, and
(iii) K_3 contains all objects from K_2 and the objects $(k,n+1)$ for $k < 0$ and $(k,0)$ for $k \leq 0$.

The inclusion i_4 thus adds the remaining objects in the negative k-direction. By Proposition 3.23, associated to these fully faithful functors there are fully faithful Kan extension functors

$$\mathscr{D}^{\vec{A}_n} \xrightarrow{(i_1)_*} \mathscr{D}^{K_1} \xrightarrow{(i_2)_!} \mathscr{D}^{K_2} \xrightarrow{(i_3)_!} \mathscr{D}^{K_3} \xrightarrow{(i_4)_*} \mathscr{D}^{M_n}.$$

These Kan extension morphisms can be analyzed in turn, and the conclusion is that $(i_1)_*$ adds zero objects on the boundary stripes in the positive direction, $(i_2)_!$ adds bicartesian squares in the positive direction, $(i_3)_!$ adds zero objects in on the boundary stripes in the negative direction, and finally $(i_4)_*$ fills up by bicartesian squares in the negative direction. Thus the essential image agrees with $\mathscr{D}^{M_n,\mathrm{ex}}$, and this concludes the construction of an equivalence $F_{\vec{A}_n}\colon \mathscr{D}^{\vec{A}_n} \xrightarrow{\sim} \mathscr{D}^{M_n,\mathrm{ex}}$ which is inverse to i^*. Again, since we only used restrictions and sufficiently finite Kan extension morphisms, both i^* and its inverse $F_{\vec{A}_n}$ are pseudo-natural with respect to exact morphisms (Remark 3.61). □

Variants of this theorem for different orientations of A_n-quivers lead to the following result.

THEOREM 4.16. *Let $n \geq 0$ be fixed. All A_n-quivers are strongly stably equivalent.*

PROOF. We sketch very roughly the main idea of the proof, and for this purpose we consider an arbitrarily oriented A_n-quiver Q. For every such Q there are admissible embeddings $i_Q\colon Q \to M_n$ and the corresponding restriction morphisms $(i_Q)^*\colon \mathscr{D}^{M_n} \to \mathscr{D}^Q$ again restrict to equivalences $(i_Q)^*\colon \mathscr{D}^{M_n,\mathrm{ex}} \xrightarrow{\sim} \mathscr{D}^Q$. In this case it is more tricky to write down in closed form the inverse equivalence F_Q (which depends on Q and the choice of the admissible embedding i_Q).

To illustrate this step, we content ourselves by one example, but we invite the reader to come up with additional examples. In the case of $n = 3$ and the source of valence two $Q = (1 \leftarrow 2 \to 3)$, we can consider the admissible embedding $i_Q\colon Q \to M_3$ given by

$$2 \mapsto (0,2), \quad 1 \mapsto (1,1), \quad \text{and} \quad 2 \mapsto (0,3).$$

The corresponding equivalence F_Q sends a representation X looking like

$$\begin{array}{ccc} x & \xrightarrow{g} & y \\ {\scriptstyle f}\downarrow & & \\ z & & \end{array}$$

to a coherent diagram as in Figure 8. Therein, the boundary stripes are populated by zero objects and all squares are bicartesian. To put it differently, the diagram $F_Q(X)$ is a refined octahedral diagram of $(Ff \to x \to y)$ (as in §2.4).

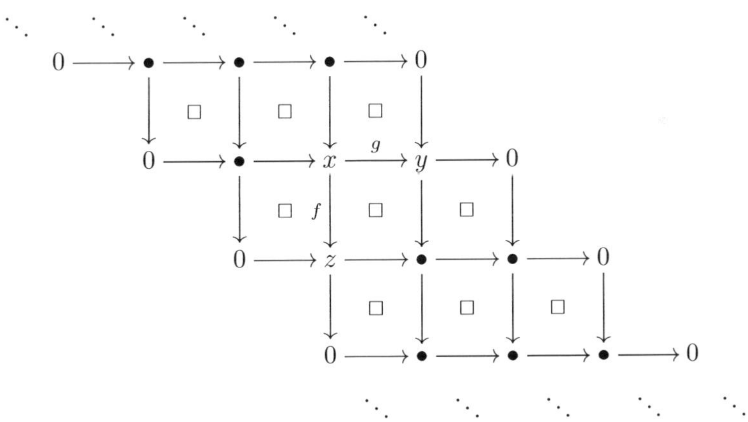

FIGURE 8. The equivalence F_Q for $Q = (\bullet \leftarrow \bullet \to \bullet)$

Now, given an additional A_n-quiver Q' we choose an admissible embedding $i_{Q'}\colon Q' \to M_n$, and obtain the corresponding inverse equivalence $F_{Q'}$ of $(i_{Q'})^*$. In order to conclude the proof it suffices to make the definition

$$\Phi_{Q',Q} = (i_{Q'})^* \circ F_Q\colon \mathscr{D}^Q \xrightarrow{\sim} \mathscr{D}^{M_n,\mathrm{ex}} \xrightarrow{\sim} \mathscr{D}^{Q'}.$$

As a composition of pseudo-natural equivalences, this defines a strong stable equivalence $\Phi_{Q',Q}\colon Q \overset{\mathrm{s}}{\sim} Q'$. □

The strong stable equivalences constructed in Theorem 4.16 arise as finite compositions from certain basic building blocks. These building blocks admit a fairly elementary description as we discuss next, and they turn out to be special cases of the more general reflection morphisms from §4.3.

CONSTRUCTION 4.17. Let $n \geq 0$, let $\mathscr{D}$be a stable derivator, and let M_n be the mesh category. Choosing two A_n-quivers Q, Q' and admissible embeddings $i_Q, i_{Q'}$ of the quivers to M_n, the resulting strong stable equivalence $\Phi_{Q',Q}\colon \mathscr{D}^Q \xrightarrow{\simeq} \mathscr{D}^{Q'}$ measures the difference between the restriction morphisms $(i_Q)^*$ and $(i_{Q'})^*$. The above-mentioned basic building blocks arise when the two embeddings "differ by one square only". Instead of making this precise by explicit combinatorial formulas, we illustrate this by an example related to Figure 8.

As in the proof of Theorem 4.16, let $Q = (\bullet \leftarrow \bullet \rightarrow \bullet)$ and let $i_Q\colon Q \to M_3$ be the embedding indicated by Figure 8. If $Q' = (\bullet \rightarrow \bullet \leftarrow \bullet)$ is the sink of valence two, then there is an obvious embedding of $i_{Q'}\colon Q' \to M_3$ which has two objects in common with i_Q but differs from it by one square. In this case, the resulting strong stable equivalence $\Phi_{Q',Q}$ agrees with the one from Example 4.9. Alternatively, we can consider the linearly oriented quiver $\vec{A}_3 = (\bullet \rightarrow \bullet \rightarrow \bullet)$ with the standard embedding $i\colon \vec{A}_3 \to M_3$ as in (4.14). Also these embeddings differ by one square only and we reproduce the strong stable equivalence from Example 4.10. Finally, there is the additional vertical embedding $j\colon \vec{A}_3 \to M_3$ and in this case the strong stable equivalence $Q \overset{s}{\sim} \vec{A}_3$ forms the fiber of the morphism labeled by g in Figure 8.

All these three cases are special cases of reflection functors. Given an A_n-quiver Q and a source $a \in Q$, let $\sigma_a Q$ be the A_n-quiver obtained by reorienting all edges adjacent to a such that the vertex $a \in \sigma_a Q$ now is a sink. By Theorem 4.16 the quivers Q and $Q' = \sigma_a Q$ are strongly stably equivalent, and a preferred such equivalence is constructed as follows. Let $i_Q\colon Q \to M_n$ be an admissible embedding and let $i_{Q'} = i_{\sigma_a Q}\colon Q' = \sigma_a Q \to M_n$ be the induced admissible embedding which differs from i_Q by one square only. The strong stable equivalence

$$s_a^- = \Phi_{Q',Q} = (i_{Q'})^* \circ F_Q\colon \mathscr{D}^Q \xrightarrow{\simeq} \mathscr{D}^{Q'}$$

and its inverse

$$s_a^+ = \Phi_{Q,Q'} = (i_Q)^* \circ F_{Q'}\colon \mathscr{D}^{Q'} \xrightarrow{\simeq} \mathscr{D}^Q$$

are **reflection functors**. It is easy to see that these morphisms do not depend on the choice of the embeddings i_Q and $i_{Q'}$ as long as they differ by precisely one square.

The strong stable equivalences from Theorem 4.16 can be described as compositions of the above reflection morphisms, and by the following remark the order of the reflections does not matter.

REMARK 4.18. Let Q be an A_n-quiver and let $a_1, a_2 \in Q$ be two distinct sinks. In that case the vertex a_2 is also a sink in the reflected quiver $\sigma_{a_1} Q$ and we can hence iterate the reflection thereby obtaining $\sigma_{a_2}\sigma_{a_1} Q$. A straightforward argument shows that at the level of the quivers the order of the reflections is irrelevant and definition

$$\sigma_{\{a_1,a_2\}} Q = \sigma_{a_1}\sigma_{a_2} Q = \sigma_{a_2}\sigma_{a_1} Q$$

hence makes sense. Playing a bit with the corresponding embeddings, we note that for the corresponding reflection functors there are natural isomorphisms

$$s^+_{a_1}s^+_{a_2} \cong s^+_{a_2}s^+_{a_1} : \mathscr{D}^Q \xrightarrow{\sim} \mathscr{D}^{\sigma\{a_1,a_2\}Q}.$$

Of course there are similar results for sinks instead of sources.

DEFINITION 4.19. An **admissible sequence of sinks** in a finite quiver Q is a total ordering $(a_1, \ldots, a_n)$ of all vertices of Q such that a_1 is a sink in Q and a_i is a sink in $\sigma_{a_{i-1}} \cdots \sigma_{a_1} Q$ for all $2 \leq i \leq n$. **Admissible sequences of sources** are defined dually.

It turns out that every finite, acyclic quiver admits an admissible sequence of sources and sinks [**BGP73**, Lemma 1.2(1)]. Moreover, one can show that the quiver $\sigma_{a_n} \cdots \sigma_{a_1} Q$ agrees with the original quiver Q and similarly in the case of sources. In what follows we only need these results for Dynkin quivers of type A, in which case the arguments are more straightforward. While admissible sequences always exist, they are by no means unique as simple examples show. However, at the level of iterated reflection functors there is the following result.

CONSTRUCTION 4.20. Let $n \geq 0$ and let Q be an A_n-quiver. For every stable derivator $\mathscr{D}$and every admissible sequence of sinks $(a_1, \ldots, a_n)$ in Q there is the iterated reflection functor

$$\Phi^+_{(a_1,\ldots,a_n)} = s^+_{a_n} \circ \ldots \circ s^+_{a_1} : \mathscr{D}^Q \to \mathscr{D}^Q.$$

A combination of Remark 4.18 with some combinatorial arguments imply that up to natural isomorphism this iterated reflection functor is independent of $(a_1, \ldots, a_n)$ (see the proof of [**BGP73**, Lemma 1.2(3)]). Of course, the same observation applies to iterated reflections $\Phi^-_{(b_1,\ldots,b_n)}$ at admissible sequences of sources.

By Construction 4.20 the following functors are well-defined up to natural isomorphisms.

DEFINITION 4.21. Let $\mathscr{D}$be a stable derivator and let Q be an A_n-quiver.

(i) The **Coxeter functor** $\Phi^+ = \Phi^+_Q \colon \mathscr{D}^Q \to \mathscr{D}^Q$ is $\Phi^+ = \Phi^+_{(a_1,\ldots,a_n)}$ for some admissible sequence of sinks $(a_1, \ldots, a_n)$ in Q.

(ii) The **Coxeter functor** $\Phi^- = \Phi^-_Q \colon \mathscr{D}^Q \to \mathscr{D}^Q$ is $\Phi^- = \Phi^-_{(b_1,\ldots,b_n)}$ for some admissible sequence of sources $(b_1, \ldots, b_n)$ in Q.

We illustrate these reflections and Coxeter functors by an example.

EXAMPLE 4.22. The linearly oriented A_3-quiver $\vec{A_3} = (1 \to 2 \to 3)$ has a unique admissible sequence of sinks $(3, 2, 1)$. Consequently, for every stable derivator $\mathscr{D}$the Coxeter functor is given by

$$\Phi^+ = s^+_1 \circ s^+_2 \circ s^+_3 : \mathscr{D}^{\vec{A_3}} \xrightarrow{\sim} \mathscr{D}^{\vec{A_3}}.$$

Given a representation $X = (x \xrightarrow{f} y \xrightarrow{g} z)$ in $\mathscr{D}$, to calculate $\Phi^+(X)$ we first determine $s^+_3(X)$ by considering the diagram on the left in

$$\begin{array}{ccccc} & & Fg & \longrightarrow & 0 \\ & & \downarrow & \square & \downarrow \\ x & \xrightarrow[f]{} & y & \xrightarrow[g]{} & z, \end{array} \qquad \begin{array}{ccccc} F(gf) & \longrightarrow & Fg & \longrightarrow & 0 \\ \downarrow & \square & \downarrow & \square & \downarrow \\ x & \xrightarrow[f]{} & y & \xrightarrow[g]{} & z. \end{array}$$

In order to describe $s_2^+s_3^+(X)$ we next reflect $(x \to y \leftarrow Fg)$ at the vertex decorated by y. This is achieved by extending the above diagram by one more bicartesian square. Since the pasting of the squares is again bicartesian, the upper left corner is indeed populated by $F(gf)$. As a final step, it remains to reflect the representation $(x \leftarrow F(gf) \to Fg)$ at the vertex decorated by x. For this purpose we extend the diagram by one more fiber square in order to obtain

$$\begin{array}{ccccccc} \Omega z & \longrightarrow & F(gf) & \longrightarrow & Fg & \longrightarrow & 0 \\ \downarrow & \square & \downarrow & \square & \downarrow & \square & \downarrow \\ 0 & \longrightarrow & x & \xrightarrow[f]{} & y & \xrightarrow[g]{} & z. \end{array}$$

Again, the outer most square is bicartesian and hence a loop square, and the upper left corner consequently agrees with Ωz. As an upshot we conclude that the Coxeter functor Φ^+ is given by

$$\Phi^+(x \xrightarrow{f} y \xrightarrow{g} z) = \big(\Omega z \to F(gf) \to Fg\big). \tag{4.23}$$

Similarly, the quiver $\vec{A}_2 = (1 \to 2)$ has $(2,1)$ as unique admissible sequence of sinks. We leave it to the reader to check that in this simpler case the Coxeter functor Φ^+ is given by

$$\Phi^+ \cong \mathsf{fib}^2\colon \mathscr{D}^{\vec{A}_2} \xrightarrow{\sim} \mathscr{D}^{\vec{A}_2}. \tag{4.24}$$

EXAMPLE 4.25. Let $\mathscr{D}$ be a stable derivator and let $Q = (1 \leftarrow 2 \to 3)$ be the source of valence two. This quiver has $(1,3,2)$ and $(3,1,2)$ as admissible sequences of sinks. In order to calculate the Coxeter functor $\Phi^+\colon \mathscr{D}^Q \xrightarrow{\sim} \mathscr{D}^Q$ we consider an arbitrary representation $X = (z \leftarrow x \to y)$ and extend it to the diagram

$$\begin{array}{ccccc} w & \longrightarrow & Fg & \longrightarrow & 0 \\ \downarrow & \boxed{2} & \downarrow & \boxed{3} & \downarrow \\ Ff & \longrightarrow & x & \xrightarrow{g} & y \\ \downarrow & \boxed{1} & \downarrow{\scriptstyle f} & \boxed{0} & \downarrow \\ 0 & \longrightarrow & z & \longrightarrow & y \cup_x z \end{array}$$

consisting of bicartesian cubes. Ignoring the lower right square for the moment, each of the remaining three squares will be used to define a reflection at a sink and the labels match the corresponding vertices. This illustrates again the commutativity of reflections at different sinks, since independently of the order we have

$$s_1^+s_3^+(X) \cong s_3^+s_1^+(X) \cong (Ff \to x \leftarrow Fg).$$

To determine $\Phi^+(X)$ it suffices to understand the upper left corner. For this purpose we also drew the remaining auxiliary cocartesian square $\boxed{0}$, since it allows us to conclude that the total pasting of these squares is a loop square. Consequently, there is a canonical isomorphism $w \cong \Omega(y \cup_x z)$ and we obtain

$$\Phi^+(z \xleftarrow{f} x \xrightarrow{g} y) \cong (Ff \leftarrow \Omega(y \cup_x z) \to Fg). \tag{4.26}$$

Motivated by the classical situation (see [**RVdB02**, §I.2]) we make the following definition.

DEFINITION 4.27. Let $\mathscr{D}$ be a stable derivator and let Q be an A_n-quiver. The **Serre functor** of $\mathscr{D}^Q$ is

$$S = \Sigma\Phi^+ \cong \Phi^+\Sigma\colon \mathscr{D}^Q \to \mathscr{D}^Q.$$

In order to put this definition into context we recall the notion of a Serre functor. Serre functors have been formalized in various contexts and references include [**BK89**, **BO01**] in the framework of k-linear categories, [**Che11**] in the context of k-linear triangulated categories, and [**LM08**] in the realm of (pretriangulated) A_∞-categories (see also [**BK90**, **KS09**, **BLM08**]). Here we are mainly interested in the case of sufficiently well-behaved triangulated categories (see [**RVdB02**, §I.1] and [**BK89**]).

DEFINITION 4.28. Let k be a field and let $\mathcal{T}$ be a k-linear triangulated category with finite-dimensional mapping vector spaces. An exact auto-equivalence $S\colon \mathcal{T} \to \mathcal{T}$ is a **Serre functor** if it is endowed with an isomorphism

$$\hom_{\mathcal{T}}(x,y) \xrightarrow{\cong} \big(\hom_{\mathcal{T}}(y,Sx)\big)^*$$

which is natural in x and y.

Here, $(-)^*$ of course denotes the vector space duals. As a consequence of the Yoneda lemma we see that Serre functors are essentially unique if they exist, so the existence of a Serre functor is really a property of the triangulated category. Let us recall the following two typical examples of Serre functors.

EXAMPLES 4.29.

(i) The example which is of key interest to here is given by bounded derived categories of finitely generated modules over finite dimensional algebras of finite global dimension ([**Hap87**, §3.6], [**KL06**]). In that case, Serre functors exist and they are given by *derived Nakayama functors* [**Hap87**, §3.6]. We will come back to this explicit description in §4.5.

(ii) Let k be an algebraically closed field, let X be a smooth, projective variety of dimension n over k, and let $D^b(X)$ be the bounded derived category of coherent sheaves on X. The derived tensor product with the shifted canonical line bundle

$$S_X = \omega_X[n] \otimes -\colon D^b(X) \xrightarrow{\sim} D^b(X)$$

is a Serre equivalence on $D^b(X)$ [**Huy06**, Thm. 3.12]. Relevant classical references include [**BK89**, **BO01**, **Ser55**] and also [**Muk81**, **Huy06**].

Let us now return to the situation in abstract representation theory, and let us revisit Example 4.22 and Example 4.25.

EXAMPLES 4.30. There are the following examples of Serre functors.

(i) The Serre functor of $\vec{A}_2$ is $S \simeq \mathsf{cof}\colon \mathscr{D}^{\vec{A}_2} \to \mathscr{D}^{\vec{A}_2}$ as it easily follows from (4.24) and (3.52).

(ii) In the case of the linearly oriented A_3-quiver $\vec{A}_3$ it follows from (4.23) that the Serre functor is given by

$$S(x \xrightarrow{f} y \xrightarrow{g} z) = \big(z \to C(gf) \to Cg\big).$$

(iii) Similarly, in the case of $Q = (1 \leftarrow 2 \to 3)$ we can recycle (4.26) in order to conclude that the Serre functor has the description

$$S(z \xleftarrow{f} x \xrightarrow{g} y) \cong (Cf \leftarrow y \cup_x z \to Cg).$$

Remark 4.31. The justification for the terminology Serre functors in abstract representation theory is as follows. Specialized to the derivator $\mathscr{D}_k$ of a field, these Serre functors induce on underlying categories the classical Serre functors from representation theory. For an additional example of Serre functors in abstract representation we refer to [**BG18**]. Jointly with Falk Beckert, we intend to study these abstract Serre functors more systematically elsewhere.

For more details on the abstract representation theory of Dynkin quivers of type A we refer to [**GŠ16a**]. In particular, therein is also a discussion of an abstract fractional Calabi–Yau property [**Kel05a, Kel08, KS12, vR12**] which plays a key role in the context of spectral Picard groups. These Picard groups will be briefly discussed in Remark 4.109. We conclude this subsection by a remark on *higher triangulations.*

Remark 4.32. Despite the fact that triangulated categories are extremely useful in plenty of situations, from the early days on people were pointing out defects of the theory (see for instance the introduction to the early [**Hel68**]). Besides the enhancements in the sense of stable model categories, stable ∞-categories or differential-graded categories, there are the attempts to fix some of the deficiencies of triangulated categories by also keeping track of *higher triangles.* The idea of considering higher triangles already appears in [**BBD82**, Remark 1.1.14], while a more recent definition of strong triangulations is in [**Mal05a**] (see also the closely related notion in [**Bal11**] as well as the thesis of Beckert [**Bec18**]).

We sketch the rough idea and consider an additive category $\mathcal{A}$ together with an auto-equivalence $\Sigma\colon \mathcal{A} \xrightarrow{\simeq} \mathcal{A}$. An *n-angle* in $(\mathcal{A}, \Sigma)$ is essentially an ordinary diagram $X\colon M_n \to \mathcal{A}$ which vanishes on the boundary stripes and such that the triangular fundamental domains match up to suitable powers of the suspension Σ. Thus, 2-angles correspond to the usual triangles in the sense of Puppe and Verdier, while 3-angles are closely related to octahedral diagrams; we refer to [**Mal05a**] for more details. A *strong triangulation* or *∞-triangulation* on $(\mathcal{A}, \Sigma)$ is given by classes of distinguished n-angles for $n \geq 2$ subject to various axioms. By [**GŠ16a**, Thm. 13.6] the values of a strong, stable derivator $\mathscr{D}$admit canonical strong triangulations. In fact, analogously to Remark 3.72, there is a lift of $\mathscr{D}$to the 2-category of strong triangulated categories (as follows from the details in [**GŠ16a**, §13] and [**Gro16**]). Moreover, the coherent versions of these distinguished n-triangles (thus objects in $\mathscr{D}^{M_n,\mathrm{ex}}$) in combination with the Mayer–Vietoris triangles from [**GPS14b**] give rise to many additional triangles in the underlying category $\mathscr{D}(*)$.

4.3. Reflection functors. In this subsection we discuss additional examples of strong stable equivalences given by reflection functors. Happel [**Hap87**, §1.7] showed that for every finite, acyclic quiver reflection functors at sources or sinks induce derived equivalences between the corresponding category algebras. In [**GŠ15**] these derived equivalences were shown to be shadows of corresponding strong stable equivalences (even without the finiteness or acyclicity assumption). We illustrate these reflection functors by some important special cases.

Let us begin this subsection by the following obstruction against the existence of strong stable equivalence.

Proposition 4.33. *If the finite acyclic quivers Q and Q' are strongly stably equivalent, then the underlying unoriented graphs are isomorphic.*

PROOF. This is [**GŠ14**, Prop. 5.3] which in turn relies essentially on [**Hap87**]. □

Thus, in the context of finite acyclic quivers all that can happen by strong stable equivalences are reorientations of some of the edges. These can not be carried out arbitrary as the case of oriented cycles shows (see Corollary 4.46). However, a valid reorientation is the classical reflection of (acyclic) quivers at sources and sinks (see, for instance, [**ASS06**, §VII.5], or [**Kra08**] for more detail). Let us recall this construction.

CONSTRUCTION 4.34. Let Q be a quiver and let $q \in Q$. The vertex q is a **source** if there are no edges in Q with q as target, and there is dual notion of a **sink** of a quiver. Given a source q in a quiver, the **reflected quiver** $\sigma_q Q$ is obtained from Q by turning the source into a sink. Thus, both quivers have the same vertices and their is a bijection between the edges. The only difference is that the orientations of all edges in Q that start at the source $q \in Q$ have been reversed in $\sigma_q Q$ so that they now end at $q \in \sigma_q Q$.

Special cases of such reflections already occurred in §4.2. For instance, in the case of $\vec{A}_3$-quivers, Example 4.9 and Example 4.10 provide examples of strong stable equivalences

$$(1 \leftarrow 2 \rightarrow 3) \overset{s}{\sim} \sigma_2(1 \leftarrow 2 \rightarrow 3) = (1 \rightarrow 2 \leftarrow 3)$$

and

$$(1 \leftarrow 2 \rightarrow 3) \overset{s}{\sim} \sigma_1(1 \leftarrow 2 \rightarrow 3) = (1 \rightarrow 2 \rightarrow 3).$$

These reflections at the level of quivers are accompanied by the following reflection functors.

CONSTRUCTION 4.35. Let Q be a finite quiver, let $q_0 \in Q$ be a source, and let k be a field. For every representation $M\colon Q \to \mathrm{Mod}(k)$ we obtain an induced representation $s^-_{q_0} M\colon \sigma_{q_0} Q \to \mathrm{Mod}(k)$ of the reflected quiver as follows. For all vertices $q' \neq q_0 \in \sigma_{q_0} Q$ we set $(s^-_{q_0} M)_{q'} = M_{q'}$. To describe the remaining component, let $q_1, \ldots, q_n$ denote the finitely many vertices of Q such that there are edges $q_0 \to q_i$ for $i = 1, \ldots, n$. Then we define

$$(s^-_{q_0} M)_{q_0} = \mathrm{cok}(M_{q_0} \to M_{q_1} \oplus \ldots \oplus M_{q_n}),$$

and the canonical homomorphisms yield the required maps $(s^-_{q_0} M)_{q_i} \to (s^-_{q_0} M)_{q_0}$. This construction is clearly functorial in M, and we hence obtain the **reflection functor** $s^-_{q_0}\colon \mathrm{Mod}(kQ) \to \mathrm{Mod}(k\sigma_{q_0} Q)$ associated to Q and the source q_0. Dually, if q_0 is a sink, then there is a reflection functor $s^+_{q_0}\colon \mathrm{Mod}(kQ) \to \mathrm{Mod}(k\sigma_{q_0} Q)$.

At the level of abelian categories of modules these reflection functors fail to be equivalences (see, for example, [**ASS06**, Thm. VII.5.3]), but there is the following positive derived version of it.

THEOREM 4.36. *Let k be a field, let Q be a finite acyclic quiver, let $q_0 \in Q$ be a source, and let $Q' = \sigma_{q_0} Q$ be the reflected quiver. The reflection functors induce an exact equivalence of bounded derived categories*

$$(Ls^-_{q_0}, Rs^+_{q_0})\colon D^b(kQ) \overset{\Delta}{\simeq} D^b(kQ').$$

PROOF. This is due to Happel and can be found in [**Hap87**, §1.7]. □

It can be shown that these reflection functors extend to abstract stable homotopy theories, and by now there are two approaches to such a generalization. Let us recall from Construction 4.34 and Construction 4.35, that these constructions rely on some "local modifications at the level of shapes and representations". Correspondingly, the idea of gluing shapes and representations is central to this business. Depending on the formal framework, there are different ways of dealing with it.

(i) Similar to the case of triangulated categories, gluing is not available in the 2-category $\mathcal{DER}_{\mathrm{St,ex}}$ of stable derivators. But in that framework (and this project began in this framework of derivators), one can partially sail around this defect by more involved combinatorial arguments. This approach was taken in [**GŠ14, GŠ16b, GŠ15**], where representation-theoretic techniques were developed (one-point extensions and homotopical epimorphisms).

(ii) In contrast to this, the more flexible language of stable ∞-categories allows, among many other things, for gluing. At the price of quoting results from the more involved machinery [**Lur09, Lur14**], a very elegant construction of reflection functors in stable ∞-categories was worked out by Dyckerhoff, Jasso, and Walde in [**DJW19**].

Because of its importance, we single out the special case of a source and a sink of finite valence. We begin by recalling some notation from [**GŠ16b**] and [**BG18**].

NOTATION 4.37. For $n \geq 1$ we denote by $\square^n$ the n-cube which is to say the n-fold product

$$\square^n = [1] \times \ldots \times [1].$$

The n-cube $\square^n$ can be realized as the power set of $\{1, \ldots, n\}$ with the ordering given by inclusion. Correspondingly, the n-cube admits the following filtration by full subcategories. For $0 \leq l \leq n$ we denote by $\square^n_{0\leq l}$ the full subposet spanned by the subsets of cardinality at most l. The subposet $\square^n_{k\leq l}$ is defined similarly, and there are fully faithful inclusion functors $\iota_{k,l} \colon \square^n_{k\leq l} \to \square^n$. As special cases, we obtain the inclusion of the source and the sink of valence n,

$$\iota_{0,1} \colon \square^n_{0\leq 1} \to \square^n \quad \text{and} \quad \iota_{n-1\leq n} \colon \square^n_{n-1\leq n} \to \square^n.$$

Similarly, there are the inclusions of the two possible punctured n-cubes

$$\iota_{0,n-1} \colon \square^n_{0\leq n-1} \to \square^n \quad \text{and} \quad \iota_{1\leq n} \colon \square^n_{1\leq n} \to \square^n.$$

DEFINITION 4.38. Let $\mathscr{D}$ be a derivator, let $n \geq 1$, and let $X \in \mathscr{D}(\square^n)$. The n-cube X is **cocartesian** if it lies in the essential image of $(\iota_{0,n-1})_! \colon \mathscr{D}(\square^n_{0\leq n-1}) \to \mathscr{D}(\square^n)$. The n-cube X is **strongly cocartesian** if it lies in the essential image of the functor $(\iota_{0,1})_! \colon \mathscr{D}(\square^n_{0\leq 1}) \to \mathscr{D}(\square^n)$.

Of course, there are dual notions of **(strongly) cartesian n-cubes**. For squares there is obviously no difference between the two notions in Definition 4.38. In higher dimension every strongly cocartesian n-cube is cocartesian since $\iota_{0,1} \colon \square^n_{0\leq 1} \to \square^n$ factors through $\square^n_{0\leq n-1}$. But there is no converse to this and, in fact, there is the following more precise result.

THEOREM 4.39. *Let $\mathscr{D}$ be a derivator, let $n \geq 2$, and let $X \in \mathscr{D}(\square^n)$. The n-cube X is strongly cocartesian if and only if every subsquare of it is cocartesian.*

PROOF. This is an entirely combinatorial problem and the details can be found in [**GŠ14**, Thm. 8.4]. $\square$

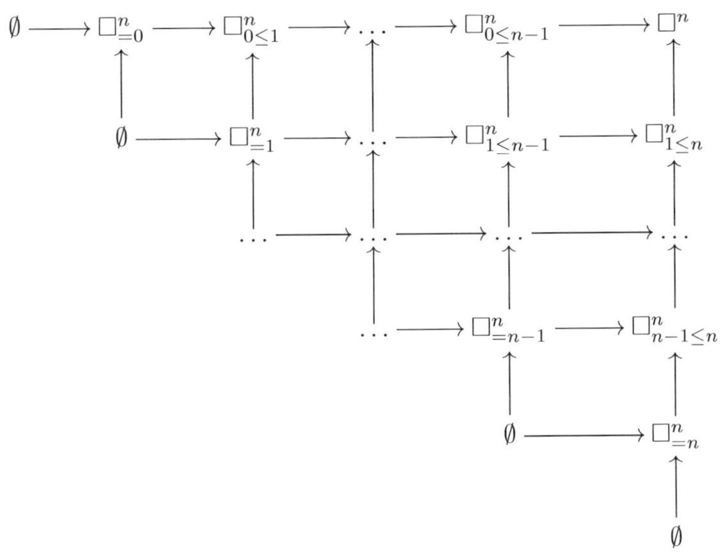

FIGURE 9. The cardinality filtration of the n-cube $\square^n$

COROLLARY 4.40. *Let $\mathscr{D}$ be a stable derivator and let $n \geq 2$. An n-cube in $\mathscr{D}$ is strongly cocartesian if and only if it is strongly cartesian.*

PROOF. This follows immediately from Theorem 4.39 and Definition 3.30. $\square$

We refer to this common class of n-cubes as the class of **strongly bicartesian** n-cubes.

COROLLARY 4.41. *For every fixed $n \geq 2$ the source and the sink of of valence n are strongly stably equivalent,*

$$\square^n_{0\leq 1} \overset{s}{\sim} \square^n_{n-1\leq n}.$$

PROOF. It is straightforward to adapt the proof of Example 4.9 to our current situation (invoking Corollary 4.40 instead of Definition 3.30). $\square$

REMARK 4.42. The way we have presented Corollary 4.41, it is an immediate consequence of the calculus of n-cubes in stable derivators. This cubical calculus was developed more systematically in [**BG18**], and here we want to mention some of the key results. Using suggestive notation (compare to Notation 4.37), the cardinality filtration of the n-cube

$$\emptyset \longrightarrow \square^n_{=0} \longrightarrow \square^n_{\leq 1} \longrightarrow \ldots \longrightarrow \square^n_{\leq n-1} \longrightarrow \square^n$$

leads to an interpolation between cocartesian and strongly cocartesian n-cubes. The differences between these individual filtration layers are given by the full subposets $\square^n_{k\leq l} \subseteq \square^n, 0 \leq k \leq l \leq n$, which we refer to as **chunks of n-cubes**. The chunks and the inclusion functors between them are neatly organized by Figure 9. In *loc. cit.* the authors study in detail the calculus of iterated partial cofiber constructions of n-cubes and their interaction with functors in Figure 9. As a generalization of Corollary 4.41 there is a strong stable equivalence

$$\square^n_{k\leq l} \overset{\mathrm{s}}{\sim} \square^n_{n-l\leq n-k};$$

see [**BG18**, Thm. 10.15]. Moreover, for every fixed n these strong stable equivalences for the various $0 \leq k \leq l \leq n$ are suitably compatible with each other [**BG18**, Thm. 12.15].

An additional interesting symmetry comes from a further incarnation of Serre equivalences. For every $n \geq 1$ and every $0 \leq k \leq l \leq n$ there is a **Serre equivalence**

$$S = S_{k,l}\colon \square^n_{k\leq l} \overset{\mathrm{s}}{\sim} \square^n_{k\leq l}.$$

Again, also these Serre equivalences turn out to be compatible with the morphisms between the chunks. More specifically, in this situation, for every inclusion functor $i\colon \square^n_{k'\leq l'} \to \square^n_{k\leq l}$ in Figure 9 there is a canonical isomorphism

$$S_{k,l} \circ i_! \cong i_* \circ S_{k',l'};$$

see [**BG18**, Thm. 12.6]. To put this into plain English, in this case left and right Kan extensions match up to a conjugation by Serre equivalences. As a special case ([**BG18**, Cor. 12.12]), for every stable derivator $\mathscr{D}$there are canonical isomorphisms

$$\mathrm{colim}_{\square^n_{k\leq l}} \cong \lim_{\square^n_{k\leq l}} \circ S_{k,l}\colon \mathscr{D}^{\square^n_{k\leq l}} \to \mathscr{D}.$$

These compatibilities of local Serre equivalences with Kan extensions are the starting point of an on-going project on **global Serre dualities**. Jointly with Falk Beckert we intend to come back to this more systematically elsewhere.

Let us now return to the case of acyclic quivers. The approach to reflection functors for acyclic quivers in stable derivators taken in [**GŠ14, GŠ16b, GŠ15**] relies largely on the special case in Corollary 4.41. More precisely, the classical construction of reflection functors uses finite biproduct in vector spaces (Construction 4.35). In stable derivators finite biproducts and all the related inclusion and projection maps can be organized in suitable cubical diagrams. For instance, in the case of two objects $x, y \in \mathscr{D}(*)$ in the underlying category of a stable derivator $\mathscr{D}$, there is the associated diagram

$$\begin{array}{ccccc}
0 & \longrightarrow & x & \longrightarrow & 0 \\
\downarrow & \square & \downarrow & \square & \downarrow \\
y & \longrightarrow & x \oplus y & \longrightarrow & y \\
\downarrow & \square & \downarrow & \square & \downarrow \\
0 & \longrightarrow & x & \longrightarrow & 0,
\end{array}$$

in which all squares are bicartesian. Given a finite acyclic quiver and a source, such cubes can be glued to the quiver in order to then mimic the construcion of reflection functors in Construction 4.35. One of the main results of [**GŠ15**] is the following.

THEOREM 4.43. *Let Q be a finite acyclic quiver, let $q_0 \in Q$ be a source, and let $Q' = \sigma_{q_0} Q$ be the reflected quiver. The reflection functors define a strong stable equivalence*

$$(s^-_{q_0}, s^+_{q_0})\colon Q \overset{\mathrm{s}}{\sim} Q'.$$

PROOF. We refer the reader to [**GŠ15**] for a detailed construction of these reflection functors. The theorem appears as [**GŠ15**, Thm. 10.3] as a specializiation of a more general result ([**GŠ15**, Thm. 9.11]). □

COROLLARY 4.44. *Let T be a finite oriented tree and let T' be an arbitrary reorientation of T. There is a strong stable equivalence*

$$T \overset{\mathrm{s}}{\sim} T'.$$

PROOF. It is a purely combinatorial argument that a reorientation of a finite tree can be obtained by a finite number of reflections at sources and sinks ([**BGP73**, Thm. 1.2(1)]). The claim hence follows from Theorem 4.43. □

In fact, there is also a converse to this result as follows from a specialization to the derivator $\mathscr{D}_k$ of a field k.

REMARK 4.45. Theorem 4.43 is obtained by specialization of the following more general result. Let A be a small category and let $q_1, \ldots, q_n \in A$ be finitely many, not necessarily pairwisely different objects. The two categories which are obtained from A by freely gluing a source or a sink of valence n to the given objects are strongly stable equivalent [**GŠ15**, Thm. 9.11]. In particular, this shows that a variant of Theorem 4.43 also works for quivers which fail to be finite or acyclic.

In contrast to the case of trees, for acyclic quivers the underlying graphs to not determine the strong stable equivalence type. To illustrate this, we revisit the following classical example of orientations Q of an unoriented n-cycle

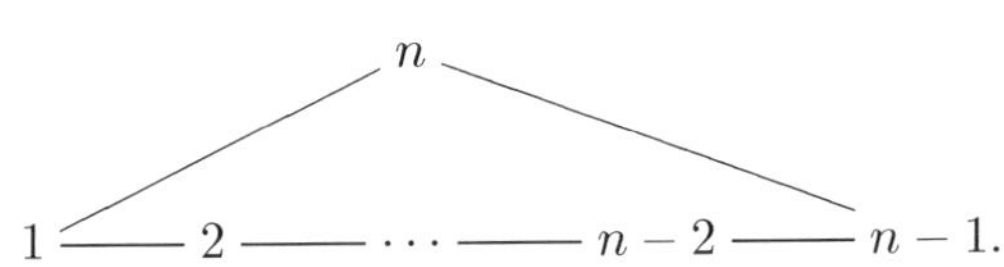

Such quivers are called **Euclidean** or **extended Dynkin** quivers of type $\widetilde{A}_{n-1}$ [**Rin84, SS07**]. For every orientation Q of the n-cycle we write $c(Q) = \{p, q\}$, if p arrows in Q are oriented clockwise and q are oriented counterclockwise.

COROLLARY 4.46. *If Q and Q' are n-cycles with $c(Q) = c(Q')$, then there is a strong stable equivalence*

$$Q \overset{\mathrm{s}}{\sim} Q'.$$

PROOF. This is [**GŠ15**, Prop. 10.5]. □

Again, the case of the derivator of a field can be used to show that the converse implication also holds, and $c(Q)$ hence determines the strong stable equivalence type of an n-cycle Q.

REMARK 4.47. Ladkani studied additional interesting examples of reflection functors for representations of posets with values in arbitrary abelian categories. For these universal derived equivalences we refer to [**Lad07a, Lad07c, Lad07b**]. Dyckerhoff, Jasso, and Walde [**DJW19**] obtain a general construction of reflection functors in stable ∞-categories which also generalizes the examples of Ladkani to the homotopical setting.

4.4. Digression: Monoidal derivators. In this subsection we include a digression on monoidal derivators, the calculus of tensor products of functors, and enriched derivators. We briefly discuss the universality of the derivator of spectra and the resulting canonical enrichment of stable derivators over spectra [**Hel88, Hel97, Cis08, Tab08, CT11, CT12**]. These techniques are central to the construction of universal tilting modules in §4.5.

In this subsection we follow the notional conventions from [**GPS14a**] and also refer to that paper for a more detailed discussion of monoidal and stable, monoidal derivators. For basics on monoidal categories see for instance [**ML98**, §VII].

The starting point of this theory is the following elementary observation.

REMARK 4.48. The 2-category $\mathcal{DER}$ of derivators admits (2-)products. In more detail, given two derivators $\mathscr{D}_1$ and $\mathscr{D}_2$ the pointwise product

$$\mathscr{D}_1 \times \mathscr{D}_2 \colon \mathcal{C}at^{\mathrm{op}} \to \mathcal{CAT} \colon A \mapsto \mathscr{D}_1(A) \times \mathscr{D}_2(A)$$

is again a derivator. The derivator $\mathscr{D}_1 \times \mathscr{D}_2$ enjoys the universal property of a product.

This simple observation allows us to speak about pseudo-monoid objects and their pseudo-actions in the world of derivators. For instance, a first attempt to define a monoidal derivator would be as a pseudo-monoid object in $\mathcal{DER}$, i.e., a derivator $\mathscr{V}$ which is endowed with coherently associative and unital multiplications $\otimes \colon \mathscr{V} \times \mathscr{V} \to \mathscr{V}$. Such a pseudo-monoid structure is equivalent to a lift of $\mathscr{V}$ to the 2-category of monoidal categories, strong monoidal functors and monoidal transformations. However, since we are mostly interested in monoidal structures which are compatible with colimits, we want to build this into the basic notion. Consequently, we begin by making this compatibility precise and, with later applications in mind, we immediately consider arbitrary morphisms of two variables.

REMARK 4.49. Let $\mathscr{D}_1, \mathscr{D}_2$, and $\mathscr{D}_3$ be derivators. A **morphism of two variables** is a morphism of derivators $\otimes \colon \mathscr{D}_1 \times \mathscr{D}_2 \to \mathscr{D}_3$. By definition of the product $\mathscr{D}_1 \times \mathscr{D}_2$ such a morphism has components

$$\otimes_A \colon \mathscr{D}_1(A) \times \mathscr{D}_2(A) \to \mathscr{D}_3(A), \qquad A \in \mathcal{C}at. \tag{4.50}$$

These components are related by pseudo-naturality isomorphisms. In particular, given diagrams $X \in \mathscr{D}_1(A)$ and $Y \in \mathscr{D}_2(A)$, there are canonical isomorphisms

$$(X \otimes_A Y)_a \cong X_a \otimes Y_a, \qquad a \in A.$$

We refer to this description of a morphism of two variables as the **internal version**.

Given two categories A and B, we can restrict diagrams along the projections $A \times B \to A$ and $A \times B \to B$ and then apply $\otimes_{A \times B}$. This way we obtain functors

$$\otimes \colon \mathscr{D}_1(A) \times \mathscr{D}_2(B) \to \mathscr{D}_3(A \times B), \qquad A, B \in \mathcal{C}at, \tag{4.51}$$

which again are endowed with suitable coherence isomorphisms. For instance, for diagrams $X \in \mathscr{D}_1(A)$ and $Y \in \mathscr{D}_2(B)$, these specialize to canonical isomorphisms

$$(X \otimes Y)_{(a,b)} \cong X_a \otimes Y_b, \qquad a \in A, b \in B.$$

One checks that $\otimes \colon \mathscr{D}_1 \times \mathscr{D}_2 \to \mathscr{D}_3$ can be recovered from this datum by also invoking restrictions along diagonal functors $\Delta \colon A \to A \times A$. In fact, these two descriptions are equivalent [**GPS14a**, Theorem 3.11], and we refer to the second one as the **external version** of the morphism of two variables. Note that we distinguish these two versions notationally.

One point of this external reformulation is that it allows for the following simple definition of cocontinuity.

DEFINITION 4.52. Let $\mathscr{D}_1, \mathscr{D}_2$ and $\mathscr{D}_3$ be derivators and let $\otimes\colon \mathscr{D}_1 \times \mathscr{D}_2 \to \mathscr{D}_3$ be a morphism of two variables. The morphism $\otimes$ **preserves left Kan extensions along** $u\colon A_1 \to A_2$ **in the first variable** if the canonical morphisms

$$(4.53) \qquad (u \times 1)_!(X \otimes Y) \xrightarrow{\sim} u_!(X) \otimes Y, \qquad X \in \mathscr{D}_1(A_1), Y \in \mathscr{D}_2(B),$$

are isomorphisms.

This notion has various relevant variants. In particular, we say that $\otimes$ is **cocontinuous in the first variable** if (4.53) are invertible for all $u\colon A_1 \to A_2$. Similarly to [**Gro13**, Prop. 2.3], one only has to verify that colimits are preserved. With this preparation we now make the following definition.

DEFINITION 4.54. A **monoidal derivator** is a pseudo-monoid object $(\mathscr{V}, \otimes, \mathbb{S})$ in $\mathcal{DER}$such that the monoidal structure $\otimes\colon \mathscr{V} \times \mathscr{V} \to \mathscr{V}$ preserves left Kan extensions in both variables separately.

Just to stress this, our convention is that a monoidal derivator is obtained by categorifying the notion of a monoid *and* imposing a cocontinuity condition. There are obvious variants of **braided** or **symmetric monoidal derivators** (see also [**ML98**, §XI] for the classical case).

EXAMPLES 4.55. There are the following expected examples of monoidal derivators.

(i) Let $\mathcal{V}$ be a complete and cocomplete category together with a monoidal structure $\otimes\colon \mathcal{V} \times \mathcal{V} \to \mathcal{V}$ which preserves colimits in both variables separately (this is for example the case when $\mathcal{V}$ is closed monoidal). Then the represented derivator $y_{\mathcal{V}}$ inherits a monoidal structure. In the external version this monoidal structure sends diagrams $X\colon A \to \mathcal{V}$ and $Y\colon B \to \mathcal{V}$ to
$$(X \otimes Y)_{(a,b)} = X_a \otimes Y_b, \qquad a \in A, b \in B.$$
(ii) The homotopy derivator of a monoidal model category is a monoidal derivator. For a more general statements related to Quillen adjunctions of two variables see [**GPS14a**, Example 3.23].
(iii) Let us again consider a commutative ring R. The category $\mathrm{Ch}(R)$ of unbounded chain complexes over R endowed with the projective model structure [**Hov99**, §2.3] yields the derivator $\mathscr{D}_R$ of the ring. With respect to the usual tensor product $\otimes_R$ of chain complexes, $\mathrm{Ch}(R)$ is a stable, symmetric monoidal model category. We conclude that the derivator $\mathscr{D}_R$ together with the derived tensor product is stable and symmetric monoidal. This applies, in particular, to the derivator $\mathscr{D}_k$ of a field k.
(iv) There are various Quillen equivalent stable, symmetric monoidal closed model categories of spectra such as the ones in [**HSS00**, **EKMM97**, **MMSS01**]. Taking any of these as a model for the derivator of spectra $\mathcal{S}p$, we conclude that $\mathcal{S}p$endowed with the derived smash product is a stable, symmetric monoidal derivator. Similarly, also the derivators $\mathcal{S}$ and $\mathcal{S}_*$ are symmetric monoidal. (One can show that there is an intrinsic approach to these monoidal structures based on universal properties of the derivators under consideration.)

Besides the internal and the external versions of morphisms of two variables, there also is the *canceling version* which we discuss next. These specialize to a categorification of the usual tensor products of bimodules over rings and play a key role in §4.5. The classical tensor product in algebra is obtained by coequalizing left and right module structures. The categorification of this construction is given by *coends*, so we begin by extending coends to derivators. Recall that in classical category theory there are various equivalent ways of defining coends [**ML98**, §IX]. In homotopical situations one has to be a bit careful which one to choose, since not all of them lead to the good notion (see [**BG18**, Rmk. 13.21]). The following approach works perfectly well (see [**GPS14a**, §5] and [**GPS14a**, Appendix A] for more details).

CONSTRUCTION 4.56. Let A be a small category. The **twisted arrow category** $\mathrm{tw}(A)$ of A has as objects all morphisms $f\colon a\to b$ in A. A morphism $f_1\to f_2$ in $\mathrm{tw}(A)$ is a commutative diagram

$$\begin{array}{ccc} a_1 & \xrightarrow{f_1} & b_1 \\ \uparrow & & \downarrow \\ a_2 & \xrightarrow[f_2]{} & b_2, \end{array}$$

and the "twist" is the reorientation of the first coordinate. To put this in plain english, a morphism $f_1\to f_2$ is a 2-sided factorization of f_2 through f_1. We note that $\mathrm{tw}(A)$ is simply the category of elements of $\hom_A\colon A^{\mathrm{op}}\times A\to \mathrm{Set}$.

The category $\mathrm{tw}(A)$ comes with a functor

$$(s,t)\colon\ \mathrm{tw}(A)\to A^{\mathrm{op}}\times A$$

given by the source and target functors. As in the case of ordinary category theory, in order to define the coend construction we need essentially the opposite of this functor, namely the composition

$$(t^{\mathrm{op}},s^{\mathrm{op}})\colon\ \mathrm{tw}(A)^{\mathrm{op}}\overset{(s,t)^{\mathrm{op}}}{\to}(A^{\mathrm{op}}\times A)^{\mathrm{op}}\cong A^{\mathrm{op}}\times A.$$

DEFINITION 4.57. Let $\mathscr{D}$ be a derivator, let $A\in\mathcal{C}at$, and let $X\in\mathscr{D}(A^{\mathrm{op}}\times A)$. The **coend** $\int^A X\in\mathscr{D}(*)$ of X is given by

$$\int^A X=\operatorname{colim}_{\mathrm{tw}(A)^{\mathrm{op}}}(t^{\mathrm{op}},s^{\mathrm{op}})^*X.$$

As a composition of functors, the coend functor $\int^A\colon\mathscr{D}(A^{\mathrm{op}}\times A)\to\mathscr{D}(*)$ is

$$\int^A\colon\mathscr{D}(A^{\mathrm{op}}\times A)\overset{(t^{\mathrm{op}},s^{\mathrm{op}})^*}{\to}\mathscr{D}(\mathrm{tw}(A)^{\mathrm{op}})\overset{\mathrm{colim}}{\to}\mathscr{D}(*).$$

EXAMPLES 4.58. In the case of a represented derivator, this definition of coends reduces to a formula in [**ML98**]. Hence, the notion reproduces the classical one in a (complete and) cocomplete category. There is also a different classical description of coends as certain coequalizers. This, however, does not extend that directly to homotopical frameworks. In fact, in that case coequalizers have to be replaced by geometric realizations of simplicial bar constructions and for an extension of this reformulation to derivators we refer to [**GPS14a**, Appendix A].

The motivation for us to discuss coends here is the following construction.

CONSTRUCTION 4.59. Let $\mathscr{D}_1, \mathscr{D}_2$, and $\mathscr{D}_3$ be derivators and let $\otimes\colon \mathscr{D}_1 \times \mathscr{D}_2 \to \mathscr{D}_3$ be a morphism of two variables. Based on the coend, there is the following third version of such a morphism. Given a small category A and diagrams $X \in \mathscr{D}_1(A^{\mathrm{op}})$ and $Y \in \mathscr{D}_2(A)$, the external product $X \otimes Y$ lives in $\mathscr{D}_3(A^{\mathrm{op}} \times A)$, and qualifies as an input for the coend. The **canceling version** of $\otimes$ is

$$X \otimes_{[A]} Y = \int^A X \otimes Y, \qquad X \in \mathscr{D}_1(A^{\mathrm{op}}), Y \in \mathscr{D}_2(A).$$

Thus, as a functor $\otimes_{[A]}$ is the composition

$$\otimes_{[A]}\colon \mathscr{D}_1(A^{\mathrm{op}}) \times \mathscr{D}_2(A) \xrightarrow{\otimes} \mathscr{D}_3(A^{\mathrm{op}} \times A) \xrightarrow{\int^A} \mathscr{D}_3(*), \tag{4.60}$$

and we refer to it as the **(canceling) tensor product of functors**. Note that we distinguish this canceling version notationally from the internal version (4.50) and the external one (4.51). For the canceling one the subscript $[A]$ is added in order to indicate which "category is canceled".

In order to develop some first intuition for this calculus, we unravel the notion of a twisted arrow category for posets.

LEMMA 4.61. *Let P be a small poset. The functor $(s,t)\colon \mathrm{tw}(P) \to P^{\mathrm{op}} \times P$ induces an isomorphism between $\mathrm{tw}(P)$ and the up-set generated by the diagonal $\Delta_P \subseteq P^{\mathrm{op}} \times P$.*

PROOF. The functor (s,t) is injective on objects and fully faithful, so it induces an isomorphism onto its image. By reflexivity the diagonal $\Delta_P = \{(p,p) \mid p \in P\}$ lies in the image. Let $f\colon p \to q$ be an object in $\mathrm{tw}(P)$, which is to say that $p, q \in P$ and $p \leq q$. Then in $P^{\mathrm{op}} \times P$ we have $(p,p), (q,q) \leq (p,q)$, and we deduce that the image of (s,t) lies in the up-set generated by the diagonal. Conversely, let $p, q_1, q_2 \in P$ be such that $(p,p) \leq (q_1, q_2)$ in $P^{\mathrm{op}} \times P$. Then $q_1 \leq p$ and $p \leq q_2$ in P, and hence $q_1 \to q_2$ defines an object in $\mathrm{tw}(P)$ which under (s,t) is mapped to (q_1, q_2). □

In the calculation of coends we essentially use the opposite functor of (s,t). Thus in order to calculate tensor products of functors over posets we first form the external product, then restrict the diagram to the *down-set* generated by the diagonal and finally calculate the colimit. We illustrate this by the following prominent example of Construction 4.59. Therein and in what follows, our drawing convention for diagrams of two variable is that the first coordinate is drawn horizontally and the second one vertically.

EXAMPLE 4.62. Let $\mathscr{D}_1, \mathscr{D}_2$, and $\mathscr{D}_3$ be derivators and let $\otimes\colon \mathscr{D}_1 \times \mathscr{D}_2 \to \mathscr{D}_3$ be a morphism of two variables. Given coherent morphisms $X = (f\colon x \to y) \in \mathscr{D}_1([1]^{\mathrm{op}})$ and $X' = (f'\colon x' \to y') \in \mathscr{D}_2([1])$, the external product $X \otimes X' \in \mathscr{D}_3([1]^{\mathrm{op}} \times [1])$ takes by pseudo-naturality the form

$$\begin{array}{ccc} y \otimes x' & \xleftarrow{f \otimes \mathrm{id}} & x \otimes x' \\ {\scriptstyle \mathrm{id} \otimes f'} \downarrow & & \downarrow {\scriptstyle \mathrm{id} \otimes f'} \\ y \otimes y' & \xleftarrow[f \otimes \mathrm{id}]{} & x \otimes y'. \end{array}$$

The restriction of $X \otimes X'$ to the down-set generated by the diagonal is the span $(y \otimes x' \leftarrow x \otimes x' \to x \otimes y')$, and the tensor product of functors $X \otimes_{[[1]]} X' \in \mathscr{D}_3(*)$ hence sits in a defining cocartesian square

$$\begin{array}{ccc} y \otimes x' & \xleftarrow{f \otimes \mathrm{id}} & x \otimes x' \\ \downarrow & & \downarrow{\scriptstyle \mathrm{id} \otimes f'} \\ X \otimes_{[[1]]} X' & \longleftarrow & x \otimes y'. \end{array}$$

In this special case the universal object

$$X \otimes_{[[1]]} X' = y \otimes x' \cup_{x \otimes x'} x \otimes y'$$

is often referred to as the **pushout product** of the two morphisms.

In §4.5 we illustrate the calculus of pushout products by specializing to more specific morphisms. Here, we content ourselves by pointing out that this calculus also plays a key role in refined axioms for tensor triangulated categories or monoidal triangulated categories (see Remark 4.72 and the references therein).

REMARK 4.63. Let $\mathscr{D}_1$, $\mathscr{D}_2$, and $\mathscr{D}_3$ be derivators and let $\otimes \colon \mathscr{D}_1 \times \mathscr{D}_2 \to \mathscr{D}_3$ be a morphism of two variables. The internal (4.50), external (4.51), and canceling versions (4.60) of $\otimes$ can be suitably combined. For instance there is the functor

$$\mathscr{D}_1(A \times B^{\mathrm{op}}) \times \mathscr{D}_2(A \times B \times C) \to \mathscr{D}_3(A \times C)$$

which treats the A-variable internally and the C-variable externally, while the B-variable is canceled by means of a coend. Taking the philosophy of derivators serious, we should not be happy with such functors but instead ask for corresponding morphisms of derivators. In the first two cases this leads to parametrized internal and parametrized external tensor products [**GPS14a**, §3],

$$\otimes_A \colon \mathscr{D}_1^A \times \mathscr{D}_2^A \to \mathscr{D}_3^A \qquad \text{and} \qquad \otimes \colon \mathscr{D}_1^A \times \mathscr{D}_2^B \to \mathscr{D}_3^{A \times B}.$$

As a preparation for the remaining case, one notes that for every derivator $\mathscr{D}$there are parametrized coends

$$\int^A \colon \mathscr{D}^{A^{\mathrm{op}}} \times \mathscr{D}^A \to \mathscr{D}.$$

These are derivatorish versions of the usual coends with parameters [**ML98**, §IX.7]. With this in place, the parametrized canceling tensor products

$$\otimes_{[A]} \colon \mathscr{D}_1^{A^{\mathrm{op}}} \times \mathscr{D}_2^A \to \mathscr{D}_3$$

can now be defined by the same formula as in Construction 4.59.

For every monoidal derivator $\mathscr{V}$, the monoidal structure $\otimes \colon \mathscr{V} \times \mathscr{V} \to \mathscr{V}$ is associative and unital up to coherence isomorphisms. This is the case by definition for the internal version and it is straightforward to also check this for the external version. We now turn towards a similar result for the canceling tensor product.

CONSTRUCTION 4.64. Let $\mathscr{V}$be a monoidal derivator and let $A, B \in \mathcal{C}at$. We also call objects in $\mathscr{V}(A \times B^{\mathrm{op}})$ **bimodules** in $\mathscr{V}$ or (A, B)-**bimodules** in case we want to be more specific. Given a third small category C, we can define a **composition functor** (over B) as the parametrized canceling tensor product

$$\otimes_{[B]} \colon \mathscr{V}(A \times B^{\mathrm{op}}) \times \mathscr{V}(B \times C^{\mathrm{op}}) \to \mathscr{V}(A \times C^{\mathrm{op}}) \tag{4.65}$$

which treats A and C^{op} externally. For every small category B there is also a preferred (B,B)-bimodule $\mathbb{I}_B$ (which somehow corresponds to the regular bimodule over a ring in algebra). To build towards these bimodules, let us recall that, as part of the monoidal structure, $\mathscr{V}$ is endowed with a monoidal unit $\mathbb{S}$. This is a pseudo-functor $\mathbb{S}\colon y_* = * \to \mathscr{V}$ defined on the terminal derivator. More concretely, $\mathbb{S}$ amounts to a pseudo-functorial choice of objects $\mathbb{S}_A \in \mathscr{V}(A), A \in \mathcal{C}at$, all of which are monoidal units with respect to the internal tensor products $\otimes_A$. The pseudo-functoriality constraint for the projection $\pi_A\colon A \to *$ gives a preferred isomorphism $\mathbb{S}_A \cong \pi_A^*(\mathbb{S}_*)$, which is to say that $\mathbb{S}_A$ is constant on the monoidal unit $\mathbb{S}_*$ of the underlying category $\mathscr{V}(*)$. With this preparation we can define the **Yoneda bimodule** or **identity profunctor** of B as the bimodule

$$\mathbb{I}_B = (t,s)_!\mathbb{S}_{\mathrm{tw}(B)} \in \mathscr{V}(B \times B^{\mathrm{op}}). \tag{4.66}$$

Here $(t,s)\colon \mathrm{tw}(B) \to B \times B^{\mathrm{op}}$ sends $(f\colon a \to b)$ to (b,a) (see Construction 4.56). By an application of the pointwise formulas for left Kan extensions one checks that for $a,b \in B$ there is a canonical isomorphism

$$(\mathbb{I}_B)_{(b,a)} \cong \coprod_{\hom_B(a,b)} \mathbb{S}_*.$$

In plain English, the value at (b,a) is the coproduct of copies of the unit $\mathbb{S}_*$ parametrized by the morphism set $\hom_B(a,b)$, and this justifies the terminology *Yoneda bimodule.*

The composition of bimodules in monoidal derivators is associative and unital in the following sense. For basic terminology on bicategories we refer the reader to [**Bén67**], [**Bor94a**, §7.7], or [**ML98**, §§XII.6-7].

THEOREM 4.67. *For every monoidal derivator $\mathscr{V}$ there is a bicategory $\mathcal{P}rof(\mathscr{V})$ with the following description. Its objects are small categories, the category of homomorphisms from A to B is the category of bimodules $\mathscr{V}(A \times B^{\mathrm{op}})$, the composition functors are given by* (4.65) *and the identity 1-cells are the Yoneda bimodules* (4.66). *We refer to $\mathcal{P}rof(\mathscr{V})$ as the* ***bicategory of bimodules*** *in $\mathscr{V}$.*

PROOF. The proof of the associativity of composition functors is a fairly direct consequence of a derivatorish version of the Fubini lemma on iterated coends [**GPS14a**, Lem. 5.3]. It turns out that the proof of the unitality is more involved [**GPS14a**, Appendix B]. For details we refer the reader to [**GPS14a**, Theorem 5.9]. □

In [**GPS14a**] we refer to $\mathcal{P}rof(\mathscr{V})$ as the *bicategory of profunctors* (and this explains the notation), while here we prefer the terminology of bimodules. For later reference we include the following example of Yoneda bimodules.

EXAMPLE 4.68. Let $\mathscr{V}$ be a pointed, monoidal derivator and let P be a poset. The Yoneda bimodule $\mathbb{I}_P \in \mathcal{V}(P \times P^{\mathrm{op}})$ restricted to the up-set generated by the diagonal Δ_P is constant with value the monoidal unit $\mathbb{S}_* \in \mathscr{V}(*)$ and it vanishes on the complement [**GŠ16a**, Lem. 7.4].

By now we have a reasonably solid understanding of the basic formalism of tensor products of functors in monoidal derivators (more examples will be discussed in §4.5). As a final preparation for the study of tensor-hom-adjunctions we extend the notion of adjunctions of two variables to derivators. There are various ways to make this precise [**GPS14a**, §8], and here we choose the following definition.

DEFINITION 4.69. Let $\mathscr{D}_1, \mathscr{D}_2$, and $\mathscr{D}_3$ be derivators. A morphism of two variables $\otimes\colon \mathscr{D}_1 \times \mathscr{D}_2 \to \mathscr{D}_3$ is a **two-variable left adjoint** if

(i) the external components $\otimes\colon \mathscr{D}_1(A) \times \mathscr{D}_2(B) \to \mathscr{D}_3(A \times B)$ are two-variable left adjoints for all $A, B \in \mathcal{C}at$ and
(ii) the morphism $\otimes\colon \mathscr{D}_1 \times \mathscr{D}_2 \to \mathscr{D}_3$ is cocontinuous in both variables separately.

Thus, for every $A, B \in \mathcal{C}at$ we ask for the existence of functors

$$\rhd_{[B]}\colon \mathscr{D}_2(B)^{\mathrm{op}} \times \mathscr{D}_3(A \times B) \to \mathscr{D}_1(A) \quad \text{and} \quad \lhd_{[A]}\colon \mathscr{D}_3(A \times B) \times \mathscr{D}_1(A)^{\mathrm{op}} \to \mathscr{D}_2(B)$$

and for natural isomorphisms

$$\hom_{\mathscr{D}_3(A\times B)}(X \otimes Y, Z) \cong \hom_{\mathscr{D}_1(A)}(X, Y \rhd_{[B]} Z) \cong \hom_{\mathscr{D}_2(B)}(Y, Z \lhd_{[A]} X).$$

The notational convention follows [**GPS14a**] and it has the feature that it preserves the cyclic structure of the arguments X, Y, and Z. The notation of the functors $\rhd_{[B]}$ and $\lhd_{[A]}$ again indicated which "category is canceled". One can show that there are morphisms of derivators

$$\lhd\colon \mathscr{D}_2^{\mathrm{op}} \times \mathscr{D}_3 \to \mathscr{D}_1 \qquad \text{and} \qquad \rhd\colon \mathscr{D}_3 \times \mathscr{D}_1^{\mathrm{op}} \to \mathscr{D}_2$$

and that the functors $\lhd_{[B]}$ and $\rhd_{[A]}$ arise from these by certain *ends* (see [**GPS14a**, §§8-9]).

DEFINITION 4.70. A **closed monoidal derivator** is a monoidal derivator such that the monoidal structure is a two-variable left adjoint.

There is the obvious variant of symmetric closed monoidal derivators.

EXAMPLES 4.71. (i) Derivators represented by closed monoidal categories are also closed monoidal. More generally, two-variable adjunctions between complete and cocomplete categories induce two-variable adjunctions of represented derivators [**GPS14a**, Ex. 8.9].
(ii) Homotopy derivators of cofibrantly-generated monoidal model categories are closed monoidal [**GPS14a**, Thm. 9.11]. Two-variable Quillen left adjoint functors between combinatorial model categories induces two-variable adjunctions between homotopy derivators [**GPS14a**, Ex. 8.11].
(iii) The derivator $\mathscr{D}_R$ of a commutative ring R, the derivator $\mathcal{S}$ of spaces, the derivator $\mathcal{S}_*$ of pointed spaces, and the derivator $\mathcal{S}p$ of spectra are symmetric closed monoidal.
(iv) Let $\otimes\colon \mathscr{D}_1 \times \mathscr{D}_2 \to \mathscr{D}_3$ be a two-variable left adjoint morphism of derivators. Then also the internal, external, and canceling versions of $\otimes$ from Remark 4.63 are two-variable left adjoints [**GPS14a**, §8]. In particular, for every closed monoidal derivator $\mathscr{V}$the bicategory $\mathcal{P}rof(\mathscr{V})$ of bimodules is closed.

REMARK 4.72. Many interesting examples of triangulated categories arising in nature come with an additional monoidal structure (see for instance [**Bal10**, §1]). A natural question then is what kind of axioms should be imposed on a category which is simultaneously endowed with a triangulation and a monoidal structure. The search for good compatibility axioms to be imposed on such categories has already some history and references include [**Mar83, HPS97, May01, KN02**]. Of course, the answer to this question depends on ones purposes (for instance for the discussion of spectra of tensor triangulated categories relatively simple axioms are good enough; see [**Bal10**] and the many references therein).

A different purpose one might have is the study of *duality phenomena* (see [**DP80**] as well as [**BG99**] and the references there). Let $\mathcal{T}$ be monoidal, triangulated category and let $x \in \mathcal{T}$ be a dualizable object. Every endomorphism $\varphi\colon x \to x$ has a trace $\mathrm{tr}(\varphi)\colon \mathbb{S} \to \mathbb{S}$ which is an endomorphism of the monoidal unit. Morally, one would expect that traces of dualizable objects are additive with respect to distinguished triangles. More precisely, given an endomorphism of a distinguished triangle

$$\begin{array}{ccccccc} x & \longrightarrow & y & \longrightarrow & z & \longrightarrow & \Sigma x \\ {\scriptstyle\varphi_x}\downarrow & & {\scriptstyle\varphi_y}\downarrow & & \downarrow{\scriptstyle\varphi_z} & & \downarrow{\scriptstyle\Sigma\varphi_x} \\ x & \longrightarrow & y & \longrightarrow & z & \longrightarrow & \Sigma x \end{array}$$

such that x, y, and z are dualizable, the formula

$$\mathrm{tr}(\varphi_x) + \mathrm{tr}(\varphi_z) = \mathrm{tr}(\varphi_y)$$

should hold. However, as a further reminiscence of the non-functoriality of cones, such a formula does not hold at the level of triangulated categories [**Fer06**]. And, in fact, this failure was part of the original motivation of Grothendieck to come up with the notion of a derivator in the first place.

May [**May01**] proposed very carefully chosen compatibility axioms for monoidal, triangulated categories which can be used to establish additivity of Euler characteristics (traces of identities), and a representation theoretic perspective on these appears in work of Keller–Neeman [**KN02**]. In [**GPS14a**] the basic formalism of monoidal, stable derivators is developed, and canonical monoidal triangulations in the sense of May are constructed for such derivators. As an additional application of these techniques, it is shown that traces of coherent morphisms are additive with respect to cofiber sequences [**GPS14a**].

The credo suggested by this is that all the compatibility is already encoded in a stable, monoidal derivator (or stable, monoidal ∞-category). Hence, instead of imposing more and more complicated *axioms* on monoidal, triangulated categories, one should try to prove *lemmas* about, say, stable, monoidal derivators and the corresponding calculus of bimodules. This was illustrated successfully in the papers [**PS16, PS14, GAdS14**] where the above additivity formula was extended to homotopy finite colimits of dualizable objects (see also the closely related [**JY18**]).

Definition 4.73. Let $\mathscr{V}$ be a monoidal derivator. A **$\mathscr{V}$-module** is a cocontinuous pseudo-module over $\mathscr{V}$. A **$\mathscr{V}$-enriched derivator** is a $\mathscr{V}$-module such that the action is a two-variable left adjoint.

Thus, a $\mathscr{V}$-module is a derivator $\mathscr{D}$ together with a morphism of two variables

$$\otimes\colon \mathscr{V} \times \mathscr{D} \to \mathscr{D}, \tag{4.74}$$

the *action* or *module structure*, such that

(i) the action is associative and unital up to coherence isomorphisms and
(ii) the action is cocontinuous in both variables separately.

If $\mathscr{V}$ and $\mathscr{D}$ are stable derivators, then these are derivatorish versions of actions of triangulated categories (as in [**Ste13**]). For a $\mathscr{V}$-enriched derivator the action (4.74) is supposed to be a two-variable left adjoint (Definition 4.69). Thus, for a

$\mathscr{V}$-enriched derivator $\mathscr{D}$besides the action morphisms we also have **internal homs** or **internal mapping objects**

$$\rhd\colon \mathscr{D}^{\mathrm{op}} \times \mathscr{D} \to \mathscr{V}$$

and **cotensors**

$$\lhd\colon \mathscr{D} \times \mathscr{V}^{\mathrm{op}} \to \mathscr{D}.$$

EXAMPLES 4.75. (i) If $\mathcal{V}$ is complete and cocomplete closed symmetric monoidal category and $\mathcal{C}$ is a complete and cocomplete $\mathcal{V}$-enriched category in the classical sense [**Kel05b**], then $y_{\mathcal{C}}$ is a $y_{\mathcal{V}}$-enriched derivator (Examples 4.71).

(ii) If $\mathcal{V}$ is symmetric monoidal model category and $\mathcal{M}$ is a $\mathcal{V}$-model category in the sense of [**Hov99**], then $\mathscr{H}o_{\mathcal{M}}$ is a $\mathscr{H}o_{\mathcal{V}}$-enriched derivator.

(iii) Any closed monoidal derivator is enriched over itself.

This gives rise to many interesting examples of enriched derivators. We invite the reader to come up with additional closure properties of enriched derivator such as the passage to shifted derivators or the behavior under monoidal adjunctions. We conclude this subsection by a discussion of the universality of the classical homotopy theories of spaces, pointed spaces, and spectra, and the resulting canonical enrichments of stable derivators over spectra. These are deep theorems and they are crucial to our applications in §4.5. Related references include [**Hel88, Hel97, Fra96, Cis08, Tab08, CT11, CT12**].

REMARK 4.76. The homotopy derivator of spaces $\mathcal{S}$ is freely generated under colimits by the singleton Δ^0 [**Cis08**, Theorem 3.24]. In more detail, given derivators $\mathscr{D}, \mathscr{E}$ we denote by $\mathrm{Hom}_!(\mathscr{D}, E)$ the category of cocontinuous morphisms $\mathscr{D} \to \mathscr{E}$ and all natural transformations. In particular, given a cocontinuous morphism $F\colon \mathcal{S} \to \mathscr{D}$, we can evaluate the underlying functor $\mathcal{S}(*) \to \mathscr{D}(*)$ at Δ^0. The universality of spaces is made precise by the statement that the evaluation induces an equivalence of categories

$$\mathrm{Hom}_!(\mathcal{S}, \mathscr{D}) \xrightarrow{\sim} \mathscr{D}(*)\colon F \mapsto F(\Delta^0).$$

This universal property explains the ubiquity of spaces in abstract homotopy theory. A formally correct proof of this result is highly non-trivial, and we refer the reader to [**Hel88**] and [**Cis08**]. But to provide some evidence for this result let us recall the following.

(i) Every topological space has a CW-approximation.
(ii) CW-complexes are constructed from n-cells under coproducts, pushouts and countable colimits.

Since the n-cells are all weakly contractible, this means that every space can be built from Δ^0 using homotopy colimits only. These heuristics make the universality at least quite plausible.

There are variants of this result for pointed spaces and spectra. The derivator of pointed spaces $\mathcal{S}_*$ is the free pointed derivator generated by the 0-sphere $\mathbb{S}^0$. And the derivator $\mathcal{S}p$ is the free stable derivator generated by the sphere spectrum $\mathbb{S}$ [**CT11**, Theorem A.11]. Thus, for every stable derivator $\mathscr{D}$the evaluation

$$\mathrm{Hom}_!(\mathcal{S}p, \mathscr{D}) \xrightarrow{\sim} \mathscr{D}(*)\colon F \mapsto F(\mathbb{S}).$$

is an equivalence of categories.

THEOREM 4.77. *Every stable derivator is enriched over the derivator $\mathcal{S}p$ of spectra.*

PROOF. The proof is essentially a consequence of the above-mentioned universal property of the derivator of spectra, and we refer the reader to [**CT11**, Appendix A.3] for details. It turns out that the action

$$\otimes\colon \mathcal{S}p \times \mathscr{D} \to \mathscr{D}$$

which belongs to this enrichment is characterized by two properties. First, the action preserves colimits in both variables separately, and, second, the sphere spectrum acts trivially (which is to say $\mathbb{S}\otimes - \cong \mathrm{id}_{\mathscr{D}}$ as it is the case for all actions). □

In §4.5 we use this canonical enrichment to construct universal tilting modules. Let us close this subsection by the following obvious remark just to underline the usefulness of mapping spectra.

REMARK 4.78. Let $\mathscr{D}$be a stable derivator, let $A \in \mathcal{C}at$, and let $X, Y \in \mathscr{D}(A)$. We refer to $X \rhd_{[A]} Y \in \mathcal{S}p(*)$ as the **mapping spectrum** of X, Y. This is a higher-structured version of the categorical morphisms $\hom_{\mathscr{D}(A)}(X,Y)$. In fact, these can be recovered as

$$\pi_0(X \rhd_{[A]} Y) \cong \hom_{\mathcal{S}p(*)}(\mathbb{S}, X \rhd_{[A]} Y) \cong \hom_{\mathscr{D}(A)}(\mathbb{S}\otimes X, Y) \cong \hom_{\mathscr{D}(A)}(X,Y).$$

Similarly, the extension groups can be described as homotopy groups of the mapping spectrum. The formalism of two-variable adjunctions allows us to calculate the mapping spectrum $X \rhd_{[A]} Y$ as an end which is useful in many situations.

4.5. Universal tilting modules. In this section we discuss universal tilting modules: certain explicitly constructed spectral bimodules realizing our strong stable equivalences. These universal tilting modules are spectral refinements of the classical tilting complexes and they are non-trivial invertible elements in the bicategory of spectral bimodules. Considered this way, abstract representation contributes to the calculation of spectral Picard groupoids.

Thus, analogously to the situation in tilting theory, here the focus shifts from the equivalences themselves to the representing bimodules.

We begin this section by defining the class of morphisms which are associated to spectral bimodules.

DEFINITION 4.79. Let $A, B \in \mathcal{C}at$ and $M \in \mathcal{S}p(B \times A^{\mathrm{op}})$. For every stable derivator $\mathscr{D}$, the **weighted colimit** with weight M is the morphism

$$M \otimes_{[A]} -\colon \mathscr{D}^A \to \mathscr{D}^B.$$

There is the dual notion of a **weighted limit** (using cotensors and ends instead of tensors end coends). In this subsection we will always focus on the case of weighted colimits.

REMARK 4.80. Let $\mathscr{D}$be a stable derivator. We say that a morphism $F\colon \mathscr{D}^A \to \mathscr{D}^B$ is a weighted colimit if there is a spectral bimodule $M \in \mathcal{S}p(B \times A^{\mathrm{op}})$ and a natural isomorphism

$$F \cong M \otimes_{[A]} -\colon \mathscr{D}^A \to \mathscr{D}^B.$$

If we do not merely have a morphism $F\colon \mathscr{D}^A \to \mathscr{D}^B$ for a fixed stable derivator, but a pseudo-natural transformation

$$F\colon (-)^A \to (-)^B\colon \mathcal{DER}_{\mathrm{St,ex}} \to \mathcal{DER}$$

between the corresponding abstract representation theories, then the weight is uniquely determined by F. In fact, it suffices to consider the universal stable derivator $\mathscr{D} = \mathcal{S}p$ of spectra. The natural isomorphism $F_{\mathcal{S}p} \cong M \otimes_{[A]} -$ specializes to a natural isomorphism of functors

$$\mathcal{S}p^A(A^{\mathrm{op}}) = \mathcal{S}p(A \times A^{\mathrm{op}}) \to \mathcal{S}p^B(A^{\mathrm{op}}) = \mathcal{S}p(B \times A^{\mathrm{op}}).$$

Plugging in the Yoneda bimodule $\mathbb{I}_A$, we can invoke the unitality constraint from Theorem 4.67 in order to obtain

$$F(\mathbb{I}_A) \cong M \otimes_{[A]} \mathbb{I}_A \cong M.$$

This remark shows that in favorable cases the weight is obtained if one evaluates the morphism on a suitable Yoneda bimodule. This idea is also central to the following two results. The first of these two results also justifies the terminology of weighted *colimits.*

THEOREM 4.81. *Let $\mathscr{D}$ be a stable derivator and let $u \colon A \to B$ be a functor.*

(i) *The restriction morphism $u^* \colon \mathscr{D}^B \to \mathscr{D}^A$ is a weighted colimit.*
(ii) *The left Kan extension morphism $u_! \colon \mathscr{D}^A \to \mathscr{D}^B$ is a weighted colimit. In particular, the colimit morphism $\mathrm{colim}_A \colon \mathscr{D}^A \to \mathscr{D}$ is a weighted colimit.*

PROOF. The stable derivator $\mathscr{D}$ is by Theorem 4.77 a $\mathcal{S}p$-enriched derivator, and let us denote the corresponding adjunction of two variables by

$$\otimes \colon \mathcal{S}p \times \mathscr{D} \to \mathscr{D}.$$

Moreover, for arbitrary categories C, D, E it follows from Remark 4.63 that also the external-canceling version of this action

$$\otimes_{[D]} \colon \mathcal{S}p^{C \times D^{\mathrm{op}}} \times \mathscr{D}^{D \times E^{\mathrm{op}}} \to \mathscr{D}^{C \times E^{\mathrm{op}}}$$

is a left adjoint of two variables. In particular, the morphism $\otimes_{[D]}$ is compatible with restrictions and left Kan extensions in both variables separately. To conclude the proof it suffices to specialize this to the two situations under consideration.

In the first case, let $X \in \mathscr{D}(B)$ and let $\mathbb{I}_B \in \mathcal{S}p(B \times B^{\mathrm{op}})$ be the Yoneda bimodule. By left unitality and pseudo-naturality we obtain

$$u^*(X) \cong u^*(\mathbb{I}_B \otimes_{[B]} \otimes X) \cong \big((u \times \mathrm{id}_{B^{\mathrm{op}}})^* \mathbb{I}_B\big) \otimes_{[B]} X,$$

and this defines the intended natural isomorphism

$$u^* \cong \big((u \times \mathrm{id}_{B^{\mathrm{op}}})^* \mathbb{I}_B\big) \otimes_{[B]} - \colon \mathscr{D}^B \to \mathscr{D}^A, \tag{4.82}$$

exhibiting the restriction morphism as a weighted colimit.

Similarly, in the second case, let $X \in \mathscr{D}(A)$ and let $\mathbb{I}_A \in \mathcal{S}p(A \times A^{\mathrm{op}})$ be the Yoneda bimodule. Using the fact that the morphism $- \otimes_{[A]} X$ preserves colimits and hence left Kan extensions, we obtain the isomorphisms

$$u_!(X) \cong u_!(\mathbb{I}_A \otimes_{[A]} X) \cong \big((u \times \mathrm{id}_{A^{\mathrm{op}}})_! \mathbb{I}_A\big) \otimes_{[A]} X.$$

This yields the intended natural isomorphism

$$u_! \cong \big((u \times \mathrm{id}_{A^{\mathrm{op}}})_! \mathbb{I}_A\big) \otimes_{[A]} - \colon \mathscr{D}^A \to \mathscr{D}^B \tag{4.83}$$

exhibiting left Kan extensions as weighted colimits. Of course, the case of colimits is obtained by specializing to $u = \pi_A$. □

This first result generalizes to more general enriched derivators in not necessarily stable situations. In contrast to this, Theorem 4.85 relies crucially on stability.

REMARK 4.84. Let us recall that a left exact morphism of derivators is defined as a morphism which preserves terminal objects and cartesian squares (Examples 3.56). By Theorem 3.59 such morphisms preserve homotopy finite limits. Since right Kan extensions in derivators can be calculated pointwise in terms of limits (by (Der4)), it follows that right exact morphisms also preserve *right homotopy finite* right Kan extension (see [**Gro16**, Thm. 9.14]). This includes those right Kan extensions for which the corresponding slice categories are homotopy finite, and the following result hence covers a large class of examples.

THEOREM 4.85. *Let $\mathscr{D}$ be a stable derivator and let $u\colon A \to B$ be a functor.*

(i) *If u is a sieve, then the right Kan extension morphism $u_*\colon \mathscr{D}^A \to \mathscr{D}^B$ is a weighted colimit.*
(ii) *If $u\colon A \to B$ is right homotopy finite, then the right Kan extension morphism $u_*\colon \mathscr{D}^A \to \mathscr{D}^B$ is a weighted colimit. In particular, for every homotopy finite category A the limit morphism $\lim_A\colon \mathscr{D}^A \to \mathscr{D}$ is a weighted colimit.*

PROOF. As in the proof of Theorem 4.81 for small categories C, D, E the action of spectra yields a left adjoint of two variables

$$\otimes_{[D]}\colon \mathcal{S}p^{C\times D^{\mathrm{op}}} \times \mathscr{D}^{D\times E^{\mathrm{op}}} \to \mathscr{D}^{C\times E^{\mathrm{op}}}$$

between *stable* derivators. In particular, for a fixed object $X \in \mathscr{D}(A)$ the morphism

$$-\otimes_{[A]} X\colon \mathcal{S}p^{A^{\mathrm{op}}} \to \mathscr{D}$$

is exact and hence preserves right Kan extensions along sieves and, more generally, right homotopy finite right Kan extensions ([**Gro16**, Thm. 9.17]). Thus, in both situations we obtain isomorphisms

$$u_*(X) \cong u_*(\mathbb{I}_A \otimes_{[A]}) X) \cong \big((u\times \mathrm{id}_{A^{\mathrm{op}}})_*\mathbb{I}_A\big) \otimes_{[A]} X.$$

Letting the diagram X vary we obtain the natural isomorphism

$$u_* \cong \big((u\times \mathrm{id}_{A^{\mathrm{op}}})_*\mathbb{I}_A\big) \otimes_{[A]} -\colon \mathscr{D}^A \to \mathscr{D}^B \tag{4.86}$$

showing that these particular *right* Kan extensions are weighted colimits. □

REMARK 4.87. We want to include a short discussion of these two theorems.

(i) The proofs of Theorem 4.81 and Theorem 4.85 are constructive in that they offer formulas which allow us to calculate the weights for restrictions and suitable Kan extensions. In fact, as detailed by the formulas (4.82), (4.83), and (4.86), in all cases we start with a suitable Yoneda bimodule. One of the variables is bound by the canceling tensor product (or the corresponding coend), and we simply apply the corresponding operation to the remaining free variable. This allows us to explicitly calculate the representing weights as we illustrate a bit further below.
(ii) Restrictions and all sufficiently finite left and right Kan between stable derivators are weighted colimits, and there are dual statements for *weighted limits*. The corresponding representing spectral bimodules admit a fairly rich calculus, and we refer the reader to [**GPS14a**, **GŠ16a**, **GS17**, **Shu08**, **PS16**] for more details.

(iii) The motivation to break up the above examples of weighted colimits into two classes is twofold. First, Theorem 4.81 extends to more general $\mathscr{V}$-modules and $\mathscr{V}$-enriched derivator, while only in Theorem 4.85 we rely on additional exactness properties. In fact, the result about right Kan extensions along sieves is also valid in modules over pointed monoidal derivators. Second, the results in Theorem 4.85 indicate that in stable land the distinction between constructions on the left and constructions on the right is blurred to some extent (sufficiently finite limits are weighted colimits). It turns out that this is a defining feature of stability and this perspective offers an interesting invitation to a formal study of abstract stability (see [**GS17**] for first steps along these lines).

EXAMPLE 4.88. For every diagram of spectra $M \in \mathcal{S}p(A)$ and every stable derivator the action

$$M \otimes -\colon \mathscr{D} \to \mathscr{D}^A$$

is a weighted colimit. As a special case, for the suspension of the sphere spectrum $\Sigma\mathbb{S} \in \mathcal{S}p(*)$ and $X \in \mathscr{D}(B)$ we have

$$(\Sigma\mathbb{S}) \otimes X \cong \Sigma(\mathbb{S} \otimes X) \cong \Sigma X.$$

Consequently, the suspension morphism $\Sigma\colon \mathscr{D} \to \mathscr{D}$ is a weighted colimit with weight $\Sigma\mathbb{S}$,

$$\Sigma\mathbb{S} \otimes - \cong \Sigma\colon \mathscr{D} \to \mathscr{D}.$$

There is a similar description of $\Sigma^n, n \in \mathbb{Z}$ as weighted colimit with weight $\Sigma^n\mathbb{S}$.

In the following examples we study special cases of weighted colimits of morphisms. In those cases we start from the Yoneda bimodule $\mathbb{I}_{[1]}$ and follow the above recipe to construct weights (see Remark 4.87). To also get more used to the calculus of coends from §4.4, we double-check that the weights really represent the constructions we started with. Let us recall that our drawing convention for bimodules is that the first coordinate is drawn horizontally and the second one vertically. As a special case of Example 4.68, there is the following example.

EXAMPLE 4.89. The Yoneda module $\mathbb{I}_{[1]} \in \mathcal{S}p([1] \times [1]^{\mathrm{op}})$ looks like

$$\begin{array}{ccc} \mathbb{S} & \xrightarrow{\mathrm{id}} & \mathbb{S} \\ \uparrow & & \uparrow{\scriptstyle \mathrm{id}} \\ \mathbb{S} & \longrightarrow & \mathbb{S}. \end{array}$$

EXAMPLE 4.90. For every stable derivator the evaluation morphism $0^*\colon \mathscr{D}^{[1]} \to \mathscr{D}$ is a weighted colimit (Theorem 4.81). The weight P_0 is obtained from $\mathbb{I}_{[1]}$ by evaluation at 0 in the covariant variable and is hence given by

$$P_0 = (0 \times \mathrm{id}_{[1]^{\mathrm{op}}})^* \mathbb{I}_{[1]} = (\mathbb{S} \leftarrow 0) \in \mathcal{S}p([1]^{\mathrm{op}}).$$

To double-check this result, for $X = (f\colon x \to y) \in \mathscr{D}([1])$ we calculate the canceling tensor product $P_0 \otimes_{[[1]]} X = (\mathbb{S} \leftarrow 0) \otimes_{[[1]]} (x \to y)$. By Example 4.62 this is simply the pushout-product of these two morphisms whose calculation is displayed in the following diagram

$$\begin{array}{ccc} \mathbb{S}\otimes x & \longleftarrow & 0\otimes x \\ \downarrow & & \downarrow\cong \\ \mathbb{S}\otimes y & \longleftarrow & 0\otimes y, \end{array} \qquad \begin{array}{ccc} x & \longleftarrow & 0 \\ \cong\downarrow & \square & \downarrow\cong \\ x & \leftarrow - - & 0. \end{array}$$

Since isomorphisms are stable under cobase change [**Gro13**, Prop. 3.12], the desired pushout is simply x. This calculation hence confirms that there is a natural isomorphism

$$0^* \cong P_0 \otimes_{[[1]]} - = (\mathbb{S} \leftarrow 0) \otimes_{[[1]]} -\colon \mathscr{D}^{[1]} \to \mathscr{D}.$$

Similarly, the weight P_1 for the evaluation morphisms $1^*\colon \mathscr{D}^{[1]} \to \mathscr{D}$ is given by

$$P_1 = (1 \times \mathrm{id}_{[1]^{\mathrm{op}}})^*\mathbb{I}_{[1]} = (\mathbb{S} \leftarrow \mathbb{S}) \in \mathcal{S}p([1]^{\mathrm{op}}).$$

Analogously to the previous case, in order to calculate $P_1 \otimes_{[[1]]} X$ we contemplate the diagrams

$$\begin{array}{ccc} \mathbb{S}\otimes x & \xleftarrow{\cong} & \mathbb{S}\otimes x \\ \downarrow & & \downarrow \\ \mathbb{S}\otimes y & \xleftarrow[\cong]{} & \mathbb{S}\otimes y, \end{array} \qquad \begin{array}{ccc} x & \xleftarrow{\cong} & x \\ \downarrow & \square & \downarrow f \\ y & \leftarrow \underset{\cong}{-} - & y. \end{array}$$

Again, this calculation confirms the above claim and we obtain a natural isomorphism

$$1^* \cong P_1 \otimes_{[[1]]} - = (\mathbb{S} \leftarrow \mathbb{S}) \otimes_{[[1]]} -\colon \mathscr{D}^{[1]} \to \mathscr{D}.$$

This example generalizes as follows.

REMARK 4.91. Let $\mathscr{D}$be a stable derivator, $A \in \mathcal{C}at$, and $a \in A$. The evaluation morphism $a^*\colon \mathscr{D}^A \to \mathscr{D}$ is a weighted colimit with weight $P_a \in \mathcal{S}p(A^{\mathrm{op}})$, the free diagram generated at a by $\mathbb{S} \in \mathcal{S}p(*)$. More formally, let $a\colon * \to A^{\mathrm{op}}$ be the functor classifying the object a, then P_a is given by

$$P_a \cong a_!(\mathbb{S}) \in \mathcal{S}p(A^{\mathrm{op}}).$$

This conclusion also holds in arbitrary derivators, and for this it suffices to consider the corresponding weight in spaces instead of in spectra.

EXAMPLE 4.92. Let $\mathscr{D}$be a stable derivator and let $C\colon \mathscr{D}^{[1]} \to \mathscr{D}$ be the cone morphism. We recall from Construction 3.24 that C is a finite composition of suitably finite Kan extensions and evaluation morphisms. By Theorem 4.81 and Theorem 4.85 the cone morphism is a weighted colimit. Moreover, the weight is obtained from the Yoneda bimodule $\mathbb{I}_{[1]} \in \mathcal{S}p([1] \times [1]^{\mathrm{op}})$ by an application of the cone with parameters in $[1]^{\mathrm{op}}$. The isomorphisms

$$C(0 \to \mathbb{S}) \cong \mathbb{S} \qquad \text{and} \qquad C(\mathbb{S} \xrightarrow{\sim} \mathbb{S}) \cong 0$$

imply that the weight M of C is isomorphic to $(0 \leftarrow \mathbb{S}) \in \mathcal{S}p([1]^{\mathrm{op}})$. We again verify this description. Given a morphism $X = (f\colon x \to y) \in \mathscr{D}([1])$ the following calculation

$$\begin{array}{ccc} 0\otimes x & \longleftarrow & \mathbb{S}\otimes x \\ \downarrow & & \downarrow \\ 0\otimes y & \longleftarrow & \mathbb{S}\otimes y, \end{array} \qquad \begin{array}{ccc} 0 & \longleftarrow & x \\ \downarrow & \square & \downarrow f \\ Cf & \leftarrow - - & y, \end{array}$$

whose details are left to the reader double-checks the claim (compare again to Construction 3.24). Thus, we obtain the intended natural isomorphism

$$C \cong (0 \leftarrow \mathbb{S}) \otimes_{[[1]]} - : \mathscr{D}^{[1]} \to \mathscr{D}.$$

REMARK 4.93. This description of cones as weighted colimits generalizes to pointed derivators. In fact, in that case the derivator $\mathcal{S}_*$ of pointed spaces is universal Remark 4.76, and every pointed derivator is a $\mathcal{S}_*$-module. Of course, in contrast to this the cone is *not an ordinary colimit* but merely a weighted colimit. This is easy to see already for represented derivators because in that case the cone agrees with the usual cokernel (Examples 3.27).

EXAMPLE 4.94. For every stable derivator $\mathscr{D}$there is the morphism

$$G\colon \mathscr{D}^{[1]} \to \mathscr{D}^{[1]\times[2]}$$

which sends a morphism to a (coherent) cofiber sequence. Let us recall from the proof of Theorem 3.42 that G is a finite composition of sufficiently finite homotopy Kan extensions (see Proposition 2.39 for the classical case which serves as a blueprint for arbitrary stable derivators). Consequently, G is a weighted colimit whose weight M lives in $\mathcal{S}p(([1]\times[2])\times[1]^{\mathrm{op}})$. As one easily verifies by calculations similar to the previous cases, the representing weight M looks like Figure 10. In that figure the paper plane corresponds to the coordinates in $[1]\times[2]$ while the diagonal direction corresponds to the copy of $[1]^{\mathrm{op}}$. Evaluating at $(0,0), (0,1)$, and $(1,1) \in [1]\times[2]$, we obtain the coherent morphisms

$$M_{0,0} = (\mathbb{S} \leftarrow 0), \quad M_{0,1} = (\mathbb{S} \leftarrow \mathbb{S}), \quad \text{and} \quad M_{1,1} = (0 \leftarrow \mathbb{S}) \in \mathcal{S}p([1]^{\mathrm{op}})$$

which already occured as weights for $0^*, 1^*$, and C, respectively (see Example 4.90 and Example 4.92). Let us only double-check the remaining non-trivial evaluation $M_{1,2} \in \mathcal{S}p([1]^{\mathrm{op}})$ which is given by

$$(\Sigma\mathbb{S} \leftarrow 0) \cong (\Sigma\mathbb{S}) \otimes (\mathbb{S} \leftarrow 0).$$

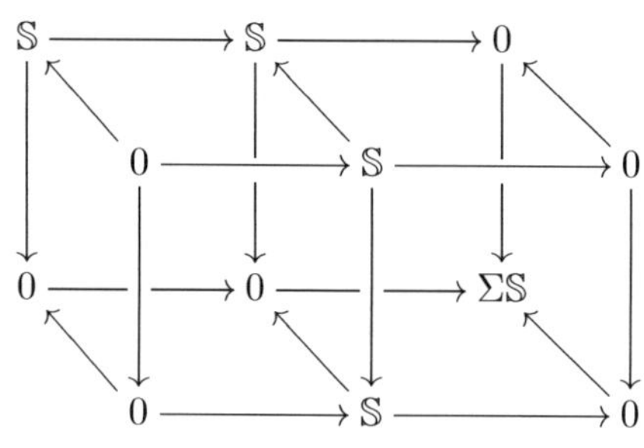

FIGURE 10. The universal constructor for cofiber sequences.

Invoking Example 4.88 and Example 4.90, the corresponding weighted colimit sends $X = (f\colon x \to y) \in \mathscr{D}([1])$ to

$$\begin{aligned}(\Sigma\mathbb{S} \leftarrow 0) \otimes_{[[1]]} X &\cong \big((\Sigma\mathbb{S}) \otimes (\mathbb{S} \leftarrow 0)\big) \otimes_{[[1]]} X\\ &\cong (\Sigma\mathbb{S}) \otimes \big((\mathbb{S} \leftarrow 0) \otimes_{[[1]]} X\big)\\ &\cong (\Sigma\mathbb{S}) \otimes x\\ &\cong \Sigma x,\end{aligned}$$

as it is supposed to be the case. Since M is a representing weight for G, we also refer to M as the **universal constructor for cofiber sequences**.

If we restrict M along the functor $i\colon [1] \to [1]\times[2]$ which classifies the morphism $(0,1) \to (1,1)$, then we obtain the bimodule $N = i^*M \in \mathcal{S}p([1]\times[1]^{\mathrm{op}})$ looking like

$$\begin{array}{ccc} \mathbb{S} & \longrightarrow & 0 \\ \uparrow & & \uparrow \\ \mathbb{S} & \longrightarrow & \mathbb{S}. \end{array}$$

The corresponding weighted colimit is the morphism $\mathsf{cof}\colon \mathscr{D}^{[1]} \to \mathscr{D}^{[1]}$,

$$N \otimes_{[[1]]} - \cong \mathsf{cof}\colon \mathscr{D}^{[1]} \to \mathscr{D}^{[1]},$$

which at the level of canonical triangulations corresponds to the rotation of triangles.

From these considerations we can also calculate the weight which describes the fiber morphism $\mathsf{fib}\colon \mathscr{D}^{[1]} \to \mathscr{D}^{[1]}$ as a weighted *colimit*. In fact, by (3.52) there is a natural isomorphism $\mathsf{cof}^3 \cong \Sigma\colon \mathscr{D}^{[1]} \to \mathscr{D}^{[1]}$, and in combination with Theorem 3.31 this implies

$$\mathsf{fib} \cong \Omega \circ \mathsf{cof}^2.$$

The weight for cof^2 can be calculated as $N \otimes_{[[1]]} N$. More efficiently, it can simply be read off from M (see Figure 10) by restriction along the functor $j\colon [1] \to [1]\times[2]$ classifying the morphism $(1,1) \to (1,2)$. By Example 4.88 and Theorem 3.31 we conclude that the weight for fib is given by $\Omega(j^*M)$ and looks like

$$\begin{array}{ccc} 0 & \longrightarrow & \mathbb{S} \\ \uparrow & \square & \uparrow \\ \Omega\mathbb{S} & \longrightarrow & 0. \end{array}$$

This example concludes the little detour which was included in order to develop some first feeling for the calculus of weighted colimits and canceling tensor products (for additional examples we refer to [**GPS14a**] or [**GŠ16a**]). We now turn to the case which is of particular interest to abstract representation theory namely the case of weights such that the corresponding weighted colimits are equivalences. After all our goal is to describe strong stable equivalences by means of spectral bimodules.

PROPOSITION 4.95. *Let $A, B \in \mathcal{C}at$ and let $M \in \mathcal{S}p(B\times A^{\mathrm{op}}), N \in \mathcal{S}p(A^{\mathrm{op}}\times B)$ be spectral bimodules. The following are equivalent.*

(i) *There are isomorphisms $M \otimes_{[A]} N \cong \mathbb{I}_B$ and $N \otimes_{[B]} M \cong \mathbb{I}_A$.*
(ii) *The weighted colimits $M \otimes_{[A]} -\colon \mathscr{D}^A \to \mathscr{D}^B$ and $N \otimes_{[B]} -\colon \mathscr{D}^B \to \mathscr{D}^A$ are inverse equivalences for all stable derivators $\mathscr{D}$.*
(iii) *The weighted colimits $M \otimes_{[A]} -\colon \mathcal{S}p^A \to \mathcal{S}p^B$ and $N \otimes_{[B]} -\colon \mathcal{S}p^B \to \mathcal{S}p^A$ are inverse equivalences.*

In this situation we say that M is an ***invertible spectral bimodule*** *and that N is an* ***inverse bimodule*** *of M.*

PROOF. Statement (ii) follows easily from (i) since weighted colimits associated to Yoneda bimodules are naturally isomorphic to the corresponding identity morphisms. Clearly, (iii) is a consequence of (ii), and it remains to show that (iii) implies (i). This implication is immediate from the uniqueness of representing weights for weighted colimits (see Remark 4.80). $\square$

REMARK 4.96. This proposition of course only makes explicit the notion of equivalences internal to the bicategory $\mathcal{P}rof(\mathcal{S}p)$ of spectral bimodules. The point is that the strong stable equivalences discussed in this paper give rise to such representing invertible spectral bimodules. In that context we also refer to these bimodules as **universal tilting modules** for the following two reasons.

(i) The word "universal" alludes to the fact that invertible bimodules realize simultaneous symmetries in all stable derivators.

(ii) These spectral bimodules are spectral refinements of the more classical tilting complexes. In more detail, for every commutative ring k extensions of scalars along $\mathbb{Z} \to k$ defines a cocontinuous monoidal morphism $k \otimes -\colon \mathscr{D}_{\mathbb{Z}} \to \mathscr{D}_k$. Moreover, let $H\mathbb{Z}$ be the integral Eilenberg–MacLane spectrum (realized as a symmetric ring spectrum) and let $\mathscr{D}_{H\mathbb{Z}}$ be the derivator of $H\mathbb{Z}$-module spectra. There is a zigzag of weakly monoidal Quillen equivalences relating $\mathrm{Ch}(\mathbb{Z})$ and $\mathrm{Mod}(H\mathbb{Z})$ [**Shi07**], and this induces a monoidal equivalence of derivators $\mathscr{D}_{H\mathbb{Z}} \simeq \mathscr{D}_{\mathbb{Z}}$. As a final ingredient, inducing up along $\mathbb{S} \to H\mathbb{Z}$ yields a cocontinuous monoidal morphism $H\mathbb{Z} \wedge -\colon \mathcal{S}p \to \mathscr{D}_{H\mathbb{Z}}$, and we end up with a cocontinuous monoidal morphism of derivators

$$k \otimes -\colon \mathcal{S}p \xrightarrow{H\mathbb{Z}\wedge -} \mathscr{D}_{H\mathbb{Z}} \simeq \mathscr{D}_{\mathbb{Z}} \xrightarrow{k\otimes -} \mathscr{D}_k. \tag{4.97}$$

In the case of a field k, an application of this morphism to invertible spectral bimodules yields tilting complexes as in tilting theory (see for example [**APR79, BB80, HR82, Ric91, Kel94**] or the survey articles in [**AHHK07**]).

We illustrate this notion of invertible spectral bimodules by some examples.

EXAMPLE 4.98. For every stable derivator the morphisms $\Sigma^n\colon \mathscr{D} \to \mathscr{D}$ for $n \in \mathbb{Z}$ are invertible (Theorem 3.31) and weighted colimits (Example 4.88). By Proposition 4.95 the corresponding weights $\Sigma^n\mathbb{S} \in \mathcal{S}p(*)$ are invertible.

EXAMPLE 4.99. For every stable derivator $\mathscr{D}$the iterated cofiber morphisms

$$\mathsf{cof}^n\colon \mathscr{D}^{[1]} \to \mathscr{D}^{[1]}$$

are invertible for $n \in \mathbb{Z}$ (Theorem 3.31) and these are by Example 4.94 weighted colimits. In more detail, the weights for cof and fib are respectively given by

$$\begin{array}{ccc} \mathbb{S} & \longrightarrow & 0 \\ \uparrow & & \uparrow \\ \mathbb{S} & \longrightarrow & \mathbb{S}, \end{array} \qquad\qquad \begin{array}{ccc} 0 & \longrightarrow & \mathbb{S} \\ \uparrow & \square & \uparrow \\ \Omega\mathbb{S} & \longrightarrow & 0, \end{array}$$

and these bimodules are inverse to each other (Proposition 4.95).

EXAMPLE 4.100. By Corollary 4.11 all A_3-quivers are strongly stably equivalent. The proof of this statement shows that these strong stable equivalences are weighted colimits which by Proposition 4.95 give rise to invertible spectral bimodules. To consider a specific example, let $Q = \ulcorner = (\bullet \leftarrow \bullet \to \bullet)$ be the source of valence two and let $\vec{A}_3 = (\bullet \to \bullet \to \bullet)$ be the linearly oriented A_3-quiver. We explicitly describe the universal tilting modules associated to the strong stable equivalence $\Phi\colon \vec{A}_3 \overset{s}{\sim} Q$ and its inverse as constructed in Example 4.10.

The universal tilting module $T_{Q,\vec{A}_3} \in \mathcal{S}p(Q \times \vec{A}_3^{\mathrm{op}})$ is obtained from the Yoneda bimodule $\mathbb{I}_{\vec{A}_3}$ by an application of the strong stable equivalence $\Phi\colon \mathcal{S}p^{\vec{A}_3} \to \mathcal{S}p^Q$ (Remark 4.87). The bimodule $\mathbb{I}_{\vec{A}_3}$ takes by Example 4.68 the form as shown on the left in:

$$\mathbb{I}_{\vec{A}_3}\colon\quad \begin{array}{ccccc} \mathbb{S} & \longrightarrow & \mathbb{S} & \longrightarrow & \mathbb{S} \\ \uparrow & & \uparrow & & \uparrow \\ 0 & \longrightarrow & \mathbb{S} & \longrightarrow & \mathbb{S} \\ \uparrow & & \uparrow & & \uparrow \\ 0 & \longrightarrow & 0 & \longrightarrow & \mathbb{S} \end{array} \qquad\qquad T_{Q,\vec{A}_3}\colon\quad \begin{array}{ccccc} 0 & \longleftarrow & \mathbb{S} & \longrightarrow & \mathbb{S} \\ \uparrow & & \uparrow & & \uparrow \\ \mathbb{S} & \longleftarrow & \mathbb{S} & \longrightarrow & \mathbb{S} \\ \uparrow & & \uparrow & & \uparrow \\ 0 & \longleftarrow & 0 & \longrightarrow & \mathbb{S} \end{array}$$

The strong stable equivalence Φ simply forms the cofiber of the first morphism, and this amounts in the three cases to

$$\begin{array}{ccccc} \mathbb{S} & \longrightarrow & \mathbb{S} & \longrightarrow & \mathbb{S} \\ \downarrow & \square & \downarrow & & \\ 0 & \longrightarrow & 0, & & \end{array} \qquad \begin{array}{ccccc} 0 & \longrightarrow & \mathbb{S} & \longrightarrow & \mathbb{S} \\ \downarrow & \square & \downarrow & & \\ 0 & \longrightarrow & \mathbb{S}, & & \end{array} \qquad \begin{array}{ccccc} 0 & \longrightarrow & 0 & \longrightarrow & \mathbb{S} \\ \downarrow & \square & \downarrow & & \\ 0 & \longrightarrow & 0. & & \end{array}$$

Consequently, for every stable derivator $\mathscr{D}$ there is a natural isomorphism

$$\Phi_{\mathscr{D}} \cong T_{Q,\vec{A}_3} \otimes_{[\vec{A}_3]} -\colon \mathscr{D}^{\vec{A}_3} \xrightarrow{\sim} \mathscr{D}^Q,$$

where $T_{Q,\vec{A}_3}$ takes the form as displayed in the above diagram on the right.

Similarly, also the inverse strong stable equivalence $\Phi^{-1}\colon Q \overset{s}{\sim} \vec{A}_3$ is given by a universal tilting module $T_{\vec{A}_3,Q} \in \mathcal{S}p(\vec{A}_3 \times Q^{\mathrm{op}})$. This invertible spectral bimodule is obtained from $\mathbb{I}_Q$ by an application of Φ^{-1} which amounts to forming fibers of the morphisms pointing from right to left. We leave it to the reader to double-check that $\mathbb{I}_Q$ and $T_{\vec{A}_3,Q}$ look like:

$$\mathbb{I}_Q\colon\quad \begin{array}{ccccc} \mathbb{S} \; (& & 0 & \longrightarrow & 0 \\ \downarrow & & \downarrow & & \downarrow \\ \mathbb{S} & \longleftarrow & \mathbb{S} & \longrightarrow & \mathbb{S} \\ \uparrow & & \uparrow & & \uparrow \\ 0 & \longleftarrow & 0 & \longrightarrow & \mathbb{S} \end{array} \qquad\qquad T_{\vec{A}_3,Q}\colon\quad \begin{array}{ccccc} \Omega\mathbb{S} & \longrightarrow & 0 & \longrightarrow & 0 \\ \downarrow & & \downarrow & & \downarrow \\ 0 & \longrightarrow & \mathbb{S} & \longrightarrow & \mathbb{S} \\ \uparrow & & \uparrow & & \uparrow \\ 0 & \longrightarrow & 0 & \longrightarrow & \mathbb{S} \end{array}$$

These universal tilting modules $T_{Q,\vec{A}_3}$ and $T_{\vec{A}_3,Q}$ are inverse to each other.

Additional interesting examples of universal tilting bimodules arise from the Coxeter and the Serre functors discussed in §4.2.

EXAMPLE 4.101. For every stable derivator $\mathscr{D}$ the Serre functor $S\colon \mathscr{D}^{\vec{A}_2} \xrightarrow{\sim} \mathscr{D}^{\vec{A}_2}$ is naturally isomorphic to the cofiber morphism cof (Examples 4.30). The universal tilting bimodule is by Example 4.99 given by

$$\begin{array}{ccc} \mathbb{S} & \longrightarrow & 0 \\ \uparrow & & \uparrow \\ \mathbb{S} & \longrightarrow & \mathbb{S}. \end{array}$$

We invite the reader to directly calculate the universal tilting bimodules for the Serre functors in the cases $\vec{A}_3$ and $Q = (\bullet \leftarrow \bullet \to \bullet)$ (see Example 4.22 and Example 4.25). Here, instead we turn to the following different perspective on these bimodules. Let us recall from our discussion of opposite derivators that the Yoneda bimodule $\mathbb{I}_A \in \mathcal{S}p(A \times A^{\mathrm{op}})$ can also be considered as an object in

$$\mathcal{S}p(A \times A^{\mathrm{op}})^{\mathrm{op}} \cong \mathcal{S}p^{\mathrm{op}}((A \times A^{\mathrm{op}})^{\mathrm{op}}) \cong \mathcal{S}p^{\mathrm{op}}(A^{\mathrm{op}} \times A).$$

The point of this reformulation is that the sphere spectrum $\mathbb{S}$ defines a morphism $- \rhd \mathbb{S} \colon \mathcal{S}p^{\mathrm{op}} \to \mathcal{S}p$.

DEFINITION 4.102. For every small category A the **canonical duality bimodule** $D_A \in \mathcal{S}p(A \times A^{\mathrm{op}})$ is the image of the Yoneda bimodule $\mathbb{I}_A \in \mathcal{S}p(A \times A^{\mathrm{op}})$ under

$$\mathcal{S}p^{\mathrm{op}}(A^{\mathrm{op}} \times A) \overset{-\rhd\mathbb{S}}{\to} \mathcal{S}p(A^{\mathrm{op}} \times A) \overset{\cong}{\to} \mathcal{S}p(A \times A^{\mathrm{op}}).$$

EXAMPLE 4.103. For every poset P the canonical duality bimodule D_P admits the following description. Let $P \times P^{\mathrm{op}}$ be endowed with the product order and let $\Delta_P \subseteq P \times P^{\mathrm{op}}$ be the diagonal. The bimodule $D_P \in \mathcal{S}p(P \times P^{\mathrm{op}})$ restricted to the down-set generated by Δ_P is constant with value the sphere spectrum $\mathbb{S}$ and it vanishes on the complement [**GŠ16a**, Lem. 7.4].

We illustrate this class of examples in the following specific cases. As in previous cases, for the bimodules $D_P \in \mathcal{S}p(P \times P^{\mathrm{op}})$ the P-coordinate is drawn horizontally and the P^{op}-coordinate vertically.

EXAMPLE 4.104. The canonical duality bimodule $D_{\vec{A}_2} \in \mathcal{S}p(\vec{A}_2 \times \vec{A}_2^{\mathrm{op}})$ looks like

$$\begin{array}{ccc} \mathbb{S} & \longrightarrow & 0 \\ \uparrow & & \uparrow \\ \mathbb{S} & \longrightarrow & \mathbb{S}. \end{array}$$

EXAMPLE 4.105. The canonical duality bimodule $D_{\vec{A}_3} \in \mathcal{S}p(\vec{A}_3 \times \vec{A}_3^{\mathrm{op}})$ takes the form

$$\begin{array}{ccccc} \mathbb{S} & \longrightarrow & 0 & \longrightarrow & 0 \\ \uparrow & & \uparrow & & \uparrow \\ \mathbb{S} & \longrightarrow & \mathbb{S} & \longrightarrow & 0 \\ \uparrow & & \uparrow & & \uparrow \\ \mathbb{S} & \longrightarrow & \mathbb{S} & \longrightarrow & \mathbb{S}. \end{array}$$

EXAMPLE 4.106. Let $Q = (\bullet \leftarrow \bullet \to \bullet)$ be the source of valence two. The canonical duality bimodule $D_Q \in \mathcal{S}p(Q \times Q^{\mathrm{op}})$ is given by

$$\begin{array}{ccccc} \mathbb{S} & \longleftarrow & \mathbb{S} & \longrightarrow & 0 \\ \downarrow & & \downarrow & & \downarrow \\ 0 & \longleftarrow & \mathbb{S} & \longrightarrow & 0 \\ \uparrow & & \uparrow & & \uparrow \\ 0 & \longleftarrow & \mathbb{S} & \longrightarrow & \mathbb{S} \end{array}$$

DEFINITION 4.107. Let $\mathscr{D}$ be a stable derivator. The **Nakayama functor** associated to $A \in \mathcal{C}at$ is

$$D_A \otimes_{[A]} -\colon \mathscr{D}^A \to \mathscr{D}^A.$$

This definition is of course motivated by the corresponding situation in representation theory (see for instance [**Hap88**, §4.6]). Note that in the case of $\vec{A}_2$ the Serre functor and the Nakayama functor are naturally isomorphic (Example 4.101 and Example 4.104). This holds in full generality.

THEOREM 4.108. *Let $\mathscr{D}$ be a stable derivator and let Q be a Dynkin quiver of type A. The Serre functor and the Nakayama functor are naturally isomorphic*

$$S \cong D_Q \otimes_{[Q]} -\colon \mathscr{D}^Q \xrightarrow{\sim} \mathscr{D}^Q.$$

PROOF. This theorem is in [**GŠ16a**]. The special case of linearly oriented Dynkin quivers is established by direct calculation as [**GŠ16a**, Thm. 7.7]. And the general case is deduced from it as an application of the general calculus of bimodules [**GŠ16a**, Thm. 10.9]. □

Additional universal tilting bimodules arise from more general reflection functors (as in §4.3). We conclude this paper by the following remark offering a different perspective on the project on abstract representation theory.

REMARK 4.109. A different way of thinking of abstract representation theory is that it contributes to the calculation of spectral Picard groups (or groupoids). For every closed monoidal derivator $\mathscr{V}$ there is the bicategory $\mathcal{P}rof(\mathscr{V})$ of bimodules (Theorem 4.67). In particular, for every small category A we obtain the monoidal category

$$\mathcal{P}rof(\mathscr{V})(A,A) = \mathscr{V}(A \times A^{\mathrm{op}})$$

of (A,A)-bimodules with values in $\mathscr{V}$. The monoidal structure is the canceling tensor product $\otimes_{[A]}$ and the monoidal unit is the Yoneda bimodule $\mathbb{I}_A$. Correspondingly, we define the **Picard group** $\mathrm{Pic}_{\mathscr{V}}(A)$ **of** A **relative to** $\mathscr{V}$ as the Picard group of the monoidal category $\mathcal{P}rof(\mathscr{V})(A,A)$,

$$\mathrm{Pic}_{\mathscr{V}}(A) = \mathrm{Pic}\big(\mathcal{P}rof(\mathscr{V})(A,A)\big).$$

Elements are simply isomorphism classes of invertible bimodules and the isomorphism class of $\mathbb{I}_A$ is the neutral element. Particularly interesting cases arise for the stable, closed monoidal derivators $\mathcal{S}p, \mathscr{D}_{\mathbb{Z}}$, and $\mathscr{D}_k$ for arbitrary commutative rings k. Correspondingly, we obtain the **spectral Picard group** $\mathrm{Pic}_{\mathcal{S}p}(A)$, the **integral Picard group** $\mathrm{Pic}_{\mathscr{D}_{\mathbb{Z}}}(A)$, and also the **$k$-linear Picard group** $\mathrm{Pic}_{\mathscr{D}_k}(A)$. In particular, in the case of a field k for special choices of A the group $\mathrm{Pic}_{\mathscr{D}_k}(A)$ agrees with the *derived Picard group* of Miyachi and Yekutieli [**MY01**]. For every stable, closed monoidal derivator $\mathscr{V}$ the suspension of the monoidal unit $\Sigma\mathbb{S}$ defines an element in $\mathrm{Pic}_{\mathscr{V}}(*)$ (Example 4.88).

To the best of the knowledge of the author, the only shape for which the spectral Picard group is known is $A = *$. In that case, the Picard group $\mathrm{Pic}_{\mathcal{S}p}(*)$ is the Picard group of the stable homotopy category which is isomorphic to $\mathbb{Z}$ with $\Sigma\mathbb{S}$ as generator ([**HMS94**] or [**Str92**, Thm. 2.2]). Every construction of a strong stable equivalence $A \overset{s}{\sim} A$ and its universal tilting bimodule $T_{A,A}$ yields an element in $\mathrm{Pic}_{\mathcal{S}p}(A)$, and the hope is that at least for some shapes A this leads to a calculation of the spectral Picard group.

One nice feature of these Picard groups is their functoriality in $\mathscr{V}$. In fact, every cocontinuous, monoidal morphism $F\colon \mathscr{V} \to \mathscr{W}$ of monoidal derivators induces a morphism of bicategories $F\colon \mathcal{P}rof(\mathscr{V}) \to \mathcal{P}rof(\mathscr{W})$ (as follows from the fact that cocontinuous morphisms preserve coends). In particular, for every $A \in \mathcal{C}at$ we obtain a monoidal functor $\mathcal{P}rof(\mathscr{V})(A,A) \to \mathcal{P}rof(\mathscr{W})(A,A)$ and hence a group homomorphism $F\colon \mathrm{Pic}_{\mathscr{V}}(A) \to \mathrm{Pic}_{\mathscr{W}}(A)$. A particularly interesting case is given by the morphism (4.97). Hence, associated to every commutative ring k a group homomorphism

$$k \otimes -\colon \mathrm{Pic}_{\mathcal{S}p}(A) \to \mathrm{Pic}_{\mathscr{D}_k}(A).$$

This functoriality of Picard groups is calculationally useful. For instance,in combination with calculations of derived Picard groups by Miyachi and Yekutieli [**MY01**, Theorem 4.1] and an abstract fractionally Calabi–Yau property of Dynkin quivers of type A [**GŠ16a**, Cor. 5.20], this was used to show that for every such quiver Q the homomorphism

$$k \otimes -\colon \mathrm{Pic}_{\mathcal{S}p}(Q) \to \mathrm{Pic}_{\mathscr{D}_k}(Q)$$

is a split epimorphism [**GŠ16a**, Thm. 12.6]. Conjecturally, there is no kernel, and, jointly with Jan Šťovíček, we intend to come back to this and related calculations elsewhere.

Acknowledgments

It is a great pleasure to begin by thanking my Ph.D. supervisor Stefan Schwede for having suggested to me to think about abstract homotopy theory in the first place. My understanding of the subject was sharpened through various more or less directly related cooperations, hence special thanks go to my coauthors Falk Beckert [**BG18**], David Gepner and Thomas Nikolaus [**GGN15**], Kate Ponto and Mike Shulman [**GPS14b, GPS14a**], again Mike Shulman [**GS17**] and, last but not least (watch the alphabetic order!), Jan Šťovíček [**GŠ14, GŠ16b, GŠ16a, GŠ15**]. During the last years many colleagues and friends gave me (willingly or not) the opportunity to discuss this project and/or to have a good time. It is a great pleasure to thank Dimitri Ara, Peter Arndt, Denis-Charles Cisinski, Ivo Dell'Ambrogio, David Gepner, Drew Heard, Gustavo Jasso, André Joyal, Philipp Jung, Bernhard Keller, Steffen König, Henning Krause, Rosie Laking, Georges Maltsiniotis, Thomas Nikolaus, Justin Noël, Eric Peterson, Mona Rahn, George Raptis, Ulrich Schlickewei, Timo Schürg, Stefan Schwede, Mike Shulman, Johan Steen, Greg Stevenson, and Jan Šťovíček.

The author also thanks the organizers of the International Conference on Representation theory of Algebras in Prague in August 2018 (ICRA2018). On occasion of that conference the author offered a series of three talks aiming for an overview of this project. The sections of this account are in obvious bijection with the talks given at ICRA2018, but here we intend to draw a slightly more complete picture.

References

[ABG+09] Matthew Ando, Andrew J. Blumberg, David Gepner, Michael J. Hopkins, and Charles Rezk, *Units of ring spectra, orientations and Thom spectra via rigid infinite loop space theory*, J. Topol. **7** (2014), no. 4, 1077–1117, DOI 10.1112/jtopol/jtu009. MR3286898

[ABG10] Matthew Ando, Andrew J. Blumberg, and David Gepner, *Twists of K-theory and TMF*, Superstrings, geometry, topology, and C^*-algebras, Proc. Sympos. Pure Math., vol. 81, Amer. Math. Soc., Providence, RI, 2010, pp. 27–63, DOI 10.1090/pspum/081/2681757. MR2681757

[AHHK07] Lidia Angeleri Hügel, Dieter Happel, and Henning Krause (eds.), *Handbook of tilting theory*, London Mathematical Society Lecture Note Series, vol. 332, Cambridge University Press, Cambridge, 2007. MR2385175

[And79] D. W. Anderson, *Axiomatic homotopy theory*, Algebraic topology, Waterloo, 1978 (Proc. Conf., Univ. Waterloo, Waterloo, Ont., 1978), Lecture Notes in Math., vol. 741, Springer, Berlin, 1979, pp. 520–547. MR557184

[APR79] Maurice Auslander, María Inés Platzeck, and Idun Reiten, *Coxeter functors without diagrams*, Trans. Amer. Math. Soc. **250** (1979), 1–46, DOI 10.2307/1998978. MR530043

[ASS06] Ibrahim Assem, Daniel Simson, and Andrzej Skowroński, *Elements of the representation theory of associative algebras. Vol. 1*, London Mathematical Society Student Texts, vol. 65, Cambridge University Press, Cambridge, 2006. Techniques of representation theory. MR2197389

[Bal10] Paul Balmer, *Tensor triangular geometry*, Proceedings of the International Congress of Mathematicians. Volume II, Hindustan Book Agency, New Delhi, 2010, pp. 85–112. MR2827786

[Bal11] Paul Balmer, *Separability and triangulated categories*, Adv. Math. **226** (2011), no. 5, 4352–4372, DOI 10.1016/j.aim.2010.12.003. MR2770453

[BB80] Sheila Brenner and M. C. R. Butler, *Generalizations of the Bernstein-Gel′fand-Ponomarev reflection functors*, Representation theory, II (Proc. Second Internat. Conf., Carleton Univ., Ottawa, Ont., 1979), Lecture Notes in Math., vol. 832, Springer, Berlin-New York, 1980, pp. 103–169. MR607151

[BBD82] A. A. Beĭlinson, J. Bernstein, and P. Deligne, *Faisceaux pervers* (French), Analysis and topology on singular spaces, I (Luminy, 1981), Astérisque, vol. 100, Soc. Math. France, Paris, 1982, pp. 5–171. MR751966

[BCR97] D. J. Benson, Jon F. Carlson, and Jeremy Rickard, *Thick subcategories of the stable module category*, Fund. Math. **153** (1997), no. 1, 59–80. MR1450996

[Bec14] Hanno Becker, *Models for singularity categories*, Adv. Math. **254** (2014), 187–232, DOI 10.1016/j.aim.2013.11.016. MR3161097

[Bec18] Falk Beckert. *The bivariant parasimplicial $S_\bullet$-construction.* PhD thesis, Bergische Universität Wuppertal, 2018.

[Bek00] Tibor Beke, *Sheafifiable homotopy model categories*, Math. Proc. Cambridge Philos. Soc. **129** (2000), no. 3, 447–475, DOI 10.1017/S0305004100004722. MR1780498

[Bén67] Jean Bénabou, *Introduction to bicategories*, Reports of the Midwest Category Seminar, Springer, Berlin, 1967, pp. 1–77. MR0220789

[Ber07] Julia E. Bergner, *A model category structure on the category of simplicial categories*, Trans. Amer. Math. Soc. **359** (2007), no. 5, 2043–2058, DOI 10.1090/S0002-9947-06-03987-0. MR2276611

[Ber10] Julia E. Bergner, *A survey of $(\infty,1)$-categories*, Towards higher categories, IMA Vol. Math. Appl., vol. 152, Springer, New York, 2010, pp. 69–83, DOI 10.1007/978-1-4419-1524-5_2. MR2664620

[BF78] A. K. Bousfield and E. M. Friedlander, *Homotopy theory of Γ-spaces, spectra, and bisimplicial sets*, Geometric applications of homotopy theory (Proc. Conf., Evanston, Ill., 1977), II, Lecture Notes in Math., vol. 658, Springer, Berlin, 1978, pp. 80–130. MR513569

[BG99] James C. Becker and Daniel Henry Gottlieb, *A history of duality in algebraic topology*, History of topology, North-Holland, Amsterdam, 1999, pp. 725–745, DOI 10.1016/B978-044482375-5/50026-2. MR1721120

[BG18] Falk Beckert and Moritz Groth. Abstract cubical homotopy theory. `https://arxiv.org/abs/1803.06022`, 2018. Preprint.

[BGP73] I. N. Bernšteĭn, I. M. Gel′fand, and V. A. Ponomarev, *Coxeter functors, and Gabriel's theorem* (Russian), Uspehi Mat. Nauk **28** (1973), no. 2(170), 19–33. MR0393065

[BIK11] David J. Benson, Srikanth B. Iyengar, and Henning Krause, *Stratifying modular representations of finite groups*, Ann. of Math. (2) **174** (2011), no. 3, 1643–1684, DOI 10.4007/annals.2011.174.3.6. MR2846489

[BK72] A. K. Bousfield and D. M. Kan, *Homotopy limits, completions and localizations*, Lecture Notes in Mathematics, Vol. 304, Springer-Verlag, Berlin-New York, 1972. MR0365573

[BK89] A. I. Bondal and M. M. Kapranov, *Representable functors, Serre functors, and reconstructions* (Russian), Izv. Akad. Nauk SSSR Ser. Mat. **53** (1989), no. 6, 1183–1205, 1337, DOI 10.1070/IM1990v035n03ABEH000716; English transl., Math. USSR-Izv. **35** (1990), no. 3, 519–541. MR1039961

[BK90] A. I. Bondal and M. M. Kapranov, *Framed triangulated categories* (Russian), Mat. Sb. **181** (1990), no. 5, 669–683, DOI 10.1070/SM1991v070n01ABEH001253; English transl., Math. USSR-Sb. **70** (1991), no. 1, 93–107. MR1055981

[BLM08] Yu. Bespalov, V. Lyubashenko, and O. Manzyuk, *Pretriangulated A_∞-categories*, Pratsī Īnstitutu Matematiki Natsīonal′noï Akademīï Nauk Ukraïni. Matematika ta īï Zastosuvannya [Proceedings of Institute of Mathematics of NAS of Ukraine. Mathematics and its Applications], vol. 76, Natsīonal′na Akademīya Nauk Ukraïni, Īnstitut Matematiki, Kiev, 2008. MR2513383

[BO01] Alexei Bondal and Dmitri Orlov, *Reconstruction of a variety from the derived category and groups of autoequivalences*, Compositio Math. **125** (2001), no. 3, 327–344, DOI 10.1023/A:1002470302976. MR1818984

[Boa64] Michael Boardman. *Stable Homotopy Theory and Some Applications*. PhD thesis, University of Cambridge, 1964.

[Bor94a] Francis Borceux, *Handbook of categorical algebra. 1*, Encyclopedia of Mathematics and its Applications, vol. 50, Cambridge University Press, Cambridge, 1994. Basic category theory. MR1291599

[Bor94b] Francis Borceux, *Handbook of categorical algebra. 2*, Encyclopedia of Mathematics and its Applications, vol. 51, Cambridge University Press, Cambridge, 1994. Categories and structures. MR1313497

[BV73] J. M. Boardman and R. M. Vogt, *Homotopy invariant algebraic structures on topological spaces*, Lecture Notes in Mathematics, Vol. 347, Springer-Verlag, Berlin-New York, 1973. MR0420609

[Cam13] Omar Antolín Camarena, *A whirlwind tour of the world of $(\infty,1)$-categories*, Mexican mathematicians abroad: recent contributions, Contemp. Math., vol. 657, Amer. Math. Soc., Providence, RI, 2016, pp. 15–61, DOI 10.1090/conm/657/13088. MR3466443

[CD09] Denis-Charles Cisinski and Frédéric Déglise, *Local and stable homological algebra in Grothendieck abelian categories*, Homology Homotopy Appl. **11** (2009), no. 1, 219–260. MR2529161

[Che11] Xiao-Wu Chen, *Generalized Serre duality*, J. Algebra **328** (2011), 268–286, DOI 10.1016/j.jalgebra.2010.08.022. MR2745566

[Cis03] Denis-Charles Cisinski, *Images directes cohomologiques dans les catégories de modèles* (French, with French summary), Ann. Math. Blaise Pascal **10** (2003), no. 2, 195–244. MR2031269

[Cis04] Denis-Charles Cisinski, *Le localisateur fondamental minimal* (French, with English summary), Cah. Topol. Géom. Différ. Catég. **45** (2004), no. 2, 109–140. MR2072934

[Cis06] Denis-Charles Cisinski, *Les préfaisceaux comme modèles des types d'homotopie* (French, with English and French summaries), Astérisque **308** (2006), xxiv+390. MR2294028

[Cis08] Denis-Charles Cisinski, *Propriétés universelles et extensions de Kan dérivées* (French, with English summary), Theory Appl. Categ. **20** (2008), No. 17, 605–649. MR2534209

[Cis10] Denis-Charles Cisinski, *Catégories dérivables* (French, with English and French summaries), Bull. Soc. Math. France **138** (2010), no. 3, 317–393, DOI 10.24033/bsmf.2592. MR2729017

[CS02] Wojciech Chachólski and Jérôme Scherer, *Homotopy theory of diagrams*, Mem. Amer. Math. Soc. **155** (2002), no. 736, x+90, DOI 10.1090/memo/0736. MR1879153

[CT11] Denis-Charles Cisinski and Gonçalo Tabuada, *Non-connective K-theory via universal invariants*, Compos. Math. **147** (2011), no. 4, 1281–1320, DOI 10.1112/S0010437X11005380. MR2822869

[CT12] Denis-Charles Cisinski and Gonçalo Tabuada, *Symmetric monoidal structure on non-commutative motives*, J. K-Theory **9** (2012), no. 2, 201–268, DOI 10.1017/is011011005jkt169. MR2922389

[DJW18] Tobias Dyckerhoff, Gustavo Jasso, and Tashi Walde, *Simplicial structures in higher Auslander-Reiten theory*, Adv. Math. **355** (2019), 106762, 73, DOI 10.1016/j.aim.2019.106762. MR3994443

[DJW19] Tobias Dyckerhoff, Gustavo Jasso, and Tashi Walde. Generalised BGP reflection functors via the Grothendieck construction. `http://arxiv.org/pdf/1901.06993`, 2019. preprint.

[DP80] Albrecht Dold and Dieter Puppe, *Duality, trace, and transfer*, Proceedings of the International Conference on Geometric Topology (Warsaw, 1978), PWN, Warsaw, 1980, pp. 81–102. MR656721

[Dug01a] Daniel Dugger, *Combinatorial model categories have presentations*, Adv. Math. **164** (2001), no. 1, 177–201, DOI 10.1006/aima.2001.2015. MR1870516

[Dug01b] Daniel Dugger, *Universal homotopy theories*, Adv. Math. **164** (2001), no. 1, 144–176, DOI 10.1006/aima.2001.2014. MR1870515

[EE05] Edgar Enochs and Sergio Estrada, *Relative homological algebra in the category of quasi-coherent sheaves*, Adv. Math. **194** (2005), no. 2, 284–295, DOI 10.1016/j.aim.2004.06.007. MR2139915

[EH62] B. Eckmann and P. J. Hilton, *Group-like structures in general categories. I. Multiplications and comultiplications*, Math. Ann. **145** (1961/62), 227–255, DOI 10.1007/BF01451367. MR136642

[EJ00] Edgar E. Enochs and Overtoun M. G. Jenda, *Relative homological algebra*, De Gruyter Expositions in Mathematics, vol. 30, Walter de Gruyter & Co., Berlin, 2000. MR1753146

[EKMM97] A. D. Elmendorf, I. Kriz, M. A. Mandell, and J. P. May, *Rings, modules, and algebras in stable homotopy theory*, Mathematical Surveys and Monographs, vol. 47, American Mathematical Society, Providence, RI, 1997. With an appendix by M. Cole. MR1417719

[Fer06] Daniel Ferrand. On the non-additivity of traces in derived categories. arXiv:0506589, 2006.

[Fra96] Jens Franke. Uniqueness theorems for certain triangulated categories with an Adams spectral sequence, 1996. Preprint.

[Fre09] Benoit Fresse, *Modules over operads and functors*, Lecture Notes in Mathematics, vol. 1967, Springer-Verlag, Berlin, 2009. MR2494775

[Gab72] Peter Gabriel, *Unzerlegbare Darstellungen. I* (German, with English summary), Manuscripta Math. **6** (1972), 71–103; correction, ibid. 6 (1972), 309, DOI 10.1007/BF01298413. MR332887

[GAdS14] Martin Gallauer Alves de Souza, *Traces in monoidal derivators, and homotopy colimits*, Adv. Math. **261** (2014), 26–84, DOI 10.1016/j.aim.2014.03.029. MR3213295

[Gar05] Grigory Garkusha, *Systems of diagram categories and K-theory. II*, Math. Z. **249** (2005), no. 3, 641–682, DOI 10.1007/s00209-004-0726-0. MR2121745

[Gar06] G. Garkusha, *Systems of diagram categories and K-theory. I* (Russian, with Russian summary), Algebra i Analiz **18** (2006), no. 6, 131–186, DOI 10.1090/S1061-0022-07-00978-8; English transl., St. Petersburg Math. J. **18** (2007), no. 6, 957–996. MR2307357

[GGN15] David Gepner, Moritz Groth, and Thomas Nikolaus, *Universality of multiplicative infinite loop space machines*, Algebr. Geom. Topol. **15** (2015), no. 6, 3107–3153, DOI 10.2140/agt.2015.15.3107. MR3450758

[Gil11] James Gillespie, *Model structures on exact categories*, J. Pure Appl. Algebra **215** (2011), no. 12, 2892–2902, DOI 10.1016/j.jpaa.2011.04.010. MR2811572

[GM03] Sergei I. Gelfand and Yuri I. Manin, *Methods of homological algebra*, 2nd ed., Springer Monographs in Mathematics, Springer-Verlag, Berlin, 2003. MR1950475

[Goo92] Thomas G. Goodwillie, *Calculus. II. Analytic functors*, K-Theory **5** (1991/92), no. 4, 295–332, DOI 10.1007/BF00535644. MR1162445

[GPS14a] Moritz Groth, Kate Ponto, and Michael Shulman, *The additivity of traces in monoidal derivators*, J. K-Theory **14** (2014), no. 3, 422–494, DOI 10.1017/is014005011jkt262. MR3349323

[GPS14b] Moritz Groth, Kate Ponto, and Michael Shulman, *Mayer-Vietoris sequences in stable derivators*, Homology Homotopy Appl. **16** (2014), no. 1, 265–294, DOI 10.4310/HHA.2014.v16.n1.a15. MR3211746

[Gro] Alexander Grothendieck. Les dérivateurs. `http://www.math.jussieu.fr/~maltsin/groth/Derivateurs.html`. Manuscript.

[Gro83] A. Grothendieck. Pursuing stacks. manuscript, to appear in Documents Mathématics, 1983.

[Gro10] Moritz Groth, *A short course on ∞-categories*, in Handbook of homotpoy theory (Haynes Miller, ed.). CRC Press, Boca Raton, FL, 2020, 549–618.

[Gro13] Moritz Groth, *Derivators, pointed derivators and stable derivators*, Algebr. Geom. Topol. **13** (2013), no. 1, 313–374, DOI 10.2140/agt.2013.13.313. MR3031644

[Gro16] Moritz Groth, *Revisiting the canonicity of canonical triangulations*, Theory Appl. Categ. **33** (2018), Paper No. 14, 350–389. MR3806332

[Gro18] Moritz Groth. The theory of derivators, 2018. Book project in preparation. Draft versions of parts of it are available under `http://www.math.uni-bonn.de/~mgroth/monos.htmpl`.

[GŠ14] Moritz Groth and Jan Šťovíček, *Tilting theory via stable homotopy theory*, J. Reine Angew. Math. **743** (2018), 29–90, DOI 10.1515/crelle-2015-0092. MR3859269

[GŠ15] Moritz Groth and Jan Šťovíček, *Abstract tilting theory for quivers and related categories*, Ann. K-Theory **3** (2018), no. 1, 71–124, DOI 10.2140/akt.2018.3.71. MR3695365

[GŠ16a] Moritz Groth and Jan Šťovíček, *Abstract representation theory of Dynkin quivers of type A*, Adv. Math. **293** (2016), 856–941, DOI 10.1016/j.aim.2016.02.018. MR3474336

[GŠ16b] Moritz Groth and Jan Šťovíček, *Tilting theory for trees via stable homotopy theory*, J. Pure Appl. Algebra **220** (2016), no. 6, 2324–2363, DOI 10.1016/j.jpaa.2015.11.009. MR3448799

[GS17] Moritz Groth and Mike Shulman. Generalized stability for abstract stable homotopy theories. available at arXiv:1704.08084, 2017. to appear in Annals of K-Theory.

[GZ67] P. Gabriel and M. Zisman, *Calculus of fractions and homotopy theory*, Ergebnisse der Mathematik und ihrer Grenzgebiete, Band 35, Springer-Verlag New York, Inc., New York, 1967. MR0210125

[Hap87] Dieter Happel, *On the derived category of a finite-dimensional algebra*, Comment. Math. Helv. **62** (1987), no. 3, 339–389, DOI 10.1007/BF02564452. MR910167

[Hap88] Dieter Happel, *Triangulated categories in the representation theory of finite-dimensional algebras*, London Mathematical Society Lecture Note Series, vol. 119, Cambridge University Press, Cambridge, 1988. MR935124

[Hel68] Alex Heller, *Stable homotopy categories*, Bull. Amer. Math. Soc. **74** (1968), 28–63, DOI 10.1090/S0002-9904-1968-11871-3. MR224090

[Hel88] Alex Heller, *Homotopy theories*, Mem. Amer. Math. Soc. **71** (1988), no. 383, vi+78, DOI 10.1090/memo/0383. MR920963

[Hel97] Alex Heller, *Stable homotopy theories and stabilization*, J. Pure Appl. Algebra **115** (1997), no. 2, 113–130, DOI 10.1016/S0022-4049(96)00116-8. MR1431157

[Hin97] Vladimir Hinich, *Homological algebra of homotopy algebras*, Comm. Algebra **25** (1997), no. 10, 3291–3323, DOI 10.1080/00927879708826055. MR1465117

[Hir03] Philip S. Hirschhorn, *Model categories and their localizations*, Mathematical Surveys and Monographs, vol. 99, American Mathematical Society, Providence, RI, 2003. MR1944041

[HJR10] Thorsten Holm and Peter Jørgensen, *Triangulated categories: definitions, properties, and examples*, Triangulated categories, London Math. Soc. Lecture Note Ser., vol. 375, Cambridge Univ. Press, Cambridge, 2010, pp. 1–51, DOI 10.1017/CBO9781139107075.002. MR2681706

[HMS94] Michael J. Hopkins, Mark Mahowald, and Hal Sadofsky, *Constructions of elements in Picard groups*, Topology and representation theory (Evanston, IL, 1992), Contemp. Math., vol. 158, Amer. Math. Soc., Providence, RI, 1994, pp. 89–126, DOI 10.1090/conm/158/01454. MR1263713

[Hov99] Mark Hovey, *Model categories*, Mathematical Surveys and Monographs, vol. 63, American Mathematical Society, Providence, RI, 1999. MR1650134

[Hov01] Mark Hovey, *Model category structures on chain complexes of sheaves*, Trans. Amer. Math. Soc. **353** (2001), no. 6, 2441–2457, DOI 10.1090/S0002-9947-01-02721-0. MR1814077

[Hov02] Mark Hovey, *Cotorsion pairs, model category structures, and representation theory*, Math. Z. **241** (2002), no. 3, 553–592, DOI 10.1007/s00209-002-0431-9. MR1938704

[HPS97] Mark Hovey, John H. Palmieri, and Neil P. Strickland, *Axiomatic stable homotopy theory*, Mem. Amer. Math. Soc. **128** (1997), no. 610, x+114, DOI 10.1090/memo/0610. MR1388895

[HR82] Dieter Happel and Claus Michael Ringel, *Tilted algebras*, Trans. Amer. Math. Soc. **274** (1982), no. 2, 399–443, DOI 10.2307/1999116. MR675063

[HS01] A. Hirschowitz and C. Simpson. Descente pour les n-champs. arXiv:math/9807049v3 [math.AG], 2001.

[HSS00] Mark Hovey, Brooke Shipley, and Jeff Smith, *Symmetric spectra*, J. Amer. Math. Soc. **13** (2000), no. 1, 149–208, DOI 10.1090/S0894-0347-99-00320-3. MR1695653

[Huy06] D. Huybrechts, *Fourier-Mukai transforms in algebraic geometry*, Oxford Mathematical Monographs, The Clarendon Press, Oxford University Press, Oxford, 2006. MR2244106

[Jar87] J. F. Jardine, *Simplicial presheaves*, J. Pure Appl. Algebra **47** (1987), no. 1, 35–87, DOI 10.1016/0022-4049(87)90100-9. MR906403

[Jar00] J. F. Jardine, *Motivic symmetric spectra*, Doc. Math. **5** (2000), 445–552. MR1787949

[Joy08] André Joyal. The theory of quasi-categories and its applications. Lectures at the CRM (Barcelona). Preprint, 2008.

[JY18] Fangzhou Jin and Enlin Yang. Kunneth formulas for motives and additivity of traces. arXiv:1812.06441, 2018.

[Kel91] Bernhard Keller, *Derived categories and universal problems*, Comm. Algebra **19** (1991), no. 3, 699–747, DOI 10.1080/00927879108824166. MR1102982

[Kel94] Bernhard Keller, *Deriving DG categories*, Ann. Sci. École Norm. Sup. (4) **27** (1994), no. 1, 63–102. MR1258406

[Kel05a] Bernhard Keller, *On triangulated orbit categories*, Doc. Math. **10** (2005), 551–581. MR2184464

[Kel05b] G. M. Kelly, *Basic concepts of enriched category theory*, Repr. Theory Appl. Categ. **10** (2005), vi+137. Reprint of the 1982 original [Cambridge Univ. Press, Cambridge; MR0651714]. MR2177301

[Kel07a] Bernhard Keller, *Appendice: Le dérivateur triangulé associé à une catégorie exacte* (French, with French summary), Categories in algebra, geometry and mathematical physics, Contemp. Math., vol. 431, Amer. Math. Soc., Providence, RI, 2007, pp. 369–373, DOI 10.1090/conm/431/08355. MR2342837

[Kel07b] Bernhard Keller, *Derived categories and tilting*, Handbook of tilting theory, London Math. Soc. Lecture Note Ser., vol. 332, Cambridge Univ. Press, Cambridge, 2007, pp. 49–104, DOI 10.1017/CBO9780511735134.005. MR2384608

[Kel08] Bernhard Keller, *Calabi-Yau triangulated categories*, Trends in representation theory of algebras and related topics, EMS Ser. Congr. Rep., Eur. Math. Soc., Zürich, 2008, pp. 467–489, DOI 10.4171/062-1/11. MR2484733

[KL06] Henning Krause and Jue Le, *The Auslander-Reiten formula for complexes of modules*, Adv. Math. **207** (2006), no. 1, 133–148, DOI 10.1016/j.aim.2005.11.008. MR2264068

[KM08] Bruno Kahn and Georges Maltsiniotis, *Structures de dérivabilité* (French, with English and French summaries), Adv. Math. **218** (2008), no. 4, 1286–1318, DOI 10.1016/j.aim.2008.03.010. MR2419385

[KN02] Bernhard Keller and Amnon Neeman, *The connection between May's axioms for a triangulated tensor product and Happel's description of the derived category of the quiver D_4*, Doc. Math. **7** (2002), 535–560. MR2015053

[Kra08] Henning Krause. Representations of quivers via reflection functors. `http://arxiv.org/abs/math/0804.1428`, 2008. Course notes.

[KS74] G. M. Kelly and Ross Street, *Review of the elements of 2-categories*, Category Seminar (Proc. Sem., Sydney, 1972/1973), Springer, Berlin, 1974, pp. 75–103. Lecture Notes in Math., Vol. 420. MR0357542

[KS09] M. Kontsevich and Y. Soibelman, *Notes on A_∞-algebras, A_∞-categories and non-commutative geometry*, Homological mirror symmetry, Lecture Notes in Phys., vol. 757, Springer, Berlin, 2009, pp. 153–219. MR2596638

[KS12] Bernhard Keller and Sarah Scherotzke, *The integral cluster category*, Int. Math. Res. Not. IMRN **12** (2012), 2867–2887, DOI 10.1093/imrn/rnr129. MR2942712

[Lac10] Stephen Lack, *A 2-categories companion*, Towards higher categories, IMA Vol. Math. Appl., vol. 152, Springer, New York, 2010, pp. 105–191, DOI 10.1007/978-1-4419-1524-5_4. MR2664622

[Lad07a] Sefi Ladkani. Universal derived equivalences of posets. arXiv:0705.0946, 2007.

[Lad07b] Sefi Ladkani. Universal derived equivalences of posets of cluster tilting modules. arXiv:0710.2860, 2007.

[Lad07c] Sefi Ladkani. Universal derived equivalences of posets of tilting modules. arXiv:0708.1287, 2007.

[Lad08] Sefi Ladkani. *Homological Properties of Finite Partially Ordered Sets*. PhD thesis, The Hebrew University of Jerusalem, 2008.

[Len10] Fabian Lenhardt, *Stable frames in model categories*, J. Pure Appl. Algebra **216** (2012), no. 5, 1080–1091, DOI 10.1016/j.jpaa.2011.10.024. MR2875328

[Len17] Tobias Lenz, *Homotopy (pre)derivators of cofibration categories and quasicategories*, Algebr. Geom. Topol. **18** (2018), no. 6, 3601–3646, DOI 10.2140/agt.2018.18.3601. MR3868229

[LM08] Volodymyr Lyubashenko and Oleksandr Manzyuk, *A_∞-bimodules and Serre A_∞-functors*, Geometry and dynamics of groups and spaces, Progr. Math., vol. 265, Birkhäuser, Basel, 2008, pp. 565–645, DOI 10.1007/978-3-7643-8608-5_14. MR2402416

[LMSM86] L. G. Lewis Jr., J. P. May, M. Steinberger, and J. E. McClure, *Equivariant stable homotopy theory*, Lecture Notes in Mathematics, vol. 1213, Springer-Verlag, Berlin, 1986. With contributions by J. E. McClure. MR866482

[Lur09] Jacob Lurie, *Higher topos theory*, Annals of Mathematics Studies, vol. 170, Princeton University Press, Princeton, NJ, 2009. MR2522659

[Lur14] Jacob Lurie. Higher algebra. `http://www.math.harvard.edu/~lurie/`, 2014. Preprint.

[Mal01a] Georges Maltsiniotis. Groupe de travail sur les dérivateurs. `http://people.math.jussieu.fr/~maltsin/textes.html`, 2001. Seminar in Paris.

[Mal01b] Georges Maltsiniotis, *La théorie de l'homotopie de Grothendieck* (French, with English and French summaries), Astérisque **301** (2005), vi+140. MR2200690

[Mal05a] Georges Maltsiniotis. Catégories triangulées supérieures. `http://people.math.jussieu.fr/~maltsin/textes.html`, 2005. Preprint.

[Mal05b] Georges Maltsiniotis, *La théorie de l'homotopie de Grothendieck* (French, with English and French summaries), Astérisque **301** (2005), vi+140. MR2200690

[Mal07] Georges Maltsiniotis, *La K-théorie d'un dérivateur triangulé* (French, with French summary), Categories in algebra, geometry and mathematical physics, Contemp. Math., vol. 431, Amer. Math. Soc., Providence, RI, 2007, pp. 341–368, DOI 10.1090/conm/431/08280. MR2342836

[Mal12] Georges Maltsiniotis, *Carrés exacts homotopiques et dérivateurs* (French, with English and French summaries), Cah. Topol. Géom. Différ. Catég. **53** (2012), no. 1, 3–63. MR2951712

[Mar83] H. R. Margolis, *Spectra and the Steenrod algebra: Modules over the Steenrod algebra and the stable homotopy category*, North-Holland Mathematical Library, vol. 29, North-Holland Publishing Co., Amsterdam, 1983. MR738973

[Mat76] Michael Mather, *Pull-backs in homotopy theory*, Canadian J. Math. **28** (1976), no. 2, 225–263, DOI 10.4153/CJM-1976-029-0. MR402694

[May01] J. P. May, *The additivity of traces in triangulated categories*, Adv. Math. **163** (2001), no. 1, 34–73, DOI 10.1006/aima.2001.1995. MR1867203

[ML98] Saunders Mac Lane, *Categories for the working mathematician*, 2nd ed., Graduate Texts in Mathematics, vol. 5, Springer-Verlag, New York, 1998. MR1712872

[MM02] M. A. Mandell and J. P. May, *Equivariant orthogonal spectra and S-modules*, Mem. Amer. Math. Soc. **159** (2002), no. 755, x+108, DOI 10.1090/memo/0755. MR1922205

[MMSS01] M. A. Mandell, J. P. May, S. Schwede, and B. Shipley, *Model categories of diagram spectra*, Proc. London Math. Soc. (3) **82** (2001), no. 2, 441–512, DOI 10.1112/S0024611501012692. MR1806878

[MS06] J. P. May and J. Sigurdsson, *Parametrized homotopy theory*, Mathematical Surveys and Monographs, vol. 132, American Mathematical Society, Providence, RI, 2006. MR2271789

[MSS07] Fernando Muro, Stefan Schwede, and Neil Strickland, *Triangulated categories without models*, Invent. Math. **170** (2007), no. 2, 231–241, DOI 10.1007/s00222-007-0061-2. MR2342636

[Muk81] Shigeru Mukai, *Duality between $D(X)$ and $D(\hat{X})$ with its application to Picard sheaves*, Nagoya Math. J. **81** (1981), 153–175. MR607081

[MV99] Fabien Morel and Vladimir Voevodsky, *$\mathbf{A}^1$-homotopy theory of schemes*, Inst. Hautes Études Sci. Publ. Math. **90** (1999), 45–143 (2001). MR1813224

[MV15] Brian A. Munson and Ismar Volić, *Cubical homotopy theory*, New Mathematical Monographs, vol. 25, Cambridge University Press, Cambridge, 2015. MR3559153

[MY01] Jun-ichi Miyachi and Amnon Yekutieli, *Derived Picard groups of finite-dimensional hereditary algebras*, Compositio Math. **129** (2001), no. 3, 341–368, DOI 10.1023/A:1012579131516. MR1868359

[Nee01] Amnon Neeman, *Triangulated categories*, Annals of Mathematics Studies, vol. 148, Princeton University Press, Princeton, NJ, 2001. MR1812507

[Pog17] Thomas Poguntke. Higher Segal structures in algebraic K-theory. `https://arxiv.org/abs/1709.06510`, 2017. Preprint.

[PS14] Kate Ponto and Michael Shulman, *The linearity of fixed point invariants*, New directions in homotopy theory, Contemp. Math., vol. 707, Amer. Math. Soc., Providence, RI, 2018, pp. 89–120, DOI 10.1090/conm/707/14256. MR3807744

[PS16] Kate Ponto and Michael Shulman, *The linearity of traces in monoidal categories and bicategories*, Theory Appl. Categ. **31** (2016), Paper No. 23, 594–689. MR3518981

[Pup58] Dieter Puppe, *Homotopiemengen und ihre induzierten Abbildungen. I* (German), Math. Z. **69** (1958), 299–344, DOI 10.1007/BF01187411. MR100265

[Pup67] Dieter Puppe, *Stabile Homotopietheorie. I* (German), Math. Ann. **169** (1967), 243–274, DOI 10.1007/BF01362348. MR211400

[Qui67] Daniel G. Quillen, *Homotopical algebra*, Lecture Notes in Mathematics, No. 43, Springer-Verlag, Berlin-New York, 1967. MR0223432

[Qui73] Daniel Quillen, *Higher algebraic K-theory. I*, Algebraic K-theory, I: Higher K-theories (Proc. Conf., Battelle Memorial Inst., Seattle, Wash., 1972), Springer, Berlin, 1973, pp. 85–147. Lecture Notes in Math., Vol. 341. MR0338129

[RB06] Andrei Radulescu-Banu. Cofibrations in Homotopy Theory. `http://arxiv.org/abs/math/0610009`, 2006. Preprint.

[Ren09] Olivier Renaudin, *Plongement de certaines théories homotopiques de Quillen dans les dérivateurs* (French, with French summary), J. Pure Appl. Algebra **213** (2009), no. 10, 1916–1935, DOI 10.1016/j.jpaa.2009.02.014. MR2526867

[Rez01] Charles Rezk, *A model for the homotopy theory of homotopy theory*, Trans. Amer. Math. Soc. **353** (2001), no. 3, 973–1007, DOI 10.1090/S0002-9947-00-02653-2. MR1804411

[Ric91] Jeremy Rickard, *Derived equivalences as derived functors*, J. London Math. Soc. (2) **43** (1991), no. 1, 37–48, DOI 10.1112/jlms/s2-43.1.37. MR1099084

[Ric97] Jeremy Rickard, *Idempotent modules in the stable category*, J. London Math. Soc. (2) **56** (1997), no. 1, 149–170, DOI 10.1112/S0024610797005309. MR1462832

[Rin84] Claus Michael Ringel, *Tame algebras and integral quadratic forms*, Lecture Notes in Mathematics, vol. 1099, Springer-Verlag, Berlin, 1984. MR774589

[RV] Emily Riehl and Dominic Verity, *The 2-category theory of quasi-categories*, Adv. Math. **280** (2015), 549–642, DOI 10.1016/j.aim.2015.04.021. MR3350229

[RV17a] Emily Riehl and Dominic Verity, *Fibrations and Yoneda's lemma in an ∞-cosmos*, J. Pure Appl. Algebra **221** (2017), no. 3, 499–564, DOI 10.1016/j.jpaa.2016.07.003. MR3556697

[RV17b] Emily Riehl and Dominic Verity, *Kan extensions and the calculus of modules for ∞-categories*, Algebr. Geom. Topol. **17** (2017), no. 1, 189–271, DOI 10.2140/agt.2017.17.189. MR3604377

[RVdB02] I. Reiten and M. Van den Bergh, *Noetherian hereditary abelian categories satisfying Serre duality*, J. Amer. Math. Soc. **15** (2002), no. 2, 295–366, DOI 10.1090/S0894-0347-02-00387-9. MR1887637

[Sch10] Stefan Schwede, *Algebraic versus topological triangulated categories*, Triangulated categories, London Math. Soc. Lecture Note Ser., vol. 375, Cambridge Univ. Press, Cambridge, 2010, pp. 389–407, DOI 10.1017/CBO9781139107075.010. MR2681714

[Sch13] Stefan Schwede, *The p-order of topological triangulated categories*, J. Topol. **6** (2013), no. 4, 868–914, DOI 10.1112/jtopol/jtt018. MR3145143

[Seg74] Graeme Segal, *Categories and cohomology theories*, Topology **13** (1974), 293–312, DOI 10.1016/0040-9383(74)90022-6. MR353298

[Ser55] Jean-Pierre Serre, *Un théorème de dualité* (French), Comment. Math. Helv. **29** (1955), 9–26, DOI 10.1007/BF02564268. MR67489

[Shi07] Brooke Shipley, *$H\mathbb{Z}$-algebra spectra are differential graded algebras*, Amer. J. Math. **129** (2007), no. 2, 351–379, DOI 10.1353/ajm.2007.0014. MR2306038

[Shu08] Michael Shulman, *Framed bicategories and monoidal fibrations*, Theory Appl. Categ. **20** (2008), No. 18, 650–738. MR2534210

[Sim12] Carlos Simpson, *Homotopy theory of higher categories*, New Mathematical Monographs, vol. 19, Cambridge University Press, Cambridge, 2012. MR2883823

[SS00] Stefan Schwede and Brooke E. Shipley, *Algebras and modules in monoidal model categories*, Proc. London Math. Soc. (3) **80** (2000), no. 2, 491–511, DOI 10.1112/S002461150001220X. MR1734325

[SS03] Stefan Schwede and Brooke Shipley, *Stable model categories are categories of modules*, Topology **42** (2003), no. 1, 103–153, DOI 10.1016/S0040-9383(02)00006-X. MR1928647

[SS07] Daniel Simson and Andrzej Skowroński, *Elements of the representation theory of associative algebras: Tubes and concealed algebras of Euclidean type. Vol. 2*, London Mathematical Society Student Texts, vol. 71, Cambridge University Press, Cambridge, 2007. MR2360503

[Ste13] Greg Stevenson, *Support theory via actions of tensor triangulated categories*, J. Reine Angew. Math. **681** (2013), 219–254. MR3181496

[Šťo14] Jan Šťovíček, *Exact model categories, approximation theory, and cohomology of quasi-coherent sheaves*, Advances in representation theory of algebras, EMS Ser. Congr. Rep., Eur. Math. Soc., Zürich, 2013, pp. 297–367. MR3220541

[Str92] N. P. Strickland, *On the p-adic interpolation of stable homotopy groups*, Adams Memorial Symposium on Algebraic Topology, 2 (Manchester, 1990), London Math. Soc. Lecture Note Ser., vol. 176, Cambridge Univ. Press, Cambridge, 1992, pp. 45–54, DOI 10.1017/CBO9780511526312.010. MR1232198

[Tab08] Gonçalo Tabuada, *Higher K-theory via universal invariants*, Duke Math. J. **145** (2008), no. 1, 121–206, DOI 10.1215/00127094-2008-049. MR2451292

[Tho80] R. W. Thomason, *Cat as a closed model category*, Cahiers Topologie Géom. Différentielle **21** (1980), no. 3, 305–324. MR591388

[Ver67] Jean-Louis Verdier. *Des catégories dérivées des catégories abéliennes.* PhD thesis, Université de Paris, 1967.

[Ver96] Jean-Louis Verdier, *Des catégories dérivées des catégories abéliennes* (French, with French summary), Astérisque **239** (1996), xii+253 pp. (1997). With a preface by Luc Illusie; Edited and with a note by Georges Maltsiniotis. MR1453167

[Voe98] Vladimir Voevodsky, *$\mathbf{A}^1$-homotopy theory*, Proceedings of the International Congress of Mathematicians, Vol. I (Berlin, 1998), Doc. Math. **Extra Vol. I** (1998), 579–604. MR1648048

[Vog70] Rainer Vogt, *Boardman's stable homotopy category*, Lecture Notes Series, No. 21, Matematisk Institut, Aarhus Universitet, Aarhus, 1970. MR0275431

[vR12] Adam-Christiaan van Roosmalen, *Abelian hereditary fractionally Calabi-Yau categories*, Int. Math. Res. Not. IMRN **12** (2012), 2708–2750, DOI 10.1093/imrn/rnr118. MR2942708

[Wal85] Friedhelm Waldhausen, *Algebraic K-theory of spaces*, Algebraic and geometric topology (New Brunswick, N.J., 1983), Lecture Notes in Math., vol. 1126, Springer, Berlin, 1985, pp. 318–419, DOI 10.1007/BFb0074449. MR802796

[Wei94] Charles A. Weibel, *An introduction to homological algebra*, Cambridge Studies in Advanced Mathematics, vol. 38, Cambridge University Press, Cambridge, 1994. MR1269324

Rheinische Friedrich-Wilhelms-Universität Bonn, Mathematisches Institut, Endenicher Allee 60, 53115 Bonn, Germany

Current address: Institut für Mathematik, Johannes Gutenberg-Universität Mainz, Staudingerweg 9, 55128 Mainz, Germany

Email address: moritz.rahn@uni-mainz.de

Conference Proceedings Papers

Contemporary Mathematics
Volume **758**, 2020
https://doi.org/10.1090/conm/758/15235

Cohomology of some local selfinjective algebras

Karin Erdmann

Abstract. We show that the cohomology for some 2-generated selfinjective algebras is finitely generated. This applies in particular to the algebras $A5(\beta)$ for $\beta \neq 0$ in the classificationof connected Hopf algebras of dimension p^3 over characteristic p, by Nguyen-Wang-Wang.

1. Introduction

Assume A is a finite-dimensional Hopf algebra, this has a trivial module k, defined using the counit, and the cohomology of A is the algebra $H^*(A) = \mathrm{Ext}^*_A(k,k)$. More generally, a selfinjective algebra may have a trivial module, and then there also is a cohomology. One would like to know whether this cohomology is finitely generated; this is known for some classes of algebras, but is open in general.

In this note we show that for some selfinjective local algebras the cohomology is finitely generated. The novelty is that we only use some identities for generators, rather than a complete presentation of the algebra. Moreover, the proofs are completely elementary. This applies in particular to 2-generated q-complete intersections as studied in [**1**], [**2**], [**3**], [**7**], and also to group algebras of 2-generated finite p-groups over fields of prime characteristic (though for group algebras, cohomology is known to be finitely generated [**5**].)

This is motivated by results in [**6**] and [**4**]. The work in [**6**] classifies connected Hopf algebras of dimension p^3 over characteristic p, the result is a list of 24 families of algebras. These algebras and their cohomology are analysed in [**4**]. Perhaps surprisingly, the algebra structures occuring are quite divers; and by exploiting different ideas, in all but one family it could be proved in [**4**] that the cohomology is finitely generated. The only case left was the algebra with label $A5$ where $\mathrm{char}(k) \geq 3$, with presentation

$$k\langle z,y\rangle/([y,z]^p,\ y^p,\ [[y,z]y],\ [[y,z],z],\ z^p + y[y,z]^{p-1} - \beta[y,z])$$

with $\beta \in k$, here $[y,z] = yz - zy$. For $\beta \neq 0$ we had mentioned in [**4**] that its cohomology is finitely generated. Our general result of this note shows that this is indeed the case. When $\beta = 0$ the situation is different.

In Section 2 we describe the input. Section 3 computes the first few syzygies and discusses further input. In Section 4 we compute an explicit minimal projective

---

2010 *Mathematics Subject Classification.* Primary 16G10, 16T05, 20G20.
Key words and phrases. Cohomology, finite generation, selfinjective algebras, Hopf algebras.

resolution of k, and in Section 5 we show that the cohomology is finitely generated. The final section deals with the case of the algebra $A5$ for $\beta \neq 0$.

Our proof only shows finite generation, but does not compute a presentation of the cohomology. Such a presentation depends on the input, as will be clear from illustrations in Sections 5 and 6.

2. The input

Let Λ be a local selfinjective algebra, of the form

$$\Lambda = k\langle x, y\rangle/I,$$

where x and y are independent generators of the radical $J = J(\Lambda)$ of Λ. This has a trivial module, $k = \Lambda/J$, and the cohomology of Λ is the algebra $H^*(\Lambda) := \mathrm{Ext}^*_\Lambda(k,k)$.

We impose conditions on the ideal I, see below. This includes several classes of algebras: group algebras of abelian p-groups of rank 2 over fields of characteristic p, also quantum complete intersections. As well this deals with the algebra $A5$ above when $\beta \neq 0$ and $\mathrm{char}(k) \geq 3$.

ASSUMPTION 2.1. We impose three conditions on I, the first two are as follows.

(1) There is a minimal relation $\psi_1 x + \psi_2 y = 0$ with ψ_1, ψ_2 independent modulo J^2.
(2) There are minimal relations $\sigma_1 x + \sigma_2 y = 0$ and $\theta_1 x + \theta_2 y = 0$. Moreover if $\sigma := (\sigma_1, \sigma_2)$ and $\theta := (\theta_1, \theta_2)$ then σ and θ are independent modulo $\Lambda\psi$ where $\psi := (\psi_1, \psi_2)$. For example, such relations might be given via $x^n = 0$ and $y^m = 0$.

The third condition will be formulated in Section 3, in Assumption 3.4.

EXAMPLE 2.2. Let

$$A = \langle x, y\rangle/(x^n, y^m, xy - qyx)$$

with $0 \neq q \in k$. This includes group algebras of rank 2 finite abelian p-groups (when k has characteristic p and $q = 1$, and m, n are powers of p) but also q-complete intersections when $q \neq 1$ as studied for example in [**2**] and [**1**]. These algebras satisfy (1) and (2) of Assumption 2.1, taking

$$\psi = (-qy, x), \quad \sigma = (x^{n-1}, 0), \quad \theta = (0, y^{m-1}).$$

We use this algebra A as an illustration throughout.

3. The first few syzygies of k

We work with left modules, and we write maps to the right. First, $\Omega(k) = \Lambda x + \Lambda y$. Next, we take the second syzygy of k in the form

$$\Omega^2(k) = \{(a, b) \in \Lambda^2 \mid ax + by = 0\}.$$

By conditions (1) and (2) of Assumption 2.1, this module contains the independent generators:

$$\sigma = (\sigma_1, \sigma_2), \ \psi = (\psi_1, \psi_2), \ \theta = (\theta_1, \theta_2).$$

The following will show that these in fact generate all of $\Omega^2(k)$. Let $N := \dim \Lambda$.

LEMMA 3.1. *We have $\Omega^2(k) = \langle \sigma, \psi, \theta\rangle$ as a left Λ-module.*

PROOF. By conditions (1) and (2) we have one inclusion, and we observe that $\dim \Omega^2(k) = N+1$. Now, since the two components of ψ are generators for J, and since A is selfinjective, we have $\Lambda\psi \cong \Omega^{-1}(k)$ as a Λ-module, in particular it has dimension $N-1$. By assumption the elements σ and θ are independent modulo $A\psi$. Hence $k\sigma + k\theta + \Lambda\psi$ has dimension $N+1$, and the claim follows. □

REMARK. For future reference, the proof shows the following: The submodule $\langle \sigma, \psi \rangle$ of $\Omega^2(k)$ has dimension N, and so does the submodule $\langle \psi, \theta \rangle$.

We have the following immediate consequence.

LEMMA 3.2. *A presentation of* $\Omega^2(k)$ *is determined completely by specifying four elements* ρ_i *of* Λ *such that*

$$x\sigma + \rho_1\psi = 0, \qquad y\sigma + \rho_2\psi = 0$$
$$x\theta + \rho_3\psi = 0, \qquad y\theta + \rho_4\psi = 0.$$

The module $\Omega^3(k)$ has has dimension $2N-1$.
We take the generators σ, ψ, θ of $\Omega^2(k)$ in this order. Then we identify

$$\Omega^3(k) = \{(a_i) \in A^3 \mid a_1\sigma + a_2\psi + a_3\theta = 0\}.$$

Using the identities in Lemma 3.2, we identify the following four elements in $\Omega^3(k)$,

$$f_1^{(3)} = (x, \rho_1, 0)$$
$$f_2^{(3)} = (y, \rho_2, 0)$$
$$f_3^{(3)} = (0, \rho_3, x)$$
$$f_4^{(3)} = (0, \rho_4, y).$$

The aim is to show that these are (independent) generators of $\Omega^3(k)$, by using only the conditions (1) and (2) of Assumption 2.1

LEMMA 3.3. *We have* $\Omega^3(k) = \sum_{i=1}^4 \Lambda f_i^{(3)}$.

PROOF. In the proof we omit the label $(-)^{(3)}$. Let $U_1 = \Lambda f_1 + \Lambda f_2$ and let $U_2 = \Lambda f_3 + \Lambda f_4$, then $U_1 + U_2 \subseteq \Omega^3(k) \subseteq \Lambda^3$. Let γ_i be the projection from $\Omega^3(k)$ onto the i-th component for $1 \leq i \leq 3$.

(a) The image $\gamma_1(U_1)$ is equal to $\Lambda x + \Lambda y = \Omega(k)$, hence has dimension $N-1$. Similarly $\gamma_3(U_2) = \Omega(k)$ of dimension $N-1$. Directly, the intersection $U_1 \cap U_2$ is contained in $(0, \Lambda, 0)$ and hence is in the kernel of γ_1 and γ_3. We have

$$\dim(U_1 + U_2) \leq \dim \gamma_1(U_1) + \dim \gamma_3(U_2) = 2N-2$$

which is one less than $\dim \Omega^3(k)$. Assume for a contradiction that $U_1 \cap U_2 = 0$, then $U_1 + U_2 = U_1 \oplus U_2$, a maximal submodule of $\Omega^3(k)$, which is indecomposable (since Λ is selfinjective). As well $\mathrm{soc}\Omega^3(k) \cong \mathrm{top}\Omega^2(k)$ and is 3-dimensional. So the socle of Λ^3 must be contained in $\Omega^3(k)$ and hence there is an element $m := (0, \omega, 0) \in \Omega^3(k)$ with ω an element spanning the socle of Λ. It follows that $\Omega^3(k) = U_1 \oplus U_2 \oplus \langle m \rangle$, and is decomposable, a contradiction. This shows that $U_1 \cap U_2$ is non-zero and then must be 1-dimensional and spanned by $(0, \omega, 0)$ with ω in the socle of Λ. The Lemma follows. □

REMARK. With the notation as in the proof, the kernel of γ_1 is

$$\{(0, a_1\rho_1 + a_2\rho_2, 0) \mid (a_1, a_2) \in \Omega^2(k)\}$$

and similarly we can write down the kernel of γ_2. Hence we have that if $\rho = \binom{\rho_1}{\rho_2}$ or $\binom{\rho_3}{\rho_4}$ then $\sigma\cdot\rho$, $\psi\cdot\rho$ and $\theta\cdot\rho$ lie in the socle of Λ. So far we only use Assumption 2.1. In the two main types of algebras we want to deal with, we have more information. We make an assumption (which possibly is redundant).

ASSUMPTION 3.4. We give now the third condition on the algebra Λ, in addition to the conditions in Assumption 2.1. We assume that there are elements $\rho_i \in \Lambda$ satisfying Lemma 3.2 and in addition

$$\begin{aligned} \sigma_1\rho_1 + \sigma_2\rho_2 &= 0 \qquad & \psi_1\rho_3 + \psi_2\rho_4 &= 0 \\ \psi_1\rho_1 + \psi_2\rho_2 &= 0 \qquad & \theta_1\rho_3 + \theta_2\rho_4 &= 0 \\ \theta_1\rho_1 + \theta_2\rho_2 &= \omega \end{aligned}$$

where $0 \neq \omega \in \mathrm{soc}\Lambda$, and

$$\theta_1\rho_1 + \theta_2\rho_2 + c(\sigma_1\rho_3 + \sigma_2\rho_4) = 0$$

for some $0 \neq c \in k$.

EXAMPLE 3.5. Let $\Lambda = A$ as in Example 2.2. We may take $\rho_1 = \rho_4 = 0$ and $\rho_2 = q^{-n}x^{n-1}$, $\rho_3 = -q^{m-1}y^{m-1}$ in Lemma 3.2. Then 3.4 is satisfied:

LEMMA 3.6. *Assume $\Lambda = A$. Then:*

$$a_1\rho_1 + a_2\rho_2 = \begin{cases} 0 & (a_1, a_2) = \sigma \\ 0 & (a_1, a_2) = \psi \\ q^{-n}y^{m-1}x^{n-1} & (a_1, a_2) = \theta. \end{cases}$$

$$a_3\rho_3 + a_4\rho_4 = \begin{cases} -q^{m-1}x^{n-1}y^{m-1} & (a_3, a_4) = \sigma \\ 0 & (a_3, a_3) = \psi \\ 0 & (a_3, a_4) = \theta. \end{cases}$$

For the general case, we summarize the identities which have formulated. In the following these are all we will use.

(1)

$$\begin{aligned} \sigma_1 x + \sigma_2 y &= 0 \\ \psi_1 x + \psi_2 y &= 0 \\ \theta_1 x + \theta_2 y &= 0 \end{aligned}$$

(2)

$$\begin{aligned} x\sigma_1 + \rho_1\psi_1 &= 0 \qquad & x\theta_1 + \rho_3\psi_1 &= 0 \\ x\sigma_2 + \rho_1\psi_2 &= 0 \qquad & x\theta_2 + \rho_3\psi_2 &= 0 \end{aligned}$$

$$\begin{aligned} y\sigma_1 + \rho_2\psi_1 &= 0 \qquad & y\theta_1 + \rho_4\psi_1 &= 0 \\ y\sigma_2 + \rho_2\psi_2 &= 0 \qquad & y\theta_2 + \rho_4\psi_2 &= 0 \end{aligned}$$

(3)

$$\begin{aligned} \sigma_1\rho_1 + \sigma_2\rho_2 &= 0 & \sigma_1\rho_3 + \sigma_2\rho_4 &= (-c)\omega \\ \psi_1\rho_1 + \psi_2\rho_2 &= 0 & \psi_1\rho_3 + \psi_2\rho_4 &= 0 \\ \theta_1\rho_1 + \theta_2\rho_2 &= \omega & \theta_1\rho_3 + \theta_2\rho_4 &= 0 \end{aligned}$$

Here ω is non-zero in the socle of Λ, and c is a non-zero scalar, and we have

$$\theta_1\rho_1 + \theta_2\rho_2 + c\sigma_1\rho_3 + c\sigma_2\rho_4 = 0.$$

4. A minimal projective resolution

As before, we use left Λ-modules, and we write maps to the right and compose from left to right. We will construct an exact sequence

$$(\dagger) \qquad \cdots \to P_{r+1} \overset{d_{r+1}}{\to} P_r \overset{d_r}{\to} \cdots \overset{d_3}{\to} P_2 \overset{d_2}{\to} P_1 \overset{d_1}{\to} P_0 \to k \to 0$$

with P_r projective so that the image of d_r is $\Omega^r(k)$.

We define $P_r := \Lambda^{r+1}$ (for $r = 0, 1, \ldots$). Then we define maps d_r as matrices, where the matrix d_r has $r+1$ rows and r columns, and the action is given by applying d_r to the standard generators of Λ^{r+1}, and express the answer in terms of the standard generators of Λ^r.

First by the results of Section 3 we may take

$$d_1 = \begin{pmatrix} x \\ y \end{pmatrix}, \quad d_2 = \begin{pmatrix} \sigma_1 & \sigma_2 \\ \psi_1 & \psi_2 \\ \theta_1 & \theta_2 \end{pmatrix}, d_3 = \begin{pmatrix} x & \rho_1 & \\ y & \rho_2 & \\ & \rho_3 & x \\ & \rho_4 & y \end{pmatrix}, \quad d_4 = \begin{pmatrix} \sigma_1 & \sigma_2 & & \\ \psi_1 & \psi_2 & & \\ \theta_1 & \theta_2 & c\sigma_1 & c\sigma_2 \\ & & \psi_1 & \psi_2 \\ & & \theta_1 & \theta_2 \end{pmatrix}.$$

so that the image of d_1 is precisely the radical of Λ, as a submodule of P_0, the image of d_2 is $\Omega^2(k)$, a submodule of Λ^2, and so on.

4.1. The matrix d_r for $r \geq 5$ and r odd. We define this recursively as follows. Let $r = 2t-1$. Then take for d_r the matrix with $r+1$ rows and r columns, which we write in block form,

$$d_r = \begin{pmatrix} d_{r-2} & S_r \\ 0 & T \end{pmatrix}$$

Here S_r is the matrix with $r-1$ rows and two columns, given as

$$S_r := \begin{pmatrix} S_r' \\ S_r'' \end{pmatrix}, \quad S_r' = 0_{(r-3)\times 2}, \quad S_r'' = \begin{pmatrix} \rho_1 c^{t-2} & 0 \\ \rho_2 c^{t-2} & 0 \end{pmatrix}.$$

(That is, S_r' is the zero matrix of size $(r-3) \times 2$). Moreover, T is the 2×2 matrix

$$T = \begin{pmatrix} \rho_3 & x \\ \rho_4 & y \end{pmatrix}.$$

4.2. The matrix d_r for $r \geq 4$ and r even. Let $r = 2t$. We define d_r to be the matrix with $r+1$ rows and r columns which we write in block form:

$$d_r = \begin{pmatrix} d_{r-2} & U_r \\ 0 & V \end{pmatrix}.$$

Here U_r is the matrix with $r-1$ rows and two columns, given as

$$U_r := \begin{pmatrix} U_r' \\ U_r'' \end{pmatrix}, \quad U_r' = 0_{(r-3)\times 2}, \quad U_r'' = \begin{pmatrix} 0 & 0 \\ c^{t-1}\sigma_1 & c^{t-1}\sigma_2 \end{pmatrix}.$$

Moreover, V is the matrix

$$V = \begin{pmatrix} \psi_1 & \psi_2 \\ \theta_1 & \theta_2 \end{pmatrix}.$$

We show first that $d_{r+1}d_r = 0$ for each r. This will show that the sequence (†) is a complex.

LEMMA 4.1. *The composition $d_{r+1}d_r$ is zero for each $r \geq 1$.*

PROOF. For $r = 1, 2, 3$ one checks this directly. We continue by induction on r.

Assume first that r is odd, and $2t - 1 = r \geq 5$. Assume the lemma holds for odd s smaller than r. We compute $d_{r+1}d_r$ using the block shape, this is equal to

$$\begin{pmatrix} d_{r-1}d_{r-2} & d_{r-1}S_r + U_{r+1}T \\ 0 & VT \end{pmatrix}.$$

One checks that $VT = 0$, using identities from (3) and (1). By the inductive hypothesis, $d_{r-1}d_{r-2} = 0$. It remains to show that $d_{r-1}S_r + U_{r+1}T = 0$. This is a matrix with r rows and two columns. First,

$$d_{r-1}S_r = \begin{pmatrix} d_{r-3} & U_{r-1} \\ 0 & V \end{pmatrix} \cdot \begin{pmatrix} S_r' \\ S_r'' \end{pmatrix} = \begin{pmatrix} U_{r-1}S_r'' \\ VS_r'' \end{pmatrix}.$$

We find

$$VS_r'' = \begin{pmatrix} 0 & 0 \\ c^{t-2}(\theta_1\rho_1 + \theta_2\rho_2) & 0 \end{pmatrix}.$$

Next,

$$U_{r-1}S_r'' = \begin{pmatrix} U_{r-1}' \\ U_{r-1}'' \end{pmatrix} \cdot S_r'' = \begin{pmatrix} 0 \\ U_{r-1}''S_r'' \end{pmatrix} = 0$$

using that $\sigma_1\rho_1 + \sigma_2\rho_2 = 0$. Next,

$$U_{r+1}T = \begin{pmatrix} U_{r+1}' \\ U_{r+1}'' \end{pmatrix} T = \begin{pmatrix} 0 \\ U_{r+1}''T \end{pmatrix}$$

and

$$U_{r+1}''T = \begin{pmatrix} 0 & 0 \\ c^{t-1}(\sigma_1\rho_3 + \sigma_2\rho_4) & 0 \end{pmatrix}$$

We deduce that $d_{r-1}S_r + U_{r+1}T$ is the matrix of size $r \times 2$ with all entries zero, except possibly the $r1$, bu this is equal to

$$c^{t-2}(\theta_1\rho_1 + \theta_2\rho_2 + c\sigma_1\rho_3 + c\sigma_2\rho_4) = 0$$

Hence $d_{r+1}d_r = 0$ for r odd.

Now assume $r = 2t \geq 6$, and assume true for even s smaller than r. Then $d_{r+1}d_r$ is given by

$$\begin{pmatrix} d_{r-1}d_{r-2} & d_{r-1}U_r + S_{r+1}V \\ 0 & TV \end{pmatrix}.$$

One checks that $TV = 0$, and by the inductive hypothesis also $d_{r-1}d_{r-2} = 0$. We show now that $d_{r-1}U_r + S_{r+1}V = 0$. This is a matrix with r rows and two columns. We have

$$d_{r-1}U_r = \begin{pmatrix} d_{r-3} & S_{r-1} \\ 0 & T \end{pmatrix} \cdot \begin{pmatrix} U_r' \\ U_r'' \end{pmatrix} = \begin{pmatrix} S_{r-1}U_r'' \\ TU_r'' \end{pmatrix}$$

We calculate

$$TU_r'' = \begin{pmatrix} c^{t-1}x\sigma_1 & c^{t-1}x\sigma_2 \\ c^{t-1}y\sigma_1 & c^{t-1}y\sigma_2 \end{pmatrix}.$$

The matrix $S_{r-1}U_r''$ is zero. Next, we compute $S_{r+1}V$, the first $r-2$ rows are zero and the last two rows are

$$\begin{pmatrix} \rho_1\psi_1 c^{t-1} & \rho_1\psi_2 c^{t-1} \\ \rho_2\psi_1 c^{t-1} & \rho_2\psi_2 c^{t-1} \end{pmatrix}.$$

Hence $d_{r-1}U_r + S_{r+1}V$ is the $r \times 2$ matrix with $r-2$ rows of zeros, and where the last two rows are

$$c^{t-1}\begin{pmatrix} x\sigma_1 + \rho_1\psi_1 & x\sigma_2 + \rho_1\psi_1 \\ y\sigma_1 + \rho_2\psi_1 & y\sigma_2 + \rho_2\psi_2 \end{pmatrix}.$$

This is zero, by the identities at the end of Section 3. □

This shows that (†) is a complex, and next we will show that it is exact.

PROPOSITION 4.2. *Let $r = 2t$ or $2t-1$ for $t \geq 1$. We have* $\mathrm{im}(d_r) = \Omega^r(k)$ *of dimension $tN+1$ if r is even, or $tN-1$ if r is odd.*

PROOF. This is true for $r = 1, 2, 3$ by Section 3, and let $r = 2t$ or $2t-1$ where $r \geq 4$. As an inductive hypothesis, assume true for all $m \leq r-1$. Then $\mathrm{im}(d_{r-1}) = \Omega^{r-1}(k)$ of dimension $tN-1$ or $(t-1)N+1$, depending on the parity of r. Since $P_{r-1} = \Lambda^r$, it follows that $\Omega^r(k)$ has dimension $tN+1$ or $tN-1$; and moreover we know $\mathrm{Im}(d_r) \subseteq \Omega^r(k)$ since $d_r d_{r-1} = 0$. We are done if we show that $\mathrm{Im}(d_r)$ has the same dimension as $\Omega^r(k)$.

(1) Assume first r is even, so that $r = 2t \geq 4$. We must show that $\mathrm{im}(d_r)$ has dimension $tN+1$. By the above, the dimension is at most $tN+1$. Let $\pi : \Lambda^{r+1} \to \Lambda^{r-1}$ be the projection onto the first $r-1$ coordinates. This takes f_i^r to f_i^{r-2} for $1 \leq i \leq r-1$, and it takes f_r^r and f_{r+1}^r to zero. Hence $(\mathrm{Im}(d_r))\pi = \mathrm{Im}(d_{r-2})$ which has dimension $(t-1)N+1$. Furthermore, the submodule $\langle f_r^r, f_{r+1}^r \rangle$ contained in the kernel of π is isomorphic to $\langle \psi, \theta \rangle$ and has dimension N (see the remark below 3.1). So the dimension of $\mathrm{Im}(d_r)$ is at least $N + (t-1)N + 1$, and then equality follows.

(2) Now assume r is odd so that $r = 2t-1$, we must show that $\mathrm{im}(d_r)$ has dimension $tN-1$. By the above, the dimension is at most $tN-1$. Write $\mathrm{Im}(d_r) = U + V \subseteq \Lambda^r$ where U is the submodule generated by $f_1^r, \cdots, f_{r-1}^r$ and V is the submodule generated by f_r^r, f_{r+1}^r.

We find the dimension of V. Let $\pi : \Lambda^r \to \Lambda$ be the projection onto the last coordinate. Then $(V)\pi = \Lambda x + \Lambda y$ and hence has dimension $N-1$. We consider the kernel of π restricted to V. As a vector space it is isomorphic to

$$\{a_1\rho_3 + a_2\rho_4 \mid (a_1, a_2) \in \Omega^2(k)\}.$$

This is computed in Section 3, and it is just the socle of Λ (concentrated in the $(r-1)$-th coordinate), it is 1-dimensional, and hence $\dim V = N$.

Now we find the dimension of U. Let $\pi' : \Lambda^r \to \Lambda^{r-2}$ be the projection onto the first $r-2$ components. We have $(f_i^r)\pi' = f_i^{r-2}$ for $1 \leq i \leq r-1$.

So $(U)\pi' = \mathrm{Im}(d_{r-2})$ of dimension $(t-1)N-1$, and hence $\dim U$ is at least $(t-1)N-1$. We can identify the kernel of π' restricted to U. Note that π' restricted to $\langle f_i^r \mid 1 \le i \le r-4\rangle$ is injective, we find that $\ker(\pi'|_U)$ is isomorphic to

$$\{a_1\rho_1 + a_2\rho_2 \mid (a_1, a_2) \in \Omega^2(k)\}.$$

This is again the socle of Λ (in the $(r-1)$-th coordinate) and is 1-dimensional. So $\dim U = (t-1)N$. Moreover, we see that $U \cap V$ is 1-dimensional and then $\dim(U+V) = tN-1$ as required. □

5. The cohomology ring

By Section 3, for each $r \ge 1$, the module $\Omega^r(k) \subseteq \Lambda^r$ is generated by the $r+1$ rows of the matrix d_r. Write f_i^r for the i-th row of this matrix, for $1 \le i \le r+1$. Then the f_i^r form a minimal set of generators for $\Omega^r(k)$, and hence we can take as a vector space basis for $H^r(\Lambda) = \mathrm{Hom}_\Lambda \Omega^r(k), k)$ the set $\{\varphi_i^r\}$ which is the dual basis for the set $\{f_i^r\}$.

If f, g are homogeneous elements of degrees s, t (say), then we take the product $f \cdot g$ as the class of $\Omega^t(f) \cdot g$ (recall that we apply maps to the right).

We will show that for $t \ge 1$ we have

$$H^{2t}(\Lambda) \times H^1(\Lambda) = H^{2t+1}(\Lambda),$$
$$H^{2t}(\Lambda) \times H^2(\Lambda) = H^{2t+2}(\Lambda).$$

In this section we write just H^m meaning $H^m(\Lambda)$.

LEMMA 5.1. *We have $H^{2t} \times H^1 = H^{2t+1}$. More precise, for $0 \le s \le t$, the following hold*

$$\varphi_{2s+1}^{2t} \cdot \varphi_1^1 = \varphi_{2s+1}^{2t+1}, \quad \varphi_{2s+1}^{2t} \cdot \varphi_2^1 = \varphi_{2s+2}^{2t+1}.$$

That is, $H^{2t} \times H^1$ contains a basis of H^{2t+1}

PROOF. The space H^{2t} has basis $\{\varphi_i^{2t} \mid 1 \le i \le 2t+1\}$. The product $\varphi_i^{2t} \cdot \varphi_j^1$ is then the class of

$$\Omega(\varphi_i^{2t})\varphi_j^1 \ \ (1 \le j \le 2).$$

The diagram to compute $\Omega(\varphi_i^{2t})'s$ is of the form

$$\begin{array}{ccccccc}
\Omega^{2t+1}(k) & \longrightarrow & P_{2t} & \xrightarrow{d_{2t}} & \Omega^{2t}(k) & \longrightarrow & 0 \\
{\scriptstyle \Omega(\varphi_i^{2t})}\downarrow & & {\scriptstyle h_i}\downarrow & & {\scriptstyle \varphi_i^{2t}}\downarrow & & \\
\Omega(k) & \longrightarrow & \Lambda & \xrightarrow{d_0} & k & \longrightarrow & 0
\end{array}$$

The composition $d_{2t}\varphi_i^{2t}$ is equal to ε_i^*, the dual of the standard basis element. We can therefore take

$$h_i = \varepsilon_i^T$$

(the ith element in the standard basis of column vectors). Then $\Omega(\varphi_i^{2t})$ is the restriction of h_i to $\Omega^{2t+1}(k)$, which is the submodule of Λ^{2t+1} generated by the rows of d_{2t+1}, which we call $f_1^{2t+1}, f_2^{2t+1}, \cdots, f_{2t+2}^{2t+1}$.

Let $i = 2s+1$ for $0 \le s \le t$. Then h_i takes

$$f_i^{2t+1} \mapsto x$$
$$f_{i+1}^{2t+1} \mapsto y$$
$$f_u^{2t+1} \mapsto 0 (u \ne i, i+1)$$

From this the claim follows. □

Recall that the space H^{2t} has basis $\{\varphi_i^{2t} \mid 1 \leq i \leq 2t+1\}$.

LEMMA 5.2. *We have $H^{2t} \times H^2 = H^{2t+2}$. More precisely, for $1 \leq s \leq t$ we have*

$$\varphi_{2s+1}^{2t} \cdot \varphi_1^2 = c^s \varphi_{2s+1}^{2t+2},\ \varphi_{2s+1}^{2t} \cdot \varphi_2^2 = \varphi_{2s+2}^{2t+2} \text{ and } \varphi_{2s+1}^{2t} \cdot \varphi_3^2 = \varphi_{2s+3}^{2t+2}.$$

Hence $H^{2t} \times H^2$ contains a basis for H^{2t+2}.

PROOF. Let $1 \leq i \leq 2t+1$. The product $\varphi_i^{2t} \cdot \varphi_j^2$ is the class of

$$\Omega^2(\varphi_i^{2t}) \circ \varphi_j^2 \quad (1 \leq j \leq 3).$$

We take as the diagram to compute $\Omega^2(\varphi_i^{2t})'s$ the extension of the previous diagram, that is we take

$$\begin{array}{ccccccccc}
\Omega^{2t+2}(k) & \longrightarrow & P_{2t+1} & \xrightarrow{d_{2t+1}} & P_{2t} & \xrightarrow{d_{2t}} & \Omega^{2t}(k) & \longrightarrow & 0 \\
{\scriptstyle \Omega^2(\varphi_i^{2t})}\downarrow & & {\scriptstyle \ell_i}\downarrow & & {\scriptstyle h_i}\downarrow & & {\scriptstyle \varphi_i^{2t}}\downarrow & & \\
\Omega^2(k) & \longrightarrow & \Lambda^2 & \xrightarrow{d_1} & \Lambda & \xrightarrow{d_0} & k & \longrightarrow & 0
\end{array}$$

As before, we take $h_i := \varepsilon_i^T$ (the standard column vector). Now assume i is an odd number $\leq 2t+1$, write $i = 2s+1$ $(0 \leq s \leq t)$. Let ℓ_i be the $(2t+2) \times 2$ matrix whose row i is $(1\ 0)$ and whose row $i+1$ is $(0\ 1)$, and where all other entries are zero. Then $\ell_i \cdot d_1$ is the column vector $x\varepsilon_i + y\varepsilon_{i+1}$ and this is the same as $d_{2t+1} \circ h_i$. So the map given by ℓ_i lifts h_i. Therefore $\Omega^2(\varphi_i^{2t})$ can be taken as the restriction of ℓ_i to $\Omega^{2t+2}(k)$, ie the submodule of Λ^{2t+2} generated by the rows of the matrix d_{2t+2}.

Thus ℓ_i takes

$$\begin{aligned}
f_i^{2t+2} &\mapsto c^s\sigma, \\
f_{i+1}^{2t+2} &\mapsto \psi \\
f_{i+3}^{2t+2} &\mapsto \theta \\
f_j^{2t+2} &\mapsto 0 \text{ else}
\end{aligned}$$

Recalling that $f_1^2 = \sigma, f_2^2 = \psi$ and $f_3^2 = \theta$, we obtain the statement of the Lemma. □

THEOREM 5.3. *The algebra $H^*(\Lambda)$ is finitely generated, and is generated by elements in degrees ≤ 2.*

This follows directly from lemmas 5.1 and 5.2, by induction on t. We note that this only uses little information on the algebra Λ. To get a presentation of the cohomology, requires further detailed information, which one can see from the following illustrations.

5.1. Products of two elements of degree 1. We have $H^1(\Lambda) = \mathrm{Hom}_\Lambda(\Omega(k), k)$. We have $f_1^1 = x$ and $f_2^1 = y$ as minimal generators of $\Omega(k)$, and then φ_i^1 are corresponding dual basis elements of $H^1(\Lambda)$. We must compute $\Omega(\varphi_i^1)$ for $i = 1, 2$. Hence we need to find h_i lifting φ_i^1 and then $\Omega(\varphi_i^1)$ is the restriction of h_i to $\Omega^2(k)$.

The relevant diagram is

$$\begin{array}{ccccccc}
\Omega^2(k) & \longrightarrow & P_1 = \Lambda^2 & \xrightarrow{d_1} & \Omega(k) & \longrightarrow & 0 \\
{\scriptstyle \Omega(\varphi_i^1)}\downarrow & & {\scriptstyle h_i}\downarrow & & {\scriptstyle \varphi_i^1}\downarrow & & \\
\Omega(k) & \longrightarrow & \Lambda & \xrightarrow{d_0} & k & \longrightarrow & 0
\end{array}$$

The composition $d_1 \circ \varphi_i^1$ is the dual ε_i^* of the standard basis element for Λ^2. We can take $h_i := \varepsilon_i^T$, the column vector with 1 in place i and zero elsewhere. Then $\Omega(\varphi_i^1)$, the restriction of h_i to $\Omega^2(k)$, takes each generator of $\Omega^2(k)$ to its ith coordinate. The generators of $\Omega^2(k)$ are

$$f_1^2 = \sigma, \quad f_2^2 = \psi, \quad f_3^2 = \theta.$$

The $\sigma_i, \psi_i, \theta_i$ are in the radical of Λ, so there are expressions

$$\begin{aligned}
\sigma_i &= u_{ix}x + u_{iy}y \\
\psi_i &= v_{ix}x + v_{iy}y \\
\theta_i &= w_{ix}x + w_{iy}y
\end{aligned}$$

with coefficients in Λ.

Let $\overline{(-)} : \Lambda \to k = \Lambda/\mathrm{rad}\Lambda$ be the canonical map. Then for any $p \in \mathrm{rad}\Lambda$ if we write $p = cx + dy$, then

$$\varphi_1^1(p) = \overline{c}, \quad \varphi_2^1(p) = \overline{d}.$$

Therefore

$$\begin{aligned}
\varphi_1^1 \cdot \varphi_1^1 &= \overline{u_{1x}}\varphi_1^2 + \overline{v_{1x}}\varphi_2^2 + \overline{w_{1x}}\varphi_3^2 \\
\varphi_1^1 \cdot \varphi_2^1 &= \overline{u_{1y}}\varphi_1^2 + \overline{v_{1y}}\varphi_2^2 + \overline{w_{1y}}\varphi_3^2 \\
\varphi_2^1 \cdot \varphi_1^1 &= \overline{u_{2x}}\varphi_1^2 + \overline{v_{2x}}\varphi_2^2 + \overline{w_{2x}}\varphi_3^2 \\
\varphi_2^1 \cdot \varphi_2^1 &= \overline{u_{2y}}\varphi_1^2 + \overline{v_{2y}}\varphi_2^2 + \overline{w_{2y}}\varphi_3^2.
\end{aligned}$$

Since the ψ_1, ψ_2 are independent generators of the radical, and so are x, y, we would know that the matrix $\begin{pmatrix} \overline{v_{1x}} & \overline{v_{1y}} \\ \overline{v_{2x}} & \overline{v_{2y}} \end{pmatrix}$ is invertible. With this we would get φ_2^2 in H^2, but we cannot say much more without knowing the algebra.

EXAMPLE 5.4. Take $\Lambda = A$. Then

$$\varphi_1^1 \cdot \varphi_1^1 = \begin{cases} \varphi_1^2 & n = 2 \\ 0 & \text{else} \end{cases}$$

Similarly $\varphi_2^2 \cdot \varphi_2^2 = \varphi_2^2$ if $m = 2$, and is zero otherwise. Furthermore, $\varphi_1^2 \cdot \varphi_2^1 = -q\varphi_2^2$ and $\varphi_2^1 \cdot \varphi_1^1 = \varphi_2^2$.

5.2. The products $H^2 \times H^2$. In Lemma 5.2 we have dealt with the products $\varphi_i^2 \cdot \varphi_j^2$ where i is odd. We consider products with $i = 2$, so we must find a suitable map ℓ_2 such that $\ell_2 \cdot d_1 = d_3 \cdot h_2$.

The map $d_3 \cdot h_2$ takes $\varepsilon_j \mapsto \rho_j$ for $1 \leq j \leq 4$. Since ρ_j is in the radical of Λ, there are elements $u_j, v_j \in \Lambda$ such that $\rho_j = u_j x + v_j y$. We fix such elements, and define

$$\ell_2 := \begin{pmatrix} u_1 & v_1 \\ u_2 & v_2 \\ u_3 & v_3 \\ u_4 & v_4 \end{pmatrix}.$$

Then $\ell_2 \cdot d_1 = d_3 \cdot h_2$. Now $\Omega^2(\varphi_2^2)$ is the restriction of ℓ_2 to the submodule $\langle f_j^4 \mid 1 \le i \le 5 \rangle$ of Λ^4. The elements $f_i^4 \cdot \ell_2$ must be Λ-combinations of σ, ψ, θ. In general, no easy description can be expected. We will discuss this for the algebra $\Lambda = A5$ in the next section.

We observe that in general the cohomology is not graded commutative, in fact if $c \neq 1$ then the even part is not commutative.

6. The Hopf algebra $A5$ for $p \ge 3$ with $\beta \neq 0$

We consider the algebra, $A5(\beta)$, as in the Introduction. Here z plays the role of x (and y is kept). We have $\Lambda = k\langle z, y\rangle / I$, and we have the following. We set $a := yz - zy$, it is central, and with this, the ideal I is determined by

$$y^p = 0, \ a^p = 0, \ z^p + a^{p-1}y - \beta a = 0.$$

We use freely that a is central and $a^p = 0$. In particular we will use the fact that $a^{p-1}yz = a^{p-1}zy$.

We define

$$\sigma := (z^{p^2-1}, 0) \quad \psi = (z^{p-1} - \beta y, a^{p-1} + \beta z), \quad \theta = (0, y^{p-1}).$$

We will show that these satisfy Assumptions 2.1 and 3.4.

We start with some identities. By definition $z^p = \beta a - a^{p-1}y$. The first observation is that

$$z^p a^{p-1} = 0.$$

Next, $(z^p)^2 = \beta^2 a^2$ and hence $(z^p)^r = \beta^r a^r$ for $2 \le r$ and $z^{p^2} = 0$. Moreover

$$z^{p^2-2} = \beta^{p-1}a^{p-1}z^{p-2} \quad \text{and} \quad z^{p^2-1} = (z^p)^{p-1}z^{p-1} = \beta^{p-1}a^{p-1}z^{p-1} \neq 0.$$

Hence $z^{p^2-1}a = 0$ and z^{p^2-1} commutes with y. For computations, we will use the following formula which allows one to interchange powers of y with powers of z. This can be proved by induction from $a = yz - zy$.

LEMMA 6.1. *We have*

$$y^b z^c = \sum_{r \le b,c} r! \binom{b}{r} \binom{c}{r} a^r z^{c-r} y^{b-r}$$

Next, we describe a basis for the algebra.

LEMMA 6.2. *The algebra Λ has k-basis*

$$\{a^i y^j z^k \mid 0 \le i, j, k \le p-1\}.$$

The element $a^i y^j z^k$ in this set belongs to $J^m \setminus J^{m+1}$ where $m = pi + j + k$. The socle of Λ is spanned by $a^{p-1}y^{p-1}z^{p-1}$.

This is straightforward. Next, we show that the Assumption 2.1 is satisfied.

LEMMA 6.3. *The elements σ and θ are not in $\Lambda\psi$, and are linearly independent.*

PROOF. Clearly σ and θ are linearly independent over k. Assume for a contradiction that σ belongs to $\Lambda\psi$. That is, there is some $\gamma \in \Lambda$ such that $\sigma = \gamma\psi$. That is

$$z^{p^2-1} = \gamma(z^{p-1} - \beta y), \quad 0 = \gamma(a^{p-1} + \beta z).$$

Consider the first identity, $z^{p-1} - \beta y \in J \setminus J^1$ and the product is in J^{p^2-1}. This means that γ must be in $J^{p^2-2} \setminus J^{p^2-3}$. A basis for this quotient consists of the cosets of elements

$$a^i y^j z^k, \quad pi + j + k = p^2 - 2.$$

Solving this, shows that $i = 1$ or 2 and γ has highest order term

$$a^{p-1} y^j z^k, \quad j + k = p - 2 \quad \text{or} \quad a^{p-2} y^{p-1} z^{p-1}.$$

In both cases there is no solution for $\gamma(z^{p-1} - \beta y) = \beta^{p-1}(a^{p-1} z^{p-1})$, a contradiction.

Next, assume for a contradiction that there is $\delta \in \Lambda$ such that $\theta = \delta\psi$. That is

$$0 = \delta\psi_1 = \delta(z^{p-1} - \beta y), \quad y^{p-1} = \delta(a^{p-1} + \beta z).$$

Consider the second identity. Modulo the ideal $I := a^{p-1}\Lambda$ we require $\beta\delta z = y^{p-1}$, which has no solution, a contradiction. □

LEMMA 6.4. *Let*

$$\rho_1 = 0 = \rho_4,$$
$$\rho_2 = \beta^{-1} z^{p^2-1}$$
$$\rho_3 = -\beta^{-1} y^{p-1} - \beta^{-2}(y^{p-2} z^{p-1}) + \beta^{-3}(a^{p-1} y^{p-2} z^{p-2}).$$

These satisfy Lemma 3.2.

PROOF. It is clear that ρ_1 and ρ_4 satisfy Lemma 3.2

(1) To show that ρ_2 satisfies Lemma 3.2, we need
   (i) $yz^{p^2-1} + \beta^{-1} z^{p^2-1}(z^{p-1} - \beta y) = 0$ and
   (ii) $\beta^{-1} z^{p^2-1}(a^{p-1} + \beta z) = 0$.

   For (i), we have $z^{p^2-1} z^{p-1} = 0$ and hence (i) becomes

$$yz^{p^2-1} - z^{p^2-1} y = 0.$$

   To prove (ii), we have $z^{p^2-1} a = 0$ and hence (ii) becomes $z^{p^2} = 0$.

(2) We show now that ρ_3 satisfies Lemma 3.2, that is we must show
   (i) $\rho_3 \psi_1 = 0$ and
   (ii) $zy^{p-1} + \rho_3 \psi_2 = 0$.

   (i) We compute the product of each of the three terms of ρ_3 with ψ_1, ie with $z^{p-1} - \beta y$.

$$-\beta^{-1} y^{p-1}(z^{p-1} - \beta y) = -\beta^{-1} y^{p-1} z^{p-1}$$

   To compute the second product, observe

$$z^{p-1} z^{p-1} = (\beta a - a^{p-1} y) z^{p-2} \quad \text{and} \quad z^{p-1} y = yz^{p-1} + az^{p-2}$$

   So

$$\begin{aligned}
&-\beta^{-1} y^{p-2} z^{p-1}(z^{p-1} - \beta y) \\
&= -\beta^{-2} y^{p-2}(\beta a - a^{p-1} y) z^{p-2} + \beta^{-1} y^{p-2}(yz^{p-1} + az^{p-2}) \\
&= -\beta^{-1} a y^{p-2} z^{p-2} + \beta^{-2} a^{p-1} y^{p-1} z^{p-2} + \beta^{-1} y^{p-1} z^{p-1} + \beta^{-1} a y^{p-2} z^{p-2} \\
&= \beta^{-2} a^{p-1} y^{p-1} z^{p-2} + \beta^{-1} y^{p-1} z^{p-1}.
\end{aligned}$$

For the third product,

$$\beta^{-3}a^{p-1}y^{p-2}z^{p-2}(z^{p-1}-\beta y)=0-\beta^{-2}a^{p-1}y^{p-1}z^{p-2}.$$

The sum of the three terms we obtained is equal to zero, as stated.

(ii) We compute each of the three terms of $\rho_3\psi_2$. First

$$-\beta^{-1}y^{p-1}\psi_2=-\beta^{-1}a^{p-1}y^{p-1}-y^{p-1}z$$

For the next product, we use $y^{p-2}z^p=\beta ay^{p-2}-a^{p-1}y^{p-1}$ and get

$$-\beta^{-2}y^{p-2}z^{p-1}\psi_2=-\beta^{-2}a^{p-1}y^{p-2}z^{p-1}-ay^{p-2}+\beta^{-1}a^{p-1}y^{p-1}$$

Finally

$$\begin{aligned}\beta^{-3}a^{p-1}y^{p-2}z^{p-2}\psi_2&=\beta^{-3}a^{p-1}y^{p-2}z^{p-2}a^{p-2}+\beta^{-2}a^{p-1}y^{p-2}z^{p-1}\\&=\beta^{-2}a^{p-1}y^{p-2}z^{p-1}.\end{aligned}$$

Adding these we see

$$\rho_3\psi_2=-y^{p-1}z-ay^{p-2}.$$

Using the formula 6.1 we have

$$y^{p-1}z=zy^{p-1}-ay^{p-2}$$

and $\rho_3\psi_2+zy^{p-1}=0$ as required. □

It remains to verify the identities in Assumption 3.4.

Three of these are easy,

$$\begin{aligned}&\sigma_1\rho_1+\sigma_2\rho_2=0\\&\theta_1\rho_3+\theta_2\rho_4=0\\&\psi_1\rho_1+\psi_2\rho_2=(a^{p-1}+\beta z)\beta^{-1}z^{p^2-1}=0.\end{aligned}$$

As well

$$\theta_1\rho_1+\theta_2\rho_2=\beta^{-1}y^{p-1}z^{p^2-1}=\beta^{p-2}(a^{p-1}y^{p-1}z^{p-1})$$

which is a non-zero element in the socle of Λ. Moreover

$$\sigma_1\rho_3+\sigma_4\rho_4=\beta^{p-1}a^{p-1}z^{p-1}\rho_3=-\beta^{p-2}a^{p-1}y^{p-1}z^{p-1}$$

where the last equality follows by noting that $(a^{p-1}z^{p-1})$ is annihilated by a and by z and that z,y commute on elements which have a factor a^{p-1}. This gives the last equation in Assumption 3.4, where the scalar c is equal to 1.

It remains to show that $\psi_1\rho_3=0$. We compute each of the three terms. The first term of $\psi_1\rho_3$ is equal to

$$(\dagger)\qquad (z^{p-1}-\beta y)(-\beta^{-1}y^{p-1})=-\beta^{-1}(z^{p-1}y^{p-1})$$

The second product is equal to

$$\beta^{-2}(z^{p-1}y^{p-2}z^{p-1})+\beta^{-1}(y^{p-1}z^{p-1}).$$

We write (I) for the first summand and (II) for the second. Consider (I), this needs

$$\begin{aligned} z^{p-1}y^{p-2}z^{p-1} =& z^{p-1}\sum_{r\leq p-2} r!\binom{p-2}{r}\binom{p-1}{r}a^r z^{p-1-r}y^{p-2-r} \\ =& z^p \sum_{r\leq p-2} r!\binom{p-2}{r}\binom{p-1}{r}a^r z^{p-2-r}y^{p-2-r} \\ =& \beta[\sum_{r\leq p-2} r!\binom{p-2}{r}\binom{p-1}{r}a^{r+1}z^{p-2-r}y^{p-2-r}] \ - a^{p-1}yz^{p-2}y^{p-2} \\ =& \beta(\Sigma_1) \ - a^{p-1}z^{p-2}y^{p-1} \end{aligned}$$

defining Σ_1 to be the sum.

Similarly (II) is equal to

$$\beta^{-1}y^{p-1}z^{p-1} = \beta^{-1}(\sum_{r\leq p-1} r!\binom{p-1}{r}^2 a^r z^{p-1-r}y^{p-1-r}) =: \beta^{-1}\Sigma_2.$$

The third product is equal to

$$(z^{p-1} - \beta y)\beta^{-3}a^{p-1}y^{p-2}z^{p-2} = -\beta^{-2}a^{p-1}y^{p-1}z^{p-2}.$$

In total we have

$$\psi_1\rho_3 = (\dagger) - \beta^{-2}\beta\Sigma_1 + \beta^{-2}a^{p-1}z^{p-2}y^{p-1} + \beta^{-1}\Sigma_2 - \beta^{-1}a^{p-1}y^{p-1}z^{p-2}.$$

The term $(\dagger)$ cancels agains the term for $r = 0$ in Σ_2. The two monomials with β^{-2} cancel. This leaves

$$\Delta := -\beta^{-1}\Sigma_1 + \beta^{-1}\sum_{r=1^{p-1}} r!\binom{p-1}{r}^2 a^r z^{p-1-r}y^{p-1-r}.$$

Changing variables in the second sum, we can combine the terms and get

$$\Delta = \beta^{-1}\sum_{r=0}^{p-2} c_r a^{r+1}z^{p-2-r}y^{p-2-r}$$

where

$$c_r = -r!\binom{p-2}{r}\binom{p-1}{r} + (r+1)!\binom{p-1}{r+1}^2.$$

By an elementary calculation one gets that this is zero in k. This completes the proof that the algebra $A5$ satisfies the identities in 3.4. □

6.1. Products in $H^2 \times H^2$ for the algebra $A5$. We will continue with the notation of Lemma 5.2, in particular it will confirm that the even part of $H^*(\Lambda)$ is indeed commutative (which is the case for cohomology rings of Hopf algebras).

LEMMA 6.5. *For each $1 \leq i, j \leq 3$ we have $\varphi_i^2 \cdot \varphi_j^2 = \varphi_j^2 \cdot \varphi_i^2$.*

PROOF. Most of this is proved in Lemma 5.2. Note that we do not need to deal with squares. We are left to show that

$$\varphi_2^2 \cdot \varphi_1^2 = \varphi_2^4 \quad \text{and } \varphi_2^2 \cdot \varphi_3^2 = \varphi_4^4$$

Recall $\rho_1 = \rho_4 = 0$ and we have

$$\ell_2 = \begin{pmatrix} 0 & 0 \\ u_2 & v_2 \\ u_3 & v_3 \\ 0 & 0 \end{pmatrix}.$$

where $\rho_j = u_j z + v_j y$ for $j = 2, 3$. Then for each j we know that

$$f_j^4 \cdot \ell_2 = \alpha_j \sigma + \beta_j \psi + \gamma_j \theta$$

where $\alpha_j, \gamma_j \in k$ are unique (from Lemma 6.3), and where $\beta_j \in \Lambda$. We will only need to identify the α_j and γ_j.

First we have $f_1^4 \cdot \ell_2 = 0$ and $f_5^4 \cdot \ell_2 = 0$. Moreover we see directly that $f_2^4 \cdot \ell_2 = \sigma$, so $\alpha_2 = 1$ and $\gamma_2 = 0$, as required.

Next, we need

$$f_3^4 \cdot \ell_2 = (y^{p-1}(\beta^{-1} z^{p^2-2}) + z^{p^2-1} u_3, 0 + z^{p^2-1} v_3)$$

Since u_3 and v_3 are in the radical, it is clear that there is no non-zero summand $\alpha_3 \sigma$ or $\gamma_3 \theta$. That is $\alpha_3 = 0$ and $\gamma_3 = 0$. Similarly we see $\gamma_4 = 1$ and $\alpha_4 = 0$. □

References

[1] David J. Benson, Karin Erdmann, and Miles Holloway, *Rank varieties for a class of finite-dimensional local algebras*, J. Pure Appl. Algebra **211** (2007), no. 2, 497–510, DOI 10.1016/j.jpaa.2007.02.006. MR2341266

[2] Petter Andreas Bergh and Karin Erdmann, *Homology and cohomology of quantum complete intersections*, Algebra Number Theory **2** (2008), no. 5, 501–522, DOI 10.2140/ant.2008.2.501. MR2429451

[3] Karin Erdmann and Magnus Hellstrøm-Finnsen, *Hochschild cohomology of some quantum complete intersections*, J. Algebra Appl. **17** (2018), no. 11, 185–215, 22, DOI 10.1142/S0219498818502158. MR3879091

[4] Karin Erdmann, Øyvind Solberg, and Xingting Wang, *On the structure and cohomology ring of connected Hopf algebras*, J. Algebra **527** (2019), 366–398, DOI 10.1016/j.jalgebra.2019.02.030. MR3924439

[5] Leonard Evens, *The cohomology ring of a finite group*, Trans. Amer. Math. Soc. **101** (1961), 224–239, DOI 10.2307/1993372. MR137742

[6] Van C. Nguyen, Linhong Wang, and Xingting Wang, *Classification of connected Hopf algebras of dimension p^3 I*, J. Algebra **424** (2015), 473–505, DOI 10.1016/j.jalgebra.2014.09.022. MR3293230

[7] Steffen Oppermann, *Hochschild cohomology and homology of quantum complete intersections*, Algebra Number Theory **4** (2010), no. 7, 821–838, DOI 10.2140/ant.2010.4.821. MR2776874

Mathematical Institute, ROQ, Oxford OX2 6GG, United Kingdom
Email address: erdmann@maths.ox.ac.uk

Contemporary Mathematics
Volume **758**, 2020
https://doi.org/10.1090/conm/758/15236

On the cohomological Hall algebra of the Kronecker quiver

H. Franzen and M. Reineke

ABSTRACT. We give a short introduction to Cohomological Hall algebras of quivers and describe the semistable Cohomological Hall algebra of central slope of the Kronecker quiver in terms of generators and relations.

1. Generalities on cohomological Hall algebras

The Cohomological Hall algebra was introduced by Kontsevich and Soibelman in [**9**]. There are various levels of generality in which a Cohomological Hall algebra can be defined. There is the critical Cohomological Hall algebra of a quiver with a potential which is also introduced in [**9**] and has been studied for example in [**2**], [**12**, **13**], [**15**]. There are also versions of the Cohomological Hall algebra of sheaves on a curve or a surface, as studied in [**8**] or of Higgs sheaves [**11**].

In this work, we concentrate on (semistable) Cohomological Hall algebras of quivers. The aim of this section is to recall their definition as well as some structural results.

1.1. Definition of the cohomological Hall algebra of a quiver. Let Q be a finite quiver with set of vertices Q_0 and set of arrows Q_1. For an arrow $\alpha \in Q_1$ the source and target of α are denoted by $s(\alpha)$ and $t(\alpha)$, respectively. For a dimension vector $\mathbf{d} \in \Lambda^+ = \mathbb{N}Q_0 \subseteq \Lambda = \mathbb{Z}Q_0$, we define

$$R_{\mathbf{d}} = \bigoplus_{\alpha \in Q_1} \operatorname{Hom}(\mathbb{C}^{d_{s(\alpha)}}, \mathbb{C}^{d_{t(\alpha)}})$$

which we regard as an affine space. Its points correspond to complex representations of Q of dimension vector $\mathbf{d}$. The linear algebraic group $G_{\mathbf{d}} = \prod_{i \in Q_0} \mathrm{GL}_{d_i}(\mathbb{C})$ acts on $R_{\mathbf{d}}$ via base change, so that the orbits for this action correspond to the isomorphism classes of representations of dimension vector $\mathbf{d}$. On the Λ_+-graded $\mathbb{Q}$-vector space of equivariant cohomology groups

$$H(Q) = \bigoplus_{\mathbf{d} \in \Lambda^+} H^*_{G_{\mathbf{d}}}(R_{\mathbf{d}}; \mathbb{Q})$$

a product, denoted $*$, can be defined using pull-back and push-forward operations. Together with this product, $H(Q)$ is an associative algebra, naturally graded by Λ^+. It is called the Cohomological Hall algebra. The product resembles the convolution product of the Ringel-Hall algebra of a quiver of [**10**]. As we will not use the definition of this convolution-like product we refer to [**9**, 2.2] for details and

2010 *Mathematics Subject Classification.* Primary 14N35, 16G20, 17B37.

instead recall the algebraic description of the multiplication of [**9**, Thm. 2]. The homogeneous component of degree $\mathbf{d}$ of the algebra $H(Q)$ is given by

$$H(Q)_{\mathbf{d}} \cong \bigotimes_{i\in Q_0} \mathbb{Q}[x_{i,1},\ldots,x_{i,d_i}]^{S_{d_i}},$$

and the multiplication is given by a shuffle product with kernel: For $f = f(x_{i,k} : i \in Q_0,\ k \le d_i') \in H(Q)_{\mathbf{d}'}$ and $g = g(x_{i,k} : i \in Q_0,\ k \le d_i'') \in H(Q)_{\mathbf{d}''}$, the product $(f * g)(x_{i,k} : i \in Q_0,\ k \le d_i' + d_i'')$ equals

$$\sum_{(\sigma_i)} f(x_{i,\sigma_i(k)},\ i \in Q_0,\ k \le d_i')g(x_{i,\sigma_i(d_i'+l)},\ i \in Q_0,\ l \le d_i'')$$

$$\times \prod_{i,j\in Q_0}\prod_{k=1}^{d_i'}\prod_{l=1}^{d_j''}(x_{j,\sigma_j(d_j'+l)} - x_{i,\sigma_i(k)})^{-\langle \mathbf{e}_i,\mathbf{e}_j\rangle},$$

where the sum ranges over all tuples $(\sigma_i)_{i\in Q_0}$ such that σ_i is a (d_i', d_i'')-shuffle permutation. In the above formula $\langle \_, \_\rangle$ denotes the Euler form of Q. This is the $\mathbb{Z}$-bilinear form on $\mathbb{Z}^{Q_0}$ given by $\langle \mathbf{a},\mathbf{b}\rangle = \sum_{i\in Q_0} a_i b_i - \sum_{\alpha\in Q_1} a_{s(\alpha)}b_{t(\alpha)}$. Moreover $\mathbf{e}_i \in \mathbb{Z}^{Q_0}$ is the ith coordinate vector.

1.2. Symmetric case. The Cohomological Hall algebra has particularly nice properties if the quiver Q is symmetric, which means that its Euler form is a symmetric bilinear form. In this case, the algebra $H(Q)$ can be made into a $\Gamma = \Lambda^+ \times \mathbb{Z}$-graded algebra by defining

$$H(Q)_{(\mathbf{d},k)} = H_{G_d}^{k-\langle \mathbf{d},\mathbf{d}\rangle}(R_{\mathbf{d}};\mathbb{Q}).$$

It is possible to find a bilinear form ψ on $\mathbb{Z}/2\mathbb{Z}^{Q_0}$ in such a way that the twisted multiplication $f \star g := (-1)^{\psi(\mathbf{d}',\mathbf{d}'')} f * g$ is graded commutative in the sense that $f \star g = (-1)^{kl} g \star f$ for $f \in H(Q)_{(\mathbf{d}',k)}$ and $g \in H(Q)_{(\mathbf{d}'',l)}$, see [**9**, 2.6]. Efimov proves in [**5**] a conjecture of Kontsevich and Soibelman that $H(Q)$, equipped with the twisted multiplication, is isomorphic to the graded symmetric algebra $\mathrm{Sym}^*(V)$ for a Γ-graded vector space V which is of the form $V = V^{\text{prim}} \otimes \mathbb{Q}[z]$, where z lives in degree $(0, 2)$. The Poincaré series of V^{prim} encodes the motivic Donaldson-Thomas invariants of Q. In fact, this was the first proof of the positivity conjecture for Donaldson–Thomas invariants of a symmetric quiver.

1.3. Semistable cohomological Hall algebra. For general Q, the structure of $H(Q)$ can be analyzed using semistable Cohomological Hall algebras: We choose a stability function $\Theta \in \Lambda^*$. For $\mathbf{d} \in \Lambda^+ \setminus \{0\}$ the rational number $\mu(\mathbf{d}) = \Theta(d)/\sum_{i\in Q_0} d_i$ is called the slope of $\mathbf{d}$. A representation M of Q is called Θ-semistable (resp. Θ-stable) if $\mu(\mathbf{dim}\, M') \le \mu(\mathbf{dim}\, M)$ (resp. $\mu(\mathbf{dim}\, M') < \mu(\mathbf{dim} M)$) for every non-zero proper subrepresentation $M' \subseteq M$. Let $R_{\mathbf{d}}^{\Theta-\text{sst}} \subseteq R_{\mathbf{d}}$ be the Zariski-open subset of Θ-semistable points. For a fixed $\mu \in \mathbb{Q}$, we consider the set $\Lambda_\mu^+ = \{\mathbf{d} \in \Lambda^+ \setminus \{0\} \mid \mu(\mathbf{d}) = \mu\} \cup \{0\}$. We then define

$$H_\mu^\Theta(Q) = \bigoplus_{\mathbf{d}\in\Lambda_\mu^+} H_{G_{\mathbf{d}}}^*(R_{\mathbf{d}}^{\Theta-\text{sst}};\mathbb{Q})$$

with the convolution structure induced by the one on $H(Q)$, called the Θ-semistable Cohomological Hall algebra of Q for slope μ. It is shown in [**6**, Thm. 6.2] that the

Harder-Narasimhan filtration of representations of Q induces an isomorphism of Λ^+-graded vector spaces

$$H(Q) \cong \overrightarrow{\bigotimes_{\mu \in \mathbb{Q}}} H_\mu^\Theta(Q).$$

In case the Euler form $\langle \_, \_ \rangle_Q$ of Q is symmetric when restricted to Λ_μ^+, good structural properties of $H_\mu^\Theta(Q)$ are expected; however, even in simple cases it is no longer (graded) commutative. In [**3**, Thm. C], Davison and Meinhardt prove (in a more general context) a PBW type theorem for $H_\mu^\Theta(Q)$. They show the existence of a filtration (the perverse filtration) on $H_\mu^\Theta(Q)$ whose associated graded is a graded symmetric algebra.

1.4. Generators. There are several geometric interpretations of the generators of the Cohomological Hall algebra. For instance, Chen identifies in [**1**] the primitive part of the Cohomological Hall algebra of a symmetric quiver with the invariant part under a Weyl group action of the cohomology of a variety introduced in [**7**]. Another interpretation can be given in terms of Chow groups (using a version of the Comological Hall algebra built on Chow groups as developed and studied in [**6**]) as follows. Denoting by $R_{\mathbf{d}}^{\Theta-\mathrm{st}}(Q)$ the locus of stable representations, it is shown in [**6**, Thm. 9.1] that

$$A_{G_{\mathbf{d}}}^*(R_{\mathbf{d}}^{\Theta-\mathrm{st}}(Q))_{\mathbb{Q}} \cong H_{\mathbf{d}}^\Theta(Q)/\sum H_\mu^\Theta(Q)_{\mathbf{d}'} * H_\mu^\Theta(Q)_{\mathbf{d}''},$$

where the sum ranges over all decompositions $\mathbf{d} = \mathbf{d}' + \mathbf{d}''$ such that $\mu(\mathbf{d}) = \mu(\mathbf{d}'')$. This describes the quotient of the algebra $H_\mu^\Theta(Q)$ by the square of the augmentation ideal $\bigoplus_{\mathbf{d} \in \Lambda_\mu^+ \setminus \{0\}} H_\mu^\Theta(Q)_{\mathbf{d}}$ as the equivariant Chow ring $A_{G_d}^*(R_{\mathbf{d}}^{\Theta-\mathrm{st}}(Q))_{\mathbb{Q}}$ of the stable locus in the sense of [**4**].

1.5. Presentation. The key to describing the semi-stable Cohomological Hall algebra explicitly in terms of generators and relations (which will be worked out in the case of the Kronecker quiver in the following sections) is the following tautological presentation for $H_\mu^\Theta(Q)$ proven in [**6**, Thm. 8.1]:

$$H_\mu^\Theta(Q)_{\mathbf{d}} \cong H(Q)_{\mathbf{d}}/\sum H(Q)_{\mathbf{d}'} * H(Q)_{\mathbf{d}''},$$

where the sum ranges over all decompositions $\mathbf{d} = \mathbf{d}' + \mathbf{d}''$ such that $\mu(\mathbf{d}') > \mu(\mathbf{d}'')$.

1.6. Examples. There are three examples in which an explicit description of the Cohomological Hall algebra in terms of generators and relations is known, namely for the quivers A_1 (one vertex and no arrows), L_1 (one vertex, one arrow), and $\tilde{A}_{2,0}$ (two vertices and two arrows of opposite orientation). The first two examples were already discussed in [**9**, 2.5], the latter in [**6**, 10.1]. All these quivers are symmetric whence we know by Efimov's theorem that their Cohomological Hall algebra is a graded symmetric algebra.

For $Q = A_1$, the Cohomological Hall algebra as a vector space is $H(A_1) = \bigoplus_{d \geq 0} \mathbb{Q}[x_1, \ldots, x_d]^{S_d}$. The Euler form of A_1 is $\langle d, e \rangle = de$. The $\Lambda^+ \times \mathbb{Z} = \mathbb{N} \times \mathbb{Z}$-grading of $H(A_1)$ is hence given by assigning to each x_i the degree 2 and letting

$$H(A_1)_{(d,k)} = \mathbb{Q}[x_1, \ldots, x_d]_{k-d^2}^{S_d} = \bigoplus_{2n_1 + \ldots + 2dn_d = k - d^2} \mathbb{Q} \cdot e_1^{n_1} \ldots e_d^{n_d}.$$

The multiplication of $H(A_1)$ is given by

$$(f*g)(x_1,\ldots,x_d)$$
$$=\sum f(x_{\sigma(1)},\ldots,x_{\sigma(d')})g(x_{\sigma(d'+1)},\ldots,x_{\sigma(d)})\frac{1}{\prod_{r=1}^{d'}\prod_{s=1}^{d''}(x_{\sigma(d'+s)}-x_{\sigma(r)})}.$$

The sum ranges over all (d',d'')-shuffle permutations, $f\in H(A_1)_{d'}$, $g\in H(A_1)_{d''}$, and $d=d'+d''$. We see that this multiplication is anti-commutative, whence $f*f=0$. This induces an algebra homomorphism $\bigwedge^* H(A_1)_1\to H(A_1)$. Let ψ_i be the polynomial $\psi_i(x)=x^i\in H(A_1)_{(1,2i+1)}$ (the i in the exponent refers to the usual multiplication of polynomials and not to the multiplication in the Cohomological Hall algebra). The elements $\psi_0,\psi_1,\psi_2,\ldots$ form a homogeneous basis of $H(A_1)$. It is easy to see that for integers $0\leq k_1<\ldots k_d$ we get

$$\psi_{k_1}*\ldots*\psi_{k_d}=s_\lambda(x_1,\ldots,x_d)$$

where $\lambda=(k_d-d+1,\ldots,k_2-1,k_1)$ and s_λ is the Schur function associated with the partition λ. This shows that the induced homomorphism $\bigwedge^*(\psi_0,\psi_1,\psi_2,\ldots)\to H(A_1)$ is surjective and a comparison of the generating functions of these two algebras yields that the surjection must be an isomorphism. So

$$H(A_1)\cong \mathrm{Sym}^*(\mathbb{Q}(1,1)\otimes\mathbb{Q}[z])$$

as $\mathbb{N}\times\mathbb{Z}$-graded algebras, where $\mathbb{Q}(d,i)$ denotes a one-dimensional vector space in degree (d,i) and z is an element of degree $(0,2)$.

The Cohomological Hall algebra of the quiver $Q=L_1$ with one loop is as a vector space again $H(L_1)=\bigoplus_{d\geq 0}\mathbb{Q}[x_1,\ldots,x_d]^{S_d}$ but as the Euler form vanishes the $\mathbb{N}\times\mathbb{Z}$-grading is given by

$$H(A_1)_{(d,k)}=\mathbb{Q}[x_1,\ldots,x_d]_k^{S_d}=\bigoplus_{2n_1+\ldots+2dn_d=k}\mathbb{Q}\cdot e_1^{n_1}\ldots e_d^{n_d}.$$

The multiplication is given by

$$(f*g)(x_1,\ldots,x_d)=\sum f(x_{\sigma(1)},\ldots,x_{\sigma(d')})g(x_{\sigma(d'+1)},\ldots,x_{\sigma(d)}).$$

The multiplication is commutative which shows that we obtain a homomorphism $S^*(H(L_1)_1)\to H(L_1)$. Again we denote $\psi_i(x)=x^i\in H(L_1)_{(1,2i)}$. We can see that for integers $k_1\geq\ldots\geq k_d\geq 0$

$$\psi_{k_1}*\ldots*\psi_{k_d}=c_\lambda m_\lambda(x_1,\ldots,x_d)$$

where m_λ is the monomial symmetric function associated with the partition $\lambda=(k_1,\ldots,k_d)$ and c_λ is a positive integer. This shows surjectivity of $S^*(\psi_0,\psi_1,\psi_2,\ldots)\to H(L_1)$. As the generating series of these algebras agree the map is an isomorphism. Therefore

$$H(L_1)\cong\mathrm{Sym}^*(\mathbb{Q}(1,0)\otimes\mathbb{Q}[z])$$

as $\mathbb{N}\times\mathbb{Z}$-graded algebras. Again z is of degree $(0,2)$.

For the quiver $\tilde{A}_{2,0}$ of affine type $\tilde{A}_1$ with the cyclic orientation, an explicit presentation can be determined by decomposing it into its semistable parts. the choice of an appropriate stability condition yields a tensor product decomposition (as $\Lambda^+\times\mathbb{Z}$-graded algebras) into three factors, two of which are isomorphic to the Cohomological Hall algebra of A_1 and the other being isomorphic to $H(L_1)$. Concretely

$$H(\tilde{A}_{2,0}) \cong \mathrm{Sym}^*\left(\left(\mathbb{Q}((1,0),1) \oplus \mathbb{Q}((1,1),0) \oplus \mathbb{Q}((0,1),1)\right) \otimes \mathbb{Q}[z]\right)$$

as $\Lambda^+ \times \mathbb{Z}$-graded algebras and z of degree $(0,2)$.

2. The case of the Kronecker quiver

We now specialize the above results to the Kronecker quiver $Q = K_2$ with vertices i and j and two arrows from i to j for the stability $\Theta(d_i, d_j) = d_i$, recall some further results from [**6**, Section 10.2.], and state our main result.

It is known [**6**, 10.2] that

$$H^{\Theta}_{\mu(\mathbf{d})}(K_2) \cong \Lambda^*(\mathbb{Q}(d,1) \otimes \mathbb{Q}[z])$$

is isomorphic to a countably generated exterior algebra for $\mathbf{d}$ of the form $(n, n+1)$ or $(n+1, n)$ for $n \geq 0$ (because in any of these cases, the category of semistable representations is equivalent to representations of a trivial quiver) and that $H^{\Theta}_{\mu}(K_2)$ is zero if μ is neither zero nor of the form $n/(2n \pm 1)$ for $n \geq 1$ (because in any of these cases, there are no semistables).

It thus remains to determine the structure of $H^{\Theta}_{1/2}(K_2)$. We denote this algebra simply by A; it is $\mathbb{N}$-graded by $A_n = H^{\Theta}_{1/2}(K_2)_{(n,n)}$. It is already known [**6**, 10.2] that there is an isomorphism of $\mathbb{N}$-graded vector spaces

$$A \cong \mathrm{Sym}^*(H^*(\mathbb{P}^1)[z])$$

with the generating vector space in degree 1. Our main result is:

THEOREM 1. *The algebra A is generated in degree 1 by generators e_n for $n \geq 0$ and f_n for $n \geq 1$, subject to the relations*

$$[E(X), E(Y)]_* - 2(Y \quad X)\frac{YE(Y) - XE(X)}{Y - X} * \frac{YF(Y) - XF(X)}{Y - X},$$

$$[E(X), F(Y)]_* = (Y - X)\left(\frac{YF(Y) - XF(X)}{Y - X}\right)^{*2},$$

$$[F(X), F(Y)]_* = 0$$

in terms of generating series

$$E(X) = \sum_{n \geq 0} e_n X^n, \quad F(X) = \sum_{n \geq 0} f_{n+1} X^n.$$

More explicitly, these identities are equivalent to the following relations:

$$\begin{aligned}
[e_p, e_q]_* &= 2(e_p * f_q + e_{p+1} * f_{q-1} + \ldots + e_{q-1} * f_{p+1}) && \textit{for } p < q,\\
[e_p, f_{q+1}]_* &= f_{p+1} * f_q + \ldots + f_q f_{p+1} && \textit{for } p < q,\\
[e_p, f_{q+1}]_* &= -f_{q+1} * f_p - \ldots - f_p * f_{q+1} && \textit{for } p > q,\\
[f_{p+1}, f_{q+1}]_* &= 0 && \textit{for } p < q.
\end{aligned}$$

From these defining relation we see that A is a filtered algebra, with the filtration induced by the degree function $\deg(e_i) = 1$ and $\deg(f_i) = 0$ for all i, such that the associated graded is the symmetric algebra of A_1.

3. Proof of Theorem 1

3.1. Idea of the proof. Since $R^{\Theta-\mathrm{st}}_{(n,n)}(K_2) = \emptyset$ for $n \geq 2$, the tautological presentation of Section 1.4 yields that

$$A_n = \sum_{k+l=n} A_k * A_l$$

for $n \geq 2$. By induction over n, we find that A_n is generated in degree 1. We will prove the following property in the next section:

There exists an ordered basis $(b_i)_{i\in I}$ for A_1 such that

$$b_j * b_i = \sum_{k\leq l} c^{i,j}_{k,l} \cdot b_k * b_k \text{ for all } i < j.$$

This implies that every A_n is linearly generated by the monomials $b_{i_1} \cdot \ldots \cdot b_{i_n}$ for $i_1 \leq \ldots \leq i_n$ in I. By the above result, this implies that these monomials already form a linear basis of A_n. This in turn implies that A is generated by the b_i subject to the relations

$$b_j * b_i - \sum_{k\leq l} c^{i,j}_{k,l} \cdot b_k * b_l \text{ for all } i < j.$$

3.2. Computations. To compute relations in A, we first describe some of the multiplication maps in $A(K_2)$ more explicitly using the above algebraic description as a modified shuffle product:

As above, we identify

$$H(K_2)_{(d,e)} \cong \mathbb{Q}[x_1, \ldots, x_d]^{S_d} \otimes \mathbb{Q}[y_1, \ldots, y_e]^{S_e}.$$

LEMMA 2. *We have:*

- *The multiplication map* $H(K_2)_{(1,0)} \otimes H(K_2)_{(0,1)} \to H(K_2)_{(1,1)}$ *is given by*
$$(f * g)(x, y) = f(x)g(y)(y - x)^2.$$
- *The multiplication map* $H(K_2)_{(1,0)} \otimes H(K_2)_{(1,2)} \to H(K_2)_{(2,2)}$ *is given by*
$$(f * g)(x_1, x_2, y_1, y_2) = \frac{1}{x_2 - x_1} \cdot \big(f(x_1)g(x_2, y_1, y_2)(y_1 - x_1)^2(y_2 - x_1)^2 \\ - f(x_2)g(x_1, y_1, y_2)(y_1 - x_2)^2(y_2 - x_2)^2\big).$$
- *The multiplication map* $H(K_2)_{(1,1)} \otimes H(K_2)_{(1,1)} \to H(K_2)_{(2,2)}$ *is given by*
$$(f * g)(x_1, x_2, y_1, y_2) = \frac{1}{(x_2 - x_1)(y_2 - y_1)} \\ \cdot \big(f(x_1, y_1)g(x_2, y_2)(y_2 - x_1)^2 - f(x_1, y_2)g(x_2, y_1)(y_1 - x_1)^2 \\ - f(x_2, y_1)g(x_1, y_2)(y_2 - x_2)^2 + f(x_2, y_2)g(x_1, y_1)(y_1 - x_2)^2\big).$$

We thus find $A_1 \cong \mathbb{Q}[x, y]/(y - x)^2$, for which we choose the basis $e_n = x^n$ for $n \geq 0$ and $f_n = x^{n-1}(y - x)$ for $n \geq 1$. We define generating series

$$E(X) = \sum_{n\geq 0} e_n X^n = \frac{1}{1 - xX} \in A_1[X], \quad F(X) = \sum_{n\geq 0} f_{n+1} X^n = \frac{y - x}{1 - xX} \in A_1[X].$$

LEMMA 3. *The following relations hold in* $A_2[X,Y]$ *(where* X *and* Y *are viewed as variables commuting with each other and with* A_1 *and* A_2*):*

$$
\begin{aligned}
[E(X),E(Y)]_* &= 2\frac{(YE(Y)-XE(X))*(YF(Y)-XF(X))}{Y-X},\\
[E(X),F(Y)]_* &= \frac{(YF(Y)-XF(X))^{*2}}{Y-X},\\
[F(X),F(Y)]_* &= 0.
\end{aligned}
$$

PROOF. We first note that

$$\frac{YE(Y)-XE(X)}{Y-X}=\frac{1}{(1-xX)(1-xY)},$$

and similarly for the series F.

We have

$$
\begin{aligned}
&\frac{1}{1-xX}*\frac{1}{1-xY}\\
&=\frac{1}{(x_2-x_1)(y_2-y_1)}\cdot\left(\frac{(y_2-x_1)^2}{(1-x_1X)(1-x_2Y)}-\frac{(y_1-x_1)^2}{(1-x_1X)(1-x_2Y)}\right.\\
&\qquad\left.-\frac{(y_2-x_2)^2}{(1-x_2X)(1-x_1Y)}+\frac{(y_1-x_2)^2}{(1-x_2X)(1-x_1Y)}\right)\\
&=\frac{1}{x_2-x_1}\cdot\left(\frac{y_1+y_2-2x_1}{(1-x_1X)(1-x_2Y)}-\frac{y_1+y_2-2x_2}{(1-x_2X)(1-x_1Y)}\right).
\end{aligned}
$$

Exchanging the variables X and Y, we find

$$\frac{1}{1-xY}*\frac{1}{1-xX}=\frac{1}{x_2-x_1}\cdot\left(\frac{y_1+y_2-2x_1}{(1-x_2X)(1-x_1Y)}-\frac{y_1+y_2-2x_2}{(1-x_1X)(1-x_2Y)}\right).$$

Thus

$$
\begin{aligned}
\left[\frac{1}{1-xX},\frac{1}{1-xY}\right]_* &= \frac{2(y_1+y_2-x_1-x_2)}{x_2-x_1}\\
&\qquad\cdot\left(\frac{1}{1-x_1X)(1-x_2Y)}-\frac{1}{(1-x_2X)(1-x_1Y)}\right)\\
&=\frac{2(y_1+y_2-x_1-x_2)(Y-X)}{(1-x_1X)(1-x_2X)(1-x_1Y)(1-x_2Y)}.
\end{aligned}
$$

On the other hand, we have

$$
\begin{aligned}
&\frac{YE(Y)-XE(X)}{Y-X}*\frac{YF(Y)-XF(X)}{Y-X}\\
&=\frac{1}{(1-xX)(1-xY)}*\frac{y-x}{(1-xX)(1-xY)}\\
&=\frac{(y_2-x_2)(y_2-x_1)^2-(y_1-x_2)(y_1-x_1)^2-(y_2-x_1)(y_2-x_2)^2+(y_1-x_1)(y_1-x_2)^2}{(x_2-x_1)(y_2-y_1)(1-x_1X)(1-x_1Y)(1-x_2X)(1-x_2Y)}\\
&=\frac{y_1+y_2-x_1-x_2}{(1-x_1X)(1-x_1Y)(1-x_2X)(1-x_2Y)},
\end{aligned}
$$

from which the first claimed relation follows.

The second relation is proved similarly, so we omit some intermediate steps. On the one hand, we have

$$[E(X),F(Y)]_* = \left[\frac{1}{1-xX},\frac{y-x}{1-xY}\right]_* = \frac{(y_1+y_2-x_1-x_2)^2(Y-X)}{(1-x_1X)(1-x_2X)(1-x_1Y)(1-x_2Y)},$$

and on the other hand

$$\begin{aligned}(\frac{YF(Y)-XF(X)}{Y-X})^{*2} &= (\frac{y-x}{(1-xX)(1-yY)})^{*2}\\ &= \frac{(y_1+y_2-x_1-x_2)^2}{(1-x_1X)(1-x_2X)(1-x_1Y)(1-x_2Y)},\end{aligned}$$

and the claimed relation follows.

To prove the third relation, we first proceed similarly to the previous computations to arrive at

$$[F(X),F(Y)]_* = \frac{(y_1+y_2-x_1-x_2)((x_1+x_2)(y_1+x_2)-x_1^2-x_2^2-2y_1y_2)(Y-X)}{(1-x_1X)(1-x_2X)(1-x_1Y)(1-x_2Y)}.$$

This element is non-zero in $A_2[X,Y] = H_{(2,2)}(K_2)[X,Y]$, but we claim that it belongs to the image of the multiplication map

$$H(K_2)_{(1,0)}[X,Y]\otimes H(K_2)_{(1,2)}[X,Y]\to H(K_2)_{(2,2)}[X,Y].$$

Namely, by the above description of this map, we find

$$\begin{aligned}&\frac{1}{(1-xX)(1-xY)} * \frac{1}{(1-xX)(1-xY)}\\ &\qquad = \frac{(y_1+y_2-x_1-x_2)(x_1^2+x_2^2+2y_1y_2-(x_1+x_2)(y_1+y_2))}{(1-x_1X)(1-x_1Y)(1-x_2X)(1-x_2Y)},\end{aligned}$$

as claimed. The lemma is proved. □

We would now like to understand which relations between the individual basis elements e_n, f_n result from the above identities of generating series. We have

$$[E(X),E(Y)]_* = \sum_{p,q\geq 0}[e_p,e_q]_* X^pY^q$$

and

$$\frac{YE(Y)-XE(X)}{Y-X} = \sum_{p,q\geq 0} e_{p+q}X^pY^q.$$

Thus

$$\begin{aligned}&\frac{YE(Y)-XE(X)}{Y-X} * (YF(Y)-XF(X))\\ &= \sum_{i,j\geq 0} e_{i+j}X^iY^j * (\sum_{k\geq 0} f_{k+1}Y^{k+1} - \sum_{k\geq 0} f_{k+1}X^{k+1})\\ &= \sum_{p,q\geq 0}(\sum_{j+k=q-1} e_{p+j}f_{k+1} - \sum_{i+k=p-1} e_{i+q}f_{k+1})X^pY^q.\end{aligned}$$

Thus for $p<q$, we find the relations

$$[e_p,e_q]_* = 2(e_p * f_q + e_{p+1} * f_{q-1} + \ldots + e_{q-1} * f_{p+1}).$$

The same calculation for the second type of relation leads to

$$[e_p, f_{q+1}]_* = f_{p+1} * f_q + \ldots + f_q * f_{p+1} \text{ for } p < q,$$
$$[e_p, f_{q+1}]_* = -f_{q+1} * f_p - \ldots - f_p * f_{q+1} \text{ for } p > q.$$

Finally, we have

$$[f_{p+1}, f_{q+1}]_* = 0$$

for all $p < q$. We can now finish the proof of the theorem: ordering the basis elements as

$$e_0, e_1, e_2, \ldots, f_1, f_2, \ldots,$$

we see that the above relations allow us to reorder any given product of the generators into a monomial in standard ordering.

4. Further descriptions

In this section, we indicate three alternative descriptions of the algebra A in order to better understand the nature of its defining relations.

4.1. Twisted symmetric algebra. For a vector space V and an operator $c \in \mathrm{End}(V \otimes V)$ such that $c^2 = \mathrm{id}$, we can define the c-twisted symmetric algebra

$$\mathrm{Sym}^c(V) = T(V)/(\mathrm{Ker}(c + \mathrm{id})).$$

In the case of the flip operator $c(v \otimes w) = w \otimes v$, we get the usual symmetric algebra. If c moreover satisfies the quantum Yang-Baxter equation

$$(c \otimes \mathrm{id})(\mathrm{id} \otimes c)(c \otimes \mathrm{id}) = (\mathrm{id} \otimes c)(c \otimes \mathrm{id})(\mathrm{id} \otimes c),$$

the algebra $\mathrm{Sym}^c(V)$ is called a braided symmetric algebra.

THEOREM 4. *The algebra A is isomorphic to $\mathrm{Sym}^c(A_1)$ for the operator c defined on the formal series $E(X)$ and $F(X)$ by*

$$c(E(X) \otimes E(Y)) = E(Y) \otimes E(X) + (Y - X)\Big(\frac{YE(Y) - XE(X)}{Y - X} \otimes \frac{YF(Y) - XF(X)}{Y - X}$$
$$+ \frac{YF(Y) - XF(X)}{Y - X} \otimes \frac{YE(Y) - XE(X)}{Y - X}\Big),$$
$$c(E(X) \otimes F(Y)) = F(Y) \otimes E(X) + (Y - X)\frac{YF(Y) - XF(X)}{Y - X} \otimes \frac{YF(Y) - XF(X)}{Y - X},$$
$$c(F(X) \otimes E(Y)) = E(Y) \otimes F(X) + (Y - X)\frac{YF(Y) - XF(X)}{Y - X} \otimes \frac{YF(Y) - XF(X)}{Y - X},$$
$$c(F(X) \otimes F(Y)) = F(Y) \otimes F(X).$$

PROOF. We abbreviate $E(X, Y) = YE(Y) - XE(X)$, and similarly $F(X, Y)$. Then

$$\begin{aligned}
&c^2(E(X) \otimes F(Y)) \\
&= c(F(Y) \otimes E(X)) + \frac{1}{(Y - X)} c(F(X, Y) \otimes F(X, Y)) \\
&= E(X) \otimes F(Y) + \frac{1}{X - Y} F(X, Y) \otimes F(X, Y) + \frac{1}{Y - X} F(X, Y) \otimes F(X, Y) \\
&= E(X) \otimes F(Y).
\end{aligned}$$

With a similar computation we find $c(E(X,Y)\otimes F(X,Y)) = F(X,Y)\otimes E(X,Y)$, which yields

$$\begin{aligned}
&c^2(E(X)\otimes E(Y))\\
&\quad = c(E(Y)\otimes E(X)) + \frac{1}{Y-X}c(E(X,Y)\otimes F(X,Y) + F(X,Y)\otimes E(X,Y))\\
&\quad = E(X)\otimes E(Y) + \frac{1}{X-Y}(E(X,Y)\otimes F(X,Y) + F(X,Y)\otimes E(X,Y))\\
&\qquad + \frac{1}{Y-X}(E(X,Y)\otimes F(X,Y) + F(X,Y)\otimes E(X,Y))\\
&= E(X)\otimes F(Y),
\end{aligned}$$

proving $c^2 = \mathrm{id}$. The defining relations of $\mathrm{Sym}^c(V)$ are then given by

$$[E(X),E(Y)] = \frac{1}{Y-X}(E(X,Y)\cdot F(X,Y) + F(X,Y)\cdot E(X,Y)),$$

$$[E(X),F(Y)] = \frac{1}{Y-X}F(X,Y)^2,\ [F(X),F(Y)] = 0.$$

Comparing this with the defining relations of A, it suffices to prove that $E(X,Y)$ and $F(X,Y)$ commute in A. But this follows from the identity

$$[E(Y),F(X)] = (X-Y)F(X,Y)^2 = -(Y-X)F(X,Y)^2 = -[E(X),F(Y)]$$

in A, finishing the proof. □

CONJECTURE 5. *The operator c given above satisfies the Quantum Yang-Baxter equation.*

4.2. Differential operators. We consider the algebra $B = \mathbb{Q}[w_1, w_2, \ldots]$ and define operators e_i and f_i on B as follows: the operator f_i acts by multiplication with w_i, and the operator e_i is a derivation defined on generators of B by

$$e_i w_{j+1} = w_{i+1}w_j + \ldots + w_{i+j}w_1 - (w_{j+1}w_i + \ldots + w_{i+j}w_1).$$

In more compact form, define

$$E(X) = \sum_{n\geq 0} e_n X^n,\quad F(X) = \sum_{n\geq 0} f_{n+1}X^n,\quad W(X) = \sum_{n\geq 0} w_{n+1}X^n.$$

Then $F(X)$ acts on $B[X]$ via multiplication by $W(X)$, and $E(X)$ acts via derivations such that

$$E(X)W(Y) = (Y-X)\left(\frac{(YW(Y)-XW(X))}{Y-X}\right)^2.$$

THEOREM 6. *The algebra A acts on B via the operators defined above.*

PROOF. We verify that the above operators on B fulfill the defining relations of A. Since $F(X)$ acts via multiplication on the commutative algebra B, the relation $[F(X),F(Y)] = 0$ is obvious.

As in the previous proof, we will abbreviate $W(X,Y) = YW(Y) - XW(X)$, and similarly for $E(X,Y)$, $F(X,Y)$. Since $E(X)$ acts by derivations, we have

$$\begin{aligned}
E(X)(F(Y)(W(Z))) &= E(X)(W(Y)W(Z))\\
&= E(X)(W(Y))W(Z) + W(Y)E(X)(W(Z))\\
&= E(X)(W(Y))W(Z) + F(Y)(E(X)(W(Z))),
\end{aligned}$$

and thus

$$[E(X), F(Y)](W(Z)) = E(X)(W(Y))W(Z) = \frac{1}{Y-X}W(X,Y)^2W(Z)$$
$$= ((Y-X)F(X,Y)^2)(W(Z)),$$

verifying the second defining relation.

The verification of the first defining relation is more involved. We have

$$E(X)(E(Y)(W(Z))) = \frac{1}{Z-Y}E(X)(W(Y,Z)^2) = \frac{2}{Z-Y}W(Y,Z)E(X)(W(Y,Z)),$$

and similarly

$$E(Y)(E(X)(W(Z))) = \frac{2}{Z-X}W(X,Z)E(Y)(W(X,Y)),$$

thus $[E(X), E(Y)](W(Z))$

$$= \frac{2}{(Z-Y)(Z-X)}((Z-X)W(Y,Z)E(X)(W(Y,Z))$$
$$- (Z-Y)W(X,Z)E(Y)(W(X,Z)))$$
$$= \frac{2}{(Z-Y)(Z-X)}(Z(Z-X)W(Y,Z)E(X)(W(Z))$$
$$- Y(Z-X)W(Y,Z)E(X)(W(Y))$$
$$- Z(Z-Y)W(X,Z)E(Y)(W(Z)) + X(Z-Y)W(X,Z)E(Y)(W(X)))$$
$$= \frac{2}{(Z-Y)(Z-X)}(ZW(Y,Z)W(X,Z)^2$$
$$- Y(Z-X)W(Y,Z)E(X)(W(Y))$$
$$- ZW(X,Z)W(Y,Z)^2 + X(Z-Y)W(X,Z)E(Y)(W(X))).$$

Using $W(X,Z) - W(Y,Z) = W(X,Y)$ and $W(Y,X) = -W(X,Y)$, this can be rewritten as

$$\frac{2}{(Z-Y)(Z-X)(Y-X)}(Z(Y-X)W(X,Z)W(Y,Z)W(X,Y)$$
$$-W(X,Y)^2(Y(Z-X)W(Y,Z) + X(Z-Y)W(X,Z)))$$
$$= \frac{2}{Y-X}W(X,Y)\frac{1}{(Z-Y)(Z-X)}(Z(Y-X)W(X,Z)W(Y,Z)$$
$$-W(X,Y)(Y(Z-X)W(Y,Z) + X(Z-Y)W(X,Z))).$$

Again using $W(X,Z) - W(Y,Z) = W(X,Y)$, we find

$$Z(Y-X)W(X,Z)W(Y,Z) - W(X,Y)(Y(Z-X)W(Y,Z) + X(Z-Y)W(X,Z))$$
$$= Y(Z-X)W(Y,Z)^2 - X(Z-Y)W(X,Z)^2,$$

and thus

$$E(X)(E(Y)(W(Z))) = \frac{2}{Y-X}W(X,Y)\frac{Y(Z-X)W(Y,Z)^2 - X(Z-Y)W(X,Z)^2}{(Z-Y)(Z-X)}$$
$$= \frac{2}{Y-X}W(X,Y)(YE(Y)(W(Z)) - XE(X)(W(Z)))$$
$$= \frac{2}{Y-X}F(X,Y)(E(X,Y)(W(Z))),$$

proving the identity $[E(X), E(Y)] = 2\frac{1}{Y-X}F(X,Y)E(X,Y)$, which by the proof of Theorem 4 is equivalent to the first defining relation of A. This finishes the proof. □

CONJECTURE 7. *The subalgebra generated by the operators e_i and f_i inside the ring of differential operators on B is isomorphic to A.*

4.3. Yangian. The category of regular representations of the Kronecker quiver and the category of torsion sheaves on the projective line being equivalent, one can expect an isomorphism of the associated cohomological Hall algebras. In the latter case, this is a special case of the class of algebras which will be described in [**14**]. The defining relations given there show that A can be viewed as a kind of Yangian algebra.

Acknowledgments

The authors would like to thank Ben Davison, Evgeny Feigin, Catharina Stroppel and Oksana Yakimova for helpful discussions on the results of this work, and Olivier Schiffmann for pointing out [**14**] and explaining the results obtained there.

References

[1] Zongbin Chen, *Geometric construction of generators of CoHA of doubled quiver* (English, with English and French summaries), C. R. Math. Acad. Sci. Paris **352** (2014), no. 12, 1039–1044, DOI 10.1016/j.crma.2014.09.025. MR3276816

[2] Ben Davison, *The critical CoHA of a quiver with potential*, Q. J. Math. **68** (2017), no. 2, 635–703, DOI 10.1093/qmath/haw053. MR3667216

[3] Ben Davison and Sven Meinhardt, *Cohomological Donaldson-Thomas theory of a quiver with potential and quantum enveloping algebras*, Invent. Math. **221** (2020), no. 3, 777–871, DOI 10.1007/s00222-020-00961-y. MR4132957

[4] Dan Edidin and William Graham, *Equivariant intersection theory*, Invent. Math. **131** (1998), no. 3, 595–634. MR1614555 (99j:14003a)

[5] Alexander I. Efimov, *Cohomological Hall algebra of a symmetric quiver*, Compos. Math. **148** (2012), no. 4, 1133–1146, DOI 10.1112/S0010437X12000152. MR2956038

[6] Hans Franzen and Markus Reineke, *Semistable Chow-Hall algebras of quivers and quantized Donaldson-Thomas invariants*, Algebra Number Theory **12** (2018), no. 5, 1001–1025, DOI 10.2140/ant.2018.12.1001. MR3840869

[7] Tamás Hausel, Emmanuel Letellier, and Fernando Rodriguez-Villegas, *Positivity for Kac polynomials and DT-invariants of quivers*, Ann. of Math. (2) **177** (2013), no. 3, 1147–1168, DOI 10.4007/annals.2013.177.3.8. MR3034296

[8] Mikhail Kapranov, Olivier Schiffmann, and Eric Vasserot, *The Hall algebra of a curve*, Selecta Math. (N.S.) **23** (2017), no. 1, 117–177, DOI 10.1007/s00029-016-0239-9. MR3595890

[9] Maxim Kontsevich and Yan Soibelman, *Cohomological Hall algebra, exponential Hodge structures and motivic Donaldson-Thomas invariants*, Commun. Number Theory Phys. **5** (2011), no. 2, 231–352, DOI 10.4310/CNTP.2011.v5.n2.a1. MR2851153

[10] Claus Michael Ringel, *Hall algebras and quantum groups*, Invent. Math. **101** (1990), no. 3, 583–591, DOI 10.1007/BF01231516. MR1062796

[11] Francesco Sala and Olivier Schiffmann, *Cohomological Hall algebra of Higgs sheaves on a curve*, Algebr. Geom. **7** (2020), no. 3, 346–376, DOI 10.14231/ag-2020-010. MR4087863

[12] Olivier Schiffmann and Eric Vasserot, *On cohomological Hall algebras of quivers: generators*, J. Reine Angew. Math. **760** (2020), 59–132, DOI 10.1515/crelle-2018-0004. MR4069884

[13] Olivier Schiffmann and Eric Vasserot, *On cohomological Hall algebras of quivers: Yangians*, 2017. Preprint. arXiv:1705.07491.

[14] Olivier Schiffmann and Eric Vasserot, *The cohomological Hall algebra of the stack of coherent sheaves on a smooth projective curve*, 2019. In preparation.

[15] Yaping Yang and Gufang Zhao, *The cohomological Hall algebra of a preprojective algebra*, Proc. Lond. Math. Soc. (3) **116** (2018), no. 5, 1029–1074, DOI 10.1112/plms.12111. MR3805051

Faculty of Mathematics, Ruhr-Universität Bochum, Universitätsstrasse 150, 44780 Bochum

Email address: `hans.franzen@rub.de`

Faculty of Mathematics, Ruhr-Universität Bochum, Universitätsstrasse 150, 44780 Bochum

Email address: `markus.reineke@rub.de`

Contemporary Mathematics
Volume **758**, 2020
https://doi.org/10.1090/conm/758/15237

The recollements of purity

X. H. Fu and I. Herzog

ABSTRACT. The theory of recollements of abelian categories is used to study purity of modules. The three equivalent definitions of purity for modules give rise to three recollements (covariant, projective quotient, and contravariant) of abelian functor categories over the module category R-Mod. The theory is motivated by 1) describing the rôle of the Snake Lemma in the theory of functor categories, and 2) a homological characterization of exactness properties of coherent functors. The classical theorems of Eilenberg and Watts are interpreted as statements about the left adjoint of the covariant recollement and the right adjoint of the contravariant recollement, and are extended in a unified way to include the middle adjoint of the projective quotient recollement.

1. Introduction

The notion of purity arises in the work of Prüfer [**16**] on abelian groups as a kind of relative divisiblity. A subgroup $A \subseteq B$ is said to be *pure* in B if for every $n \in \mathbb{Z}$,

$$nA = A \cap nB.$$

This generalizes to modules over a ring R by allowing for a matrix C over the ring R to replace n · a submodule $_RM \subseteq {}_RN$ is pure in $_RN$ if for every $m \times n$ matrix C with entries in R,

$$CM^n = M^m \cap CN^n.$$

In this general context, purity of modules plays an important rôle in the model theory of modules, and in the study of infinite-dimensional representations of algebras.

P.M. Cohn [**4**] found an equivalent condition that places purity in the context of relative homological algebra. A short exact sequence

$$(1) \qquad \Sigma : 0 \longrightarrow M \xrightarrow{\ i\ } N \xrightarrow{\ p\ } K \longrightarrow 0$$

in the category R-Mod of left R-modules is *pure exact* if the inclusion $i(M) \subseteq N$ is a pure submodule. In that case, we say that $i : M \to N$ is a pure monomorphism

2010 *Mathematics Subject Classification.* Primary 16D90, 18E99.
Key words and phrases. Recollement of abelian categories, covariant functor category, contravariant functor category, projective quotient category, canonical resolution/coresolution, tilting functor, cotilting functor.
The first author was supported in part by the National Natural Science Foundation of China, Grant No. 11671069; the second by NSF Grant DMS 12-01523.

and $p : N \to K$ a pure epimorphism. Cohn proved that a short exact sequence is pure exact if and only if it remains exact upon tensoring by a right R-module X_R :

$$X \otimes_R \Sigma : 0 \longrightarrow X \otimes_R M \xrightarrow{X \otimes i} X \otimes_R N \xrightarrow{X \otimes p} X \otimes_R K \longrightarrow 0.$$

For this to hold, it suffices that the sequence be exact for finitely presented X_R.

The equivalence of these two conditions is a curious thing. The tensor product $X \otimes_R -$ is right exact, so if we apply it to the short exact sequence (1), exactness at N and K is guaranteed, and pure exactness is therefore equivalent to the condition that $X \otimes_R -$ be exact at ${}_RM$. But observe that if C is an $m \times n$ matrix over R, then the functor $M \mapsto CM^n$ respects monomorphism and epimorphism, so that if it is applied to an arbitrary short exact sequence (1), then exactness at M and K is guaranteed and pure exactness is therefore equivalent to the condition that this functor be exact at ${}_RN$.

Well, there is a third equivalent condition [**21**]. The sequence (1) is pure exact if and only if it remains exact upon application of the functor $(W, -) := \mathrm{Hom}_R(W, -)$,

$$(W, \Sigma) : 0 \longrightarrow (W, M) \xrightarrow{(W,i)} (W, N) \xrightarrow{(W,p)} (W, K) \longrightarrow 0,$$

where ${}_RW$ is a finitely presented left R-module. The functor $(W, -)$ is left exact, so that exactness at M and N is guaranteed, and pure exactness is equivalent to the condition that the functor $(W, -)$ be exact at ${}_RK$.

Auslander's category [**2**] of coherent functors fp(R-mod, Ab) contains all three categories of functors that test for purity. The representable coherent functors $(W, -)$ are the projective objects of fp(R-mod, Ab). By the Yoneda embedding $\mathcal{Y} : W \mapsto (W, -)$, they form a subcategory equivalent to $(\text{R-mod})^{\text{op}}$. The functors of the form $X \otimes_R -$ are the injective objects of fp(R-mod, Ab); they form a subcategory equivalent to mod-R by the coYoneda embedding $\mathrm{c}\mathcal{Y} : N_R \mapsto N \otimes_R -$. The subcategory of functors $M \mapsto CM^n$, C an $m \times n$ matrix, is equivalent to the *projective quotient* category $\mathrm{PQ}(R^{\text{op}})$ of Crawley-Boevey and Sauter [**5**], defined as the homotopy category of epimorphisms $P \twoheadrightarrow X$, where X_R is finitely presented and $P_R \in$ proj-R is a finitely generated projective right R-module.

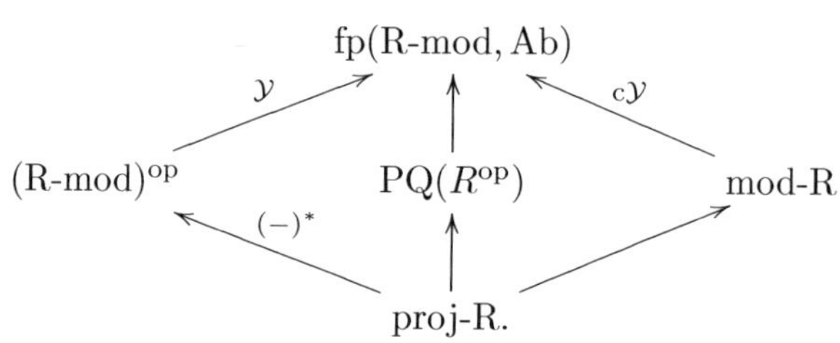

The most important coherent functor is the forgetful functor $({}_RR, -) \cong (R_R \twoheadrightarrow 0) \cong R \otimes_R -$, which belongs to all three subcategories of test functors. The smallest additive subcategory $\mathrm{add}(R \otimes_R -)$ containing the forgetful functor is equivalent to the category proj-R, via the functor $P_R \mapsto P \otimes_R -$. The functor category on $\mathrm{add}(R \otimes_R -)$ is therefore equivalent to the module category R-Mod $\cong$ (proj-R, Ab), which lies at the bottom of the diagram of restriction functors

$$\begin{array}{ccccc} & & (\mathrm{fp}(\text{R-mod},\mathrm{Ab}),\mathrm{Ab}) & & \\ & \swarrow^{(\mathcal{Y},\mathrm{Ab})} & \downarrow{\scriptstyle \mathrm{res}} & \searrow^{(c\mathcal{Y},\mathrm{Ab})} & \\ ((\text{R-mod})^{\mathrm{op}},\mathrm{Ab}) & & (\mathrm{PQ}(R^{\mathrm{op}}),\mathrm{Ab}) & & (\text{mod-R},\mathrm{Ab}) \\ & \searrow & \downarrow & \swarrow & \\ & & \text{R-Mod}. & & \end{array}$$

The important point here is (see [**17**, Example 3.12]) that such restriction functors are the localization functors of recollements of abelian categories [**8, 17**]. The recollements associated to the bottom three arrows of the diagram are the *recollements of purity.* A comprehensive treatment of the functor categories (mod-R, Ab) and [(R-mod)^{op}, Ab] is given in [**13, 14**] and their recollements are related in [**6**]; the appearence of the projective quotient category now compels a unified treatment of the three recollements of purity.

This unified treatment is based on an analysis of the functor R-Mod $\to$ (fp(R-mod, Ab), Ab), $M \mapsto \mathrm{Ev}(M)$, which sends a left R-module ${}_RM$ to the "evaluation at M" functor. In order to make this meaning precise, recall Freyd's observation [**1, 10**] that the category of coherent functors is the universal abelian category over the subcategory add$(R\otimes_R -) \cong$ proj-R. This implies that every left R-module, considered as a functor M : proj-R $\to$ Ab extends, in an essentially unique way, to an exact functor on fp(R-mod, Ab),

$$\begin{array}{ccc} \mathrm{fp}(\text{R-mod},\mathrm{Ab}) & & \\ \uparrow & \searrow^{\mathrm{Ev}(M)} & \\ \text{proj-R} & \xrightarrow{\quad M \quad} & \mathrm{Ab}. \end{array}$$

The value of $\mathrm{Ev}(M)(F)$ is given by the left Kan extension (along R-mod $\subseteq$ R-Mod) of F. For brevity, the notation $\mathrm{Ab}(R) = \mathrm{fp}(\text{R-mod}, \mathrm{Ab})$ will also be used. The functor R-Mod $\to$ (Ab(R), Ab)), $M \mapsto \mathrm{Ev}(M)$ is a fully faithful embedding of R-Mod onto the subcategory of exact functors on Ab(R).

Unlike Ab(R), the additive subcategories mod-R, PQ(R^{op}), and (R-mod)^{op} are not necessarily abelian; mod-R is only closed under cokernels (in Ab(R)); PQ(R^{op}) under images (in Ab(R)); and (R-mod)^{op} kernels (in Ab(R)). The celebrated theorems of Eilenberg and Watts [**7, 22**] assert that every right exact functor in (mod-R, Ab) is equivalent to one of the form $- \otimes_R M$, and that every left exact functor in ((R-mod)^{op}, Ab) is of the form $(-, M)$. In the context of the recollements of purity, these theorems are unified and extended to include the middle adjoint of the projective quotient recollement.

THEOREM 1. (Eilenberg-Watts) *For each of the categories* $\mathcal{A}=$mod-R, PQ(R^{op}), *and* (R-mod)^{op}, *the functor* R-Mod $\to (\mathcal{A}, \mathrm{Ab})$, $M \mapsto \mathrm{Ev}(M)|_{\mathcal{A}}$ *is a fully faithful embedding. Every structure preserving functor in* $(\mathcal{A}, \mathrm{Ab})$ *is equivalent to one of the form* $\mathrm{Ev}(M)|_{\mathcal{A}}$. *To wit,*

(1) *if* $\mathcal{A}$ = mod-R, *then the functor* $M \mapsto \mathrm{Ev}(M)|_{\text{mod-R}} = - \otimes_R M$ *is given by the left adjoint of the restriction* res : (mod-R, Ab) $\to$ R-Mod; *it is therefore right exact, i.e., preserves cokernels;*

(2) *if* $\mathcal{A} = \mathrm{PQ}(R^{\mathrm{op}})$, *then the functor* $M \mapsto \mathrm{Ev}(M)|_{\mathrm{PQ}(R^{\mathrm{op}})}$ *is given by the middle adjoint (intermediate extension functor) of the restriction* res : $(\mathrm{PQ}(R^{\mathrm{op}}), \mathrm{Ab}) \to$ R-Mod; *it therefore preserves monomorphism and epimorphisms, i.e., images; and*

(3) *if* $\mathcal{A} = (\text{R-mod})^{\mathrm{op}}$, *then the functor* $M \mapsto \mathrm{Ev}(M)|_{(\text{R-mod})^{\mathrm{op}}} = (-, M)$ *is given by the right adjoint of the restriction* res : $((\text{R-mod})^{\mathrm{op}}, \mathrm{Ab}) \to$ R-Mod; *it is therefore left exact, i.e., preserves kernels.*

We see from Theorem 1 that the evaluation bifunctor Ev: $\mathrm{Ab}(R) \times \text{R-Mod} \to \mathrm{Ab}$ restricts to the tensor bifunctor $- \otimes_R -$: mod-R $\times$ R-Mod $\to$ Ab, which is right exact in each variable, and the Hom bifunctor $\mathrm{Hom}_R(-,-) : (\text{R-mod})^{\mathrm{op}} \times \text{R-Mod} \to \mathrm{Ab}$, which is left exact in each variable, but the statement of the theorem signals the importance of the restriction to $\mathrm{PQ}(R^{\mathrm{op}}) \times$ R-Mod, which preserves images, and seems to have no classical precedent.

2. Exactness conditions

In this section, we describe how the Snake Lemma is used in the theory of functor categories. Specifically, we are interested in the usual fragments of the theory of exact functors on a skeletally small abelian category $\mathcal{A}$. Denote by $(\mathcal{A}, \mathrm{Ab})$ the category of additive functors $F : \mathcal{A} \to \mathrm{Ab}$ to the category of abelian groups, with morphisms the natural transformations between functors. Every natural transformation $\eta : F \to G$ in the functor category $(\mathcal{A}, \mathrm{Ab})$ has a kernel and a cokernel; $(\mathrm{Ker}\,\eta)(A) := \mathrm{Ker}\,\eta_A$ is the kernel of the A-component of the tranformation; the cokernel is defined dually. It is then readily verified that $(\mathcal{A}, \mathrm{Ab})$ is an abelian category and that a sequence of morphisms $F \xrightarrow{\eta} G \xrightarrow{\gamma} H$ in $(\mathcal{A}, \mathrm{Ab})$ is *exact* if every A-component sequence of abelian groups is. Similarly, the product and coproduct of a family of functors in $(\mathcal{A}, \mathrm{Ab})$ is defined componentwise. The Grothendieck structure of Ab may then be used to verify for $(\mathcal{A}, \mathrm{Ab})$ the axioms of a Grothendieck category [**20**].

2.1. Exactness conditions. An *exactness condition* on an additive functor $F : \mathcal{A} \to \mathrm{Ab}$ is obtained by choosing a subset of $\{1, 2, 3\}$ and insisting that F respect exactness of the short exact sequence

$$\Theta : 0 \longrightarrow A_1 \xrightarrow{f} A_2 \xrightarrow{g} A_3 \longrightarrow 0 \tag{2}$$

at those A_i. The Hasse diagram of the $2^3 = 8$ subcategories is displayed as

$$\begin{array}{ccccc} & & (\mathcal{A}, \mathrm{Ab}) & & \\ & \diagup & | & \diagdown & \\ \mathrm{Mono}(\mathcal{A}, \mathrm{Ab}) & & \mathrm{Hex}(\mathcal{A}, \mathrm{Ab}) & & \mathrm{Epi}(\mathcal{A}, \mathrm{Ab}) \\ | & \times & & \times & | \\ \mathrm{Lex}(\mathcal{A}, \mathrm{Ab}) & & \mathrm{Im}(\mathcal{A}, \mathrm{Ab}) & & \mathrm{Rex}(\mathcal{A}, \mathrm{Ab}) \\ & \diagdown & | & \diagup & \\ & & \mathrm{Ex}(\mathcal{A}, \mathrm{Ab}). & & \end{array}$$

The singleton subsets of $\{1, 2, 3\}$ correspond to the mono preserving, half exact, and epi preserving functors. Their complements correspond to the right exact, image preserving and left exact functors, respectively.

Recall that a subcategory of $\mathcal{W} \subseteq (\mathcal{A}, \mathrm{Ab})$ closed under extensions is a *torsion free class* if it is closed under products and subobjects. Dually, it *torsion* if it is closed under coproducts and quotient objects. It is called *torsion torsion free* (TTF) if it is both.

PROPOSITION 2. *The subcategories of* $(\mathcal{A}, \mathrm{Ab})$ *shown in the Hasse diagram above are all closed under extension, product and coproduct. The subcategory* $\mathrm{Mono}(\mathcal{A}, \mathrm{Ab}) \subseteq (\mathcal{A}, \mathrm{Ab})$ *is a torsion free class, while* $\mathrm{Epi}(\mathcal{A}, \mathrm{Ab})$ *is a torsion class.*

2.2. The snake lemma. Let $\eta : F \to G$ be a natural transformation in the functor category $(\mathcal{A}, \mathrm{Ab})$. We may complete it to an exact diagram of the form

$$\begin{array}{ccccccccc} 0 \longrightarrow \mathrm{Ker}\,\eta & \longrightarrow & F & & \xrightarrow{\quad\eta\quad} & & G & \longrightarrow & \mathrm{Coker}\,\eta \longrightarrow 0 \\ & & & \searrow & & \nearrow & & & \\ & & & & \mathrm{Im}\ \eta. & & & & \end{array}$$

Suppose that F is right exact and G left exact. Then the image of η is a quotient of an epi preserving functor, and therefore preserves epimorphism; dually, it preserves monomorphism. We conclude that $\mathrm{Im}\ \eta$ belongs to the subcategory $\mathrm{Im}(\mathcal{A}, \mathrm{Ab})$ of image preserving functors. If the top (exact) row of functors is applied to the short exact sequence Θ in (2), we find ourselves in the familiar setting of the celebrated Snake Lemma,

$$\begin{array}{ccccccccc} & & 0 & & 0 & & 0 & & \\ & & \downarrow & & \downarrow & & \downarrow & & \\ & & K(A_1) & \longrightarrow & K(A_2) & \xrightarrow{K(g)} & K(A_3) & \longrightarrow & 0 \\ & & \downarrow & & \downarrow & & \downarrow & & \\ & & F(A_1) & \longrightarrow & F(A_2) & \longrightarrow & F(A_3) & \longrightarrow & 0 \\ & & \downarrow{\scriptstyle \eta_{A_1}} & & \downarrow{\scriptstyle \eta_{A_2}} & & \downarrow{\scriptstyle \eta_{A_3}} & & \\ 0 & \longrightarrow & G(A_1) & \longrightarrow & G(A_2) & \longrightarrow & G(A_3) & & \\ & & \downarrow & & \downarrow & & \downarrow & & \\ 0 & \longrightarrow & C(A_1) & \xrightarrow{C(f)} & C(A_2) & \longrightarrow & C(A_3) & & \\ & & \downarrow & & \downarrow & & \downarrow & & \\ & & 0 & & 0 & & 0, & & \end{array}$$

where the columns and the middle two rows are exact; the kernel of η has been abbreviated to K and the cokernel to C. The Snake Lemma implies that K and C are half exact, and it provides a connecting morphism Δ that makes the sequence

$$\begin{array}{ccccccc} K(A_1) \xrightarrow{K(f)} K(A_2) \xrightarrow{K(g)} K(A_3) & & \xrightarrow{\qquad\Delta\qquad} & & C(A_1) \xrightarrow{C(f)} C(A_2) \xrightarrow{C(g)} C(A_3) \\ & \searrow & & \nearrow & \\ & & \mathrm{Im}\ \Delta & & \end{array}$$

exact. So if $\Delta = 0$, then $\mathrm{Ker}\,\eta$ is right exact and $\mathrm{Coker}\,\eta$ is left exact. Evidently, the image of Δ is a simultaneous measure of the failure of $\mathrm{Ker}\,\eta$ to be right exact and $\operatorname{Coker}\eta$ to be left exact. Let us observe that it also measures the failure of $\operatorname{Im}\ \eta$ to be exact.

THEOREM 3. (The Snake Lemma) *Let* $\eta : F \to G$ *be a natural transformation in* $(\mathcal{A}, \mathrm{Ab})$ *with* F *right exact and* G *left exact. Then* $\operatorname{Ker}\eta, \operatorname{Coker}\eta \in \operatorname{Hex}(\mathcal{A}, \mathrm{Ab})$ *and* $\operatorname{Im}\ \eta \in \operatorname{Im}(\mathcal{A}, \mathrm{Ab})$. *Furthermore, if* Σ *is a short exact sequence in* $\mathcal{A}$ *as in* (1), *there exist natural isomorphisms of homology:*

$$\operatorname{Im}\Delta \cong H_1((\operatorname{Coker}\eta)(\Theta)) \cong H_2((\operatorname{Im}\ \eta)(\Theta)) \cong H_3((\operatorname{Ker}\eta)(\Theta)).$$

PROOF. In our statement of the Snake Lemma, $H_1((\operatorname{Coker}\eta)(\Theta))$ refers to the (horizontal) homology at $C(A_1)$, while $H_3((\operatorname{Ker}\eta)(\Theta))$ refers to the homology at $K(A_3)$. We have already explained how the connecting map $\Delta : K(A_3) \to C(A_1)$ establishes a natural isomorphism $H_3(K(\Theta)) \cong \operatorname{Im}\ \Delta \cong H_1(C(\Theta))$. This isomorphism may be factored by vertically unfolding the diagram above to reveal the image of η. The lower half of the resulting diagram is given by

$$\begin{array}{ccccccccc}
 & & 0 & & 0 & & 0 & & \\
 & & \downarrow & & \downarrow & & \downarrow & & \\
0 & \longrightarrow & I(A_1) & \longrightarrow & I(A_2) & \longrightarrow & I(A_3) & \longrightarrow & 0 \\
 & & \downarrow & & \downarrow & & \downarrow & & \\
0 & \longrightarrow & G(A_1) & \longrightarrow & G(A_2) & \longrightarrow & G(A_3) & & \\
 & & \downarrow & & \downarrow & & \downarrow & & \\
0 & \longrightarrow & C(A_1) & \longrightarrow & C(A_2) & \longrightarrow & C(A_3) & & \\
 & & \downarrow & & \downarrow & & \downarrow & & \\
 & & 0 & & 0 & & 0, & &
\end{array}$$

where the columns and middle row are exact, and the image of η has been abbreviated to I. Because the columns are exact, we can view this as an exact sequence of complexes. The middle row is exact, so the long exact sequence of homology yields the natural isomorphism $H_1(C(\Theta)) \cong H_2(I(\Theta))$. $\square$

2.3. Morphisms between exact functors. The version of the Snake Lemma that we will use is the strengthened version when the natural transformation $\eta : F \to G$ is between exact functors, whose statement is given as follows.

COROLLARY 4. *Let* $\eta : F \to G$ *be a natural transformation in* $\operatorname{Ex}(\mathcal{A}, \mathrm{Ab})$. *Then* $\operatorname{Ker}\eta \in \operatorname{Lex}(\mathcal{A}, \mathrm{Ab})$, $\operatorname{Coker}\eta \in \operatorname{Rex}(\mathcal{A}, \mathrm{Ab})$, *and* $\operatorname{Im}\ \eta \in \operatorname{Im}(\mathcal{A}, \mathrm{Ab})$.

Notice the self-referential nature of Corollary 4, for it states that the kernel of a natural transformation between exact functors preserves kernels, the cokernel preserves cokernels and the image images. When the abelian category $\mathcal{A} = \operatorname{Ab}(R)$ is the free abelian category, then the exact functors are those isomorphic to one of the form $\operatorname{Ev}(M)$, $M \in R$-Mod, and the natural transformations $\eta : \operatorname{Ev}(M) \to \operatorname{Ev}(N)$ are those of the form $\operatorname{Ev}(f)$, where $f \colon {}_RM \to {}_RN$ is a morphism of R-modules.

3. The category of coherent functors

Auslander [**3**] introduced the category of coherent functors and proved that it is an abelian category with global homological dimension at most 2. In this section, we study coherent functors according to their projective dimension and characterize them in terms of 1) their relationship to the forgetful functor and 2) exactness conditions of the corresponding left Kan extension. As mentioned in the Introduction, Freyd [**10**] observed that the category of coherent functors is the free abelian category over the forgetful functor.

3.1. Definition. Let R-mod denote the category of finitely presented left R-modules. Yoneda's Lemma implies that the finitely generated projective objects of (R-mod, Ab) are those isomorphic to the representable ones $(U,-) := \mathrm{Hom}_R(U,-)$, where ${}_RU \in$ R-mod. The Yoneda functor $\mathcal{Y} : U \mapsto (U,-)$ is a fully faithful embedding $(\text{R-mod})^{\mathrm{op}} \to (\text{R-mod}, \mathrm{Ab})$ whose image is equivalent to the subcategory $\mathrm{proj}(\text{R-mod}, \mathrm{Ab}) \subseteq (\text{R-mod}, \mathrm{Ab})$ of finitely generated projective objects. A functor $F \in (\text{mod-R}, \mathrm{Ab})$ is *finitely presented* if there exists a presentation

$$(V,-) \xrightarrow{\ \eta\ } (U,-) \xrightarrow{\ \pi\ } F \longrightarrow 0 \tag{3}$$

by finitely generated projective objects. The subcategory of finitely presented functors is denoted by fp(R-mod, Ab) and is commonly known as the *category of coherent functors.*

3.2. Left Kan extensions. If F: R-mod $\to$ Ab is a coherent functor, the *left Kan extension* [**19**, Chapter 6] along the inclusion R-mod $\subseteq$ R-Mod is the functor $\tilde{F}$: R-Mod $\to$ Ab that restricts to F on R-mod and commutes with direct limits. Because every R-module is a direct limit of finitely presented modules, this determines $\tilde{F}$ up to isomorphism, and so we will abuse the language and also denote the left Kan extension by F. The reader will notice that any reference to a coherent functor is expressed in a form that extends to R-Mod in a way that commutes with direct limits. For example, the functors $\mathrm{Hom}_R(U,-)$: R-Mod $\to$ Ab, $U \in$ R-mod, commute with direct limits, and the presentation (3) defines a functor on R-Mod, which, by the exactness of direct limits, respect them. Similar considerations will hold for the functors $X \otimes_R -$ defined below, and the functors copresented by them.

The category R-mod of finitely presented modules is not necessarily abelian, so that it doesn't make sense, at first glance, to consider exactness properties of a coherent functor. But it does make sense to consider the exactness properties of the left Kan extension, so that we will denote by mono(R-mod, Ab) $\subseteq$ fp(R-mod, Ab) the subcategory of coherent functors with a left Kan extension that respects monomorphisms, by lex(R-mod, Ab) the subcategory of coherent functors whose left Kan extension is left exact, etc. This allows us to represent the relevant subcategories of $\mathrm{Ab}(R) = $ fp(R-mod, Ab), as in the previous section, by the Hasse diagram

$$\begin{array}{ccccc}
 & & \text{fp(R-mod, Ab)} & & \\
 & / & & \backslash & \\
\text{mono(R-mod, Ab)} & & & & \text{epi(R-mod, Ab)} \\
| & \backslash & & / & | \\
\text{lex(R-mod, Ab)} & & \text{im(R-mod, Ab)} & & \text{rex(R-mod, Ab)} \\
 & \backslash & | & / & \\
 & & \text{ex(R-mod, Ab).} & &
\end{array}$$

3.3. Projective dimension. Because the Yoneda functor is full, the natural transformation η in (3) is of the form $(f,-)$ where $f : {}_RU \to {}_RV$ is a morphism

R-mod. Its cokernel is finitely presented, so the exact sequence

$$U \xrightarrow{f} V \xrightarrow{g} W \longrightarrow 0$$

in R-mod, and the left exactness of the Yoneda functor, give rise to a projective resolution of F,

$$0 \longrightarrow (W,-) \xrightarrow{(g,-)} (V,-) \xrightarrow{(f,-)} (U,-) \longrightarrow F \longrightarrow 0$$

of length at most 2. It follows that the global projective dimension of the category fp(R-mod, Ab) is at most 2. If we denote by $\mathrm{proj}_n(\text{R-mod}, \mathrm{Ab}) \subseteq \mathrm{fp}(\text{R-mod}, \mathrm{Ab})$ the subcategory of coherent functors of projective dimension at most n, then $\mathrm{proj}_2(\text{R-mod}, \mathrm{Ab}) = \mathrm{fp}(\text{R-mod}, \mathrm{Ab})$ and $\mathrm{proj}_0(\text{R-mod}, \mathrm{Ab}) = \mathrm{proj}(\text{R-mod}, \mathrm{Ab})$.

In the following, we denote by $\mathrm{cogen}(R\otimes_R -) \subseteq \mathrm{fp}(\text{R-mod}, \mathrm{Ab})$ the subcategory of coherent functors F that admit a monomorphism $\mu\colon F \to (R^n, -) = (R, -)^n$ into a finite power of the forgetful functor.

PROPOSITION 5. $\mathrm{mono}(\text{R-mod}, \mathrm{Ab}) = \mathrm{proj}_1(\text{R-mod}, \mathrm{Ab}) = \mathrm{cogen}(R \otimes_R -)$.

PROOF. By [**15**, Cor. 10.2.16], every coherent functor F of projective dimension at most 1 admits an embedding into a finite power of the forgetful functor, so that $\mathrm{proj}_1(\text{R-mod}, \mathrm{Ab}) \subseteq \mathrm{cogen}(R\otimes_R -)$. A subfunctor of an exact functor certainly respects monomorphisms, so that $\mathrm{cogen}(R \otimes_R -) \subseteq \mathrm{mono}(\text{R-mod}, \mathrm{Ab})$. It remains to verify that $\mathrm{mono}(\text{R-mod}, \mathrm{Ab}) \subseteq \mathrm{proj}_1(\text{R-mod}, \mathrm{Ab})$, so let $F \in \mathrm{mono}(\text{R-mod}, \mathrm{Ab})$ and take a projective presentation (3), with $\eta = (f, -)$, $f : U \to V$. Now apply F to the monomorphism in the epi-mono factorization of $f = me : U \xrightarrow{e} I \xrightarrow{m} V$ to obtain the following commutative diagram with exact rows,

$$\begin{array}{ccccccc} (V,I) & \xrightarrow{(f,I)} & (U,I) & \xrightarrow{\pi_I} & F(I) & \longrightarrow & 0 \\ \downarrow & & \downarrow{\scriptstyle (U,m)} & & \downarrow{\scriptstyle F(m)} & & \\ (V,V) & \longrightarrow & (U,V) & \xrightarrow{\pi_V} & F(V) & \longrightarrow & 0. \end{array}$$

Then $F(m)\pi_I(e) = \pi_V(U,m)(e) = \pi_V(me) = \pi_V(f) = 0$. By hypothesis $F(m) : F(I) \to F(V)$ is a monomorphism, so that $\pi_I(e) = 0$ and therefore, there exists a $g : V \to I$ such that $e = gf = gme$. As e is an epimorphism, $gm = 1_I$ is a split monomorphism and I is a direct summand of V, which implies that I is finitely presented and the presentation of F given by $0 \longrightarrow (I,-) \xrightarrow{\eta} (U,-) \longrightarrow F \longrightarrow 0$ is a projective resolution. □

Let ${}_RU \in$ R-mod be a finitely presented module with presentation

$$(4) \qquad {}_RR^m \xrightarrow{-\cdot C} {}_RR^n \longrightarrow U \longrightarrow 0,$$

where C is an $m \times n$ matrix with entries in R. Apply the Yoneda functor to get a copresentation

$$0 \longrightarrow (U,-) \longrightarrow (R^n,-) \xrightarrow{C\cdot -} (R^m,-)$$

by finite powers of the forgetful functor of the corresponding representable functor. Denote by $\mathrm{cogen}_2(R \otimes_R -) \subseteq \mathrm{fp}(\text{R-mod}, \mathrm{Ab})$ the subcategory of coherent functors with a copresentation by finite powers of the forgetful functor.

PROPOSITION 6. $\text{lex}(\text{R-mod}, \text{Ab}) = \text{proj}(\text{R-mod}, \text{Ab}) = \text{cogen}_2(R \otimes_R -)$.

PROOF. The foregoing argument can be retraced, using Yoneda's Lemma, to show that every functor in $\text{cogen}_2(R \otimes_R -)$ is isomorphic to a representable functor $(U, -)$, $U \in$ R-mod, so that $\text{cogen}_2(R \otimes_R -) \subseteq \text{proj}(\text{R-mod}, \text{Ab})$. Every projective coherent functor is representable and so belongs to lex(R-mod, Ab), so it remains to verify that $\text{lex}(\text{R-mod}, \text{Ab}) \subseteq \text{cogen}_2(R \otimes_R -)$. But if F is left exact, then it belongs to $\text{mono}(\text{R-mod}, \text{Ab}) = \text{cogen}(R \otimes_R -)$, by Proposition 5, and admits a monomorphism $\mu \colon F \to (R^n, -)$ into a finite power of the forgetful functor. Now check that the cokernel of μ is itself a mono coherent functor and admits a monomorphism into some finite power of the forgetful functor. $\square$

3.4. Injective dimension. Consider now the finitely presented *right* R-module $X_R \in$ mod-R presented by the same $m \times n$ matrix C used in (4) to present U,

$$R^n_R \xrightarrow{C\cdot -} R^m_R \longrightarrow X_R \longrightarrow 0;$$

this is the *Auslander-Bridger transpose* of ${}_RU$, $X = \text{Tr}(U)$. The tensor product is right exact, so we obtain from this free presentation of X_R a presentation

$$R^n \otimes_R - \xrightarrow{C\cdot -} R^m \otimes_R - \longrightarrow X \otimes_R - \longrightarrow 0$$

of $X \otimes_R - \in (\text{R-mod}, \text{Ab})$ by finitely generated projective objects. It follows that $X \otimes_R -$ is a coherent functor.

Gruson and Jensen [**12**] and Auslander [**3**] independently proved the existence of a duality $D : (\text{fp}(\text{mod-R}, \text{Ab}))^{\text{op}} \to \text{fp}(\text{R-mod}, \text{Ab})$, defined by $F \mapsto DF \colon U \mapsto \text{Hom}_{(\text{mod-R},\text{Ab})}(F, - \otimes_R U)$, for $U \in$ R-mod, and whose left Kan extension is defined by the same expression, with $U \in$ R-Mod. This duality may be used to prove the following.

THEOREM 7. ([**3**,**12**]) *The category* fp(R-mod, Ab) *has enough injective objects. The* coYoneda functor $c\mathcal{Y} \colon \text{mod-R} \to \text{fp}(\text{R-mod}, \text{Ab})$, $X_R \mapsto X \otimes_R -$ *is a full and faithful embedding, and every injective coherent functor* $F \in \text{inj}(\text{R-mod}, \text{Ab}) \subseteq$ fp(R-mod, Ab) *is isomorphic to one in the image of the coYoneda functor.*[1] *Every coherent functor has injective demiension at most* 2.

Theorem 7 and the Gruson-Jensen-Auslander duality may be used to prove the statements dual to Propositions 5 and 6. The dual of Proposition 5 is

$$\text{epi}(\text{R-mod}, \text{Ab}) = \text{inj}_1(\text{R-mod}, \text{Ab}) = \text{gen}(R \otimes_R -),$$

where $\text{inj}_1(\text{R-mod}, \text{Ab}) \subseteq \text{fp}(\text{R-mod}, \text{Ab})$ is the subcategory of objects of injective dimension at most 1, and $\text{gen}(R \otimes_R -)$ is the subcategory of coherent functors F that admit an epimorphism $\eta : (R^n, -) \to F$. The dual of Proposition 6 is

$$\text{rex}(\text{R-mod}, \text{Ab}) = \text{inj}(\text{R-mod}, \text{Ab}) = \text{gen}_2(R \otimes_R -),$$

where $\text{gen}_2(R \otimes_R -) \subseteq \text{fp}(\text{R-mod}, \text{Ab})$ is the subcategory whose objects are the coherent functors that have a presentation by finite powers of the forgetful functor.

[1] These functors are not necessarily injective in (R-mod, Ab).

Propositions 5 and 6 and their duals allow us to relabel the Hasse diagram of relevant subcategories of fp(R-mod, Ab) in two different ways:

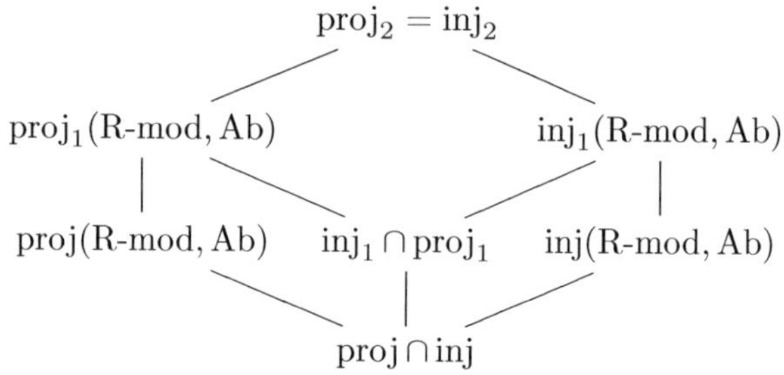

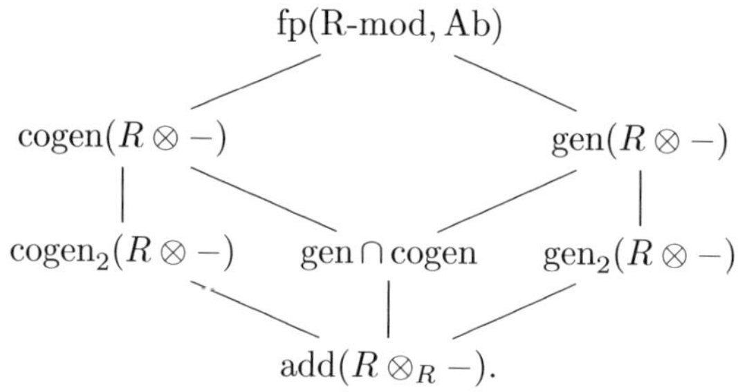

3.5. The biresolution of a matrix. Let C be the $m \times n$ matrix over R that presents (4) the finitely presented left R-module ${}_RU$ and the finitely presented right R-module $X_R = \mathrm{Tr}(U)$. We can identify the induced copresentation of $(U, -)$ and the presentation of $X \otimes_R -$ in fp(R-mod, Ab) to obtain an exact sequence
(5)

$$0 \longrightarrow (U, -) \longrightarrow R^n \otimes_R - \xrightarrow{\quad C\cdot - \quad} R^m \otimes_R - \longrightarrow X \otimes_R - \longrightarrow 0$$
$$R^n \otimes_R - \searrow \ \Omega(X \otimes -) \nearrow R^m \otimes_R -,$$

which serves, at the same time, as an injective coresolution of $(U, -)$ in fp(mod-R, Ab), and a projective resolution of $X \otimes_R -$ in (mod-R, Ab) (and fp(mod-R, Ab)). The exact sequence is obtained from its centerpiece, a morphism between finite powers of the forgetful functor, adorned by a kernel, image, and cokernel. The kernel $(U, -)$ belongs to the subcategory proj(R-mod, Ab) and every object of proj(R-mod, Ab) is isomorphic to one of the form $(U, -)$; dually, the cokernel $X \otimes_R -$ belongs to inj(R-mod, Ab) and, by Theorem 7, every object of inj(R-mod, Ab) is isomorphic to one of this form; but notice also how the syzygy $\Omega(X \otimes_R -) = \mathrm{Im}(C \cdot -)$ belongs to the subcategory $\mathrm{gen}(R \otimes_R -) \cap \mathrm{cogen}(R \otimes_R -)$ located in the middle of the Hasse diagram, and how every object in this category is also isomorphic to one of this form. Such a functor $\Omega(X \otimes_R -)$ is isomorphic to the functor $M \mapsto CM^n$, where C is a presentation matrix for X_R. These are the functors used in the original definition of purity.

3.6. The projective quotient category. A *projective quotient morphism* in mod-R is an epimorphism $P \twoheadrightarrow X_R$ in mod-R with $P_R \in$ proj-R a finitely generated projective right R-module. The *projective quotient category* $\mathrm{PQ}(R^{\mathrm{op}})$, introduced by Crawley-Boevey and Sauter [**5**], is the homotopy category of projective quotient morphisms. Let P'_R be a finitely generated projective such that $P \oplus P' = R^n_R$ is free of finite rank; then the given projective quotient morphism is isomorphic in $\mathrm{PQ}(R^{\mathrm{op}})$ to one with a free domain, $P \oplus P' \twoheadrightarrow X \oplus P'$. The category $\mathrm{PQ}(R)$ is defined dually as the homotopy category projective quotient morphisms in R-mod.

THEOREM 8. *The functor* $\mathrm{Ker}\,\mathrm{c}\mathcal{Y} : \mathrm{PQ}(R^{\mathrm{op}}) \to \mathrm{fp}(\text{R-mod}, \mathrm{Ab})$, *given by*

$$\mathrm{Ker}\,\mathrm{c}\mathcal{Y} \colon (R^n \twoheadrightarrow X_R) \mapsto \mathrm{Ker}\,(R^n \otimes_R - \longrightarrow\!\!\!\!\!\rightarrow X \otimes_R -) = \Omega(X \otimes_R -),$$

is a fully faithful embedding, whose image is dense in $\mathrm{gen}(R \otimes_R -) \cap \mathrm{cogen}(R \otimes_R -)$. *Dually, the functor* $\mathrm{Coker}\,\mathcal{Y} \colon \mathrm{PQ}(R)^{\mathrm{op}} \to \mathrm{gen}(R \otimes_R -) \cap \mathrm{cogen}(R \otimes_R -)$, *given by the rule*

$$\mathrm{Coker}\,\mathcal{Y} \colon (R^n \twoheadrightarrow {}_R U) \mapsto \mathrm{Coker}\,((U, -) \rightarrowtail (R^n, -)),$$

is also an equivalence of categories.

PROOF. The functor $\mathrm{Ker}\,\mathrm{c}\mathcal{Y}$ assigns to a projective quotient morphism $R^n \twoheadrightarrow X_R$ the object F in $\mathrm{fp}(\text{R-mod}, \mathrm{Ab})$ with injective coresolution

$$0 \longrightarrow F \longrightarrow R^n \otimes_R - \longrightarrow X_R \otimes_R - \longrightarrow 0.$$

This is well-defined, because homotopic injective coresolutions coresolve isomorphic objects. The equivalence inverse of $\mathrm{Ker}\,\mathrm{c}\mathcal{Y}$ is also well-defined, because every object in $\mathrm{cogen}(R \otimes_R -) \cap \mathrm{inj}_1(\text{R-mod}, \mathrm{Ab})$ has an injective coresolution induced by a projective quotient morphism, unique up to homotopy equivalence. The proof of the second uses a dual argument. $\square$

In view of Theorem 8 we shall identify the projective quotient category with the subcategory of coherent functors that have projective and injective dimension at most 1 or, equivalently, which are generated/cogenerated by the forgetful functor,

$$\mathrm{PQ}(R^{\mathrm{op}}) = \mathrm{gen}(R \otimes_R -) \cap \mathrm{cogen}(R \otimes_R -) = \mathrm{inj}_1(\text{R-mod}, \mathrm{Ab}) \cap \mathrm{proj}_1(\text{R-mod}, \mathrm{Ab})$$

Under this identification, the forgetful functor is identified with the projective quotient morphism $R \twoheadrightarrow 0$.

Just as the image $\mathcal{Y}(\text{R-mod}) \subseteq \mathrm{fp}(\text{R-mod}, \mathrm{Ab})$ of the Yoneda functor is closed under kernels, and the image $\mathrm{c}\mathcal{Y}(\text{mod-R}) \subseteq \mathrm{fp}(\text{R-mod}, \mathrm{Ab})$ is closed under cokernels, let us observe how the projective quotient category $\mathrm{PQ}(R^{\mathrm{op}}) \subseteq \mathrm{fp}(\text{R-mod}, \mathrm{Ab})$ is closed under images. Indeed, suppose that $\eta : \Omega(X \otimes_R -) \to \Omega(Y \otimes_R -)$ is a morphism in $\mathrm{fp}(\text{R-mod}, \mathrm{Ab})$. Both of these functors belong to $\mathrm{gen}(R \otimes_R -) \cap \mathrm{cogen}(R \otimes_R -)$, so are part of a commutative diagram

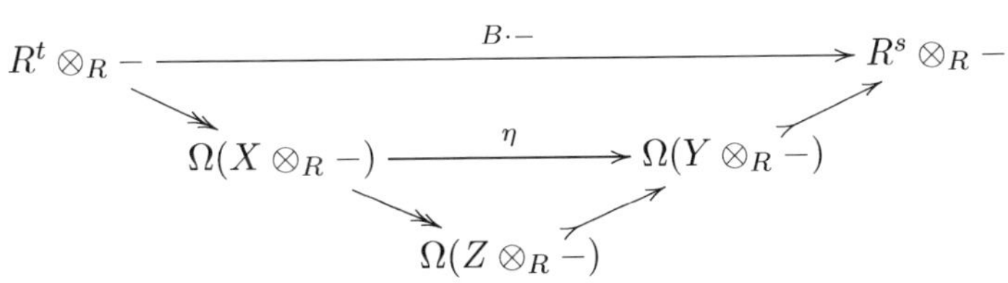

that shows the image of η to be $\Omega(Z \otimes_R -)$, where $Z_R \in$ mod-R is the module presented by the $s \times t$ matrix B.

4. Recollements of abelian categories

4.1. Definition. A *recollement* $\mathcal{R}_{\text{ab}}(\mathcal{W}, \mathcal{X}, \mathcal{Z})$ of abelian categories [**8, 18**] consists of three abelian categories together with the structural additive functors displayed in

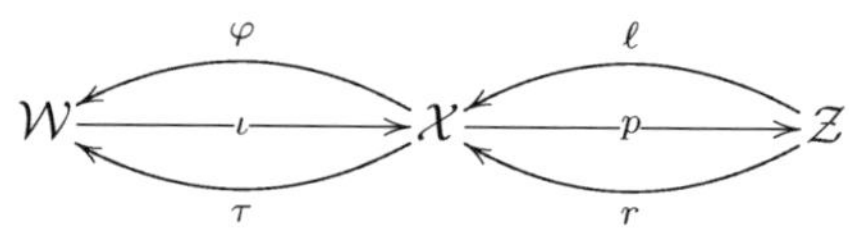

and satisfying the following properties:

(1) (φ, ι, τ) and (ℓ, p, r) are adjoint triples;
(2) the functors ι, ℓ and r are fully faithful; and
(3) the functor $\iota : \mathcal{W} \to \mathcal{X}$ is a full and faithful embedding onto the subcategory of $\mathcal{X}$ of those functors $X \in \mathcal{X}$ for which $p(X) = 0$.

We will assume that $\mathcal{X}$ has small products and coproducts. For an abelian category, this is equivalent to the existence of limits and colimits, and in the situation of a recollement, it implies that $\mathcal{W}$ and $\mathcal{Z}$ also have small products and coproducts.

Condition (3) allows us to identify $\mathcal{W}$ along ι with a (full) subcategory of $\mathcal{X}$, closed under isomorphism, and the condition that (φ, ι, τ) is an adjoint triple implies that $\mathcal{W}$ is part of a torsion triple $(\mathcal{T}, \mathcal{W}, \mathcal{F})$ in $\mathcal{X}$, where $\tau : \mathcal{X} \to \mathcal{W}$ is the torsion radical of the torsion pair $(\mathcal{W}, \mathcal{F})$ and $\varphi : \mathcal{X} \to \mathcal{W}$ is the torsion free quotient functor associated to the torsion pair $(\mathcal{T}, \mathcal{W})$.

Because $\mathcal{W}$ is a TTF class, it is a localizing subcategory of $\mathcal{X}$, in the sense of Gabriel, and the functor $p : \mathcal{X} \to \mathcal{Z} \cong \mathcal{X}/\mathcal{W}$ is the localization of $\mathcal{X}$ with respect to $\mathcal{Z}$. In a recollement situation, the codomain $\mathcal{X}/\mathcal{W}$ of the localization is defined to be the category whose objects $X_{\mathcal{W}} := X \in \mathcal{X}$ are the same as those of $\mathcal{X}$, but with morphisms given by

$$\operatorname{Hom}_{\mathcal{X}/\mathcal{W}}(X_{\mathcal{W}}, Y_{\mathcal{W}}) := \operatorname{Hom}_{\mathcal{X}}(t(X), Y/\tau(Y)),$$

where $t : \mathcal{X} \to \mathcal{T}$ is the torsion radical associated to the torsion pair $(\mathcal{T}, \mathcal{W})$.

The left and right adjoints of the localization functor $p : \mathcal{X} \to \mathcal{Z}$, denoted by $\ell : \mathcal{Z} \to \mathcal{X}$ and $r : \mathcal{Z} \to \mathcal{X}$, respectively, are a part of the given recollement structure. The *norm* of the recollement is the natural transformation $\mathcal{N} : \ell \to r$ whose Z-component $\mathcal{N}_Z : \ell(Z) \to r(Z)$, $Z \in \mathcal{Z}$, corresponds to the identity 1_Z under the canonical isomorphism $\operatorname{Hom}_{\mathcal{X}}(\ell(Z), r(Z)) \cong \operatorname{Hom}_{\mathcal{Z}}(Z, pr(Z)) = \operatorname{End}_{\mathcal{Z}} Z$. The image of the norm is commonly known as the *intermediate extension* functor, but we prefer the more suggestive terminology of a *middle adjoint*, which we denote by m,

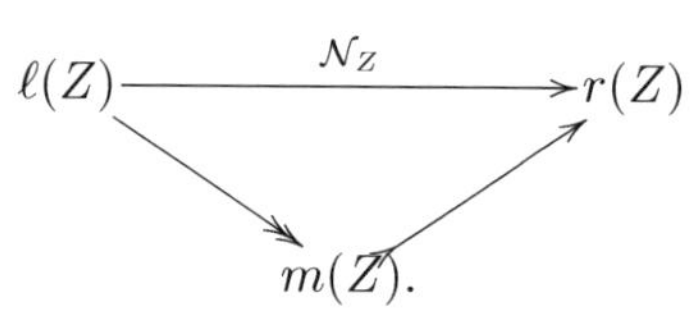

The left adjoint is right exact, the right adjoint is left exact, and the middle adjoint preserves monomorphisms and epimorphisms, but may fail to preserve exactness at the *middle* term of a short exact sequence. The images of the three adjoints form

subcategories of $\mathcal{X}$ ordered by inclusion as displayed in yet another Hasse diagram

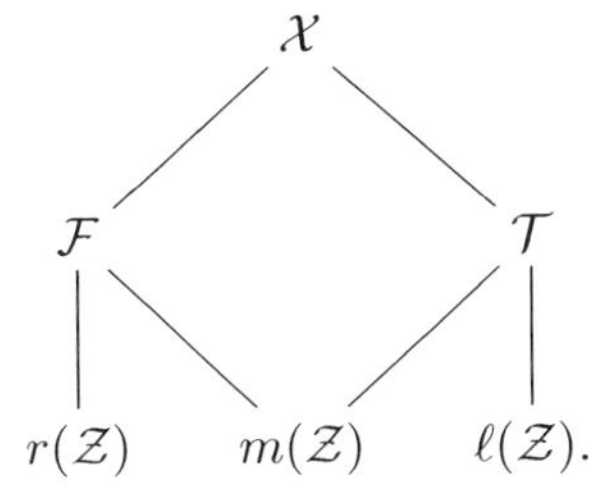

4.2. Gabriel recollements. In the latter part of Gabriel's thesis [**11**] the condition that $(\mathcal{W}, \mathcal{F})$ is cogenerated by an injective object is used to construct a right adjoint to the localization functor. Many classical examples of a recollement of abelian categories satisfy the dual condition that the torsion pair $(\mathcal{T}, \mathcal{W})$ is generated by a projective object.

DEFINITION 9. A recollement $\mathcal{R}_{\mathrm{ab}}(\mathcal{W}, \mathcal{X}, \mathcal{Z})$ is a *Gabriel* recollement if $\mathcal{Z}$ has a progenerator and an injective cogenerator.

THEOREM 10. *Let $\mathcal{R}_{\mathrm{ab}}(\mathcal{W}, \mathcal{X}, \mathcal{Z})$ be a Gabriel recollement with $P \in \mathcal{Z}$ a progenerator and $E \in \mathcal{Z}$ an injective cogenerator. The torsion pair $(\mathcal{T}, \mathcal{W})$ is generated by the projective object $\ell(P) \in \mathcal{X}$; the pair $(\mathcal{W}, \mathcal{F})$ is cogenerated by the injective object $r(E)$. The Hasse diagram above can be more explicitly relabeled as*

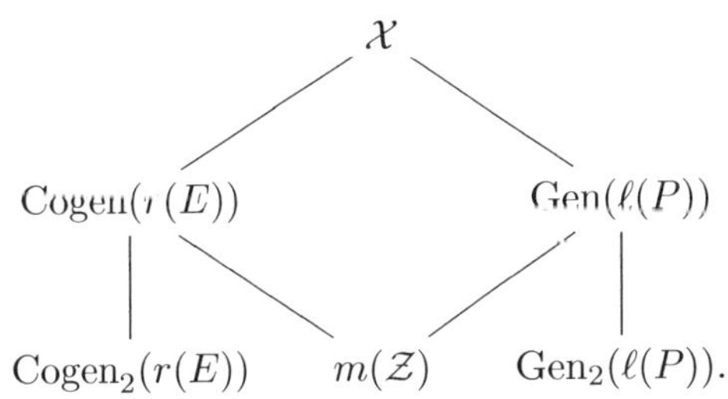

In fact, $\mathrm{Cogen}_2(r(E)) = \mathcal{F} \cap \mathcal{W}^{\perp_1}$ *and* $\mathrm{Gen}_2(\ell(P)) = \mathcal{T} \cap {}^{\perp_1}\mathcal{W}$.

PROOF. The left adjoint of an exact functor is right exact, and preserves coproducts and projective objects, so that if $P_1 \xrightarrow{f} P_0 \longrightarrow Z \longrightarrow 0$ is a projective presentation of an object in $\mathcal{Z}$, then the value of $\ell(Z)$ is determined, up to natural isomorphism, by the projective presentation $\ell(P_1) \xrightarrow{\ell(f)} \ell(P_0) \longrightarrow \ell(Z) \longrightarrow 0$ of $\ell(Z) \in \mathcal{X}$. The dual properties of the right adjoint $r : \mathcal{Z} \to \mathcal{X}$ show that an injective copresentation of $Z \in \mathcal{Z}$ lifts to one of $r(Z) \in \mathcal{X}$. The details of the proof follow from these key observations. □

Let $Z \in \mathcal{Z}$ be given by a projective presentation as in the proof of Theorem 10 and an injective copresentation $0 \longrightarrow Z \longrightarrow E_0 \xrightarrow{g} E_1$. Applying the left and right adjoints of the localization functor yields a diagram in $\mathcal{X}$ with exact rows

$$
\begin{array}{ccccccc}
\ell(P_1) & \xrightarrow{\ell(f)} & \ell(P_0) & \longrightarrow & \ell(Z) & \longrightarrow & 0 \\
 & & & & \Big\downarrow{\scriptstyle \mathcal{N}_Z} & & \\
 & & 0 & \longrightarrow & r(Z) & \longrightarrow & r(E_0) \xrightarrow{r(g)} r(E_1),
\end{array}
\tag{6}
$$

where the norm is part of an exact sequence

$$
0 \longrightarrow \tau(\ell(Z)) \longrightarrow \ell(Z) \xrightarrow{\mathcal{N}_Z} r(Z) \longrightarrow \varphi(r(Z)) \longrightarrow 0, \qquad \ell(Z) \searrow m(Z) \nearrow r(Z).
$$

4.3. Recollements of functor categories. Let $\mathcal{B} \subseteq \mathcal{A}$ be an inclusion of skeletally small additive subcategories. Given $A, B \in \mathcal{A}$, denote by $\mathcal{B}(A, B) \subseteq \mathrm{Hom}_{\mathcal{A}}(A, B)$ the subgroup of morphisms that factor through an object in $\mathcal{B}$. The quotient category $\pi : \mathcal{A} \to \mathcal{A}/\mathcal{B}$ is the category whose objects $\underline{A} := A \in \mathcal{A}$ are the same as those of $\mathcal{A}$, but whose morphisms are those of $\mathcal{A}$ modulo the ideal of morphisms that factor through an object in $\mathcal{B}$, $\mathrm{Hom}_{\mathcal{A}/\mathcal{B}}(\underline{A}, \underline{B}) := (A, B)/\mathcal{B}(A, B)$. This state of affairs is represented by a sequence of functors

$$
\mathcal{B} \xrightarrow{\ \subseteq\ } \mathcal{A} \xrightarrow{\ \pi\ } \mathcal{A}/\mathcal{B},
\tag{7}
$$

where $\pi : \mathcal{A} \to \mathcal{A}/\mathcal{B}$, $A \mapsto \underline{A}$, is a full and dense functor and $\mathcal{B} \subseteq \mathcal{A}$ is a fully faithful embedding onto the subcategory of objects for which $\pi(A) = 0$. This quotient functor is the analogue of the localization functor in the setting of additive categories.

Precomposition by the functors in the sequence (7) yields exact functors between the corresponding functor categories,

$$
(\mathcal{A}/\mathcal{B}, \mathrm{Ab}) \xrightarrow{\ \pi^*\ } (\mathcal{A}, \mathrm{Ab}) \xrightarrow{\ \mathrm{res}\ } (\mathcal{B}, \mathrm{Ab}).
$$

The functor $\mathrm{res} : (\mathcal{A}, \mathrm{Ab}) \to (\mathcal{B}, \mathrm{Ab})$, $F \mapsto F|_{\mathcal{B}}$, is given by *restriction to* $\mathcal{B}$, and the functor $\pi^* : (\mathcal{A}/\mathcal{B}, \mathrm{Ab}) \to (\mathcal{A}, \mathrm{Ab})$ is a full and faithful embedding of the quotient functor category onto the TTF class of those functors in $(\mathcal{A}, \mathrm{Ab})$ that vanish on $\mathcal{B}$.

PROPOSITION 11. *A Gabriel recollement of functor categories is given by*

$$
(\mathcal{A}/\mathcal{B}, \mathrm{Ab}) \underset{\tau}{\overset{\varphi}{\leftleftarrows}}\!\!\xrightarrow{\pi^*} (\mathcal{A}, \mathrm{Ab}) \underset{r}{\overset{\ell}{\leftleftarrows}}\!\!\xrightarrow{\mathrm{res}} (\mathcal{B}, \mathrm{Ab}),
$$

where the left adjoint of the restriction functor res *is defined for* $B \in \mathcal{B}$ *by* $\ell(B, -) = \mathrm{Hom}_{\mathcal{A}}(B, -)$.

PROOF. A progenerator of $(\mathcal{B}, \mathrm{Ab})$ is given by $\amalg_{B \in \mathcal{B}} (B, -)$ and an injective cogenerator exists in $(\mathcal{B}, \mathrm{Ab})$ because it is a Grothendieck category.[2] □

[2]Indeed, the functor category $(\mathcal{B}, \mathrm{Ab})$ is locally finitely generated, so that every nonzero object has a finitely generated subobject, and therefore a simple subquotient. It follows that the injective envelopes of simple objects form a cogenerating set, whose product is the minimal injective cogenerator.

4.4. The free abelian recollement. This is the recollement of functor categories that arises from the inclusion $\mathrm{add}(R\otimes_R -) \subseteq \mathrm{Ab}(R)$. Recall that $\mathrm{add}(R\otimes_R -) \cong$ R–Proj and that $(\text{R–Proj}, \mathrm{Ab}) \cong$ R-Mod, so that the localization functor of this recollement is given by evaluation at the forgetful functor $\mathrm{Ev}(R\otimes_R -) : (\mathrm{Ab}(R), \mathrm{Ab}) \to$ R-Mod. This recollement ca1n therefore be expressed as

$$\mathcal{R}_{\mathrm{ab}}(\mathcal{W}, (\mathrm{Ab}(R), \mathrm{Ab}), \text{R-Mod}),$$

where $\mathcal{W} \subseteq (\mathrm{Ab}(R), \mathrm{Ab})$ is the TTF class of functors that vanish on the forgetful functor.

By the universal property of the free abelian category, the map $\mathrm{Ev}(-) : \text{R-Mod} \to \mathrm{Ex}(\mathrm{Ab}(R), \mathrm{Ab})$, $M \mapsto \mathrm{Ev}(M)$, is a full and faithful section of the localization. If this section is applied to the short exact sequence Σ in (1), then the sequence

$$\mathrm{Ev}(\Sigma) : 0 \longrightarrow \mathrm{Ev}(M) \xrightarrow{\mathrm{Ev}(i)} \mathrm{Ev}(N) \xrightarrow{\mathrm{Ev}(p)} \mathrm{Ev}(K) \longrightarrow 0$$

in $(\mathrm{Ab}(R), \mathrm{Ab})$ is exact if and only if Σ is pure exact. Apart from that the functor $M \mapsto \mathrm{Ev}(M)$ does not satisfy any of the exactness conditions of §2. For this reason, it cannot be one of the adjoints of the localization functor. It will, however, satisfy the corresponding exactness conditions when it is restricted to one of the coherent subcategories of the Hasse diagrams in §3.

PROPOSITION 12. *In the free abelian recollement* $\mathcal{R}_{\mathrm{ab}}(\mathcal{W}, (\mathrm{Ab}(R), \mathrm{Ab}), \text{R-Mod})$

(1) *if* $P \in$ R-Mod *is projective, then* $\ell(P)$ *is naturally isomorphic to* $\mathrm{Ev}(P)$, *and*
(2) *if* $E \in$ R-Mod *is injective, then* $r(E)$ *is naturally isomorphic to* $\mathrm{Ev}(E)$.

PROOF. (1) Let $P \in \text{R-Mod} \cong (\text{proj-R}, \mathrm{Ab})$ be a finitely generated projective; it corresponds under the equivalence to the functor $- \otimes_R P = (P^*, -)$. The localization of the free abelian recollement is a composition of localization functors

$$(\mathrm{Ab}(R), \mathrm{Ab}) \xrightarrow{(\mathcal{Y}, \mathrm{Ab})} ((\text{R-mod})^{\mathrm{op}}, \mathrm{Ab}) \xrightarrow{((-)^*, \mathrm{Ab})} (\text{proj-R}, \mathrm{Ab}),$$

so its left adjoint is the composition of the respective left adjoints. By Proposition 11, these are given by $\ell : (P^*, -) \mapsto (-, P) \mapsto ((-, P), ?)$, which is equivalent to $\mathrm{Ev}(P)$, by Yoneda's Lemma. Thus $\ell(P) \cong \mathrm{Ev}(P)$ for all finitely generated projective modules in R-Mod. As both ℓ and Ev commute with coproducts, the equivalence holds for all projective modules in R-Mod.

(2) To prove that $r(E) \cong \mathrm{Ev}(E)$ in $(\mathrm{Ab}(R), \mathrm{Ab})$, it suffices to verify that $r(E)$ is exact. But if $F \in \mathrm{Ab}(R)$ is a coherent functor, then

$$r(E)(F) \cong \mathrm{Hom}_{(\mathrm{Ab}(R), \mathrm{Ab})}((F, -), r(E)) \cong \mathrm{Hom}_R((F, R\otimes_R -), E),$$

by Yoneda's Lemma and the adjunction. As $R\otimes_R -$ is injective in $\mathrm{Ab}(R)$ and E is injective in R-Mod, this is a composition of exact functors. □

So if ${}_RM$ is a module, with projective presentation $P_1 \xrightarrow{f} P_0 \longrightarrow M \longrightarrow 0$ and injective copresentation $0 \longrightarrow M \longrightarrow E_0 \xrightarrow{g} E_1$, Proposition 12 implies that the diagram (6) takes the form

$$
(8)\qquad
\begin{array}{ccccccccc}
\mathrm{Ev}(P_1) & \xrightarrow{\mathrm{Ev}(f)} & \mathrm{Ev}(P_0) & \longrightarrow & \ell(M) & \longrightarrow & 0 & & \\
 & & & \searrow & \downarrow & & & & \\
 & & & & m(M) & & & & \\
 & & & & \downarrow & \searrow & & & \\
 & & 0 & \longrightarrow & r(M) & \longrightarrow & \mathrm{Ev}(E_0) & \xrightarrow{\mathrm{Ev}(g)} & \mathrm{Ev}(E_1),
\end{array}
$$

which is a commutative diagram in $(\mathrm{Ab}(R), \mathrm{Ab})$, with exact rows. In the next section, we study how this diagram restricts to the various test subcategories of purity.

5. The recollements of purity: Proof of Theorem 1

This section is devoted to the proof the three statements of Theorem 1 from the Introduction. The recollements of purity are the recollements of functor categories that arise from the three embeddings proj-R $\subseteq$ mod-R, proj-R $\to \mathrm{PQ}(R^{\mathrm{op}})$, and $(-)^* :$ proj-R $\to$ $(\text{R-mod})^{\mathrm{op}}$. All three embeddings take place within the ambient free abelian category $\mathrm{Ab}(R) = \mathrm{fp}(\text{R-mod}, \mathrm{Ab})$, where the base category proj-R is replaced by the subcategory $\mathrm{add}(R \otimes_R -)$ along the equivalence $P_R \mapsto P \otimes_R - \cong (P^*, -)$. The extension categories are then given by $\mathrm{inj}(\text{R-mod}, \mathrm{Ab})$, $\mathrm{PQ}(R^{\mathrm{op}})$, and $\mathrm{proj}(\text{R-mod}, \mathrm{Ab})$, respectively. When the context is clear, we will omit the annoying restriction notation in $\mathrm{Ev}(M)|_{\mathcal{A}}$, and just write $\mathrm{Ev}(M)$.

5.1. The covariant recollement. The *covariant recollement* is the recollement of functor categories associated to the inclusion proj-R $\subseteq$ mod-R. The localization functor of this recollement is the restriction functor $(\text{mod-R}, \mathrm{Ab}) \to (\text{proj-R}, \mathrm{Ab}) = $ R-Mod, so that the TTF class in $(\text{mod-R}, \mathrm{Ab})$ is the subcategory of those functors that vanish on proj-R. The quotient category mod-R/proj-R $=$ $\underline{\text{mod}}$-R is the *stable category* of finitely presented left R-modules. The covariant recollement is therefore denoted as $\mathcal{R}_{\mathrm{ab}}((\underline{\text{mod}}\text{-R}, \mathrm{Ab}), (\text{mod-R}, \mathrm{Ab}), \text{R-Mod})$.

If $_RM$ is a left R-module, the "evaluation at M" functor $\mathrm{Ev}(M) : \mathrm{fp}(\text{R-mod}, \mathrm{Ab}) \to \mathrm{Ab}$ restricts to mod-R along the coYoneda embedding, so we get that if $X_R \in$ mod-R, then

$$\mathrm{Ev}(M)(X_R) = \mathrm{Ev}(M)(X \otimes_R -) = X \otimes_R M,$$

from which it follows that $\mathrm{Ev}(M) = - \otimes_R M$ on mod-R. By the Eilenberg-Watts Theorem, every right exact functor $G \in (\text{mod-R}, \mathrm{Ab})$ is isomorphic to $\mathrm{Ev}(M)$, where $_RM = G(R_R) = \mathrm{Ev}(R)(G)$ is given by applying the localization functor of the recollement to G.

Thus the functor $\mathrm{Ev} :$ R-Mod $\to (\text{mod-R}, \mathrm{Ab})$, $M \mapsto \mathrm{Ev}(M)$, is right exact and commutes with coproducts, just like the left adjoint $\ell :$ R-Mod $\to (\text{mod-R}, \mathrm{Ab})$ of the localization functor in the recollement. Furthermore, $\ell(P) = \mathrm{Ev}(P)$, for every projective left R-module $_RP$; this is seen by applying the left adjoint of the composition $(\mathrm{Ab}(R), \mathrm{Ab}) \to (\text{mod-R}, \mathrm{Ab}) \to$ R-Mod, which is $\mathrm{Ev}(P)$, by Proposition 12, and restricting to $(\text{mod-R}, \mathrm{Ab})$. Another application of the Eilenberg-Watts Theorem shows that Ev is isomorphic to the left adjoint.

5.2. The contravariant recollement. The *contravariant recollement* is the recollement of functor categories associated to the inclusion proj-R $\cong$ (R-proj)^{op} $\subseteq$ (R-mod)^{op}, so that the localization functor is the restriction functor ((R-mod)^{op}, Ab) $\to$ ((R-proj)^{op}, Ab), and the TTF class is the "stable" subcategory ((R-$\underline{\text{mod}}$)^{op}, Ab) $\subseteq$ ((R-mod)^{op}, Ab) of functors that vanish on R-proj. The contravariant recollement is denoted as $\mathcal{R}_{\mathrm{ab}}(((\text{R-}\underline{\text{mod}})^{\mathrm{op}}, \mathrm{Ab}), ((\text{R-mod})^{\mathrm{op}}, \mathrm{Ab}), \text{R-Mod})$.

The "evaluation at M" functor $\mathrm{Ev}(M)$: fp(R-mod, Ab) $\to$ Ab restricts to (R-mod)^{op} along the Yoneda embedding: if ${}_RU \in$ R-mod, then $\mathrm{Ev}(M)(U) = \mathrm{Ev}(M)(U, -) = (U, M)$, so that $\mathrm{Ev}(M) = (-, M)$ on (R-mod)^{op}. By the dual of the Eilenberg-Watts Theorem, every left exact functor $F \in [(\text{R-mod})^{\mathrm{op}}, \mathrm{Ab}]$ is isomorphic to $\mathrm{Ev}(M)$, where ${}_RM = F({}_RR) = \mathrm{Ev}(R)(F)$ is the value of the localization functor at F.

Shifting the emphasis on the other variable, the functor Ev : R-Mod $\to$ ((R-mod)^{op}, Ab), $M \mapsto \mathrm{Ev}(M)$, is seen to be left exact and commute with products, just like the right adjoint r : R-Mod $\to$ ((R-mod)^{op}, Ab) of the localization functor. Furthermore, $r(E) = \mathrm{Ev}(E)$ for every injective module, using Proposition 12 as above, so that another application of the dual of the Eilenberg-Watts Theorem implies that Ev is the right adjoint of the localization functor.

5.3. The projective quotient recollement. The *projective quotient recollement*, associated to the inclusion $\mathrm{add}(R \otimes_R -) \subseteq \mathrm{PQ}(R^{\mathrm{op}})$ in $\mathrm{Ab}(R)$, is expressed as

$$\mathcal{R}_{\mathrm{ab}}((\underline{\mathrm{PQ}}(R^{\mathrm{op}}), \mathrm{Ab}), (\mathrm{PQ}(R^{\mathrm{op}}), \mathrm{Ab}), \text{R-Mod}),$$

where $\underline{\mathrm{PQ}}(R^{\mathrm{op}})$ is the stable projective quotient category; it is the projective quotient category $\mathrm{PQ}(R^{\mathrm{op}})$ modulo the morphisms that factor through some finite power of the forgetful functor, so that $(\underline{\mathrm{PQ}}(R^{\mathrm{op}}), \mathrm{Ab}) \subseteq (\mathrm{PQ}(R^{\mathrm{op}}), \mathrm{Ab})$ is the subcategory of functors that vanish on $\mathrm{add}(R \otimes_R -)$.

It is clear that if $M \in$ R-Mod, then $\mathrm{Ev}(M) : \mathrm{PQ}(R^{\mathrm{op}}) \to \mathrm{Ab}$ preserves images in $\mathrm{PQ}(R^{\mathrm{op}})$; this is because these images are taken in $\mathrm{Ab}(R)$ and $\mathrm{Ev}(M)$ is the restriction to $\mathrm{PQ}(R^{\mathrm{op}})$ of an exact functor on $\mathrm{Ab}(R)$. On the other hand, a functor $F : \mathrm{PQ}(R^{\mathrm{op}}) \to \mathrm{Ab}$ that respects images must be of the form $\mathrm{Ev}(M)$, where ${}_RM = F(R \otimes_R -)$ is the module that F assigns to the forgetful functor. For if $\Omega(X \otimes_R -) \in \mathrm{PQ}(R^{\mathrm{op}})$, then it is the image of the natural transformation in the middle of (5), induced by a matrix C that presents $X_R \in$ mod-R. Thus $F(\Omega(X \otimes_R -)) \cong CM^n \cong \mathrm{Ev}(M)(\Omega(X \otimes_R -))$, naturally.

In the covariant recollement, the restriction $\mathrm{Ev} : M \mapsto \mathrm{Ev}(M)|_{\text{mod-R}} = - \otimes_R M$ was a right exact section of the localization functor; and in the contravariant recollement situation, the restriction $\mathrm{Ev}\colon M \mapsto \mathrm{Ev}(M)|_{(\text{R-mod})^{\mathrm{op}}}$ was a left exact section. For the projective quotient recollement, we have the following.

PROPOSITION 13. *The restriction* $\mathrm{Ev}\colon M \mapsto \mathrm{Ev}(M)|_{\mathrm{PQ}(R^{\mathrm{op}})}$ *is a section of the localization functor* $(\mathrm{PQ}(R^{\mathrm{op}}), \mathrm{Ab}) \to$ R-Mod *that preserves images. It is therefore naturally isomorphic to the middle adjoint of the projective quotient recollement.*

PROOF. Let us note that $\mathrm{Ev}(-)$: R-Mod $\to (\mathrm{PQ}(R^{\mathrm{op}}), \mathrm{Ab})$ preserves monomorphisms and epimorphisms, and therefore images (in R-Mod). If $f : M \to N$ is a monomorphism, look at the right square of the commutative diagram

$$\begin{array}{ccccc} M^n & \twoheadrightarrow & CM^n & \rightarrowtail & M^m \\ \downarrow{\scriptstyle f^n} & & \downarrow{\scriptstyle Cf^n} & & \downarrow{\scriptstyle f^m} \\ N^n & \twoheadrightarrow & CN^n & \rightarrowtail & N^m, \end{array}$$

and note that $f^m : M^m \to N^m$ is also a monomorphism, and therefore so is Cf^n. Thus $\mathrm{Ev}(f) : \mathrm{Ev}(M) \to \mathrm{Ev}(N)$ is a monomorphism. Use the left square to prove the dual result.

To prove the second statement, let ${}_R M \in$ R-Mod be a module and splice together a projective presentation and injective copresentation of M to obtain the diagram

$$\begin{array}{ccccccc} P_1 & \xrightarrow{f} & P_0 & \xrightarrow{\quad h \quad} & E_0 & \xrightarrow{g} & E_1 \\ & & & \searrow \quad M \quad \nearrow & & & \end{array}$$

Now look at the diagram (8) and use the fact that Ev respects images to infer that

$$\mathrm{Ev}(M) \cong \mathrm{Im}[\mathrm{Ev}(h) : \mathrm{Ev}(P_0) \to \mathrm{Ev}(E_0)] = m(M)$$

is a natural isomorphism. □

6. Equivalence of the TTF classes

In the biresolution (5) of the matrix C, the syzygy $\Omega(X \otimes_R -)$ of $X \otimes_R -$ is well-defined modulo the morphisms that factor through an object in $\mathrm{add}(R \otimes_R -)$, $\Omega : \underline{\mathrm{mod}}\text{-R} \to \underline{\mathrm{PQ}}(R^{\mathrm{op}})$. Now apply Ω to $\Omega(X \otimes_R -)$ to get the second syzygy $(U, -) = \Omega^2(X \otimes_R -)$, $U = \mathrm{Tr}\, X$, also well-defined modulo the morphisms that factor through $\mathrm{add}(R \otimes_R -)$, $\Omega : \underline{\mathrm{PQ}}(R^{\mathrm{op}}) \to (\text{R-}\underline{\mathrm{mod}})^{\mathrm{op}}$. Both of these functors have an equivalence inverse, given by the cosyzygy functor, so that precomposition by the syzygy operator yields equivalences

$$((\text{R-}\underline{\mathrm{mod}})^{\mathrm{op}}, \mathrm{Ab}) \xrightarrow{\quad \Omega^* \quad} (\underline{\mathrm{PQ}}(R^{\mathrm{op}}), \mathrm{Ab}) \xrightarrow{\quad \Omega^* \quad} (\underline{\mathrm{mod}}\text{-R}, \mathrm{Ab})$$

between the TTF classes of the three recollements of purity.

To calculate the value of Ω^*, let us return to the narrative of the Introduction, and consider a short exact sequence

$$\text{(9)} \qquad \Psi : 0 \longrightarrow M_1 \xrightarrow{\quad f \quad} M_2 \xrightarrow{\quad g \quad} M_3 \longrightarrow 0$$

of left R-modules. The evaluation functor, restricted to $(\text{R-mod})^{\mathrm{op}}$, given by the composition $\text{R-Mod} \xrightarrow{\mathrm{Ev}} (\mathrm{Ab}(R), \mathrm{Ab}) \xrightarrow{(\mathcal{Y}, \mathrm{Ab})} ((\text{R-mod})^{\mathrm{op}}, \mathrm{Ab})$ is left exact, so we get an exact sequence

$$0 \longrightarrow \mathrm{Ev}(M_1) \longrightarrow \mathrm{Ev}(M_2) \longrightarrow \mathrm{Ev}(M_3) \longrightarrow H_3^{(\text{R-mod})^{\mathrm{op}}}(\mathrm{Ev}(\Psi)) \longrightarrow 0$$

in $[(\text{R-mod})^{\mathrm{op}}, \mathrm{Ab}]$, where the homology $H_3^{(\text{R-mod})^{\mathrm{op}}}(\mathrm{Ev}(\Psi))$ of $\mathrm{Ev}(\Psi)$ at $\mathrm{Ev}(M_3)$ vanishes at the forgetful functor, $H_3^{(\text{R-mod})^{\mathrm{op}}}(\mathrm{Ev}(\Psi)) \in ((\text{R-}\underline{\mathrm{mod}})^{\mathrm{op}}, \mathrm{Ab})$.

Similarly, if we apply the evaluation functor to Ψ and restrict to $(\mathrm{PQ}(R^{\mathrm{op}}), \mathrm{Ab})$, we obtain a sequence

$$
\begin{array}{ccccccccc}
 & & & & 0 & & 0 & & \\
 & & & & \downarrow & & \downarrow & & \\
0 & \longrightarrow & \mathrm{Ev}(M_1) & \longrightarrow & \mathrm{Ker}\,\mathrm{Ev}(g) & \longrightarrow & H_2^{\mathrm{PQ}(R^{\mathrm{op}})}(\Psi) & \longrightarrow & 0 \\
 & & \| & & \downarrow & & \downarrow & & \\
0 & \longrightarrow & \mathrm{Ev}(M_1) & \xrightarrow{\mathrm{Ev}(f)} & \mathrm{Ev}(M_2) & \longrightarrow & \mathrm{Coker}\,\mathrm{Ev}(f) & \longrightarrow & 0 \\
 & & & & \downarrow{\scriptstyle \mathrm{Ev}(g)} & & \downarrow & & \\
 & & & & \mathrm{Ev}(M_3) & = & \mathrm{Ev}(M_3) & & \\
 & & & & \downarrow & & \downarrow & & \\
 & & & & 0 & & 0 & &
\end{array}
$$

in $(\mathrm{PQ}(R^{\mathrm{op}}), \mathrm{Ab}(R))$ (seen in the bottom left) that may fail to be exact at $\mathrm{Ev}(M_2)$, where the homology $H_2^{\mathrm{PQ}(R^{\mathrm{op}})}(\Psi) \in (\underline{\mathrm{PQ}}(R^{\mathrm{op}}), \mathrm{Ab})$ also vanishes on the forgetful functor.

Finally, the restriction of $\mathrm{Ev} : R\text{-Mod} \to (\mathrm{Ab}(R), \mathrm{Ab})$ to $(\text{mod-R}, \mathrm{Ab})$ along the coYoneda embedding is right exact, and we obtain an exact sequence

$$
0 \longrightarrow H_1^{\text{mod-R}}(\mathrm{Ev}(\Psi)) \longrightarrow \mathrm{Ev}(M_1) \longrightarrow \mathrm{Ev}(M_2) \longrightarrow \mathrm{Ev}(M_3) \longrightarrow 0
$$

in $(\text{mod-R}, \mathrm{Ab})$ where the homology $H_1^{\text{mod-R}}(\mathrm{Ev}(\Psi))$ of $\mathrm{Ev}(\Psi)$ at $\mathrm{Ev}(M_1)$ vanishes at the forgetful functor, $H_1^{\text{mod-R}}(\mathrm{Ev}(\Psi)) \in (\underline{\text{mod}}\text{-R}, \mathrm{Ab})$. In the notation just introduced, we have the following.

LEMMA 14. *For a short exact sequence* (9), *the equivalence* $\Omega^* \colon ((\text{R-}\underline{\text{mod}})^{\mathrm{op}}, \mathrm{Ab}) \to (\underline{\mathrm{PQ}}(R^{\mathrm{op}}), \mathrm{Ab})$ *sends the homology of* $\mathrm{Ev}(\Psi)$ *at* $\mathrm{Ev}(M_3)$ *to the homology of* $\mathrm{Ev}(\Psi)$ *at* $\mathrm{Ev}(M_2)$,

$$
H_3^{(\text{R-mod})^{\mathrm{op}}}(\mathrm{Ev}(\Psi)) \xmapsto{\Omega^*} H_2^{\mathrm{PQ}(R^{\mathrm{op}})}(\Psi).
$$

The equivalence $\Omega^* \colon (\underline{\mathrm{PQ}}(R^{\mathrm{op}}), \mathrm{Ab}) \to (\underline{\text{mod}}\text{-R}, \mathrm{Ab})$ *sends the homology of* $\mathrm{Ev}(\Psi)$ *at* $\mathrm{Ev}(M_2)$ *to the homology of* $\mathrm{Ev}(\Psi)$ *at* $\mathrm{Ev}(M_1)$, $H_2^{\mathrm{PQ}(R^{\mathrm{op}})}(\Psi) \xmapsto{\Omega^*} H_1^{\text{mod-R}}(\mathrm{Ev}(\Psi))$.

PROOF. Both proofs are the same as that of Theorem 3; we only verify the first part. Think of the short exact sequence $\Psi = M.$ as a complex of left R-modules and $\mathrm{Ev}(\Psi) = \mathrm{Ev}(M.)$ as a complex in $(\mathrm{Ab}(R), \mathrm{Ab})$ of exact functors. If we apply the complex to the short exact sequence on the left side of (5) we obtain the following short exact sequence of complexes in Ab,

$$\begin{array}{ccccccccc}
 & & 0 & & 0 & & 0 & & \\
 & & \downarrow & & \downarrow & & \downarrow & & \\
0 & \to & \mathrm{Ev}(M_1)(U,-) & \to & \mathrm{Ev}(M_2)(U,-) & \to & \mathrm{Ev}(M_3)(U,-) & \to & 0 \\
 & & \downarrow & & \downarrow & & \downarrow & & \\
0 & \to & \mathrm{Ev}(M_1)(R^m \otimes_R -) & \to & \mathrm{Ev}(M_2)(R^m \otimes_R -) & \to & \mathrm{Ev}(M_3)(R^m \otimes_R -) & \to & 0 \\
 & & \downarrow & & \downarrow & & \downarrow & & \\
0 & \to & \mathrm{Ev}(M_1)(\Omega(X \otimes_R -)) & \to & \mathrm{Ev}(M_2)(\Omega(X \otimes_R -)) & \to & \mathrm{Ev}(M_3)(\Omega(X \otimes_R -)) & \to & 0 \\
 & & \downarrow & & \downarrow & & \downarrow & & \\
 & & 0 & & 0 & & 0, & &
\end{array}$$

with exact columns and middle complex exact. The connecting map of homology gives a natural isomorphism

$$\begin{aligned}
H_2^{\mathrm{PQ}(R^{\mathrm{op}})}(\Psi)(\Omega(X \otimes_R -)) = H_2(\mathrm{Ev}(M.))(\Omega(X \otimes_R -)) &\cong H_3(\mathrm{Ev}(M.))(U,-) \\
&= H_3^{(\text{R-mod})^{\mathrm{op}}}(\mathrm{Ev}(\Psi))(U,-) \\
&= H_3^{(\text{R-mod})^{\mathrm{op}}}(\mathrm{Ev}(\Psi))(\Omega^2(X \otimes_R -)) \\
&= \Omega^* H_3^{(\text{R-mod})^{\mathrm{op}}}(\mathrm{Ev}(\Psi))(\Omega(X \otimes_R -)).
\end{aligned}$$

□

Acknowledgment

We are grateful to Shijie Zhu for sharing his insights on some early results of this project.

References

[1] Murray Adelman, *Abelian categories over additive ones*, J. Pure Appl. Algebra **3** (1973), 103–117, DOI 10.1016/0022-4049(73)90026-1. MR318265

[2] Maurice Auslander, *Coherent functors*, Proc. Conf. Categorical Algebra (La Jolla, Calif., 1965), Springer, New York, 1966, pp. 189–231. MR0212070

[3] Maurice Auslander, *Isolated singularities and existence of almost split sequences*, Representation theory, II (Ottawa, Ont., 1984), Lecture Notes in Math., vol. 1178, Springer, Berlin, 1986, pp. 194–242, DOI 10.1007/BFb0075297. MR842486

[4] P. M. Cohn, *On the free product of associative rings*, Math. Z. **71** (1959), 380–398, DOI 10.1007/BF01181410. MR106918

[5] William Crawley-Boevey and Julia Sauter, *On quiver Grassmannians and orbit closures for representation-finite algebras*, Math. Z. **285** (2017), no. 1-2, 367–395, DOI 10.1007/s00209-016-1712-z. MR3598816

[6] Samuel Dean and Jeremy Russell, *Derived recollements and generalised AR formulas*, J. Pure Appl. Algebra **223** (2019), no. 2, 515–546, DOI 10.1016/j.jpaa.2018.04.004. MR3850554

[7] Samuel Eilenberg, *Abstract description of some basic functors*, J. Indian Math. Soc. (N.S.) **24** (1960), 231–234 (1961). MR125148

[8] Vincent Franjou and Teimuraz Pirashvili, *Comparison of abelian categories recollements*, Doc. Math. **9** (2004), 41–56. MR2054979

[9] Freyd, P., *Abelian Categories*, Harper & Row, 1964.

[10] Peter Freyd, *Representations in abelian categories*, Proc. Conf. Categorical Algebra (La Jolla, Calif., 1965), Springer, New York, 1966, pp. 95–120. MR0209333

[11] Pierre Gabriel, *Des catégories abéliennes* (French), Bull. Soc. Math. France **90** (1962), 323–448. MR232821

[12] L. Gruson and C. U. Jensen, *Dimensions cohomologiques reliées aux foncteurs* $\varprojlim^{(i)}$ (French), Paul Dubreil and Marie-Paule Malliavin Algebra Seminar, 33rd Year (Paris, 1980), Lecture Notes in Math., vol. 867, Springer, Berlin-New York, 1981, pp. 234–294. MR633523

[13] Ivo Herzog, *The Ziegler spectrum of a locally coherent Grothendieck category*, Proc. London Math. Soc. (3) **74** (1997), no. 3, 503–558, DOI 10.1112/S002461159700018X. MR1434441

[14] Ivo Herzog, *Contravariant functors on the category of finitely presented modules*, Israel J. Math. **167** (2008), 347–410, DOI 10.1007/s11856-008-1052-8. MR2448029

[15] Mike Prest, *Purity, spectra and localisation*, Encyclopedia of Mathematics and its Applications, vol. 121, Cambridge University Press, Cambridge, 2009. MR2530988

[16] Heinz Prüfer, *Untersuchungen über die Zerlegbarkeit der abzählbaren primären Abelschen Gruppen* (German), Math. Z. **17** (1923), no. 1, 35–61, DOI 10.1007/BF01504333. MR1544601

[17] Chrysostomos Psaroudakis, *A representation-theoretic approach to recollements of abelian categories*, Surveys in representation theory of algebras, Contemp. Math., vol. 716, Amer. Math. Soc., Providence, RI, 2018, pp. 67–154, DOI 10.1090/conm/716/14427. MR3852400

[18] Chrysostomos Psaroudakis and Jorge Vitória, *Recollements of module categories*, Appl. Categ. Structures **22** (2014), no. 4, 579–593, DOI 10.1007/s10485-013-9323-x. MR3227608

[19] Riehl, E., *Category Theory in Context,* Dover, 2016.

[20] Bo Stenström, *Rings of quotients*, Springer-Verlag, New York-Heidelberg, 1975. Die Grundlehren der Mathematischen Wissenschaften, Band 217; An introduction to methods of ring theory. MR0389953

[21] R. B. Warfield Jr., *Purity and algebraic compactness for modules*, Pacific J. Math. **28** (1969), 699–719. MR242885

[22] Charles E. Watts, *Intrinsic characterizations of some additive functors*, Proc. Amer. Math. Soc. **11** (1960), 5–8, DOI 10.2307/2032707. MR118757

School of Mathematics and Statistics, Northeast Normal University, Changchun, People's Republic of China
Email address: fuxianhui@gmail.com

The Ohio State University at Lima, Lima, Ohio
Email address: herzog.23@osu.edu

Contemporary Mathematics
Volume **758**, 2020
https://doi.org/10.1090/conm/758/15238

Higher Auslander algebras of type $\mathbb{A}$ and the higher Waldhausen S-constructions

Gustavo Jasso

ABSTRACT. These notes are an expanded version of my talk at the ICRA 2018 in Prague, Czech Republic; they are based on joint work with Tobias Dyckerhoff and Tashi Walde [**DJW19b**]. In them we relate Iyama's higher Auslander algebras of type $\mathbb{A}$ to Eilenberg–Mac Lane spaces in algebraic topology and to higher-dimensional versions of the Waldhausen S-construction from algebraic K-theory.

1. Motivation: Eilenberg–Mac Lane spaces

Let A be an abelian group. For each positive integer m there exists a topological space $K(A,m)$ characterised, up to weak homotopy equivalence[1], by the existence of natural bijections[2]

$$[X, K(A,m)] \xrightarrow{\cong} H^m(X;A) \tag{1}$$

where X is a CW complex, see Corollary III.2.17 in [**GJ99**]. Equivalently, the topological space $K(A,m)$ is characterised by the following two properties:

- The topological space $K(A,m)$ is path connected;
- there is an isomorphism

$$\pi_n(K(A,m)) \cong \begin{cases} A & \text{if } n=m, \\ 0 & \text{otherwise.} \end{cases}$$

2010 *Mathematics Subject Classification.* Primary: 18G30; Secondary 16G70, 18E30.

Key words and phrases. Auslander–Reiten theory, Eilenberg–Mac Lane spaces, Waldhausen S-construction.

The author was funded by the Deutsche Forschungs-gemeinschaft (DFG, German Research Foundation) under Germany's Excellence Strategy - GZ 2047/1, Projekt-ID 390685813.

[1]A *weak homotopy equivalence* is a continuous map $X \to Y$ between topological spaces which induces a bijection

$$\pi_0(X) \xrightarrow{\cong} \pi_0(Y)$$

on connected components as well as group isomorphisms

$$\pi_n(X,x) \xrightarrow{\cong} \pi_n(Y, f(x))$$

for all points $x \in X$ and all $n \geq 1$. Two topological spaces are *weakly homotopy equivalent* if they are connected by a zig-zag of weak homotopy equivalences.

[2]Here $[X, K(A,m)]$ denotes the set of homotopy classes of continuous maps $X \to K(A,m)$, while $H^m(X;A)$ denotes the m-th singular cohomology group of X with coefficients in A.

A topological space satisfying the above properties is called an *Eilenberg–Mac Lane space*[3]. The bijection (1) exhibits Eilenberg–Mac Lane spaces as fundamental objects in algebraic topology while their characterisation in terms of higher homotopy groups makes their relative simplicity manifest.

Rather surprisingly, the standard construction of the Eilenberg–Mac Lane spaces can be expressed in terms of the *higher Auslander–Reiten theory* of Iyama's higher Auslander algebras of type $\mathbb{A}$. This observation, which we make precise below, is one of the starting points of our investigations.

1.1. Interlude: The simplex category. The modern approach to homotopy theory makes extensive use of simplicial methods and the standard construction of Eilenberg–Mac Lane spaces is no exception. We recall a minimal amount of terminology from simplicial homotopy theory[4] needed for describing this construction.

We remind the reader of the definition of the *simplex category*, which is commonly denoted by $\boldsymbol{\Delta}$. The objects of the simplex category are the linear posets

$$[n] := \{ 0 \to 1 \to \cdots \to n \},$$

where n ranges over all non-negative integers, with morphisms the monotone functions $[m] \to [n]$. We denote the set of morphisms $[m] \to [n]$ in $\boldsymbol{\Delta}$ by $\boldsymbol{\Delta}(m,n)$. Note that the set $\boldsymbol{\Delta}(m,n)$ has a natural poset structure[5]: Namely, $\sigma \leq \tau$ if and only if $\sigma_i \leq \tau_i$ for all $i \in [m]$. The elements of $\boldsymbol{\Delta}(m,n)$ are called *m-simplices in $\boldsymbol{\Delta}^n$*; an m-simplex is *non-degenerate* if the underlying monotone function is injective and is *degenerate* otherwise. We often identify $\boldsymbol{\Delta}(m,n)$ with the set of ordered tuples $\sigma \in \mathbb{Z}^{m+1}$ which satisfy the inequalities $0 \leq \sigma_0 \leq \sigma_1 \leq \cdots \leq \sigma_m \leq n$.

A *simplicial object*[6] in a category $\mathcal{C}$ is a functor $X \colon \boldsymbol{\Delta}^{\mathrm{op}} \to \mathcal{C}$. It is elementary to verify, using a specific presentation of $\boldsymbol{\Delta}$ in terms of generators and relations, that the data of a simplicial object in $\mathcal{C}$ is equivalent to that of a diagram in $\mathcal{C}$ of the form

$$\cdots \;\substack{\rightarrow\\ \leftarrow\\ \rightarrow\\ \leftarrow\\ \rightarrow\\ \leftarrow\\ \rightarrow}\; X_2 \;\substack{\rightarrow\\ \leftarrow\\ \rightarrow\\ \leftarrow\\ \rightarrow}\; X_1 \;\substack{\rightarrow\\ \leftarrow\\ \rightarrow}\; X_0$$

where $X_n := X([n])$ is the object of *n-simplices of X*. The morphisms

$$\{ d_i \colon X_n \to X_{n-1} \mid i \in [n] \},$$

called *face maps*, and the morphisms

$$\{ s_i \colon X_n \to X_{n+1} \mid i \in [n] \},$$

[3]The symbol $K(A,m)$ is more commonly used to denote the weak homotopy equivalence class (also known as the homotopy type) of an Eilenberg–Mac Lane space. For our purposes it will be convenient to abuse the notation and use this symbol to denote a preferred representative of this equivalence class.

[4]The reader is referred to [**GJ99**] for an introduction to simplicial homotopy theory.

[5]The poset structure on the morphism sets in $\boldsymbol{\Delta}$ allows us to promote the simplex category to a 2-category; this perspective, implicit in parts of these notes, is essential in our work [**DJW19b**].

[6]Depending on the target category one speaks of simplicial sets, simplicial topological spaces, simplicial abelian groups, etc.

called *degeneracy maps*, are subject to a number of relations commonly referred to as the *simplicial identities*[7]:

$$d_i \circ d_j = d_{j-1} \circ d_i \quad i < j, \qquad d_i \circ s_j = \begin{cases} s_{j-1} \circ d_i & i < j, \\ s_j \circ d_{i-1} & i > j+1, \\ 1 & \text{otherwise.} \end{cases}$$
$$s_i \circ s_j = s_j \circ s_{i-1} \quad i > j,$$

Thus, a simplicial object in $\mathcal{C}$ is a *contravariant representation* in the category $\mathcal{C}$ of the infinite quiver with (inadmissible) relations corresponding to the standard presentation of $\mathbf{\Delta}$.

1.2. Construction via the Dold–Kan correspondence. The *Dold–Kan correspondence* [**Dol58, Kan58**] is an explicit adjoint equivalence of categories[8]

$$C\colon \mathbf{Ab}_{\mathbf{\Delta}} \rightleftarrows \mathrm{Ch}_{\geq 0}(\mathbf{Ab}) :N$$

between the (abelian) category of simplicial abelian groups and the (abelian) category of chain complexes of abelian groups which are concentrated in non-negative homological degrees[9].

Let A be an abelian group and m a positive integer. As usual, let us denote by $A[m]$ the chain complex of abelian groups whose only non-zero component consists of A placed in homological degree m. The Eilenberg–Mac Lane space $K(A,m)$ is defined as

$$K(A,m) := |N(A[m])_\bullet|,$$

that is as the geometric realisation[10] of the underlying simplicial set of $N(A[m])_\bullet$.

1.3. The Eilenberg–Mac Lane space $K(A,1)$. Let A be an abelian group. For our purposes it is instructive to give an explicit description of the simplicial abelian group $N(A[m])_\bullet$ in the simplest case $m = 1$.

DEFINITION 1.1. The simplicial abelian group $N(A[1])_\bullet$ is defined as follows. For $n \geq 0$, let $N(A[1])_n$ be the abelian group of upper-triangular arrays

$$\begin{pmatrix} a_{00} & a_{01} & a_{02} & \cdots & a_{0,n-1} & a_{0n} \\ & a_{11} & a_{12} & \cdots & a_{1,n-1} & a_{1n} \\ & & \ddots & \ddots & \vdots & \vdots \\ & & & & a_{n-1,n-1} & a_{n-1,n} \\ & & & & & a_{nn} \end{pmatrix}$$

[7]Let X be a simplicial set. The simplicial identities allow us to think of an element of X_n as an abstract (possibly degenerate) n-simplex whose boundary consists of the $(n-1)$-simplices $\{\, d_i(x) \in X_{n-1} \mid i \in [n] \,\}$.

[8]A detailed discussion of this equivalence can be found in Section III.2 in [**GJ99**].

[9]From a representation-theoretic perspective, the Dold–Kan correspondence should be viewed as a Morita equivalence between $\mathbb{Z}\mathbf{\Delta}$, the $\mathbb{Z}$-linear envelope of $\mathbf{\Delta}$, and the $\mathbb{Z}$-linear path category of the quiver $0 \to 1 \to 2 \to \cdots$ modulo the square of the Jacobson radical (note that we work with *contravariant* modules).

[10]The geometric realisation functor is part of an adjunction [**GJ99**, p. 9]

$$|-|\colon \mathbf{Set}_{\mathbf{\Delta}} \rightleftarrows \mathbf{CGHaus} : \mathrm{Sing}$$

between the category of simplicial sets and the category of compactly generated Hausdorff spaces. In fact, this adjunction is a Quillen equivalence with respect to appropriate model category structures [**Qui67, Hov99**]. This Quillen equivalence justifies the simplicial approach to homotopy theory.

with entries in A such that for each $0 \leq i \leq n$ there is an equality $a_{ii} = 0$ and for all $0 \leq i < j < k \leq n$ the *Euler relation*

$$a_{ij} - a_{ik} + a_{jk} = 0 \tag{2}$$

is satisfied. The i-th face map

$$d_i \colon N(A[1])_n \to N(A[1])_{n-1}$$

is given by deleting the i-th row and the i-th column while the i-th degeneracy map

$$s_i \colon N(A[1])_n \to N(A[1])_{n+1}$$

is given by repeating the i-th row and the i-th column.

REMARK 1.2. An n-simplex of $N(A[1])_\bullet$ is completely determined by the n-tuple $(a_{01}, a_{02}, \dots, a_{0n})$. Indeed, the Euler relations (2) imply that for each $1 \leq j < k \leq n$ the equality

$$a_{jk} = a_{0k} - a_{0j}$$

is satisfied. We conclude that there is an isomorphism

$$N(A[1])_n \cong A^n.$$

Note, however, that the above identification makes the simplicial structure of $N(A[1])_\bullet$ less apparent.

Our next task is to relate the previous construction of the Eilenberg–Mac Lane space $K(A,1)$ to the Auslander–Reiten theory of the family of quivers

$$\mathbb{A}_n := 1 \to 2 \to \cdots \to n; \qquad\qquad n \geq 0.$$

Consider the Auslander–Reiten quiver of the category of finite-dimensional representations[11] of the quiver $\mathbb{A}_n$, which we depict as follows:

$$\begin{array}{ccccccccc}
M_{00} \to & M_{01} & \to & M_{02} & \to \cdots \longrightarrow & M_{0,n-1} & \longrightarrow & M_{0n} \\
 & \downarrow & & \downarrow & & \downarrow & & \downarrow \\
 & M_{11} & \to & M_{12} & \to \cdots \longrightarrow & M_{1,n-1} & \longrightarrow & M_{1n} \\
 & & & \downarrow & & \downarrow & & \downarrow \\
 & & & \ddots & \ddots & \vdots & & \vdots \\
 & & & & & \downarrow & & \downarrow \\
 & & & & & M_{n-1,n-1} & \to & M_{n-1,n} \\
 & & & & & & & \downarrow \\
 & & & & & & & M_{nn}
\end{array}$$

For reasons which shall become clear shortly, we have included additional vertices $M_{ii} = 0$ in the above Auslander–Reiten quiver. The Grothendieck group $K_0(\mathbb{A}_n)$ admits a presentation by generators and relations as the quotient of the free abelian group with basis

$$\{\, [M_{ij}] \,|\, 0 \leq i \leq j \leq n \,\}$$

modulo the subgroup generated by the relations $[M_{ii}] = 0$, $i \in [n]$, together with the Euler relations

$$[M_{ij}] - [M_{ik}] + [M_{jk}] = 0 \tag{3}$$

[11] We work over some fixed but unspecified field.

corresponding to short exact sequences[12]

$$0 \to M_{ij} \to M_{ik} \to M_{jk} \to 0$$

in the abelian category of finite-dimensional representations of the quiver $\mathbb{A}_n$, where $0 \leq i < j < k \leq n$.

It is immediate from the above presentation of $K_0(\mathbb{A}_n)$ that there are isomorphisms

$$N(A[1])_n \cong \operatorname{Hom}_{\mathbb{Z}}(K_0(\mathbb{A}_n), A) \cong \operatorname{Hom}_{\mathbb{Z}}(\mathbb{Z}^n, A) \cong A^n, \tag{4}$$

where $n \geq 1$. We leave it to the reader to verify that the Grothendieck groups $K_0(\mathbb{A}_n)$ assemble into a co-simplicial abelian group, that is into a functor

$$K_0(\mathbb{A}_\bullet)\colon \mathbf{\Delta} \to \mathbf{Ab},$$

where $K_0(\mathbb{A}_0) := 0$. Moreover, the isomorphisms (4) assemble into an isomorphism of simplicial abelian groups

$$N(A[1])_\bullet \cong \operatorname{Hom}_{\mathbb{Z}}(K_0(\mathbb{A}_\bullet), A) \tag{5}$$

where $\operatorname{Hom}_{\mathbb{Z}}(K_0(\mathbb{A}_\bullet), A)$ denotes the composite

$$\mathbf{\Delta}^{\mathrm{op}} \xrightarrow{K_0(\mathbb{A}_\bullet)^{\mathrm{op}}} \mathbf{Ab}^{\mathrm{op}} \xrightarrow{\operatorname{Hom}_{\mathbb{Z}}(-,A)} \mathbf{Ab},$$

and $K_0(\mathbb{A}_\bullet)^{\mathrm{op}}$ indicates the passage to the opposite functor.

REMARK 1.3. It is elementary to verify that the Euler relations (3) are generated by the *Auslander–Reiten relations*

$$[M_{ij}] - ([M_{i+1,j}] + [M_{i,j+1}]) + [M_{i+1,j+1}] = 0$$

corresponding to the almost-split sequences[13]

$$0 \to M_{ij} \to M_{i+1,j} \oplus M_{i,j+1} \to M_{i+1,j+1} \to 0,$$

where $0 \leq i < j < n$, see also [**Aus84**]. In particular, the isomorphism (5) gives a precise relationship between the simplicial abelian group $N(A[1])_\bullet$ and the Auslander–Reiten theory of the linearly oriented quivers of Dynkin type $\mathbb{A}$.

---

[12] Equivalently, the Euler relations are induced by biCartesian squares

$$\begin{array}{ccc} M_{ij} & \longrightarrow & M_{ik} \\ \downarrow & & \downarrow \\ M_{ii} & \longrightarrow & M_{jk} \end{array}$$

where $0 \leq i < j < k \leq n$.

[13] Equivalently, the Auslander–Reiten relations are induced by meshes

$$\begin{array}{ccc} M_{ij} & \longrightarrow & M_{i,j+1} \\ \downarrow & & \downarrow \\ M_{i+1,j} & \longrightarrow & M_{i+1,j+1} \end{array}$$

in the Auslander–Reiten quiver, where $0 \leq i < j < n$.

1.4. The Eilenberg–Mac Lane space $K(A,m)$. Let A be an abelian group and m a positive integer. The previous discussion extends to the Eilenberg–Mac Lane space $K(A,m)$ provided that we replace the linearly oriented quivers of type $\mathbb{A}$ by Iyama's m-dimensional Auslander algebras $\mathbb{A}_\ell^{(m)}$ of type $\mathbb{A}$, see [**Iya11**] for the definition. Indeed, the Grothendieck groups of these algebras assemble into a co-simplicial abelian group

$$K_0(\mathbb{A}_{\bullet-m+1}^{(m)})\colon \mathbf{\Delta}\to\mathbf{Ab},$$

where $K_0(\mathbb{A}_{n-m+1}^{(m)}):=0$ for $n<m$. In analogy with the case $m=1$, there is a canonical isomorphism of simplicial abelian groups

$$N(A[m])_\bullet\cong\operatorname{Hom}_{\mathbb{Z}}\left(K_0(\mathbb{A}_{\bullet-m+1}^{(m)}),A\right),$$

see Theorem 1.20 in [**DJW19b**] for a detailed description of the above isomorphism.

The above description of $N(A[m])_\bullet$ utilises the presentation of the Grothendieck group $K_0(\mathbb{A}_{n-m+1}^{(m)})$ in terms of the basis

$$\{\,[M_\sigma]\mid\sigma\in\mathbf{\Delta}(m,n)\,\}$$

labelled by the indecomposable summands of the unique basic m-cluster-tilting $\mathbb{A}_n^{(m)}$-module[14]. In analogy with the case $m=1$, one imposes relations $[M_\sigma]=0$ for all degenerate m-simplices in $\mathbf{\Delta}^n$ as well as *higher Euler relations*

$$\sum_{i=0}^{m}(-1)^i[M_{d_i(\sigma)}]=0, \tag{6}$$

where σ is a non-degenerate $(m+1)$-simplex in $\mathbf{\Delta}^n$ and

$$d_i(\sigma)=(\sigma_0,\sigma_1,\cdots,\sigma_{i-1},\widehat{\sigma}_i,\sigma_{i+1},\cdots,\sigma_m).$$

The higher Euler relations are induced by exact sequences

$$0\to M_{d_m(\sigma)}\to M_{d_{m-1}(\sigma)}\to\cdots\to M_{d_1(\sigma)}\to M_{d_0(\sigma)}\to 0$$

in the abelian category of finite-dimensional representations of the higher Auslander algebra $\mathbb{A}_{n-m+1}^{(m)}$. Finally, we note that the higher Euler relations are generated by the *higher Auslander–Reiten relations*

$$\sum_{v}(-1)^{|v|}[M_{\sigma+v}]=0, \tag{7}$$

where σ is a non-degenerate m-simplex in $\mathbf{\Delta}^n$ with $\sigma_m<n$, the tuple v ranges over all vertices of the $(m+1)$-cube

$$\underbrace{\{0,1\}\times\cdots\times\{0,1\}}_{m+1\text{ times}},$$

and $|v|:=v_0+v_1+\cdots+v_m$. Again in analogy with the case $m=1$, the relations (7) are induced by the m-almost-split sequences

$$0\to M_\sigma\to\bigoplus_{|v|=1}M_{\sigma+v}\to\cdots\to\bigoplus_{|v|=m}M_{\sigma+v}\to M_{\sigma+(1,\dots,1)}\to 0$$

in the unique m-cluster-tilting subcategory associated with $\mathbb{A}_{n-m+1}^{(m)}$, see [**Iya07, Iya11**] and [**OT12**] for definitions and further details.

[14]Implicit in this discussion is the Grothendieck group of an m-cluster-tilting subcategory, which can be defined using the ideas in [**BT14**].

1.5. On the purpose of these notes. In the remainder of these notes is to describe a *categorified* version of the above discussion where the abelian group A is replaced by a refined version of a triangulated category[15], namely a *stable ∞-category* in the sense of Lurie [**Lur17**]. As such, the simplicial objects we shall consider can be thought of as 'categorified Eilenberg–Mac Lane spaces' as originally advocated by Dyckerhoff in [**Dyc17**]. The results presented in these notes can be also seen as contributions to the abstract representation theory program of Groth and Šťovíček [**GŠ18**], see also Ladkani's Ph.D. Thesis [**Lad08**]. Indeed, the results presented here are for the most part higher-dimensional versions of results in [**GŠ16a**].

2. Rudiments of the theory of ∞-categories

The language of ∞-categories, developed by Joyal in [**Joy02**] as well as in unpublished work and by Lurie in [**Lur09, Lur17**], affords an adequate framework for our results. It is however impractical to include here a formal introduction to the theory of ∞-categories[16]. Instead we aim to provide the reader with a minimal amount of intuition regarding the theory which we hope will be sufficient to follow our exposition in the subsequent section.

2.1. What are ∞-categories? An ∞-category is a mathematical structure[17] which implements the idea of a

> 'higher-dimensional category with morphisms of every positive degree endowed with a coherently-associative composition law and such that the morphisms of degree greater than 1 are invertible'.

In other words the theory of ∞-categories is a model for the theory of $(\infty, 1)$-categories. In particular, for every pair of objects x and y in an ∞-category $\mathcal{C}$ there is a 'space' $\mathrm{Map}_{\mathcal{C}}(x, y)$ of maps $x \to y$. Within this paradigm ordinary categories are those for which the non-empty spaces of maps are homotopy equivalent to discrete spaces[18]. After suitable translation, further examples of ∞-categories are provided by differential graded categories and, more generally, A_∞-categories[19].

An important feature of the theory of ∞-categories is that it allows us to formalise the notion of a universal property which only holds 'up to coherent homotopy'. For example, an object x of an ∞-category $\mathcal{C}$ is *initial* if for each object y of $\mathcal{C}$ the space $\mathrm{Map}_{\mathcal{C}}(x, y)$ is contractible (whence in particular non-empty).

Just as in ordinary category theory, universal properties are also captured by appropriate notions of (homotopy) limit and (homotopy) colimit[20]. For example,

[15]Triangulated categories are inadequate for this purpose as the cone of a morphisms lacks the necessary functoriality.

[16]The standard reference for the theory of ∞-categories is Lurie's book [**Lur09**]. Alternatives include Groth's lecture notes [**Gro10**] and Cisinski's book [**Cis19**].

[17]More precisely, an ∞-category is a particular kind of simplicial set called a 'weak Kan complex'.

[18]See Propositions 2.3.4.5 and 2.3.4.18 in [**Lur09**].

[19]See Section 1.3 in [**Lur17**] for the case of differential graded categories and [**Fao17a, Fao17b**] for the case of A_∞-categories.

[20]In ∞-category theory it is customary to drop the qualifier 'homotopy' from the terminology as only homotopy-invariant notions make sense in this context.

given two morphisms $x \to y'$ and $x \to y''$ in an ∞-category $\mathcal{C}$, we may define the (homotopy) pushout of the diagram[21]

$$\begin{array}{ccc} x & \longrightarrow & y' \\ \big\downarrow & & \\ y'' & & \end{array}$$

by means of a suitable universal property, that is as a (homotopy) colimit diagram of the form[22]

$$\begin{array}{ccc} x & \longrightarrow & y' \\ \big\downarrow & \ulcorner & \big\downarrow \\ y'' & \longrightarrow & z \end{array}$$

The space of all possible (homotopy) (co)limit diagrams of a fixed shape is either empty (if no (co)limit of the diagram exists in $\mathcal{C}$) or contractible[23].

2.2. Stable ∞-categories. In the sequel we are mostly concerned with the following class of ∞-categories, which the reader might want to think of as refined versions of triangulated categories.

DEFINITION 2.1. An ∞-category $\mathcal{A}$ is *stable*[24] if it has the following properties:

(1) The ∞-category $\mathcal{A}$ is *pointed*, that is $\mathcal{A}$ has a zero object.

(2) For every morphism $f\colon x \to y$ in $\mathcal{A}$ there exist (coherently commutative) squares in $\mathcal{A}$ of the form

$$\begin{array}{ccc} x & \xrightarrow{f} & y \\ \big\downarrow & \ulcorner & \big\downarrow \\ 0 & \longrightarrow & z \end{array} \qquad \text{and} \qquad \begin{array}{ccc} w & \longrightarrow & x \\ \big\downarrow & \lrcorner & \big\downarrow{\scriptstyle f} \\ 0 & \longrightarrow & y \end{array}$$

which are a (homotopy) pushout square and a (homotopy) pullback square, respectively. The objects z and w are called the *cofibre of f* and the *fibre of f*, respectively[25].

(3) A diagram in $\mathcal{A}$ of the form

$$\begin{array}{ccc} x & \longrightarrow & y \\ \big\downarrow & & \big\downarrow \\ 0 & \longrightarrow & z \end{array}$$

[21] Strictly speaking, specifying a coherently commutative diagram in an ∞-category involves an infinite amount of data. For expository purposes we nonetheless display such diagrams as we would display commutative diagrams in ordinary categories.

[22] We decorate the square to indicate that it is a (homotopy) pushout square.

[23] This means that, for the purposes of ∞-category theory, (co)limit cones of a fixed diagram in an ∞-category are unique in the appropriate sense. See Proposition 1.2.12.9 and Definition 1.2.13.4 in [**Lur09**] for details.

[24] Although we have not provided formal definitions of any the notions involved, the property of being *stable* is sufficiently intuitive for it to be worth to be included in these notes. A detailed treatment of the theory of stable ∞-categories can be found in Chapter 1 in [**Lur17**].

[25] The cofibre and the fibre of a morphism in a stable ∞-category are the ∞-categorical versions of the cone and the co-cone of a morphism in a triangulated category. Following the established convention, we adopt the 'topological' terminology which further reminds us of the homotopical nature of these concepts.

is a (homotopy) pushout square if and only if it is a (homotopy) pullback square. A diagram as above which satisfies these additional properties is called a *fibre-cofibre sequence*[26].

REMARK 2.2. Stable ∞-categories are defined in terms of *properties* and not in terms of additional *structure*, the latter being the case for triangulated categories. Roughly speaking, this is the reason why stable ∞-categories enjoy better *formal properties* than triangulated categories.

Every ∞-category $\mathcal{C}$ has an associated *homotopy category* $\operatorname{Ho}(\mathcal{C})$ which is in fact an ordinary category. The following result[27] relates stable ∞-categories to triangulated categories, see Theorem 1.1.2.14 in [**Lur17**].

PROPOSITION 2.3. *Let $\mathcal{A}$ be a stable ∞-category. Then, the homotopy category $\operatorname{Ho}(\mathcal{A})$ is (canonically) a triangulated category.*

REMARK 2.4. The suspension $\Sigma(x)$ of an object x of a stable ∞-category $\mathcal{A}$ is characterised by the existence of a fibre-cofibre sequence[28]

$$\begin{array}{ccc} x & \longrightarrow & 0 \\ \downarrow & \square & \downarrow \\ 0 & \longrightarrow & \Sigma(x) \end{array}$$

Similarly, exact triangles in the homotopy category $\operatorname{Ho}(\mathcal{A})$ are induced by diagrams in $\mathcal{A}$ of the form

$$\begin{array}{ccccc} x & \longrightarrow & y & \longrightarrow & 0 \\ \downarrow & \square & \downarrow & \square & \downarrow \\ 0 & \longrightarrow & z & \longrightarrow & \Sigma(x) \end{array}$$

in which each square is a fibre-cofibre sequence.

The good formal behaviour of stable ∞-categories is illustrated by the following statement[29], see Proposition 1.1.3.1 in [**Lur17**].

PROPOSITION 2.5. *Let $\mathcal{A}$ be a stable ∞-category and K a small ∞-category. Then, the ∞-category $\operatorname{Fun}(K,\mathcal{A})$ of functors $K \to \mathcal{A}$ is also stable.*

3. The higher Waldhausen S-constructions

In this section we establish a link between the higher dimensional Auslander algebras of type $\mathbb{A}$ and the higher Waldhausen S-constructions introduced by Dyckerhoff [**Dyc17**] and Poguntke [**Pog17**], and by Hesselholt and Madsen [**HM15**] in the case $m = 2$. We begin by reminding the reader of what is known in the classical situation, that is in the case $m = 1$.

[26] Fibre-cofibre sequences in stable ∞-categories are the ∞-categorical versions of exact triangles in triangulated categories.

[27] Proposition 2.3 should be compared with Happel's theorem which states that the stable category of a Frobenius exact category is a triangulated category [**Hap88**].

[28] The decoration indicates that the square is both a (homotopy) pushout and a (homotopy) pullback. In fact, in a stable ∞-category these two notions coincide, see Proposition 1.1.3.4 in [**Lur17**].

[29] Proposition 2.5 should be compared with the elementary fact that the category of functors from a small category into an abelian category is again abelian. Indeed, (homotopy) limits and (homotopy) colimits in functor ∞-categories are computed point-wise, see Corollary 5.1.2.3 in [**Lur09**].

3.1. The Waldhausen S-construction. The following construction is due to Waldhausen [**Wal85**]. It is the main ingredient in the definition of the algebraic K-theory space of a stable ∞-category[30], see for example [**BGT13**].

DEFINITION 3.1. Let $\mathcal{C}$ be a stable ∞-category and $n \geq 0$. We let $\mathsf{S}(\mathcal{C})_n$ be the (stable) ∞-category of diagrams of the form[31]

$$\begin{array}{ccccccccc} X_{00} \to & X_{01} \to & X_{02} \to & \cdots \longrightarrow & X_{0,n-1} & \longrightarrow & X_{0n} \\ & \downarrow & \downarrow & & \downarrow & & \downarrow \\ & X_{11} \to & X_{12} \to & \cdots \longrightarrow & X_{1,n-1} & \longrightarrow & X_{1n} \\ & & \downarrow & & \downarrow & & \downarrow \\ & & \ddots & \ddots & \vdots & & \vdots \\ & & & & \downarrow & & \downarrow \\ & & & & X_{n-1,n-1} & \to & X_{n-1,n} \\ & & & & & & \downarrow \\ & & & & & & X_{nn} \end{array}$$

which satisfy the following two conditions:

- For each $i \in [n]$ the object X_{ii} is a zero object of $\mathcal{C}$;
- for each $0 \leq i < j < k \leq n$ the square

$$\begin{array}{ccc} X_{ij} & \longrightarrow & X_{ik} \\ \downarrow & \square & \downarrow \\ X_{ii} & \longrightarrow & X_{jk} \end{array}$$

  is a (homotopy) pushout diagram and a (homotopy) pullback diagram.[32].

Size issues aside, the ∞-categories $\mathsf{S}(\mathcal{C})_n$ assemble into a simplicial object $\mathsf{S}(\mathcal{C})_\bullet$, called the *Waldhausen* S*-construction of* $\mathcal{C}$, which takes values in the ∞-category of stable ∞-categories and exact functors between them.

The following elementary observation can be viewed as a categorification of the isomorphism (4). Proofs can be found in Lemma 7.3 in [**BGT13**] and Lemma 1.2.2.4 in [**Lur17**]. A version in the related framework of stable derivators is proven in Theorem 4.6 in [**GŠ16a**] by means of a version of the knitting algorithm.

PROPOSITION 3.2 (Waldhausen). *Let $\mathcal{C}$ be a stable ∞-category and $n \geq 1$. The restriction functor*

$$\mathsf{S}(\mathcal{C})_n \longrightarrow \mathrm{Fun}(01 \to 02 \to \cdots 0n, \mathcal{C}),$$

which sends an object X of $\mathsf{S}(\mathcal{C})_n$ to the diagram

$$X_{01} \to X_{02} \to \cdots \to X_{0,n-1} \to X_{0n},$$

is an equivalence of (stable) ∞-categories.

---

[30] The algebraic K-theory space is a fundamental invariant which can be seen as a substantial refinement of the Grothendieck group.

[31] Such a diagram is precisely a functor $X \colon \Delta(1,n) \to \mathcal{C}$, where we endow the set $\Delta(1,n)$ of monotone maps $[1] \to [n]$ with the natural partial order.

[32] The exactness conditions imposed on the objects of $\mathsf{S}(\mathcal{C})_n$ should be thought of as a categorification of the Euler relations (3).

The following theorem extends the foregoing proposition to arbitrary orientations of the Dynkin diagram $\mathbb{A}_n$ and can be proven using combinatorial versions of the classical reflection functors. A proof, carried out in the related framework of stable derivators, can be found in [**GŠ16a**].

THEOREM 3.3 (Groth–Šťovíček). *Let $\mathcal{C}$ be a stable ∞-category and $n \geq 1$. Let S be a slice*[33] *in the poset $\Delta(1,n)$. The restriction functor*

$$\mathsf{S}(\mathcal{C})_{\bullet} \to \mathrm{Fun}(\underline{S}, \mathcal{C})$$

is an equivalence of (stable) ∞-categories.

Our aim in this section is to provide higher-dimensional versions of Proposition 3.2 and Theorem 3.3 in terms of certain higher-dimensional versions of the Waldhausen S-construction.

3.2. The higher Waldhausen S-constructions. Before proceeding we introduce further terminology. Let $I = [1]$ be the poset $\{0 \to 1\}$ and m a non-negative integer. An *m-cube* in an ∞-category $\mathcal{C}$ is a functor $X\colon I^m \to \mathcal{C}$. The isomorphism

$$I^{m+1} \cong I \times I^m$$

together with the adjunction

$$\mathrm{Fun}(I \times I^m, \mathcal{C}) \cong \mathrm{Fun}(I, \mathrm{Fun}(I^m, \mathcal{C}))$$

allow us to view an $(m+1)$-cube in $\mathcal{C}$ as morphism in the ∞-category $\mathrm{Fun}(I^m, \mathcal{C})$ of m-cubes in $\mathcal{C}$. In the case of a stable ∞-category, this identification allows for an inductive treatment of hyper-cubes as illustrated by the following definition.

DEFINITION 3.4. We say that a 0-cube in a stable ∞-category $\mathcal{A}$, which is nothing but an object of $\mathcal{A}$, is *biCartesian* if it is a zero object of $\mathcal{A}$. Inductively[34], we say that an $(m+1)$-cube X in a stable ∞-category $\mathcal{A}$ is *biCartesian* if its cofibre (taken in the stable ∞-category $\mathrm{Fun}(I^m, \mathcal{A})$) is a biCartesian m-cube in $\mathcal{A}$. For example, a 1-cube in $\mathcal{A}$ is biCartesian if its underlying morphism is an equivalence[35] in $\mathcal{A}$.

We are now ready to state the definition of the m-dimensional Waldhausen S-construction of a stable ∞-category [**HM15, Dyc17, Pog17**].

DEFINITION 3.5. Let $\mathcal{A}$ be a stable ∞-category and m a non-negative integer. For $n \geq 0$ we denote by $\mathsf{S}^{\langle m \rangle}(\mathcal{A})_n$ the full subcategory of $\mathrm{Fun}(\Delta(m,n), \mathcal{A})$ spanned by the diagrams X satisfying the following two conditions:

- For every degenerate m-simplex σ in Δ^n the object X_σ is a zero object of $\mathcal{A}$;

[33]As left implicit above, the Hasse quiver of $\Delta(1,n)$ can be identified with the Auslander–Reiten quiver of the quiver $\mathbb{A}_n$ (with additional degenerate vertices). Slices in $\Delta(1,n)$ can then be defined in the usual way.

[34]The notion of a biCartesian hyper-cube can be defined directly as a certain (homotopy) colimit diagram, see Proposition 1.2.4.13 and Lemma 1.2.4.5 in [**Lur17**].

[35]Equivalences are the ∞-categorical analogues of isomorphisms in ordinary category theory.

- for each non-degenerate $(m+1)$-simplex in Δ^n consider the $(m+1)$-cube $q\colon I^{m+1} \to \Delta(m,n)$ given by $q(v)_i = \sigma_{i+v_i}$ for each $v \in I^{m+1}$ and $i \in [m]$. Then, the induced $(m+1)$-cube

$$X \circ q\colon I^{m+1} \to \mathcal{A}$$

is biCartesian[36].

Size issues aside, the (stable) ∞-categories $\mathsf{S}^{\langle m \rangle}(\mathcal{A})_n$, $n \geq 0$, assemble into a simplicial object $\mathsf{S}^{\langle m \rangle}(\mathcal{A})_\bullet$, called the *$m$-dimensional Waldhausen* S*-construction of* $\mathcal{A}$, which takes values in the ∞-category of stable ∞-categories and exact functors between them.

REMARK 3.6. The second exactness conditions appearing in the definition of the m-dimensional Waldhausen S-construction are equivalent to the condition that the $(m+1)$-cubes of the form $X \circ q$ are biCartesian, where

$$q(v)_i = \sigma_i + v_i, \quad i \in [m]$$

and σ ranges over all non-degenerate m-simplices in Δ^n with $\sigma_m < n$. For example, for $m = 2$ these are the cubes

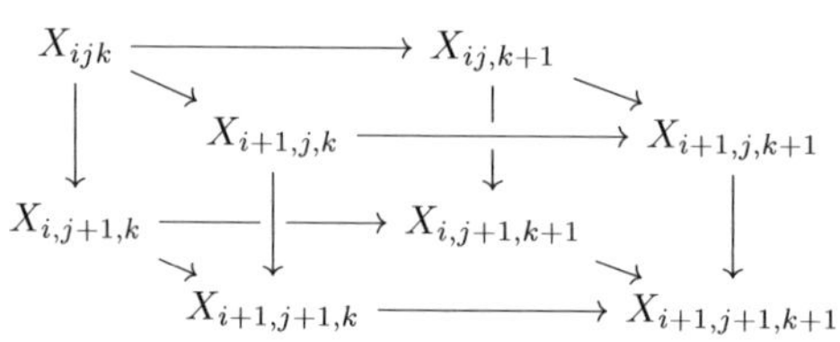

where $0 \leq i < j < k < n$. These conditions should be thought of as a categorification of the higher Auslander–Reiten relations (3).

EXAMPLE 3.7. Let $\mathcal{A}$ be a stable ∞-category, $m = 2$, and $n = 4$. After discarding redundant information pertaining additional zero objects, an object of

[36]These exactness conditions should be thought of as categorifications of the higher-dimensional Euler relations (6). For example, for $m = 2$ the cubes which are required to be biCartesian are those of the form

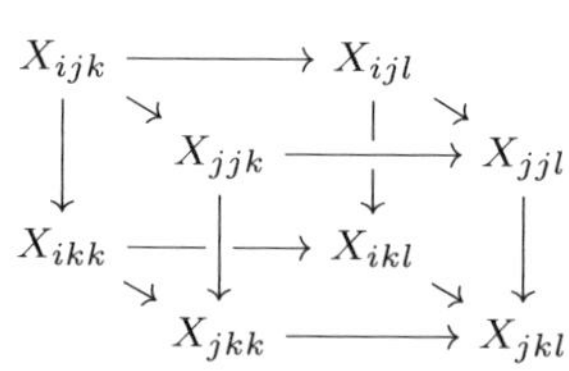

where $0 \leq i < j < k < l \leq n$; these cubes are ∞-categorical analogues of 4-term exact sequences.

$\mathsf{S}^{\langle 2\rangle}(\mathcal{A})_4$ can be identified with a diagram of the form

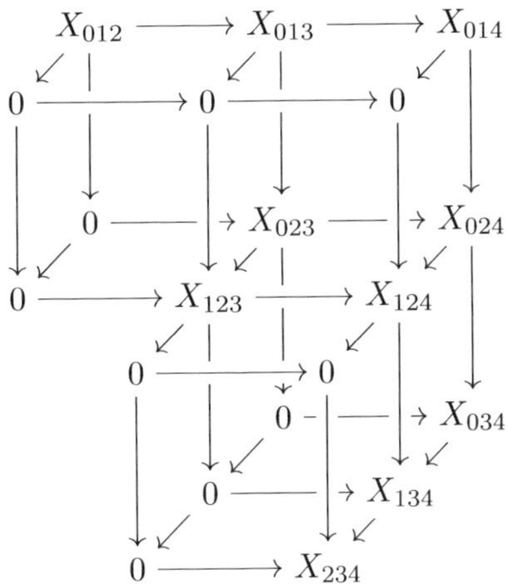

(8)

in which all 'unit cubes' are biCartesian. A standard application of the theory of Kan extensions in ∞-categories shows that such a diagram is determined by its restriction to the (coherently commutative) diagram

$$\begin{array}{ccccc} X_{012} & \longrightarrow & X_{013} & \longrightarrow & X_{014} \\ \downarrow & & \downarrow & & \downarrow \\ 0 & \longrightarrow & X_{023} & \longrightarrow & X_{024} \\ & & \downarrow & & \downarrow \\ & & 0 & \longrightarrow & X_{034} \end{array}$$

The latter diagram can be thought of as a (coherent) representation of the Auslander algebra $\mathbb{A}_3^{(2)}$ of the quiver $\mathbb{A}_3$ in the stable ∞-category $\mathcal{A}$. It is also worth noting that diagram (8) agrees with the 'higher-dimensional Auslander–Reiten quiver' of the unique cluster-tilting subcategory associated with $\mathbb{A}_3^{(2)}$, see Section 6 in [**Iya11**] or [**OT12**].

The following observation relates the m-dimensional Waldhausen S-construction to the m-dimensional Auslander algebras of type $\mathbb{A}$; it can be viewed as a higher-dimensional analogue of Proposition3.2. It can be proven either constructively, by means of a higher-dimensional version of the knitting algorithm[37], or by means of the point -wise formulas for ∞-categorical Kan extensions[38]. A proof can be found in Proposition 2.10 in [**DJW19b**].

PROPOSITION 3.8 (Dyckerhoff–J–Walde). *Let $\mathcal{A}$ be a stable ∞-category and $n \geq m \geq 1$. The restriction functor*

$$\mathsf{S}^{\langle m\rangle}(\mathcal{A})_m \longrightarrow \operatorname{Fun}_*(P(m,n),\mathcal{A})$$

[37]This approach is analogous to the proof of Theorem 4.6 in [**GŠ16a**]. The aforementioned knitting algorithm can be deduced, for example, from the proof of Theorem 5.27 in [**IO11**].

[38]This approach is analogous to the proof of Proposition 3.2 given for example in Lemma 1.2.2.4 in [**Lur17**].

is an equivalence of (stable) ∞-categories, where

$$P(m,n) := \{ \sigma \in \Delta(m,n) \mid \sigma_0 = 0 \}$$

and $\operatorname{Fun}_*(P(m,n),\mathcal{A})$ *is the full subcategory of* $\operatorname{Fun}(P(m,n),\mathcal{A})$ *spanned by those functors which send degenerate m-simplices in Δ^n to zero objects in $\mathcal{A}$.*

REMARK 3.9. After discarding further redundant information, the poset $P(m,n)$ appearing in in Proposition 3.8 can be replaced by a smaller poset which models perfectly the quiver with relations of the higher Auslander algebra $\mathbb{A}_{n-m+1}^{(m)}$. This fact is perhaps more transparent from the description of Iyama's higher Auslander algebras of type $\mathbb{A}$ given in Section 2 in [**JK16**].

As explained in [**IO11**], the following theorem can be regarded as a higher-dimensional version of Theorem 3.3. A proof can be found in Theorem 2.41 in [**DJW19b**].

THEOREM 3.10 (Dyckerhoff–J–Walde). *Let $\mathcal{A}$ be a stable ∞-category and $n \geq m \geq 1$. Let S be a slice*[39] *in the poset $\Delta(m,n)$. The restriction functor*

$$\mathsf{S}^{\langle m\rangle}(\mathcal{A})_n \to \operatorname{Fun}_*(\underline{S},\mathcal{A})$$

is an equivalence of (stable) ∞-categories, where $\underline{S}$ denotes the convex hull of S in the poset $\Delta(m,n)$.

REMARK 3.11. The proof of Theorem 3.10 relies on combinatorial versions of the derived equivalences induced by higher-dimensional reflection functors[40] in the sense of [**IO11**]; these reflection functors rely on the operation of slice mutation also introduced in [**IO11**]. In more detail, if S and S' are slices in $\Delta(m,n)$ which are mutation of each other, then there exists a slightly larger poset $\underline{S \diamond S'}$ containing both S and S' as well as a distinguished $(m+1)$-cube. This larger poset allows us to realise the aforementioned reflection functors by means of equivalences of stable ∞-categories

$$\operatorname{Fun}_*(\underline{S},\mathcal{A}) \xleftarrow{\simeq} \operatorname{Fun}_*^{ex}(\underline{S \diamond S'},\mathcal{A}) \xrightarrow{\simeq} \operatorname{Fun}_*(\underline{S'},\mathcal{A})$$

induced by the restriction functors, where the objects of the stable ∞-category $\operatorname{Fun}_*^{ex}(\underline{S \diamond S'},\mathcal{A})$ satisfy the additional requirement that their restriction along the distinguished $(m+1)$-cube in $\underline{S \diamond S'}$ is a biCartesian $(m+1)$-cube in $\mathcal{A}$, see Figure 1. The proof of Theorem 3.10 is obtained by combining these equivalences with Proposition 3.8 and the transitivity of the mutation operation on slices proven in [**IO11**]. See [**DJW19b**] for details.

3.3. Recollements. We make the following general observation, which makes the inductive nature of the m-dimensional Waldhausen S-construction of $\mathcal{A}$ readily apparent. Indeed, Proposition 3.12 is reminiscent of Iyama's inductive description of the higher Auslander algebras of type $\mathbb{A}$ by means of cones of translation quivers, see Section 6 in [**Iya11**]. A proof can be found in Proposition 2.51 in [**DJW19b**].

---

[39]The notion of a 'slice' is analogous to that introduced in [**IO11**].

[40]Combinatorial versions of classical reflection functors are investigated in [**GŠ16a**] in the case of quivers of type $\mathbb{A}$ and in [**GŠ16b**] for general trees in the related framework of stable derivators. A further generalisation, corresponding to the generalised reflection functors of Ladkani [**Lad08**], was obtained in [**DJW19a**].

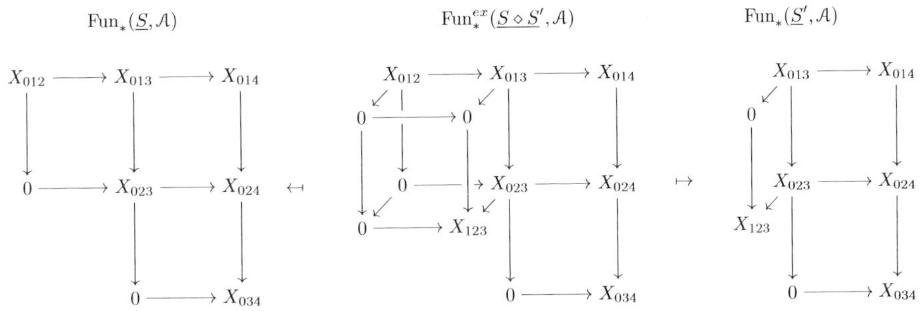

FIGURE 1. An example of slice mutation.

PROPOSITION 3.12 (Dyckerhoff–J–Walde). *Let $\mathcal{A}$ be a stable ∞-category and $n \geq m \geq 1$ integers. For each $i \in [n]$ the functor $s_i \colon \mathsf{S}^{\langle m\rangle}(\mathcal{A})_n \to \mathsf{S}^{\langle m\rangle}(\mathcal{A})_{n+1}$ is part of a recollement of stable ∞-categories*

$$\mathsf{S}^{\langle m\rangle}(\mathcal{A})_n \underset{\xleftarrow{d_{i+1}}}{\overset{\xleftarrow{d_i}}{\xrightarrow{s_i}}} \mathsf{S}^{\langle m\rangle}(\mathcal{A})_{n+1} \underset{\longleftarrow}{\overset{\longleftarrow}{\longrightarrow}} \mathsf{S}^{\langle m-1\rangle}(\mathcal{A})_n.$$

In particular, the sequence of adjunctions

$$d_0 \dashv s_0 \dashv d_1 \dashv s_1 \dashv \cdots \dashv s_n \dashv d_{n+1}$$

is part of a ladder of recollements in the sense of [**BBD82, AHKLY17**].

Acknowledgments

The author would like to acknowledge the numerous conversations with Julian Külshammer concerning the combinatorial treatment of the higher-dimensional Auslander algebras of type $\mathbb{A}$ which helped shape the author's viewpoint on the subject. The author also thanks Marius Thaule and Tashi Walde for comments on a preliminary version of this note. Versions of Proposition 3.8 and Proposition 3.12 have been obtained by Beckert in his PhD thesis [**Bec18**] in the related framework of stable derivators.

References

[AHKLY17] L. Angeleri Hügel, S. Koenig, Q. Liu, and D. Yang, *Ladders and simplicity of derived module categories*, J. Algebra **472** (2017), 15–66, DOI 10.1016/j.jalgebra.2016.10.023. MR3584869

[Aus84] M. Auslander, *Relations for Grothendieck groups of Artin algebras*, Proc. Amer. Math. Soc. **91** (1984), no. 3, 336–340, DOI 10.2307/2045297. MR744624

[BBD82] A. A. Beĭlinson, J. Bernstein, and P. Deligne, *Faisceaux pervers* (French), Analysis and topology on singular spaces, I (Luminy, 1981), Astérisque, vol. 100, Soc. Math. France, Paris, 1982, pp. 5–171. MR751966

[Bec18] F. Beckert, *The bivariant parasimplicial $S_\bullet$-construction*, Ph.d. thesis, Bergische Universität Wuppertal, July 2018.

[BGT13] A. J. Blumberg, D. Gepner, and G. Tabuada, *A universal characterization of higher algebraic K-theory*, Geom. Topol. **17** (2013), no. 2, 733–838, DOI 10.2140/gt.2013.17.733. MR3070515

[BT14] P. A. Bergh and M. Thaule, *The Grothendieck group of an n-angulated category*, J. Pure Appl. Algebra **218** (2014), no. 2, 354–366, DOI 10.1016/j.jpaa.2013.06.007. MR3120636

[Cis19] D.-C. Cisinski, *Higher categories and homotopical algebra*, Cambridge Studies in Advanced Mathematics, vol. 180, Cambridge University Press, Cambridge, 2019. MR3931682

[DJW19a] T. Dyckerhoff, G. Jasso, and T. Walde, *Generalised BGP reflection functors via the Grothendieck construction*, Int. Math. Res. Not. IMRN **11** (2019). DOI 10.1093/imrn/rnz194.

[DJW19b] T. Dyckerhoff, G. Jasso, and T. Walde, *Simplicial structures in higher Auslander–Reiten theory*, Adv. Math. **355** (2019), 106762. MR3994443

[Dol58] A. Dold, *Homology of symmetric products and other functors of complexes*, Ann. of Math. (2) **68** (1958), 54–80, DOI 10.2307/1970043. MR97057

[Dyc17] T. Dyckerhoff, *A categorified Dold–Kan correspondence*, arXiv:1710.08356 (2017).

[Fao17a] G. Faonte, *A_∞-functors and homotopy theory of dg-categories*, J. Noncommut. Geom. **11** (2017), no. 3, 957–1000, DOI 10.4171/JNCG/11-3-6. MR3713010

[Fao17b] G. Faonte, *Simplicial nerve of an A_∞-category*, Theory Appl. Categ. **32** (2017), Paper No. 2, 31–52. MR3607208

[GJ99] P. G. Goerss and J. F. Jardine, *Simplicial homotopy theory*, Progress in Mathematics, vol. 174, Birkhäuser Verlag, Basel, 1999. MR1711612

[Gro10] M. Groth, *A short course on ∞-categories*, arXiv:1007.2925 (2010).

[GŠ16a] M. Groth and J. Šťovíček, *Abstract representation theory of Dynkin quivers of type A*, Adv. Math. **293** (2016), 856–941, DOI 10.1016/j.aim.2016.02.018. MR3474336

[GŠ16b] M. Groth and J. Šťovíček, *Tilting theory for trees via stable homotopy theory*, J. Pure Appl. Algebra **220** (2016), no. 6, 2324–2363, DOI 10.1016/j.jpaa.2015.11.009. MR3448799

[GŠ18] M. Groth and J. Šťovíček, *Tilting theory via stable homotopy theory*, J. Reine Angew. Math. **743** (2018), 29–90, DOI 10.1515/crelle-2015-0092. MR3859269

[Hap88] D. Happel, *Triangulated categories in the representation theory of finite-dimensional algebras*, London Mathematical Society Lecture Note Series, vol. 119, Cambridge University Press, Cambridge, 1988. MR935124

[HM15] L. Hesselholt and I. Madsen, *Real algebraic K-theory*, Unpublished, April 2015.

[Hov99] M. Hovey, *Model categories*, Mathematical Surveys and Monographs, vol. 63, American Mathematical Society, Providence, RI, 1999. MR1650134

[IO11] O. Iyama and S. Oppermann, *n-representation-finite algebras and n-APR tilting*, Trans. Amer. Math. Soc. **363** (2011), no. 12, 6575–6614, DOI 10.1090/S0002-9947-2011-05312-2. MR2833569

[Iya07] O. Iyama, *Higher-dimensional Auslander-Reiten theory on maximal orthogonal subcategories*, Adv. Math. **210** (2007), no. 1, 22–50, DOI 10.1016/j.aim.2006.06.002. MR2298819

[Iya11] O. Iyama, *Cluster tilting for higher Auslander algebras*, Adv. Math. **226** (2011), no. 1, 1–61, DOI 10.1016/j.aim.2010.03.004. MR2735750

[JK16] G. Jasso and J. Külshammer, *Higher Nakayama algebras I: Construction*, Adv. Math. **351** (2019), 1139–1200, DOI 10.1016/j.aim.2019.05.026. With an appendix by Külshammer and Chrysostomos Psaroudakis and an appendix by Sondre Kvamme. MR3959141

[Joy02] A. Joyal, *Quasi-categories and Kan complexes*, J. Pure Appl. Algebra **175** (2002), no. 1-3, 207–222, DOI 10.1016/S0022-4049(02)00135-4. Special volume celebrating the 70th birthday of Professor Max Kelly. MR1935979

[Kan58] D. M. Kan, *Functors involving c.s.s. complexes*, Trans. Amer. Math. Soc. **87** (1958), 330–346, DOI 10.2307/1993103. MR131873

[Lad08] S. Ladkani, *Homological properties of finite partially ordered sets*, Ph.D. thesis, Hebrew University of Jerusalem, 2008.

[Lur09] J. Lurie, *Higher topos theory*, Annals of Mathematics Studies, vol. 170, Princeton University Press, Princeton, NJ, 2009. MR2522659

[Lur17] Jacob Lurie, *Higher algebra*, May 2017, Available online at `http://www.math.harvard.edu/~lurie/`.

[OT12] S. Oppermann and H. Thomas, *Higher-dimensional cluster combinatorics and representation theory*, J. Eur. Math. Soc. (JEMS) **14** (2012), no. 6, 1679–1737, DOI 10.4171/JEMS/345. MR2984586

[Pog17] T. Poguntke, *Higher Segal structures in algebraic K-theory*, arXiv:1709.06510 (2017).

[Qui67] D. G. Quillen, *Homotopical algebra*, Lecture Notes in Mathematics, No. 43, Springer-Verlag, Berlin-New York, 1967. MR0223432

[Wal85] F. Waldhausen, *Algebraic K-theory of spaces*, Algebraic and geometric topology (New Brunswick, N.J., 1983), Lecture Notes in Math., vol. 1126, Springer, Berlin, 1985, pp. 318–419, DOI 10.1007/BFb0074449. MR802796

Mathematisches Institut, Rheinische Friedrich-Wilhelms-Universität Bonn, Endenicher Allee 60, D-53115 Bonn, Germany

Email address: gjasso@math.uni-bonn.de

URL: http://gustavo.jasso.info

Contemporary Mathematics
Volume **758**, 2020
https://doi.org/10.1090/conm/758/15239

A survey on maximal green sequences

Bernhard Keller and Laurent Demonet

Abstract. Maximal green sequences appear in the study of Fomin–Zelevinsky's cluster algebras. They are useful for computing refined Donaldson–Thomas invariants, constructing twist automorphisms and proving the existence of theta bases and generic bases. We survey recent progress on their existence and properties and give a representation-theoretic proof of Greg Muller's theorem stating that full subquivers inherit maximal green sequences. In the appendix, Laurent Demonet describes maximal chains of torsion classes in terms of bricks generalizing a theorem by Igusa.

1. Introduction

A quiver is an oriented graph. Quiver mutation is an elementary operation on quivers. It is the basic combinatorial ingredient of Fomin–Zelevinsky's definition of cluster algebras [**46**]. In this definition, one recursively constructs generators for the cluster algebra by repeatedly mutating an initial seed (Q, x) consisting of a quiver Q and a set x of indeterminates associated with the vertices of Q. The construction process is recorded in a graph, the *exchange graph*, whose vertices are the seeds obtained from (Q, x) by iterated mutation and whose edges correspond to mutations. By definition, the edges of the exchange graph are unoriented. However, it was noticed early on [**89**] that there is a natural partial order on seeds whose minimal inclusions correspond to edges of the exchange graph, which thus becomes oriented. For example, for a linearly oriented quiver of type A_n, the poset of seeds is the nth Tamari lattice [**71, 109**].

Maximal green sequences were invented in [**74**] (and became part of [**77**]) but are already implicit in the work of Gaiotto–Moore–Neitzke [**51**] (published in [**52**]). A maximal green sequence is a (finite) path in the oriented exchange graph from the unique smallest element to the unique largest element. Not all quivers have maximal green sequences but they do exist for important classes of quivers and their existence has important consequences: it yields explicit formulas for Kontsevich–Soibelman's refined Donaldson–Thomas invariant associated [**80**] with the quiver, for the twist automorphism [**58**] of the associated cluster algebra and it is a sufficient condition for the existence of a theta basis [**60**] and a generic basis [**102**] in the upper cluster algebra.

2010 *Mathematics Subject Classification.* Primary 13F60; Secondary 14N35, 18E40, 18E30.
Key words and phrases. Cluster algebra, maximal green sequence, Donaldson–Thomas invariant, theta basis, generic basis, twist automorphisms, torsion class.

In section 2, we review the purely combinatorial definitions of mutation and green mutation leading to the notion of (maximal) green sequence (and, more generally, reddening sequence). In section 3, we describe the applications of maximal green sequences mentioned above. We then report on results concerning the existence and non existence of maximal green sequences (section 4). In particular, we state Greg Muller's theorem to the effect that full subquivers inherit maximal green sequences. In the final section 5, we give a proof of Muller's theorem based on recent results in the study of torsion classes [**33**]. In the appendix, Laurent Demonet establishes a bijection between maximal chains of torsion classes and maximal forward Hom-orthogonal sequences of bricks over a finite-dimensional algebra generalizing a theorem proved by Igusa [**65**] for representation-finite Jacobi algebras.

2. Mutation and green mutation

2.1. Quiver mutation. A *quiver* is an oriented graph, i.e. a quadruple $Q = (Q_0, Q_1, s, t)$ formed by a set of vertices Q_0, a set of arrows Q_1 and two maps s and t from Q_1 to Q_0 which send an arrow α respectively to its source $s(\alpha)$ and its target $t(\alpha)$. In practice, a quiver is given by a picture as in the following example

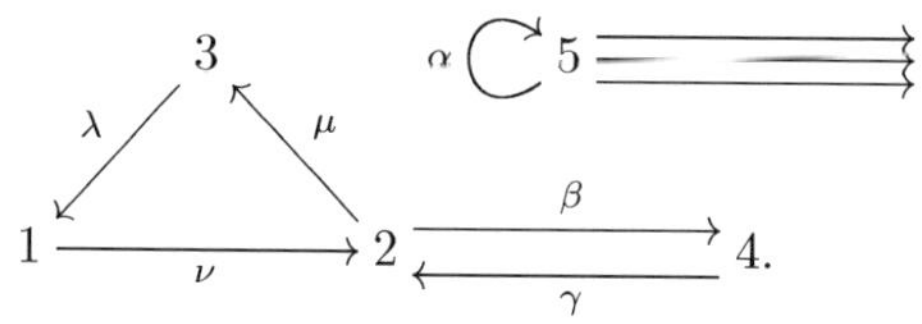

An arrow α whose source and target coincide is a *loop*; a 2-*cycle* is a pair of distinct arrows β and γ such that $s(\beta) = t(\gamma)$ and $t(\beta) = s(\gamma)$. Similarly, one defines n-*cycles* for any positive integer n. A vertex i of a quiver is a *source* (respectively a *sink*) if there is no arrow with target i (respectively with source i).

By convention, in the sequel, by a quiver, we always mean a finite quiver without loops nor 2-cycles whose set of vertices is the set of integers from 1 to n for some $n \geqslant 1$. Up to an isomorphism fixing the vertices, such a quiver Q is given by the *skew-symmetric matrix* $B = B_Q$ whose coefficient b_{ij} is the difference between the number of arrows from i to j and the number of arrows from j to i for all $1 \leqslant i, j \leqslant n$. Conversely, each skew-symmetric matrix B with integer coefficients comes from a quiver.

Let Q be a quiver and k a vertex of Q. The *mutation* $\mu_k(Q)$ is the quiver obtained from Q as follows:

1) for each subquiver $i \xrightarrow{\beta} k \xrightarrow{\alpha} j$, we add a new arrow $[\alpha\beta] : i \to j$;
2) we reverse all arrows with source or target k;
3) we remove the arrows in a maximal set of pairwise disjoint 2-cycles.

For example, if k is a source or a sink of Q, then the mutation at k simply reverses all the arrows incident with k. In general, if B is the skew-symmetric matrix associated with Q and B' the one associated with $\mu_k(Q)$, we have

$$b'_{ij} = \begin{cases} -b_{ij} & \text{if } i = k \text{ or } j = k \text{ ;} \\ b_{ij} + \operatorname{sgn}(b_{ik}) \max(0, b_{ik} b_{kj}) & \text{else.} \end{cases} \tag{2.1.1}$$

This is the *matrix mutation rule* introduced by Fomin-Zelevinsky in [**46**], cf. also [**48**]. It applies more generally to skew-symmetrizable matrices, which correspond to *valued quivers*, cf. section 3.3 of [**78**].

One checks easily that μ_k is an involution. For example, the quivers

(2.1.2)

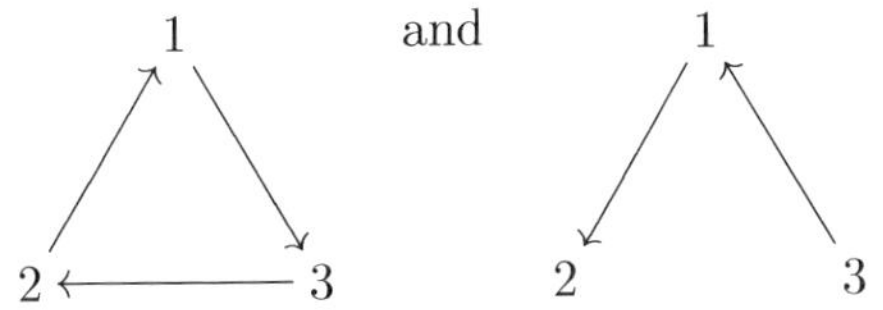

are linked by a mutation at the vertex 1. Notice that these quivers are drastically different: The first one is a cycle, the second one the Hasse diagram of a linearly ordered set.

Two quivers are *mutation equivalent* if they are linked by a finite sequence of mutations. For example, it is an easy exercise to check that any two orientations of a tree are mutation equivalent. Using the quiver mutation applet [**76**] or the Sage package [**94**] one can check that the following three quivers are mutation equivalent

(2.1.3)

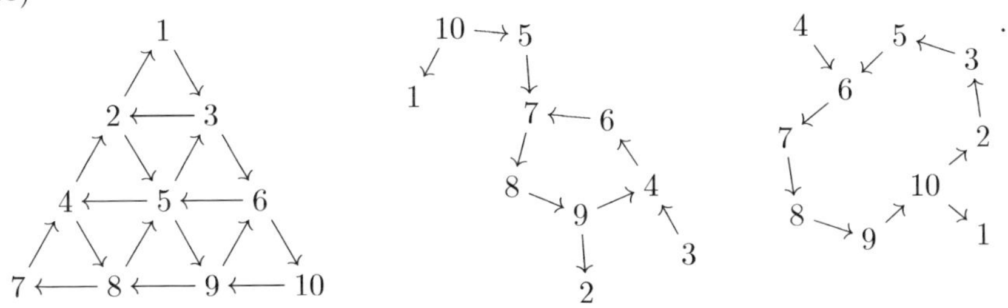

The common *mutation class* of these quivers contains 5739 quivers (up to isomorphism). The mutation class of 'most' quivers is infinite. The classification of the quivers having a finite mutation class was achieved by Felikson-Shapiro-Tumarkin [**42, 43**]: in addition to the quivers associated with triangulations of surfaces (with boundary and marked points, cf. [**45**]) the list contains 11 exceptional quivers, the largest of which is in the mutation class of the quivers (2.1.3).

2.2. Green quiver mutation. Let Q be a quiver. The *framed quiver* $\widehat{Q}$ is obtained from Q by adding, for each vertex i, a new vertex i' and a new arrow $i \to i'$. For example, if Q is the quiver $1 \to 2$, then the framed quiver $\widehat{Q}$ is

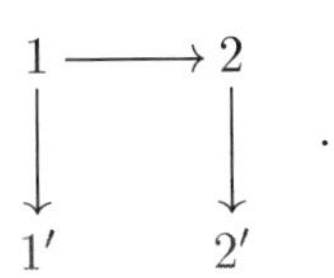

The new vertices i' are called *frozen* vertices, because we never mutate at them. Now suppose that we have transformed $\widehat{Q}$ into a quiver R by a finite sequence of mutations (at non frozen vertices). A non frozen vertex i is *green* in R if it is not the target of any arrows $j' \to i$ from frozen vertices j' in R. It is *red* if it is not the source of any arrows $i \to j'$ to frozen vertices of R, cf. Figure 1.

THEOREM 2.3 (Derksen-Weyman-Zelevinsky [**36**]). *Each non frozen vertex of R is either green or red.*

The proof is based on the theory of mutations of quivers with potential and their decorated representations developed in [**35, 36**]. Alternative proofs of the

FIGURE 1. Green ⊙ and red • vertices in R

theorem were given in [**100**] (via the cluster category) and in [**96**] (via Donaldson–Thomas theory). An important generalization to valued quivers is proved [**60**]. As shown in [**97**], the theorem is central in the theory of cluster algebras.

The *c-vector* $\alpha_i \in \mathbb{Z}^n$ associated with a non frozen vertex i of R is the integer vector whose jth component is the difference between the number of arrows from i to j' minus the number of arrows from j' to i in R. By the theorem, each c-vector has either all components $\geqslant 0$ or all components $\leqslant 0$ (sign-coherence of c-vectors). The following definition was first given in [**74, 77**]. A sequence $\mathbf{i} = (i_1, \ldots, i_N)$ of vertices is *green* if, for each $1 \leqslant t \leqslant N$, the vertex i_t is green in the partially mutated quiver

$$\mu_{i_{t-1}} \cdots \mu_{i_2} \mu_{i_1}(\widehat{Q}).$$

It is *maximal green* if moreover all non frozen vertices of $\mu_{\mathbf{i}}(\widehat{Q})$ are red. It is *reddening* [**75**] (or *green-to-red* [**92**]) if it is not necessarily green but all non frozen vertices of $\mu_{\mathbf{i}}(\widehat{Q})$ are red.

In Figure 2, we see that the quiver $\vec{A}_2$ has exactly two maximal green sequences: $(1,2)$ and $(2,1,2)$ (green vertices are encircled). The final quivers in the two sequences are isomorphic by a *frozen isomorphism* (i.e. an isomorphism which fixes the frozen vertices) to the *coframed quiver* $\check{Q}$, which is obtained from Q by adding, for each vertex i, a new vertex i' and an arrow $i' \to i$. This is a general phenomenon:

PROPOSITION 2.4 (Prop. 2.10 of [**11**]). *Suppose that Q admits a reddening sequence $\mathbf{i}$. Then there is a unique isomorphism $\mu_{\mathbf{i}}(\widehat{Q}) \xrightarrow{\sim} \check{Q}$ fixing the frozen vertices and sending a non frozen vertex i to $\sigma(i)$ for a unique permutation σ of the vertices of Q.*

For maximal green sequences of quivers in the mutation class of A_n, the permutation σ is studied in [**56, 69**]. In general, it remains mysterious.

It is a fact that large classes of quivers appearing in Lie theory and higher Teichmüller theory do admit maximal green sequences. We refer to section 4 below for classes of examples. This is interesting because of the applications sketched in the next section.

3. Applications

3.1. Refined Donaldson–Thomas invariants. The *quantum dilogarithm series* is defined by

$$\begin{aligned} \mathbb{E}(y) &= 1 + \frac{q^{1/2}}{q-1} \cdot y + \cdots + \frac{q^{n^2/2} y^n}{(q^n-1)(q^n-q)\cdots(q^n-q^{n-1})} + \cdots \\ &\in \mathbb{Q}(q^{1/2})[[y]], \end{aligned}$$

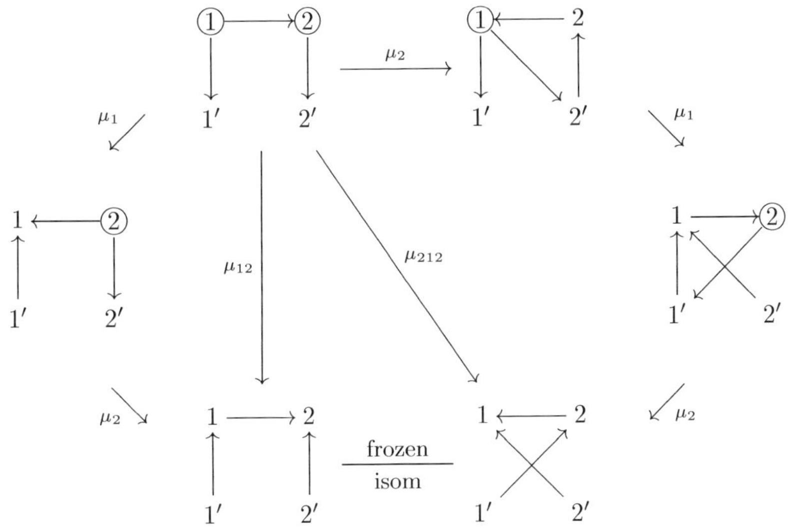

FIGURE 2. The two maximal green sequences for $\vec{A}_2$

where $q^{1/2}$ is an indeterminate whose square is denoted by q and y is an indeterminate. This series is a classical object with many remarkable properties, cf. for example [**116**].

Let Q be a quiver with n vertices. Let $\lambda_Q : \mathbb{Z}^n \times \mathbb{Z}^n \to \mathbb{Z}$ be the bilinear antisymmetric form associated with the matrix B_Q. Define the *complete quantum affine space* as the algebra

$$\widehat{\mathbb{A}}_Q = \mathbb{Q}(q^{1/2})\langle\langle y^\alpha, \alpha \in \mathbb{N}^n \mid y^\alpha y^\beta = q^{1/2\,\lambda(\alpha,\beta)} y^{\alpha+\beta} \rangle\rangle.$$

This is a slightly non commutative deformation of an ordinary commutative power series algebra in the n indeterminates $y_i = y^{e_i}$, where e_i is the ith vector of the standard basis of $\mathbb{Z}^n$. We write $\mathbb{A}_Q$ for the non completed variant of $\widehat{\mathbb{A}}_Q$. For a sequence $\mathbf{i} = (i_1, \dots, i_N)$ of vertices of Q, we define

$$\mathbb{E}_{Q,\mathbf{i}} = \mathbb{E}(y^{\varepsilon_1 \beta_1})^{\varepsilon_1} \cdots \mathbb{E}(y^{\varepsilon_N \beta_N})^{\varepsilon_N},$$

where the product is taken in $\widehat{\mathbb{A}}_Q$, the vector β_t is the c-vector associated with the vertex i_t of the partially mutated quiver

$$\mu_{i_{t-1}} \cdots \mu_{i_1}(\widehat{Q})$$

and ε_i is the common sign of its entries, $1 \leqslant t \leqslant N$. Notice that by the sign-coherence, each factor does belong to $\widehat{\mathbb{A}}_Q$.

Let $RDT_Q \in \widehat{\mathbb{A}}_Q$ denote the refined Donaldson–Thomas invariant of Q constructed by Kontsevich–Soibelman [**80, 82, 83**]. Notice that their construction has not yet been made completely rigorous due to technical difficulties arising from the fact that the potentials needed are *infinite* linear combinations of cycles. The following theorem was the motivation for [**74**]. Independently, it was discovered in the study of the BPS spectrum by Gaiotto–Moore–Neitzke in [**51**] and used in the physics literature, for example in [**2, 25**], cf. [**115**] and the references given there.

THEOREM 3.2. *If Q admits a reddening sequence* $\mathbf{i}$, *we have*

$$RDT_Q = \mathbb{E}_{Q,\mathbf{i}} \quad in \quad \widehat{\mathbb{A}}_Q.$$

The theorem is proved in section 7 of [**73**] under the assumption that RDT_Q is well-defined. It implies in particular that the right hand side is independent of the choice of $\mathbf{i}$. This was conjectured in [**95**] and can be proved rigorously using the theory of cluster algebras and their additive categorification, cf. section 7.11 of [**73**].

Thanks to the theorem, each pair $(\mathbf{i}, \mathbf{i}')$ of reddening sequences yields a quantum dilogarithm identity. For example, the two maximal green sequences of the quiver $Q = \vec{A}_2$ yield the *pentagon identity*

$$\mathbb{E}(y_1)\mathbb{E}(y_2) = \mathbb{E}(y_2)\mathbb{E}(q^{-1/2}y_1y_2)\mathbb{E}(y_1) \tag{3.2.1}$$

due to [**39**] and [**40**], cf. [**111**] for a recent account. Analogous pairs of maximal green sequences exist for all Dynkin quivers and yield generalizations of the pentagon identity due to Reineke [**104**], cf. [**77**]. These identities further generalize to square products [**79**] of Dynkin quivers, cf. [**77**]. For pairs (A_n, A_m), an alternative geometric proof of the identities is given by Allman–Rimanyi [**3**]. We refer to section 5 of [**75**] for more examples.

One may ask whether all the quantum dilogarithm identities obtained from the theorem are in fact consequences of the pentagon identity. We doubt this but it does hold if Q is an ayclic (valued) quiver, as shown by Hermes–Igusa [**62**].

In [**59**], Goncharov–Shen construct maximal green sequences for a large class of quivers appearing in higher Teichmüller theory and apply the theorem to obtain the corresponding Donaldson–Thomas invariants. Similar results for Grassmannians and double Bruhat cells are due to Weng [**112–114**] and for double Bott–Samelson cells to Shen–Weng [**108**].

3.3. Twist automorphisms. Let Q be a quiver admitting a reddening sequence $\mathbf{i}$. Let σ be the permutation of Prop. 2.4. Then the composition of the mutation sequence $\mu_{\mathbf{i}}$ with the permutation σ transforms the initial seed (Q, x) of the cluster algebra $\mathcal{A}_Q$ into a seed of the form (Q, u) and thus yields an automorphism

$$\mathsf{tw} : \mathcal{A}_Q \xrightarrow{\sim} \mathcal{A}_Q \ , \ x_i \mapsto u_i$$

called the *twist automorphism* of $\mathcal{A}_Q$.

From the independence of $\mathbb{E}_{Q,\mathbf{i}}$ of the choice of the reddening sequence $\mathbf{i}$, one can deduce that the twist automorphism tw is also independent of $\mathbf{i}$ (*cf.* section 4 of [**77**] and section 6.4 of [**78**]).

As shown by Geiss–Leclerc–Schröer [**58**], for unipotent cells of Kac–Moody groups, the twist automorphism identifies with the chamber ansatz of Berenstein–Fomin–Zelevinsky [**8**]. It has found important applications in the work of Marsh–Scott [**88**], Muller–Speyer [**93**], Rietsch–Williams [**105**], Cautis–Williams [**24**], ...

3.4. The Fock–Goncharov conjectures. Let Q be a valued quiver. The Langlands dual valued quiver Q^L is obtained by reversing all the valuations of Q (without changing the arrows). The skew-symmetrizable matrix B_{Q^L} is the opposite transpose $-B_Q^T$ of B_Q. Thus, we have $Q = Q^L$ if Q is an ordinary (i.e. equivalued) quiver. Suppose that Q is obtained from an ice quiver $\tilde{Q}$ by removing all the frozen vertices and all the arrows incident with them. Assume that the exchange matrix associated with $\widetilde{Q}$ is of maximal rank.

THEOREM 3.5 (Gross–Hacking–Keel–Kontsevich [**60**]). *If Q has a reddening sequence, then the Fock–Goncharov duality conjectures* [**44**] *hold for Q and in particular the upper cluster algebra $\mathcal{U}_{\widetilde{Q}}$ admits a basis parametrized by the tropical points of the cluster Poisson variety associated with $\widetilde{Q}^L$.*

The basis constructed by Gross–Hacking–Keel–Kontsevich is known as the *theta basis*. *Generic bases* were first considered by Dupont [**38**] for acyclic quivers. They are constructed using generic values of cluster characters [**22, 99**]. In [**58**], Geiss–Leclerc–Schröer showed that they exist for large classes of cluster algebras arising in Lie theory and coincide with Lusztig's dual semi-canonical bases. Plamondon showed in [**101**] that generic bases are also canonically parametrized by the tropical points of the cluster Poisson variety.

THEOREM 3.6 (Qin [**102**]). *Let Q be an (equivalued) quiver. If Q has a reddening sequence, then the upper cluster algebra $\mathcal{U}_{\widetilde{Q}}$ admits a generic basis parametrized by the tropical points of the cluster Poisson variety associated with $\widetilde{Q}$.*

Notice that these results concern the upper cluster algebra. It is expected that the existence of a maximal green, or reddening, sequence should have implications for the relationship between the cluster algebra and the upper cluster algebra. Notice however that this relationship depends on the choice of coefficients [**41**, Example 8.3] [**20**], whereas the oriented exchange graph, and thus the existence of a maximal green sequence, does not [**26**]. A conjecture on the precise relationship is formulated in [**90**].

4. Existence and properties

4.1. Existence. Suppose that Q is an acyclic quiver. A *source sequence* for Q is an enumeration of the vertices of Q which is increasing for the partial order defined by the existence of a path. It is easy to check that each source sequence is a maximal green sequence for Q, cf. [**11**].

Recall that a valued quiver is cluster-finite (i.e. the associated cluster algebra has only finitely many cluster variables) iff it is mutation-equivalent to an orientation of a Dynkin diagram [**47**]. As observed in in [**11**], it is immediate from [**27**] that cluster-finite valued quivers have maximal green sequences.

For valued quivers mutation-equivalent to orientations of extended Dynkin diagrams, there are only finitely many maximal green sequences [**12**]. The same holds for (equally valued) acyclic quivers [**70**].

Two canonical maximal green sequences exist for the square product of two alternating valued Dynkin quivers, cf. section 5 of [**75**].

If R is an aylic quiver and $\widetilde{w}$ a reduced expression for an element of the Coxeter group associated with the underlying graph of R, there is a canonical quiver Q associated with the pair $(R, \widetilde{w})$. It serves to obtain a cluster structure on the coordinate algebra of the unipotent cell associated with w in the Kac–Moody group determined by R, cf. [**5**] and [**57**]. It is closely related to the (upper) cluster structures on Bruhat cells obtained in [**9**]. In section 13 of [**57**], Geiss–Leclerc–Schröer exhibit a canonical reddening sequence for Q (we conjecture that it is actually maximal green).

In analogy with the definition of a full subcategory, one defines a *full subquiver* Q' of a quiver Q to be a subquiver containing all the arrows in Q between any two of its vertices.

THEOREM 4.2 (Muller [**92**]). *If Q has a maximal green (resp. reddening) sequence, then each full subquiver $Q' \subseteq Q$ has a maximal green (resp. reddening) sequence.*

REMARK 4.3. Muller's proof shows that more precisely, starting from a maximal green sequence $(k_1, \ldots, k_N)$ for Q, one obtains a maximal green sequence for Q' as follows: Let $(c_1, \ldots, c_N)$ be the sequence of c-vectors associated with the given sequence $(k_1, \ldots, k_N)$. Form the subsequence $(c'_1, \ldots, c'_M)$ of $(c_1, \ldots, c_N)$ formed by those vectors supported on Q'. Then there is a unique sequence of mutations $(k'_1, \ldots, k'_M)$ of Q' whose associated sequence of c-vectors is $(c'_1, \ldots, c'_M)$. The sequence $(k'_1, \ldots, k'_M)$ is the required maximal green sequence for Q'.

Muller's proof uses the existence and uniqueness of the scattering diagram [**61, 81**] associated with a quiver [**60**]. We will give a representation-theoretic proof in section 5.

Let Q be a quiver and Q', Q'' full subquivers. We say that Q is a *triangular extension* of Q' by Q'' if the set of vertices of Q is the disjoint union of the sets of vertices of Q' and Q'' and there are no arrows from vertices of Q'' to vertices of Q'. After pioneering work in [**56**], the following theorem was proved in [**23**] using Muller's theorem 4.2.

THEOREM 4.4. *If Q is a triangular extension of Q' by Q'', then Q has a maximal green sequence if and only if Q' and Q'' have maximal green sequences.*

We refer to [**18, 19**] for recent extensions of this theorem. The case of mutation-finite quivers is treated in section 4.9 below.

4.5. Preservation under mutations. We have the following 'rotation lemma'.

LEMMA 4.6 (Brüstle–Hermes–Igusa–Todorov [**12**]). *If $\mathbf{i} = (i_1, \ldots, i_N)$ is maximal green (resp. reddening) for a quiver Q, then $(i_2, \ldots, i_N, k)$ is maximal green (resp. reddening) for $\mu_{i_1}(Q)$, where k is the target of the unique arrow with source i'_1 in $\mu_{\mathbf{i}}(\hat{Q})$.*

THEOREM 4.7 (Muller [**92**]). *If Q admits a reddening sequence, then each quiver mutation-equivalent to Q admits a reddening sequence.*

However, the analogous statement for maximal green sequences is false, as we will see below.

4.8. Non existence. For three non negative integers a, b, c, denote by $Q_{a,b,c}$ the quiver with three vertices 1, 2, 3 and a arrows from 1 to 2, b arrows from 2 to 3 and c arrows from 3 to 1 (cf. [**107**] for a study of the case of valued quivers with 3 vertices). As shown in [**11**], the quiver $Q_{2,2,2}$ does not admit a maximal green sequence (nor does it admit a reddening sequence). Muller shows [**92**] that none of the quivers $Q_{a,b,c}$ with all three numbers a, b, $c \geqslant 2$ admits a maximal green sequence On the other hand, the quiver $Q_{2,2,3}$ is mutation-acyclic (the quiver $Q_{a,b,c}$ is mutation-acyclic iff $\min(a, b, c) < 2$ or $a^2 + b^2 + c^2 - abc > 4$ as shown in [**7**]). Thus, the existence of a maximal green sequence is not preserved under mutation [**92**].

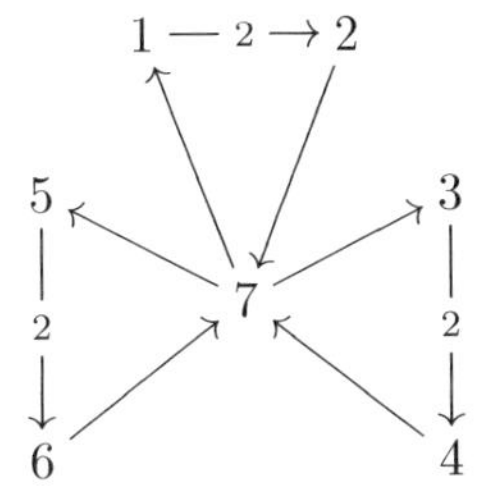

FIGURE 3. The quiver X_7 (double arrows are marked with 2)

4.9. Existence and non existence for mutation-finite quivers. Generalizing the example of $Q_{2,2,2}$, Ladkani has shown in [**86**] that the quivers associated [**45**] with once-punctured surfaces of arbitrary genus (without boundary) do not admit reddening sequences (his proof is based on the work of Labardini–Fragoso [**85**] and Corollary 5.2 below). Seven shows in [**106**] that the quivers in the mutation class of the quiver X_7 (discovered in [**34**]) do not admit maximal green sequences (and presumably no reddening sequences either).

Partial results in the direction of the following theorem were obtained in [**2,17, 21,29,56**].

THEOREM 4.10 (Mills [**91**]). *If Q is a mutation-finite quiver, it has a maximal green sequence except if it comes from a once-punctured closed surface of genus $\geqslant 1$ or is in the mutation class of X_7.*

Information on the length of the minimal length maximal green sequences for the quivers associated with annuli or punctured disks can be found in [**54,55,72**]. Maximal green sequences for minimal mutation-infinite quivers are studied in [**87**].

5. Muller's theorem via representation theory

Let Q be a quiver admitting a maximal green sequence and $Q' \subseteq Q$ a full subquiver. We wish to show that Q' has a maximal green sequence as well.

We recall the setup of Derksen–Weyman–Zelevinsky's theory of quivers with potentials and their mutations [**35**]. Let k be an uncountable field. Let kQ be the path algebra and J the two-sided ideal of kQ generated by the arrows. Let $\widehat{kQ}$ be the completed path algebra, i.e. inverse limit of the finite-dimensional quotients kQ/J^n, $n \geqslant 1$. Let W be a non degenerate potential, i.e. an element of $\widehat{kQ}$ which is an infinite linear combination of cycles of length $\geqslant 3$ such that the pair (Q, W) can be mutated indefinitely without creating 2-cycles in the quiver component Q' of the mutated quiver with potential (Q', W'). Let A be the Jacobi algebra of (Q, W), i.e. the quotient of $\widehat{kQ}$ by the closed ideal generated by the cyclic derivatives $\partial_\alpha W$, $\alpha \in Q_1$. Notice that in general, the algebra A is infinite-dimensional. Let $\mathsf{mod}\, A$ be the category of finite-dimensional (right) A-modules. Recall that all finite-dimensional A-modules are nilpotent, i.e. annihilated by a sufficiently high power of the ideal J (cf. section 10 of [**35**]). Therefore, the simple objects of $\mathsf{mod}\, A$ are the simple modules S_i associated with the vertices $i \in Q_0$. Thus, the abelian category $\mathsf{mod}\, A$ is an Ext-finite length category with finitely many simple objects, cf. [**50,84**].

Let $\mathcal{A}$ be an abelian category. For a class of objects $\mathcal{X}$, we denote by $\mathcal{X}^{\perp}$ the *right orthogonal of* $\mathcal{X}$, i.e. the full subcategory of objects Y such that $\mathsf{Hom}(X,Y) = 0$ for all $X \in \mathcal{X}$. We denote by $\mathsf{add}(\mathcal{X})$ the full subcategory formed by all direct summands of finite direct sums of objects of $\mathcal{X}$. Recall that a *torsion pair* [**37**] in $\mathcal{A}$ is a pair of full subcategories $(\mathcal{T}, \mathcal{F})$ such that $\mathsf{Hom}(T, F) = 0$ for all $T \in \mathcal{T}$ and $F \in \mathcal{F}$ and for each object M of $\mathcal{A}$, there is a short exact sequence

$$0 \longrightarrow T \longrightarrow M \longrightarrow F \longrightarrow 0$$

with $T \in \mathcal{T}$ and $F \in \mathcal{F}$. In this case, $\mathcal{T}$ is called a *torsion class.* The torsion classes of the category of finite-dimensional modules over a finite-dimensional algebra are precisely the full subcategories closed under extensions and quotients [**37**]. A *brick* is an object whose endomorphism algebra is a division algebra.

The link between torsion classes and clusters is already implicit in Marsh–Reineke–Zelevinsky's [**89**]. It is made completely explicit by Ingalls–Thomas in [**71**] (in the acyclic case). As explained in section 7.6 of [**78**], the following theorem results from Nagao's work in [**96**]. For the case of acyclic quivers, cf. [**103**].

THEOREM 5.1 (Nagao). *Each green sequence* $\mathbf{i} = (i_1, \ldots, i_N)$ *yields an ascending chain*

$$0 = \mathcal{T}_0 \subset \mathcal{T}_1 \subset \mathcal{T}_2 \subset \ldots \subset \mathcal{T}_N \subseteq \mathsf{mod}\, A$$

of torsion classes such that

$$\mathcal{T}_{t-1}^{\perp} \cap \mathcal{T}_t = \mathsf{add}(B_t) \tag{5.1.1}$$

for a unique brick B_t *with* $\mathsf{End}(B_t) = k$, $\mathsf{Ext}^1(B_t, B_t) = 0$ *and such that the dimension vector of* B_t *is the c-vector associated with the vertex* i_t, $1 \leqslant t \leqslant N$. *Moreover, the green sequence* $\mathbf{i}$ *is maximal if and only if* $\mathcal{T}_N = \mathsf{mod}\, A$.

Notice that the theorem admits an obvious generalization to arbitrary (red and green) sequences of mutations (cf. section 7.6 of [**78**]). We refer to [**53**] for applications of the theorem to the study of lattice properties of oriented exchange graphs.

COROLLARY 5.2 (Brüstle–Dupont–Pérotin [**11**]). *If* Q *admits a reddening (in particular a maximal green) sequence, then* A *is finite-dimensional.*

PROOF. By the theorem, each object of $\mathsf{mod}\, A$ admits a finite zig-zag-filtration, whose subquotients are finite direct sums of the finitely many modules B_t, $1 \leqslant t \leqslant N$. So in this case, there is a uniform bound on the Loewy length of the finite-dimensional A-modules, which implies that A itself is finite-dimensional [**49, 84**], a fact proved in a different manner in [**11**]. √

Recall that the Hasse quiver $\mathsf{Hasse}(P)$ of a poset P has as vertices the elements of P and an arrow $x \to y$ for each minimal inequality $x < y$, i.e. we have $x < y$ and whenever $x \leqslant z \leqslant y$, we have $x = z$ or $z = y$. It is immediate from the equality (5.1.1) that the inclusions $\mathcal{T}_{t-1} \subset \mathcal{T}_t$ in the chain of torsion classes associated to a green sequence are minimal inclusions. Thus, the chain yields a path starting at 0 in the Hasse quiver $\mathsf{Hasse}(\mathsf{tors}\, A)$ of the poset of torsion classes $\mathsf{tors}\, A$ of $\mathsf{mod}\, A$. The following theorem is an immediate consequence of the results of Adachi–Iyama–Reiten [**1**] and Demonet–Iyama–Jasso [**32**].

THEOREM 5.3. *Suppose* $\dim A < \infty$. *Then Nagao's map taking a green sequence* $\mathbf{i}$ *to the chain of torsion classes* $(\mathcal{T}_t)$ *is a bijection from the set of green sequences for* Q *to the set of paths starting at* 0 *in the Hasse quiver of torsion classes of* $\mathsf{mod}\, A$.

PROOF. Recall that a torsion class $\mathcal{T}$ is functorially finite if for each module M, there is a morphism $M \to T$ with $T \in \mathcal{T}$ such that each morphism $M \to T'$ with $T' \in \mathcal{T}$ factors through $M \to T$. Let us write $\mathsf{ftors}\, A$ for the poset of functorially finite torsion classes in $\mathsf{mod}\, A$. It follows from Theorem 4.1 of [**1**], cf. also Theorem 4.9 of [**16**], that Nagao's map is a bijection from the set of green sequences onto the set of paths starting at 0 in the quiver $\mathsf{Hasse}(\mathsf{ftors}\, A)$. It follows from Theorem 1.3 of [**32**] that immediate successors and predecessors in $\mathsf{Hasse}(\mathsf{tors}\, A)$ of functorially finite torsion classes are functorially finite. Thus the inclusion $\mathsf{Hasse}(\mathsf{ftors}\, A) \subset \mathsf{Hasse}(\mathsf{tors}\, A)$ induces an isomorphism of the connected component of 0 in $\mathsf{Hasse}(\mathsf{ftors}\, A)$ onto the connected component of 0 in $\mathsf{Hasse}(\mathsf{tors}\, A)$. The claim follows. $\surd$

Though it is not necessary for the proof of Muller's theorem, it is an interesting question to ask which sequences of bricks (B_t) are associated with maximal green sequences of Q. The following remarkably simple criterion is proved for maximal green sequences of cluster-finite quivers by Igusa in Corollary 2.14 of [**65**]. A more general statement concerning (possibly infinite) chains in the poset of torsion classes of a finite-dimensional algebra is proved by Demonet in Appendix A. Applications to the construction of maximal green sequences for representation-finite cluster-tilted algebras are given in the appendix to [**65**] for type A and in [**98**] for arbitrary type.

THEOREM 5.4 (Igusa [**65**], Demonet App. A). *Suppose* $\dim A < \infty$. *A sequence of bricks* $B_1, \ldots, B_N$ *is associated with a maximal green sequence for* Q *if and only if*

$$\mathsf{Hom}(B_i, B_j) = 0 \text{ for all } i < j$$

and the sequence cannot be refined keeping this condition.

To conclude the proof of Muller's theorem, we need one more recent result on torsion classes. By definition, if R and R' are quivers, a *contraction* $R \to R'$ is a functor from the path category of R to that of R' which maps each arrow to an arrow or an identity morphism.

THEOREM 5.5 (Demonet–Iyama–Reading–Reiten–Thomas [**33**]). *Let* A *be a finite-dimensional algebra and* $I \lhd A$ *a 2-sided ideal. Then the map* $\mathcal{T} \mapsto \mathcal{T} \cap \mathsf{mod}(A/I)$ *induces a contraction*

$$\mathsf{Hasse}(\mathsf{tors}\, A) \to \mathsf{Hasse}(\mathsf{tors}\, A/I).$$

PROOF. Let us first recall from Theorem 3.3 of [**33**], *cf.* also Theorem 1.0.2 of [**6**], that an inclusion of torsion classes $\mathcal{S} \subseteq \mathcal{T}$ is minimal if and only if there is a unique (up to isomorphism) brick in $\mathcal{S}^{\perp} \cap \mathcal{T}$ and that it is an equality if and only if there is no brick in $\mathcal{S}^{\perp} \cap \mathcal{T}$. Now let $\mathcal{S} \subset \mathcal{T}$ be a minimal inclusion of torsion classes in $\mathsf{mod}\, A$. Let $\mathcal{S}_I = \mathcal{S} \cap \mathsf{mod}(A/I)$ and $\mathcal{T}_I = \mathcal{T} \cap \mathsf{mod}(A/I)$. Let M be a module in $\mathcal{S}_I^{\perp} \cap \mathcal{T}_I$. Let $M_{\mathcal{S}} \subseteq M$ be its maximal submodule in $\mathcal{S}$. Then $M_{\mathcal{S}}$ clearly belongs to $\mathcal{S}_I$. Since M is right orthogonal to $\mathcal{S}_I$, we have $M_{\mathcal{S}} = 0$. Thus, the module M belongs to $\mathcal{S}^{\perp} \cap \mathcal{T}$ and we have $\mathcal{S}_I^{\perp} \cap \mathcal{T}_I \subset \mathcal{S}^{\perp} \cap \mathcal{T}$. Thus, the subcategory $\mathcal{S}_I^{\perp} \cap \mathcal{T}_I$ contains either zero or one brick and the inclusion $\mathcal{S}_I \subseteq \mathcal{S}_I$ is either an equality or minimal. $\surd$

REMARK 5.6. As explained in section 3.2 of [**33**], a brick B is associated with each minimal inclusion of torsion classes $\mathcal{S} \subset \mathcal{T}$ of $\mathsf{mod}\, A$ and the proof shows that such an inclusion is mapped to a minimal inclusion in $\mathsf{mod}(A/I)$ if and only if B belongs to $\mathsf{mod}(A/I)$ (i.e. B is annihilated by I).

We can now conclude: Suppose that Q has a maximal green sequence and that $Q' \subseteq Q$ is a full subquiver. Let e be the sum of the lazy idempotents e_i associated with the vertices not in Q'_0. Clearly, the algebra $\widehat{kQ'}$ is isomorphic to the quotient of $\widehat{kQ}$ by the two-sided ideal generated by e. Let $W' \in \widehat{kQ'}$ be the image of W under the projection $\widehat{kQ} \to \widehat{kQ'}$. By Corollary 22 of [**85**] (the published version!), the potential W' is non degenerate on Q'. Let A' be the Jacobi algebra of (Q', W'). Clearly, the algebra A' is isomorphic to the quotient of A by the two-sided ideal generated by e. By the theorem above, the map

$$\mathsf{Hasse}(\mathsf{tors}\, A) \to \mathsf{Hasse}(\mathsf{tors}\, A') \ , \ \mathcal{T} \mapsto \mathcal{T} \cap \mathsf{mod}\, A'$$

is a contraction and clearly, it takes 0 to 0 and $\mathsf{mod}\, A$ to $\mathsf{mod}\, A'$. Thus, it maps a finite path from 0 to $\mathsf{mod}\, A$ to a finite path from 0 to $\mathsf{mod}\, A'$. By Theorem 5.3, we obtain a maximal green sequence for Q'.

REMARK 5.7. By comparing remarks 4.3 and 5.6 we see that both proofs yield the same explicit recipe for constructing the induced maximal green sequence.

6. Comparing the proofs

Let us emphasize that the statement of Muller's theorem 4.2 and the proof via scattering diagrams given by him in [**92**] go through for valued quivers. In contrast, our representation-theoretic proof only works for the classes of valued quivers treated by Demonet [**30, 31**].

When trying to compare the two proofs we are naturally lead to the problem of relating torsion classes to scattering diagrams. A first step towards its solution was taken by Bridgeland [**10**] who associates a scattering diagram (with values in a motivic Hall algebra) to each finite quiver with (polynomial) relations and without loops or 2-cycles. By a result of Hua–Song [**63**], each potential on a quiver Q whose (complete) Jacobi-algebra is finite-dimensional is right-equivalent to a polynomial potential W. So if the morphism from the non complete Jacobi algebra of (Q, W) to the complete Jacobi algebra is bijective (which is not automatic even if the non complete Jacobi algebra is finite-dimensional!), then Bridgeland's construction applies to this case. In particular, it often applies when the quiver admits a reddening sequence. The problem of comparing Bridgeland's stability scattering diagram with the cluster scattering diagram of Gross–Hacking–Keel–Kontsevich [**60**] is the subject of ongoing research [**28, 102**]. Other topics closely related to the representation-theoretic proof are the investigation of chains of torsion classes induced by stability conditions [**4, 13, 14, 64, 65, 110**], the wall and chamber structure of the space of stability conditions [**15**] and the study of semi-invariants and picture groups [**66–68**].

Appendix A. Maximal chains of torsion classes, by Laurent Demonet

Let A be a finite-dimensional algebra over a field k. We consider the category $\mathsf{mod}\, A$ of finite-dimensional right A-modules and its lattice $\mathsf{tors}(A) = \mathsf{tors}(\mathsf{mod}\, A)$ of torsion classes (cf. section 5). It is a *complete* lattice, *i.e.* the associated category has all limits, called meets, and all colimits, called joins. In particular, for each set of modules $\mathcal{X} \subseteq \mathsf{mod}\, A$, there is a smallest torsion class $\mathsf{T}(\mathcal{X})$ containing $\mathcal{X}$. An element x of a complete poset is *completely join irreducible* if it is not the join of an arbitrary family of elements $< x$. For a poset P, we denote by $\mathsf{Hasse}(P)$ the *Hasse quiver* of P, whose vertices are the elements of P and which has an arrow $x \to y$ if $x < y$ and $x = z$ or $z = y$ whenever $x \leqslant z \leqslant y$ (it is opposite to the Hasse quiver of [**33**]). If P is complete, we denote by $\mathsf{cjirr}(P)$ the subposet of completely join irreducible elements.

A *brick* is an A-module whose endomorphisms form a division algebra. We write $\mathsf{brick}(A)$ for the set of isomorphism classes of bricks of $\mathsf{mod}\, A$.

A *chain of torsion classes* is a totally ordered subposet of $\mathsf{tors}(A)$. Chains of torsion classes are ordered by inclusion. Using the results of [**33**] we will describe the (possibly infinite!) maximal chains of torsion classes in terms of the bricks of $\mathsf{mod}\, A$. Notice that part of the results of [**33**] were independently obtained in [**6**].

Let I be a totally ordered set. An *I-chain of bricks* is a map

$$S_\bullet : I \to \mathsf{brick}(A)\ ,\ i \mapsto S_i$$

such that $\mathsf{Hom}(S_i, S_j) = 0$ for $i < j$. In particular, the map $S_\bullet$ induces an injection from I into the set of isomorphism classes of bricks. If I^1 and I^2 are totally ordered sets and $S^1_\bullet$, $S^2_\bullet$ are chains of bricks indexed by I^1 and I^2, we write $S^1_\bullet \leqslant S^2_\bullet$ if there is an increasing inclusion $\iota : I^1 \to I^2$ such that for all $i \in I^1$, the module S^1_i is isomorphic to $S^2_{\iota(i)}$. Suppose we have an inequality $S^1_\bullet \leqslant S^2_\bullet$ given by ι and an inequality $S^2_\bullet \leqslant S^1_\bullet$ given by κ. Then for each $i \in I^1$, the module S_i is isomorphic to $S_{\kappa(\iota(i))}$ so that $\kappa\iota(i) = i$ and ι and κ are bijective. We define the chains of bricks $S^1_\bullet$ and $S^2_\bullet$ to be *equivalent* if we have $S^1_\bullet \leqslant S^2_\bullet \leqslant S^1_\bullet$.

Let I be a totally ordered set and $S_\bullet$ an I-chain of bricks. We will associate with $S_\bullet$ a chain of torsion classes $\mathcal{T}_\bullet = \Phi(S_\bullet)$. An *ideal* of I is a subset $j \subseteq I$ such that $l \in j$ whenever $l \leqslant i$ and $i \in j$. Let J be the poset of ideals of I. For each $j \in J$, let $\mathcal{T}_j$ be the smallest torsion class containing the S_i, $i \in j$. It is the join of the torsion classes $\mathsf{T}(S_i)$, $i \in j$. Clearly, the map $j \mapsto \mathcal{T}_j$ is increasing. If we have $j \subset j'$ and i belongs to j' but not to j, then the brick S_i lies in $\mathcal{T}_{j'} \cap \mathcal{T}_j^\perp$ so that the map $j \mapsto \mathcal{T}_j$ is strictly increasing. We define $\Phi(S_\bullet)$ to be the subposet of $\mathsf{tors}(A)$ formed by the $\mathcal{T}_j$, $j \in J$.

PROPOSITION A.1. *The map Φ induces an injective morphism from the poset of equivalence classes of chains of bricks to the poset of chains of torsion classes.*

PROOF. Clearly, the map Φ is a morphism of posets. Let $S_\bullet : I \to \mathsf{brick}(A)$ and $S'_\bullet : I' \to \mathsf{brick}(A)$ be chains of bricks and $\mathcal{T}_\bullet = \Phi(S_\bullet)$ and $\mathcal{T}'_\bullet = \Phi(S'_\bullet)$ the associated chains of torsion classes. Let J and J' be the posets of ideals of I and I'. We first construct an isomorphism of posets $\pi : I \xrightarrow{\sim} I'$ such that for each $j \in J$, we have $\mathcal{T}'_{\pi(j)} = \mathcal{T}_j$. For this, notice that J and $\mathsf{tors}(A)$ are complete lattices and that the injective morphism $j \mapsto \mathcal{T}_j$ commutes with arbitrary joins. We can recover I from the set of ideals J of I as the subposet of the completely join irreducible elements. Since we have isomorphisms of posets

$$J \xrightarrow{\sim} \{\mathcal{T}_j \mid j \in J\} = \{\mathcal{T}'_{j'} \mid j' \in J'\} \xleftarrow{\sim} J'$$

we obtain an isomorphism $\pi : I \xrightarrow{\sim} I'$ and it clearly satisfies $\mathcal{T}_j = \mathcal{T}'_{\pi j}$ for all $j \in J$. Thus, we may assume that $S_\bullet$ and $S'_\bullet$ are maps $I \to \mathsf{brick}(A)$ such that for each ideal j of I, the join $\mathcal{T}_j$ of the classes $\mathsf{T}(S_i)$, $i \in j$, coincides with $\mathcal{T}'_j$. Fix an element

i of I. Let $j \subseteq I$ be the ideal of the elements $i' < i$. Let $\mathcal{U} = \mathcal{T}_j = \mathcal{T}'_j$. Let $\mathcal{T}$ be the torsion class $\mathcal{U} \vee \mathsf{T}(S_i) = \mathcal{U} \vee \mathsf{T}(S'_i)$. Since S_i belongs to $\mathcal{U}^\perp$, by part a) of Theorem 3.4 of [**33**], the torsion class $\mathcal{T}$ is completely join irreducible and the unique arrow of the Hasse quiver $\mathsf{Hasse}[\mathcal{U}, \mathsf{mod}\, A]$ ending at $\mathcal{T}$ is labelled by S_i, cf. Definition 3.5 of [**33**]. Since we also have $\mathcal{T} = \mathcal{U} \vee \mathsf{T}(S'_i)$, the arrow is also labeled by S'_i, which is therefore isomorphic to S_i. $\surd$

PROPOSITION A.2. *Let C be a maximal chain of torsion classes. Then each $\mathcal{T} \in C$ equals the join $\mathcal{V}$ of the $\mathcal{U} \in \mathsf{cjirr}(C)$ contained in $\mathcal{T}$. It also equals the join $\mathcal{V}'$ of the classes $\mathsf{T}(S_q)$, where q runs through the arrows of the Hasse quiver of $C \cap [0, \mathcal{T}]$.*

PROOF. Clearly, we have $\mathcal{V} \subseteq \mathcal{T}$. Let us show that $\mathcal{V}'$ is contained in $\mathcal{V}$. If $q : \mathcal{U}' \to \mathcal{U}$ is an arrow of the Hasse quiver of $C \cap [0, \mathcal{T}]$, it is also an arrow of $\mathsf{Hasse}(A)$, by the maximality of C. Thus, the arrow has a well-defined brick label S_q. Moreover, the brick S_q belongs to $\mathcal{U}$ and $\mathcal{U}$ is completely join irreducible. Therefore, we have $\mathsf{T}(S_q) \subset \mathcal{V}$ and $\mathcal{V}' \subset \mathcal{V}$. Let us show that the inclusion $\mathcal{V}' \subseteq \mathcal{T}$ is an equality. By Lemma 3.10 of [**33**], the modules in $\mathcal{T} \cap \mathcal{V}'^\perp$ are those admitting a filtration whose subquotients are bricks in $\mathcal{T} \cap \mathcal{V}'^\perp$. Let S be a brick of minimal dimension in $\mathcal{T} \cap \mathcal{V}'^\perp$. Consider the meet $\mathcal{W}$ of the torsion classes $\mathcal{W}' \in C$ containing $\mathcal{V}'$ and S. Since C is a maximal chain, it is stable under meets and thus contains $\mathcal{W}$. Let $\mathcal{V}'' \in C$ be a torsion class such that $\mathcal{V}' \subseteq \mathcal{V}'' \subseteq \mathcal{W}$ and $\mathcal{V}'' \neq \mathcal{W}$. Let $X \in \mathcal{V}''$. Consider a morphism $f : X \to S$. The image $\mathsf{im}(f)$ is a quotient of X and thus belongs to $\mathcal{T}$. It is also a submodule of S and thus belongs to $\mathcal{V}'^\perp$. Thus, it belongs to $\mathcal{D} \cap \mathcal{V}'^\perp$ and has a filtration whose subquotients are bricks in $\mathcal{T} \cap \mathcal{V}'^\perp$. By the minimality of the dimension of S, we have $\mathsf{im}(f) = 0$ or $\mathsf{im}(f) = S$. If we have $\mathsf{im}(f) = S$, then S is a quotient of X and belongs to $\mathcal{V}''$, which contradicts the definition of $\mathcal{W}$. Thus, we have $\mathsf{im}(f) = 0$ and $\mathcal{V}'' \subset {}^\perp S \cap \mathcal{W}$. By the maximality of C, it follows that the class ${}^\perp S \cap \mathcal{W}$ belongs to C and there is an arrow ${}^\perp S \cap \mathcal{W} \to \mathcal{S}$ in $\mathsf{Hasse}(C)$ labeled by S, *cf.* Theorem 3.4 of [**33**]. So the module S belongs to $\mathcal{V}$, which is a contradiction. $\surd$

THEOREM A.3. *The map $\Phi : S_\bullet \mapsto \mathcal{T}_\bullet$ induces a bijection from the set of equivalence classes of maximal chains of bricks to the set of maximal chains of torsion classes.*

PROOF. By Proposition A.1, it only remains to prove that Φ is surjective. Let C be a maximal chain of torsion classes. Let I be the poset $\mathsf{cjirr}(C)$. For each $i \in I$, there is a unique arrow $\mathcal{U}_i \to i$ in $\mathsf{Hasse}(C)$. By the maximality, it is also an arrow of $\mathsf{Hasse}(A)$. Let S_i be its label, *cf.* Definition 3.5 of [**33**]. We claim that the map $i \mapsto S_i$ is an I-chain of bricks. Indeed, if we have $i < j$ in I, then $i \subseteq \mathcal{U}_j \subseteq j$ and so $S_j \in \mathcal{U}_j^\perp$ and $S_i \in i \subseteq \mathcal{U}_j$ so that we have $\mathsf{Hom}(S_i, S_j) = 0$. Let $\mathcal{T}_\bullet$ be the chain of torsion classes $\Phi(S_\bullet)$. For $\mathcal{U} \in C$, consider the ideal j of I formed by the $i \in I$ contained in U. By Proposition A.2, we have $\mathcal{U} = \mathcal{T}_j$. Whence an inclusion $C \subseteq \mathcal{T}_\bullet$. By the maximality of C, we actually have an equality. $\surd$

Acknowledgments

I thank Laurent Demonet for his explanations of the results of [**33**], his help with section 5 and for contributing the appendix. I am grateful to Daniel Labardini–Fragoso for helpful discussions and for reference [**85**]. An abridged version of this note was presented at the ICRA 18 in Prague in August 2018. I thank the organisers for a wonderful conference and for considering this material for the proceedings. I am indebted to Zheng Hua for pointing out a misquotation in section 6 of a previous version of this note and to Jiarui Fei for alerting me to reference [**41**].

References

[1] Takahide Adachi, Osamu Iyama, and Idun Reiten, *τ-tilting theory*, Compos. Math. **150** (2014), no. 3, 415–452, DOI 10.1112/S0010437X13007422. MR3187626

[2] Murad Alim, Sergio Cecotti, Clay Córdova, Sam Espahbodi, Ashwin Rastogi, and Cumrun Vafa, *$\mathcal{N} = 2$ quantum field theories and their BPS quivers*, Adv. Theor. Math. Phys. **18** (2014), no. 1, 27–127. MR3268234

[3] Justin Allman and Richárd Rimányi, *Quantum dilogarithm identities for the square product of A-type Dynkin quivers*, Math. Res. Lett. **25** (2018), no. 4, 1037–1087, DOI 10.4310/MRL.2018.v25.n4.a1. MR3882154

[4] P. J. Apruzzese and Kiyoshi Igusa, *Stability Conditions for Affine Type A*, Algebr. Represent. Theory **23** (2020), no. 5, 2079–2111, DOI 10.1007/s10468-019-09926-z. MR4140066

[5] A. B. Buan, O. Iyama, I. Reiten, and J. Scott, *Cluster structures for 2-Calabi-Yau categories and unipotent groups*, Compos. Math. **145** (2009), no. 4, 1035–1079, DOI 10.1112/S0010437X09003960. MR2521253

[6] Emily Barnard, Andrew Carroll, and Shijie Zhu, *Minimal inclusions of torsion classes*, Algebr. Comb. **2** (2019), no. 5, 879–901. MR4023570

[7] Andre Beineke, Thomas Brüstle, and Lutz Hille, *Cluster-cyclic quivers with three vertices and the Markov equation*, Algebr. Represent. Theory **14** (2011), no. 1, 97–112, DOI 10.1007/s10468-009-9179-9. With an appendix by Otto Kerner. MR2763295

[8] Arkady Berenstein, Sergey Fomin, and Andrei Zelevinsky, *Parametrizations of canonical bases and totally positive matrices*, Adv. Math. **122** (1996), no. 1, 49–149, DOI 10.1006/aima.1996.0057. MR1405449

[9] Arkady Berenstein, Sergey Fomin, and Andrei Zelevinsky, *Cluster algebras. III. Upper bounds and double Bruhat cells*, Duke Math. J. **126** (2005), no. 1, 1–52, DOI 10.1215/S0012-7094-04-12611-9. MR2110627

[10] Tom Bridgeland, *Scattering diagrams, Hall algebras and stability conditions*, Algebr. Geom. **4** (2017), no. 5, 523–561, DOI 10.14231/2017-027. MR3710055

[11] Thomas Brüstle, Grégoire Dupont, and Matthieu Pérotin, *On maximal green sequences*, Int. Math. Res. Not. IMRN **16** (2014), 4547–4586, DOI 10.1093/imrn/rnt075. MR3250044

[12] Thomas Brüstle, Stephen Hermes, Kiyoshi Igusa, and Gordana Todorov, *Semi-invariant pictures and two conjectures on maximal green sequences*, J. Algebra **473** (2017), 80–109, DOI 10.1016/j.jalgebra.2016.10.025. MR3591142

[13] Thomas Brüstle, David Smith, and Hipolito Treffinger, *Stability conditions and maximal green sequences in abelian categories*, arXiv:1805.04382 [math.RT].

[14] Thomas Brüstle, David Smith, and Hipolito Treffinger, *Stability conditions, τ-tilting theory and maximal green sequences*, arXiv:1705.08227 [math.RT].

[15] Thomas Brüstle, David Smith, and Hipolito Treffinger, *Wall and chamber structure for finite-dimensional algebras*, Adv. Math. **354** (2019), 106746, 31, DOI 10.1016/j.aim.2019.106746. MR3989130

[16] Thomas Brüstle and Dong Yang, *Ordered exchange graphs*, Advances in representation theory of algebras, EMS Ser. Congr. Rep., Eur. Math. Soc., Zürich, 2013, pp. 135–193. MR3220536

[17] Eric Bucher, *Maximal green sequences for cluster algebras associated to the n-torus*, arXiv:1412.3713 [math.CO].

[18] Eric Bucher and John Machacek, *Reddening sequences for Banff quivers and the class $\mathcal{P}$*, SIGMA Symmetry Integrability Geom. Methods Appl. **16** (2020), Paper No. 049, 11, DOI 10.3842/SIGMA.2020.049. MR4108215

[19] Eric Bucher, John Machacek, Evan Runburg, Abe Yeck, and Ethan Zewde, *Building maximal green sequences via component preserving mutations*, arXiv:1902.02262 [math.CO].

[20] Eric Bucher, John Machacek, and Michael Shapiro, *Upper cluster algebras and choice of ground ring*, Sci. China Math. **62** (2019), no. 7, 1257–1266, DOI 10.1007/s11425-018-9486-6. MR3974092

[21] Eric Bucher and Matthew R. Mills, *Maximal green sequences for cluster algebras associated with the n-torus with arbitrary punctures*, J. Algebraic Combin. **47** (2018), no. 3, 345–356, DOI 10.1007/s10801-017-0778-y. MR3802321

[22] Philippe Caldero and Frédéric Chapoton, *Cluster algebras as Hall algebras of quiver representations*, Comment. Math. Helv. **81** (2006), no. 3, 595–616, DOI 10.4171/CMH/65. MR2250855

[23] Peigen Cao and Fang Li, *Uniform column sign-coherence and the existence of maximal green sequences*, J. Algebraic Combin. **50** (2019), no. 4, 403–417, DOI 10.1007/s10801-018-0861-z. MR4029226

[24] Sabin Cautis and Harold Williams, *Cluster theory of the coherent Satake category*, J. Amer. Math. Soc. **32** (2019), no. 3, 709–778, DOI 10.1090/jams/918. MR3981987

[25] Sergio Cecotti, Clay Córdova, and Cumrun Vafa, *Braids, walls and mirrors*, arXiv:1110.2115 [hep-th].

[26] Giovanni Cerulli Irelli, Bernhard Keller, Daniel Labardini-Fragoso, and Pierre-Guy Plamondon, *Linear independence of cluster monomials for skew-symmetric cluster algebras*, Compos. Math. **149** (2013), no. 10, 1753–1764, DOI 10.1112/S0010437X1300732X. MR3123308

[27] Frédéric Chapoton, Sergey Fomin, and Andrei Zelevinsky, *Polytopal realizations of generalized associahedra*, Canad. Math. Bull. **45** (2002), no. 4, 537–566, DOI 10.4153/CMB-2002-054-1. Dedicated to Robert V. Moody. MR1941227

[28] Man-Wai Cheung and Travis Mandel, *Donaldson-Thomas invariants from tropical disks*, Selecta Math. (N.S.) **26** (2020), no. 4, 57, DOI 10.1007/s00029-020-00580-8. MR4131036

[29] E. Cormier, P. Dillery, J. Resh, K. Serhiyenko, and J. Whelan, *Minimal length maximal green sequences and triangulations of polygons*, J. Algebraic Combin. **44** (2016), no. 4, 905–930, DOI 10.1007/s10801-016-0694-6. MR3566224

[30] Laurent Demonet, *Mutations of group species with potentials and their representations. Applications to cluster algebras*, arXiv:1003.5078 [math.RT].

[31] Laurent Demonet, *Categorification of skew-symmetrizable cluster algebras*, Algebr. Represent. Theory **14** (2011), no. 6, 1087–1162, DOI 10.1007/s10468-010-9228-4. MR2844757

[32] Laurent Demonet, Osamu Iyama, and Gustavo Jasso, *τ-tilting finite algebras, bricks, and g-vectors*, Int. Math. Res. Not. IMRN **3** (2019), 852–892, DOI 10.1093/imrn/rnx135. MR3910476

[33] Laurent Demonet, Osamu Iyama, Nathan Reading, Idun Reiten, and Hugh Thomas, *Lattice theory of torsion classes*, arXiv:1711.01785 [math.RT].

[34] Harm Derksen and Theodore Owen, *New graphs of finite mutation type*, Electron. J. Combin. **15** (2008), no. 1, Research Paper 139, 15. MR2465763

[35] Harm Derksen, Jerzy Weyman, and Andrei Zelevinsky, *Quivers with potentials and their representations. I. Mutations*, Selecta Math. (N.S.) **14** (2008), no. 1, 59–119, DOI 10.1007/s00029-008-0057-9. MR2480710

[36] Harm Derksen, Jerzy Weyman, and Andrei Zelevinsky, *Quivers with potentials and their representations II: applications to cluster algebras*, J. Amer. Math. Soc. **23** (2010), no. 3, 749–790, DOI 10.1090/S0894-0347-10-00662-4. MR2629987

[37] Spencer E. Dickson, *A torsion theory for Abelian categories*, Trans. Amer. Math. Soc. **121** (1966), 223–235, DOI 10.2307/1994341. MR191935

[38] G. Dupont, *Generic variables in acyclic cluster algebras*, J. Pure Appl. Algebra **215** (2011), no. 4, 628–641, DOI 10.1016/j.jpaa.2010.06.012. MR2738377

[39] L. Faddeev and A. Yu. Volkov, *Abelian current algebra and the Virasoro algebra on the lattice*, Phys. Lett. B **315** (1993), no. 3-4, 311–318, DOI 10.1016/0370-2693(93)91618-W. MR1238474

[40] L. D. Faddeev and R. M. Kashaev, *Quantum dilogarithm*, Modern Phys. Lett. A **9** (1994), no. 5, 427–434, DOI 10.1142/S0217732394000447. MR1264393

[41] Jiarui Fei, *Tensor product multiplicities via upper cluster algebras*, arXiv:1603.02521 [math.RT].

[42] Anna Felikson, Michael Shapiro, and Pavel Tumarkin, *Cluster algebras of finite mutation type via unfoldings*, Int. Math. Res. Not. IMRN **8** (2012), 1768–1804, DOI 10.1093/imrn/rnr072. MR2920830

[43] Anna Felikson, Michael Shapiro, and Pavel Tumarkin, *Skew-symmetric cluster algebras of finite mutation type*, J. Eur. Math. Soc. (JEMS) **14** (2012), no. 4, 1135–1180, DOI 10.4171/JEMS/329. MR2928847

[44] Vladimir V. Fock and Alexander B. Goncharov, *Cluster ensembles, quantization and the dilogarithm* (English, with English and French summaries), Ann. Sci. Éc. Norm. Supér. (4) **42** (2009), no. 6, 865–930, DOI 10.24033/asens.2112. MR2567745

[45] Sergey Fomin, Michael Shapiro, and Dylan Thurston, *Cluster algebras and triangulated surfaces. I. Cluster complexes*, Acta Math. **201** (2008), no. 1, 83–146, DOI 10.1007/s11511-008-0030-7. MR2448067

[46] Sergey Fomin and Andrei Zelevinsky, *Cluster algebras. I. Foundations*, J. Amer. Math. Soc. **15** (2002), no. 2, 497–529, DOI 10.1090/S0894-0347-01-00385-X. MR1887642

[47] Sergey Fomin and Andrei Zelevinsky, *Cluster algebras. II. Finite type classification*, Invent. Math. **154** (2003), no. 1, 63–121, DOI 10.1007/s00222-003-0302-y. MR2004457

[48] Sergey Fomin and Andrei Zelevinsky, *Cluster algebras. IV. Coefficients*, Compos. Math. **143** (2007), no. 1, 112–164, DOI 10.1112/S0010437X06002521. MR2295199

[49] Pierre Gabriel, *Représentations indécomposables* (French), Séminaire Bourbaki, 26e année (1973/1974), Exp. No. 444, Springer, Berlin, 1975, pp. 143–169. Lecture Notes in Math., Vol. 431. MR0485996

[50] Peter Gabriel, *Indecomposable representations. II*, Symposia Mathematica, Vol. XI (Convegno di Algebra Commutativa, INDAM, Rome, 1971), Academic Press, London, 1973, pp. 81–104. MR0340377

[51] Davide Gaiotto, Gregory W. Moore, and Andrew Neitzke, *Wall-crossing, Hitchin systems, and the WKB approximation*, Adv. Math. **234** (2013), 239–403, DOI 10.1016/j.aim.2012.09.027. MR3003931

[52] Davide Gaiotto, Gregory W. Moore, and Andrew Neitzke, *Wall-crossing, Hitchin systems, and the WKB approximation*, Adv. Math. **234** (2013), 239–403, DOI 10.1016/j.aim.2012.09.027. MR3003931

[53] Alexander Garver and Thomas McConville, *Lattice properties of oriented exchange graphs and torsion classes*, Algebr. Represent. Theory **22** (2019), no. 1, 43–78, DOI 10.1007/s10468-017-9757-1. MR3908895

[54] Alexander Garver, Thomas McConville, and Khrystyna Serhiyenko, *Minimal length maximal green sequences* (English, with English and French summaries), Sém. Lothar. Combin. **78B** (2017), Art. 16, 12. MR3678598

[55] Alexander Garver, Thomas McConville, and Khrystyna Serhiyenko, *Minimal length maximal green sequences*, Adv. in Appl. Math. **96** (2018), 76–138, DOI 10.1016/j.aam.2017.12.008. MR3767505

[56] Alexander Garver and Gregg Musiker, *On maximal green sequences for type $\mathbb{A}$ quivers*, J. Algebraic Combin. **45** (2017), no. 2, 553–599, DOI 10.1007/s10801-016-0716-4. MR3604067

[57] Christof Geiß, Bernard Leclerc, and Jan Schröer, *Kac-Moody groups and cluster algebras*, Adv. Math. **228** (2011), no. 1, 329–433, DOI 10.1016/j.aim.2011.05.011. MR2822235

[58] Christof Geiss, Bernard Leclerc, and Jan Schröer, *Generic bases for cluster algebras and the Chamber ansatz*, J. Amer. Math. Soc. **25** (2012), no. 1, 21–76, DOI 10.1090/S0894-0347-2011-00715-7. MR2833478

[59] Alexander Goncharov and Linhui Shen, *Donaldson-Thomas transformations of moduli spaces of G-local systems*, Adv. Math. **327** (2018), 225–348, DOI 10.1016/j.aim.2017.06.017. MR3761995

[60] Mark Gross, Paul Hacking, Sean Keel, and Maxim Kontsevich, *Canonical bases for cluster algebras*, J. Amer. Math. Soc. **31** (2018), no. 2, 497–608, DOI 10.1090/jams/890. MR3758151

[61] Mark Gross and Bernd Siebert, *From real affine geometry to complex geometry*, Ann. of Math. (2) **174** (2011), no. 3, 1301–1428, DOI 10.4007/annals.2011.174.3.1. MR2846484

[62] Stephen Hermes and Kiyoshi Igusa, *The no gap conjecture for tame hereditary algebras*, J. Pure Appl. Algebra **223** (2019), no. 3, 1040–1053, DOI 10.1016/j.jpaa.2018.05.013. MR3862663

[63] Zheng Hua and Gui-Song Zhou, *Noncommutative Mather–Yau theorem and its applications to Calabi–Yau algebras and birational geometry*, arXiv:1803.06128 [math.AG].

[64] Kiyoshi Igusa, *Linearity of stability conditions*, Comm. Algebra **48** (2020), no. 4, 1671–1696, DOI 10.1080/00927872.2019.1705466. MR4079339

[65] Kiyoshi Igusa, *Maximal green sequences for cluster-tilted algebras of finite representation type*, Algebr. Comb. **2** (2019), no. 5, 753–780, DOI 10.5802/alco.61. With an appendix written jointly with Gordana Todorov. MR4023565

[66] Kiyoshi Igusa, Kent Orr, Gordana Todorov, and Jerzy Weyman, *Modulated semi-invariants*, arXiv:1507.03051 [math.RT].

[67] Kiyoshi Igusa, Kent Orr, Gordana Todorov, and Jerzy Weyman, *Cluster complexes via semi-invariants*, Compos. Math. **145** (2009), no. 4, 1001–1034, DOI 10.1112/S0010437X09004151. MR2521252

[68] Kiyoshi Igusa, Gordana Todorov, and Jerzy Weyman, *Picture groups of finite type and cohomology in type A_n*, arXiv:1609.02636 [math.RT].

[69] Kiyoshi Igusa and Ying Zhou, *The formula for the permutation of mutation sequences in A_n straight orientation*, arXiv:1606.00958 [math.RT].

[70] Kiyoshi Igusa and Ying Zhou, *Maximal green sequences of quivers with multiple edges*, arXiv:1902.07375 [math.RT].

[71] Colin Ingalls and Hugh Thomas, *Noncrossing partitions and representations of quivers*, Compos. Math. **145** (2009), no. 6, 1533–1562, DOI 10.1112/S0010437X09004023. MR2575093

[72] Ryoichi Kase, *Remarks on lengths of maximal green sequences for quivers of type $\widetilde{A}_{n,1}$*, arXiv:1507.02852 [math.RT].

[73] Bernhard Keller, *Cluster algebras and derived categories*, Derived categories in algebraic geometry, EMS Ser. Congr. Rep., Eur. Math. Soc., Zürich, 2012, pp. 123–183. MR3050703

[74] Bernhard Keller, *Quantum dilogarithm identities from quiver mutations*, video of a talk given at the meeting 'Test problems for the theory of finite dimensional algebras', Banff, September 9, 2010,.

[75] Bernhard Keller, *Quiver mutation and combinatorial DT-invariants*, corrected version of a submission to FPSAC 2013, arXiv:1709.03143 [math.CO].

[76] Bernhard Keller, *Quiver mutation in Javascript and Java*, Java applet available at https://webusers.imj-prg.fr/ bernhard.keller/quivermutation/.

[77] Bernhard Keller, *On cluster theory and quantum dilogarithm identities*, Representations of algebras and related topics, EMS Ser. Congr. Rep., Eur. Math. Soc., Zürich, 2011, pp. 85–116, DOI 10.4171/101-1/3. MR2931896

[78] Bernhard Keller, *Cluster algebras and derived categories*, Derived categories in algebraic geometry, EMS Ser. Congr. Rep., Eur. Math. Soc., Zürich, 2012, pp. 123–183. MR3050703

[79] Bernhard Keller, *The periodicity conjecture for pairs of Dynkin diagrams*, Ann. of Math. (2) **177** (2013), no. 1, 111–170, DOI 10.4007/annals.2013.177.1.3. MR2999039

[80] Maxim Kontsevich and Yan Soibelman, *Stability structures, Donaldson-Thomas invariants and cluster transformations*, arXiv:0811.2435 [math.AG].

[81] Maxim Kontsevich and Yan Soibelman, *Affine structures and non-Archimedean analytic spaces*, The unity of mathematics, Progr. Math., vol. 244, Birkhäuser Boston, Boston, MA, 2006, pp. 321–385, DOI 10.1007/0-8176-4467-9_9. MR2181810

[82] Maxim Kontsevich and Yan Soibelman, *Motivic Donaldson-Thomas invariants: summary of results*, Mirror symmetry and tropical geometry, Contemp. Math., vol. 527, Amer. Math. Soc., Providence, RI, 2010, pp. 55–89, DOI 10.1090/conm/527/10400. MR2681792

[83] Maxim Kontsevich and Yan Soibelman, *Cohomological Hall algebra, exponential Hodge structures and motivic Donaldson-Thomas invariants*, Commun. Number Theory Phys. **5** (2011), no. 2, 231–352, DOI 10.4310/CNTP.2011.v5.n2.a1. MR2851153

[84] Henning Krause and Dieter Vossieck, *Length categories of infinite height*, Geometric and topological aspects of the representation theory of finite groups, Springer Proc. Math. Stat., vol. 242, Springer, Cham, 2018, pp. 213–234. MR3901161

[85] Daniel Labardini-Fragoso, *Quivers with potentials associated to triangulated surfaces*, Proc. Lond. Math. Soc. (3) **98** (2009), no. 3, 797–839, DOI 10.1112/plms/pdn051. MR2500873

[86] Sefi Ladkani, *On cluster algebras from once punctured closed surfaces*, arXiv:1310.4454 [math.RT].

[87] John W. Lawson and Matthew R. Mills, *Properties of minimal mutation-infinite quivers*, J. Combin. Theory Ser. A **155** (2018), 122–156, DOI 10.1016/j.jcta.2017.11.004. MR3741423

[88] R. J. Marsh and J. S. Scott, *Twists of Plücker coordinates as dimer partition functions*, Comm. Math. Phys. **341** (2016), no. 3, 821–884, DOI 10.1007/s00220-015-2493-7. MR3452273

[89] Robert Marsh, Markus Reineke, and Andrei Zelevinsky, *Generalized associahedra via quiver representations*, Trans. Amer. Math. Soc. **355** (2003), no. 10, 4171–4186, DOI 10.1090/S0002-9947-03-03320-8. MR1990581

[90] Matthew R. Mills, *On the relationship between green-to-red sequences, local-acyclicity, and upper cluster algebras*, arXiv:1804.00479 [math.RT].

[91] Matthew R. Mills, *Maximal green sequences for quivers of finite mutation type*, Adv. Math. **319** (2017), 182–210, DOI 10.1016/j.aim.2017.08.019. MR3695872

[92] Greg Muller, *The existence of a maximal green sequence is not invariant under quiver mutation*, Electron. J. Combin. **23** (2016), no. 2, Paper 2.47, 23. MR3512669

[93] Greg Muller and David E. Speyer, *The twist for positroid varieties*, Proc. Lond. Math. Soc. (3) **115** (2017), no. 5, 1014–1071, DOI 10.1112/plms.12056. MR3733558

[94] Gregg Musiker and Christian Stump, *A compendium on the cluster algebra and quiver package in* Sage, Sém. Lothar. Combin. **65** (2010/12), Art. B65d, 67. MR2862990

[95] Kentaro Nagao, *Quantum dilogarithm identities*, Infinite analysis 2010—Developments in quantum integrable systems, RIMS Kôkyûroku Bessatsu, B28, Res. Inst. Math. Sci. (RIMS), Kyoto, 2011, pp. 165–170. MR2885304

[96] Kentaro Nagao, *Donaldson-Thomas theory and cluster algebras*, Duke Math. J. **162** (2013), no. 7, 1313–1367, DOI 10.1215/00127094-2142753. MR3079250

[97] Tomoki Nakanishi and Andrei Zelevinsky, *On tropical dualities in cluster algebras*, Algebraic groups and quantum groups, Contemp. Math., vol. 565, Amer. Math. Soc., Providence, RI, 2012, pp. 217–226, DOI 10.1090/conm/565/11159. MR2932428

[98] Alireza Nasr-Isfahani, *Maximal forward hom-orthogonal sequences for cluster-tilted algebras of finite type*, Proc. Amer. Math. Soc. **147** (2019), no. 7, 2775–2782, DOI 10.1090/proc/14523. MR3973881

[99] Pierre-Guy Plamondon, *Cluster characters*, Homological methods, representation theory, and cluster algebras, CRM Short Courses, Springer, Cham, 2018, pp. 101–125. MR3823389

[100] Pierre-Guy Plamondon, *Cluster algebras via cluster categories with infinite-dimensional morphism spaces*, Compos. Math. **147** (2011), no. 6, 1921–1954, DOI 10.1112/S0010437X11005483. MR2862067

[101] Pierre-Guy Plamondon, *Generic bases for cluster algebras from the cluster category*, Int. Math. Res. Not. IMRN **10** (2013), 2368–2420, DOI 10.1093/imrn/rns102. MR3061943

[102] Fan Qin, *Bases for upper cluster algebras and tropical points*, arXiv:1902.09507 [math.RT].

[103] Yu Qiu, *C-sortable words as green mutation sequences*, Proc. Lond. Math. Soc. (3) **111** (2015), no. 5, 1052–1070, DOI 10.1112/plms/pdv046. MR3477229

[104] Markus Reineke, *Poisson automorphisms and quiver moduli*, J. Inst. Math. Jussieu **9** (2010), no. 3, 653–667, DOI 10.1017/S1474748009000176. MR2650811

[105] K. Rietsch and L. Williams, *Newton-Okounkov bodies, cluster duality, and mirror symmetry for Grassmannians*, Duke Math. J. **168** (2019), no. 18, 3437–3527, DOI 10.1215/00127094-2019-0028. MR4034891

[106] Ahmet I. Seven, *Maximal green sequences of exceptional finite mutation type quivers*, SIGMA Symmetry Integrability Geom. Methods Appl. **10** (2014), Paper 089, 5, DOI 10.3842/SIGMA.2014.089. MR3261855

[107] Ahmet I. Seven, *Maximal green sequences of skew-symmetrizable* 3×3 *matrices*, Linear Algebra Appl. **440** (2014), 125–130, DOI 10.1016/j.laa.2013.10.018. MR3134258

[108] Linhui Shen and Daping Weng, *Cluster structures on double Bott–Samelson cells*, arXiv:1904.07992 [math.AG].

[109] Hugh Thomas, *The Tamari lattice as it arises in quiver representations*, Associahedra, Tamari lattices and related structures, Prog. Math. Phys., vol. 299, Birkhäuser/Springer, Basel, 2012, pp. 281–291, DOI 10.1007/978-3-0348-0405-9_14. MR3221543

[110] Hipolito Treffinger, *An algebraic approach to Harder-Narasimhan filtrations*, arXiv:1810.06322 [math.CT].

[111] Alexander Yu. Volkov, *Pentagon identity revisited*, Int. Math. Res. Not. IMRN **20** (2012), 4619–4624, DOI 10.1093/imrn/rnr200. MR2989615

[112] Daping Weng, *Donaldson–Thomas transformation of double Bruhat cells in general linear groups*, arXiv:1606.01948 [math.AG].

[113] Daping Weng, *Donaldson–Thomas transformation of double Bruhat cells in semisimple Lie groups*, arXiv:1611.04186 [math.AG], to appear in Ann. Scient. ENS.

[114] Daping Weng, *Donaldson–Thomas transformation of Grassmannian*, arXiv:1603.00972 [math.RT].

[115] Dan Xie, *BPS spectrum, wall crossing and quantum dilogarithm identity*, Adv. Theor. Math. Phys. **20** (2016), no. 3, 405–524, DOI 10.4310/ATMP.2016.v20.n3.a1. MR3565860

[116] Don Zagier, *Polylogarithms, Dedekind zeta functions and the algebraic K-theory of fields*, Arithmetic algebraic geometry (Texel, 1989), Progr. Math., vol. 89, Birkhäuser Boston, Boston, MA, 1991, pp. 391–430. MR1085270

Université Paris Diderot – Paris 7, Sorbonne Université, UFR de Mathématiques, CNRS, Institut de Mathématiques de Jussieu–Paris Rive Gauche, IMJ-PRG, Bâtiment Sophie Germain, 75205 Paris Cedex 13, France

Email address: bernhard.keller@imj-prg.fr

URL: https://webusers.imj-prg.fr/~bernhard.keller/

Graduate School of Mathematics, Nagoya University, Chikusa-ku, Nagoya, 464-8602 Japan

Email address: Laurent.Demonet@normalesup.org

URL: http://www.math.nagoya-u.ac.jp/~demonet/

Contemporary Mathematics
Volume **758**, 2020
https://doi.org/10.1090/conm/758/15240

The Gerstenhaber bracket in Hochschild cohomology: Methods and examples

Andrea Solotar

ABSTRACT. Homological methods provide important information about the structure of associative algebras, revealing sometimes hidden connections amongst them. This text is about the Gerstenhaber bracket in Hochschild cohomology of unital associative algebras over a field. The first Hochschild cohomology space of an associative algebra is a Lie algebra with the Gerstenhaber bracket. The computation of Hochschild cohomology requires a resolution of the algebra considered as a bimodule over itself. The bar resolution is not very satisfactory for explicit calculations to be carried out. The use of alternative resolutions is not well adapted to the computation of the Gerstenhaber bracket. However, some results by Witherspoon-Negron, Volkov and Suárez Álvarez provide useful tools to solve this problem. I will illustrate how, using these methods, it is possible to describe the first Hochschild cohomology spaces of some families of algebras as Lie algebras, the Lie module structure of higher Hochschild cohomology spaces, and relate this to the structure of the algebras.

1. Introduction

Given a field $\Bbbk$, an associative $\Bbbk$-algebra A and an A-bimodule M, the Hochschild cohomology and homology of A with coefficients in M are respectively the graded vector spaces:

$$\mathrm{H}^*(A, M) = \mathrm{Ext}^*_{A^e}(A, M) \text{ and } \mathrm{H}_*(A, M) = \mathrm{Tor}_*^{A^e}(M, A).$$

Of course any projective resolution of A as A-bimodule can be used to compute them, and the bar resolution is always available. Unfortunately, it is not suitable for computations.

Hochschild homology and cohomology – originally defined by G. Hochschild in 1945 in [**H**] in terms of the bar resolution– are very useful tools to study the representation theory of a given associative algebra. They are invariant under Morita equivalence and more generally under derived equivalence [**R**].

Just to give a flavour of what we may learn about the algebra just knowing its Hochschild cohomology, let us mention that $\mathrm{HH}^0(A)$ is the centre of A while $\mathrm{HH}^1(A)$ is isomorphic to the quotient of the $\Bbbk$-linear derivations of A modulo the inner derivations, and $\mathrm{HH}^2(A)$ classifies infinitesimal deformations of A.

2010 *Mathematics Subject Classification.* Primary 16E40, 16D90, 16E05.
Key words and phrases. Hochschild cohomology, Gerstenhaber algebra, resolutions.
This work was supported by the projects UBACYT 20020170100613BA, PIP-CONICET 11220150100483CO and PICT 2015-0366. The author is a research member of CONICET.

The Hochschild cohomology is endowed with a very rich structure:

- it is a graded commutative algebra via the cup product $\cup$,
- it has a graded Lie bracket of degree -1 that endows it with a structure of graded Lie algebra,
- the bracket is a graded biderivation with respect to the cup product, so Hochschild cohomology becomes a Gerstenhaber algebra.
- there is an action of Hochschild cohomology on Hochschild homology via the cap product.

Since for any $p, n \in \mathbb{N}$,

$$[-,-] : \mathrm{HH}^p(A) \times \mathrm{HH}^n(A) \to \mathrm{HH}^{p+n-1}(A),$$

it turns out that $\mathrm{HH}^1(A)$ is a Lie algebra with structure given by

$$[\overline{f}, \overline{h}] = \overline{f \circ h - h \circ f}$$

for all $\overline{f}, \overline{h} \in \mathrm{HH}^1(A)$, and each $\mathrm{HH}^n(A)$ becomes an $\mathrm{HH}^1(A)$-Lie module. In other words, the Lie structure of $\mathrm{HH}^1(A)$ comes from looking at $\mathrm{Der}_{\Bbbk}(A)$ inside the Lie algebra of $\Bbbk$-linear endomorphisms of A, and passing to the quotient by the inner derivations.

The whole structure is derived invariant [**K**]. In characteristic p, it is a restricted Lie algebra, and this is also derived invariant [**Z**]. Suppose $\mathrm{char}(\Bbbk) \neq 2$. Given $i \in \mathbb{N}$, the squaring map $sq_i : \mathrm{HH}^{2i} \to \mathrm{HH}^{4i-1}$ is defined by $a \mapsto \frac{1}{2}[a, a]$. For $i = 1$, this map plays an important role since it describes obstructions to infinitesimal deformations.

Putting all this together, we get –theoretically– a lot of information. The problem is that these invariants are not at all easy to compute, since the cup product and the Gerstenhaber bracket have been initially defined in terms of the bar resolution and the Hochschild cohomology is almost never computed using it. While the cup product is well understood, both because it coincides with the Yoneda product of classes of extensions and since it can be expressed in terms of any resolution via a diagonal map, this is not yet the case of the Gerstenhaber bracket. In spite of the interpretations given by Stasheff [**St**] and by Schwede [**Sch**], it remains mysterious.

This bracket is –despite its formula – a geometrical object, as we can appreciate in the following example.

EXAMPLE 1.0.1. (Hochschild-Kostant-Rosenberg)

Let $\Bbbk$ be a perfect field and A a smooth commutative $\Bbbk$-algebra essentially of finite type. By the Hochschild–Kostant–Rosenberg Theorem [**HKR**], there is an isomorphism of algebras:

$$\Lambda_A^* \mathrm{Der}_{\Bbbk} A \cong \mathrm{HH}^*(A),$$

where the multiplication on the right hand side is the cup product.

Moreover, Ginzburg [**G**] proved that, considering on the left hand side the Schouten-Nijenhuis bracket and on the right one the Gerstenhaber bracket, it is also an isomorphism of Lie algebras.

Up to a few years ago, the Gerstenhaber bracket had only been computed in some particular cases – see for example [**SF, Str, L**]. Recently, several important advances have been made and nowadays there are different methods to attempt the calculation. These methods were obtained by Negron–Witherspoon [**NW**],

Volkov [**V**] and Suárez–Álvarez [**SA**] and allow to use other resolutions than the bar resolution without having to compute the comparison maps. The first two approaches are particularly adapted to Koszul algebras. The third one, even if it is restricted to the computation of the brackets between HH^1 and HH^n for all n, can be used for big families of algebras, and sometimes allow the computation of the whole bracket, as we shall see in Section 3.

As a consequence we now know how to describe the Gerstenhaber structure of the Hochschild cohomology of a wide variety of associative algebras, and we are at the point where we can start facing the problem of discovering what does the Gerstenhaber structure say about the given algebra. On the other hand, we can also look at which kind of Lie algebras we obtain. These two questions are the main motivation of the present work.

In Section 2, I will briefly describe the recent methods while in Section 3 I will illustrate Suárez–Álvarez's method for three families of algebras, and the conclude in Section 4 with some remarks.

2. Computing the bracket using alternative resolutions

In [**Sch**] Schwede provided a description of the Gerstenhaber bracket in Hochschild cohomology based on Retakh's description of categories of extensions. R. Hermann generalised this approach in [**Her**]. However, these results are difficult to apply while doing explicit computations. Quite recently, C. Negron and S. Witherspoon [**NW**] have developed another method that works very well for example for Koszul algebras, allowing to use the Koszul resolution. Their method is clearly explained in [**NW**, Thm. 3.2.5], stating that if $P_\bullet$ is a projective resolution of A as A-bimodule via $\epsilon_P : P_0 \to A$, satisfying the following conditions:

(1) $P_\bullet$ admits an embedding $\iota : P_\bullet \to B_\bullet$, where $B_\bullet$ is the bar resolution via $\epsilon_B : B_0 \to A$, such that $\epsilon_B \iota = \epsilon_P$,
(2) ι has a section π,
(3) the diagonal $\Delta_B : B_\bullet \to B_\bullet \otimes_A B_\bullet$ preserves $P_\bullet$ and hence induces $\Delta_P : P_\bullet \to P_\bullet \otimes_A P_\bullet$,

then, any contracting homotopy ϕ of $F_P := \epsilon_P \otimes id - id \otimes \epsilon_P$ gives a bracket on $\mathrm{Hom}_{A^e}(P_\bullet, A)$ that induces the Gerstenhaber bracket in cohomology.

The authors prove that in fact this result holds under weaker conditions. Cyclic group algebras in positive characteristic and, as mentioned, Koszul algebras, are very nice examples of how this method can be used.

Y. Volkov provided a generalisation in [**V**] and computed further examples. Volkov's method depends on the construction of an homotopy. More precisely, given any resolution $P_\bullet$ of A as A-bimodule, and a diagonal map $\Delta_P : P_\bullet \to P_\bullet \otimes_A P_\bullet$, an m-cocycle $f \in \mathrm{Hom}_{A^e}(P_m, A)$, and an A^e-linear homotopy $\psi_f : P_\bullet \to P_\bullet[1-m]$ which is an homotopy lifting of f with respect to Δ_P (see [**V**] for the definition of such an homotopy lifting), then the Gerstenhaber bracket is given by:

$$[f, g] = f\psi_g - (-1)^{(n-1)(m-1)} g\psi_f,$$

where g is any n-cocycle and ψ_g is an homotopy lifting of g with respect to Δ_P.

A different approach has been developed by M. Suárez-Álvarez in [**SA**]: he gives a method to compute the Gerstenhaber bracket of elements in $\mathrm{HH}^1(A)$ with any other cohomology class, allowing then a description of $\mathrm{HH}^*(A)$ as Lie module over $\mathrm{HH}^1(A)$ for any associative algebra A. A very easy computation for monomial algebras is detailed in [**SA**]. This method is particularly useful when A is a graded algebra, as it can be observed in [**MNPRS**], where the fact that the special biserial algebras considered there possess more than one grading allowed the whole computation of the Gerstenhaber bracket using Suárez-Álvarez's method and the graded associative algebra structure of $\mathrm{HH}^*(A)$.

Let us briefly describe his method.

Consider a $\Bbbk$-algebra B with a derivation $\delta : B \to B$. If W is a left B-module, a **δ-operator** on W is a $\Bbbk$-linear map $f : W \to W$ such that

$$f(bw) = \delta(b)w + bf(w),$$

for all $b \in B$ and $w \in W$. If $\epsilon : P_\bullet \to W$ is a projective resolution of W,

$$\cdots \longrightarrow P_2 \xrightarrow{d_2} P_1 \xrightarrow{d_1} P_0 \xrightarrow{\epsilon} W \longrightarrow 0,$$

then a **δ-lifting** of f to $P_\bullet$ is a sequence $f_\bullet = (f_n)_{n\geq 0}$ of δ-operators $f_n : P_n \to P_n$, such that the following diagram commutes:

$$\begin{array}{ccccccccc} \cdots \longrightarrow & P_2 & \xrightarrow{d_2} & P_1 & \xrightarrow{d_1} & P_0 & \xrightarrow{\epsilon} & W & \longrightarrow 0 \\ & \downarrow{\scriptstyle f_2} & & \downarrow{\scriptstyle f_1} & & \downarrow{\scriptstyle f_0} & & \downarrow{\scriptstyle f} & \\ \cdots \longrightarrow & P_2 & \xrightarrow{d_2} & P_1 & \xrightarrow{d_1} & P_0 & \xrightarrow{\epsilon} & W & \longrightarrow 0. \end{array}$$

In [**SA**], Suárez-Álvarez proves that δ-liftings exist and are unique up to an equivalence:

LEMMA 2.1 ([**SA**, Lemma 1.4]). *Let W be a left B-module and let $\epsilon : P_\bullet \twoheadrightarrow W$ be a projective resolution of W. If f is a δ-operator on W, there exists a δ-lifting $f_\bullet : P_\bullet \to P_\bullet$ of f to $P_\bullet$. Moreover, if $f_\bullet$ and $f'_\bullet$ are both δ-liftings of f to $P_\bullet$, then $f_\bullet$ and $f'_\bullet$ are B-linearly homotopic.*

Now, suppose $f : W \to W$ is a δ-operator with δ-lifting $f_\bullet : P_\bullet \to P_\bullet$ of f to $P_\bullet$. Given $n \geq 0$ and $\phi \in \mathrm{Hom}_B(P_n, W)$, there is a B-linear map $f_n^\#(\phi) : P_n \to W$ defined by setting

$$f_n^\#(\phi)(p) = f(\phi(p)) - \phi(f_n(p)),$$

for $p \in P_n$. The resulting morphism

$$f_n^\# : \mathrm{Hom}_B(P_n, W) \to \mathrm{Hom}_B(P_n, W)$$

is an endomorphism of the complex $\mathrm{Hom}_B(P_\bullet, W)$. In fact, the induced map on cohomology

$$\Delta^\bullet_{f,P_\bullet} : \mathsf{H}(\mathrm{Hom}_B(P_\bullet, W)) \to \mathsf{H}(\mathrm{Hom}_B(P_\bullet, W))$$

only depends on f and not on the choice of the lifting $f_\bullet$:

THEOREM 2.2 ([**SA**, Theorem A]). *Let W be a left B-module and f be a δ-operator on W. There is a canonical morphism*

$$\Delta^\bullet_f : \mathrm{Ext}^\bullet_B(W, W) \to \mathrm{Ext}^\bullet_B(W, W)$$

of graded vector spaces, such that for every projective resolution $\epsilon : P_\bullet \twoheadrightarrow W$ and each δ-lifting $f_\bullet : P_\bullet \to P_\bullet$ of f to $P_\bullet$, the following diagram commutes:

$$\begin{array}{ccc} \mathsf{H}(\mathrm{Hom}_B(P_\bullet, W)) & \xrightarrow{\Delta^\bullet_{f,P_\bullet}} & \mathsf{H}(\mathrm{Hom}_B(P_\bullet, W)) \\ \downarrow{\scriptstyle \cong} & & \downarrow{\scriptstyle \cong} \\ \mathrm{Ext}^\bullet_B(W, W) & \xrightarrow{\Delta^\bullet_f} & \mathrm{Ext}^\bullet_B(W, W). \end{array}$$

In what follows, let us consider an algebra A and take $B = A^e$ and $W = A$ as a left A^e-module. It is well known that any element $f \in \mathrm{HH}^1(A)$ can be represented by a derivation $f : A \to A$. We can use the above results to compute the Gerstenhaber bracket $[f, -]$ on $\mathrm{HH}^\bullet(A)$. The first step is to construct a derivation on A^e,

$$f^e := f \otimes \mathrm{id}_A + \mathrm{id}_A \otimes f : A^e \longrightarrow A^e,$$

and note that $f : A \to A$ is an f^e-operator on the A^e-module A. We can thus consider the f^e-lifting of f to the bar resolution $\mathbb{B}_\bullet(A)$ of A and get a morphism of complexes

$$f_\bullet^\# : \mathrm{Hom}_{A^e}(\mathbb{B}_\bullet(A), A) \to \mathrm{Hom}_{A^e}(\mathbb{B}_\bullet(A), A).$$

LEMMA 2.3 ([**SA**, §2.2]). *The morphism $f_\bullet^\#$ describes the action of the bracket $[f, -]$ on the complex $\mathrm{Hom}_{A^e}(\mathbb{B}_\bullet(A), A)$. In particular, the Gerstenhaber bracket $[f, -]$ on $\mathrm{HH}^\bullet(A)$ is given by*

$$\Delta^\bullet_{f,\mathbb{B}_\bullet} : \mathrm{HH}^\bullet(A) \longrightarrow \mathrm{HH}^\bullet(A).$$

REMARK 2.3.1. In the examples of the next section, we will make use of the following two general observations:

(1) By Theorem 2.2, the bracket $[f, -]$ can be computed using any projective A^e-resolution $P_\bullet$ of A, provided that we succeed to construct an f^e-lifting of f to $P_\bullet$.
(2) Let A be a $\mathbb{Z}$-graded $\mathbb{k}$-algebra with $\mathbb{Z}$-grading deg, and denote the induced grading on A^e also by deg. Write δ_{deg} for the Eulerian derivation on A defined by setting $\delta_{\mathsf{deg}}(a) = \mathsf{deg}(a)a$ for any homogeneous element $a \in A$. Suppose $P_\bullet \twoheadrightarrow A$ is a graded projective resolution of A by A^e-modules; that is, every P_n is a graded A^e-module and the differential maps preserve the grading. Then there is a $(\delta_{\mathsf{deg}})^e$-lifting $(\delta_{\mathsf{deg}})_\bullet$ of δ_{deg}, defined by setting

$$(\delta_{\mathsf{deg}})_n(p) = \mathsf{deg}_n(p)p$$

for homogeneous elements $p \in P_n$, where we denote the grading on P_n by deg_n.

3. Three examples

In this section I will use the last method to compute the complete Gerstenhaber structure of three families of algebras.

3.1. Temperley-Lieb algebras. Given $n \in \mathbb{N}$ and $\delta \in \mathbb{k}\backslash\{0\}$, the Temperley-Lieb algebra $A_n(\delta)$ is a quotient of a Hecke algebra that appears in statistical mechanics, see [**EG**] and references therein. It is defined as the $\mathbb{k}$-algebra generated by variables $t_1, \ldots, t_{n-1}$ subject to the relations

$$t_i t_j t_i = t_i \text{ if } |i-j| = 1, \qquad t_i t_j = t_j t_i \text{ if } |i-j| > 1, \qquad t_i^2 = \delta t_i,$$

for $1 \leq i, j \leq n-1$.

König and Xi proved in [**KX**] that all blocks of $A_n(\delta)$ are Morita equivalent to a $\mathbb{k}$-algebra with quiver –with m depending on the block–:

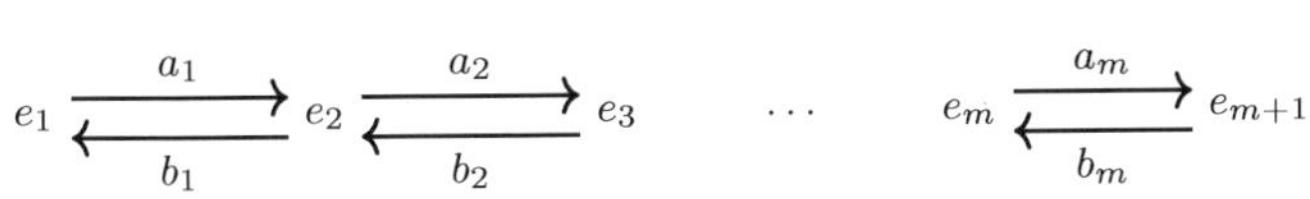

and ideal of relations generated by $a_{i+1}a_i$, b_ib_{i+1}, $a_ib_i = b_{i+1}a_{i+1}$ for all i such that $1 \leq i \leq m-1$, and a_mb_m.

It is thus sufficient to describe the Hochschild cohomology of these algebras. The dimensions of the Hochschild cohomology spaces of such a block have been computed by J. A. de la Peña and C. Xi in [**dlPX**]. They obtained that $\dim_{\Bbbk} \mathrm{HH}^0(A) = m$, $\dim_{\Bbbk} \mathrm{HH}^i(A) = 1$ for $1 \leq i \leq 2m-2$ and $\mathrm{HH}^i(A) = 0$ for $i > 2m-2$, where A denotes one of the indicated quiver algebras, where m is the number of vertices of the quiver. Afterwards Li, Chen and Xu gave generators and relations for the Hochschild cohomology as graded commutative algebra in [**LCX**]. Namely, they proved that the Hochschild cohomology is isomorphic as graded algebra to

$$\Bbbk[x_1, \ldots, x_{m-1}, y, z]/\langle x_ix_j, x_iy, x_iz, y^2, z^m, yz^{m-1}\rangle_{1\leq i,j\leq m-1},$$

where x_i is in degree 0 for all i, y is in degree 1 and z is in degree 2 (cohomological grading). For the computation of the Gerstenhaber bracket, it is important to notice that y is the class of the 1-cocycle – that we will also call y– corresponding to the grading defined in the bound quiver algebra by $|a_i| = -1$ and $|b_i| = 0$ for all i, where $|-|$ denotes the degree, and that for any r with $1 \leq r \leq m-1$,

$$\mathrm{HH}^{2r-1}(A) = \langle yz^{2r-1}\rangle \text{ and } \mathrm{HH}^{2r}(A) = \langle z^{2r}\rangle,$$

where $\langle x\rangle$ denotes the $\Bbbk$-vector space generated by x. Since $\mathrm{HH}^1(A)$ is one dimensional, it is an abelian Lie algebra. Next, let us compute the action of $\mathrm{HH}^1(A)$ on $\mathrm{HH}^i(A)$ for all i, $0 \leq i \leq 2m-2$. First notice that a direct computation gives

$$[y, x_j] = -x_j$$

for $1 \leq j \leq m-1$.

In order to use the method described in [**SA**], we need the δ-liftings of y up to degree 2. In the same way as described in [**MNPRS**], the grading provides these liftings and we easily obtain that $[y, z] = z$. Finally, using that the Gerstenhaber bracket is a graded biderivation with respect to the cup product, we obtain that for $r, s > 0$:

(1) $[y, z^r] = rz^r$,
(2) $[y, yz^r] = ryz^r$,
(3) $[z^s, z^r] = 0$,
(4) $[yz^s, yz^r] = ryz^r$.

In other words, all the generators of the Hochschild cohomology spaces are eigenvectors of $[y, -]$ and the eigenvalues equal the corresponding degrees of these generators extended to cohomology.

3.2. A family of gentle algebras. Throughout this subsection, A will be a gentle Jacobian $\Bbbk$-algebra and $char(\Bbbk) \neq 2$. Let us recall some definitions and results.

The Hochschild cohomology of some families of special biserial algebras has already been computed. In particular, M. J. Redondo and L. Román have obtained a formula for the dimension of the Hochschild cohomology spaces of quadratic string algebras – and therefore for gentle algebras– in [**RR**]. Given a connected oriented Riemann surface S with non-empty boundary ∂S, a non empty finite set M on the boundary ∂S such that there is at least one marked point in each boundary component – the elements of M are called marked points– and a triangulation $\mathbb{T}$, we consider the triple $(S; M; \mathbb{T})$, called a triangulated surface. Suppose furthermore that (S, M) is unpunctured. The finite dimensional Jacobian algebra $A = A_{\mathbb{T}}$, whose quiver $Q_{\mathbb{T}}$ was defined by Fomin, Shapiro and Thurston in [**FST**], and the potential $W_{\mathbb{T}}$ was given by D. Labardini–Fragoso in [**LF**], is gentle, see [**ABCP**].

The algebra A is isomorphic to the quotient of the path algebra of the quiver $Q_{\mathbb{T}}$ by the two–sided ideal generated by the subpaths of length 2 of each oriented 3-cycle in $Q_{\mathbb{T}}$. In her article [**Va**], Y. Valdivieso–Díaz, computed in a geometric way the dimensions of the Hochschild cohomology spaces of these algebras –usually called Jacobian gentle algebras–, and then the cyclic homology and the Hochschild homology of the same family in [**Va2**]. She proved in [**Va**] that if the triangulation $\mathbb{T}$ contains at least one internal triangle, then the cup product is non trivial, and the same holds for the Gerstenhaber bracket provided, in addition, that the characteristic of $\Bbbk$ is zero. In the main theorem of [**Va**], the author proves that for Jacobian gentle algebras, the Hochschild cohomology space $\mathrm{HH}^i(A)$ is not zero only for $i = 0, 1, 6j, 6j + 1$, with $j \in \mathbb{Z}_{>0}$, and she gives the dimensions, which for $j \geq 1$, are:

$$dim_{\Bbbk} \mathrm{HH}^{6j}(A) = dim_{\Bbbk} \mathrm{HH}^{6j+1}(A) = \#\{\text{ internal triangles of } \mathbb{T}\}.$$

In a forthcoming article with Cristian Chaparro, Sibylle Schroll and Mariano Suárez-Álvarez we will provide bases for these spaces. The precise statement follows. We first introduce some notations.

Let $A = \Bbbk Q/I$ be a quiver algebra, and let p and q be paths in Q. We will say that p is parallel to q and denote $p||q$ is both the sources and the terminus of p and q are coincident. Given sets of paths X and Y we will denote by $X||Y$ the set consisting of all pairs (p, q) in $X \times Y$ such that $p||q$.

THEOREM 3.1. *Let $A = \Bbbk Q/I$ be a Jacobian gentle algebra with basis of paths $\mathcal{B}$ and $m \geq 1$. Let Γ_r denote a basis of the space of ambiguities in the corresponding degree of Bardzell's resolution.*

(1) *For $m \equiv 0(3)$, there is a basis element of $\mathrm{HH}^{2m}(A)$ in $\Bbbk(\Gamma_{2m}||\mathcal{B})$ for each minimal complete cycle $C = a_n...a_1$ in Γ_n such that $n|2m$. Explicitly, this element is*

$$\sum_{i=1}^{n} (C_i^k, e_i)$$

where k is such that $nk = 2m$, $e_i = s(a_i), C_i = a_{i+n-1}...a_i$ and the indices are taken modulo n.

(2) *For $m \equiv 0(3)$, there are basis elements of $\mathrm{HH}^{2m+1}(A)$ in $\Bbbk(\Gamma_{2m+1}||\mathcal{B})$ for each minimal complete cycle $C = a_n...a_1$ in Γ_n such that $n|2m$. Explicitly, the pairs*

$$(a_i C_i^k, a_i)$$

where k is such that $nk = 2m$, $e_i = s(a_i), C_i = a_{i+n-1}...a_i$ and the indices are taken modulo n.

We will give a detailed proof of a more general statement in [**CSSS**].

We recall from [**Va**] that

$$dim_{\Bbbk}(\mathrm{HH}^1(A)) = 1 + |Q_1| - |Q_0| + \#\{B_1\},$$

where B_1 is the set of internal triangles of type 1. There is a basis of $\mathrm{HH}^1(A)$ containing on one hand $1 + |Q_1| - |Q_0|$ elements of type $[\alpha||\alpha]$, where $\alpha \in Q_1$, and we choose one of these elements for each fundamental cycle, and on the other hand, by $\#\{B_1\}$ elements such that each of them can be represented by the class of a shortcut $\beta||a_m...a_1$, where $\beta \in Q_1$ and $a_m...a_1$ is a path in A of length at least 2. See [**CSS20**] for the definition of shortcut.

Moreover, Valdivieso proves that the dimension of $\mathrm{HH}^0(A)$ equals:

$$1 + \#\{\text{internal triangles of type } 0\}.$$

A careful inspection of the cocycles producing Hochschild cohomology shows that $\mathrm{HH}^*(A)$ is generated as graded commutative algebra by the bases of $\mathrm{HH}^0(A)$, $\mathrm{HH}^1(A)$ and $\mathrm{HH}^6(A)$.

Again, since the Gerstenhaber bracket is a graded derivation with respect to the cup product, it is sufficient to compute the Lie algebra structure of $\mathrm{HH}^1(A)$, together with its action on $\mathrm{HH}^0(A)$ and $\mathrm{HH}^6(A)$.

THEOREM 3.2. *The Lie structure of* $\mathrm{HH}^1(A)$ *is as follows:*

(1) $[\alpha_1||\alpha_1, \alpha_2||\alpha_2] = 0$ *for all arrows* α_1, α_2,

(2) $[\beta||a_m...a_1, \beta'||a'_m...a'_1] = 0$ *for all shortcuts* $\beta||a_m...a_1$ *and* $\beta'||a'_m...a'_1$,

(3) $[\alpha||\alpha, \beta||a_m...a_1] = \begin{cases} \beta||a_m...a_1 & \text{if there exists } j \text{ such that } a_j = \alpha, \\ 0 & \text{if not.} \end{cases}$

PROOF. It follows from the fact that the Gerstenhaber bracket in $\mathrm{HH}^1(A)$ is just induced by the commutator of derivations in A. □

REMARK 3.2.1. Each element of type $[\alpha||\alpha]$ comes from a derivation of A given by $\alpha \mapsto \alpha$ and $\beta \mapsto 0$ if $\beta \in Q_1$ and $\beta \neq \alpha$. This is the Eulerian derivation associated to the grading such that $\mathsf{deg}(\alpha) = 1$ and $\mathsf{deg}(\beta) = 0$ for $\beta \in Q_1$ and $\beta \neq \alpha$. The brackets can be reinterpreted as

$$[\alpha||\alpha, \beta||a_m...a_1] = \mathsf{deg}(\beta||a_m...a_1) \cdot \beta||a_m...a_1,$$

where the degree is induced by the above mentioned grading.

We will use again Suárez-Álvarez's method [**SA**] for the computation of the brackets $[\mathrm{HH}^1(A), \mathrm{HH}^6(A)]$. As we already know, given a derivation δ of A, we need to compute a family of δ-operators δ_i lifting $\delta^e = \delta \otimes 1 + \otimes \delta$ to our resolution of A, which is in fact Bardzell's resolution, since A is a monomial algebra and this is a graded resolution. It follows from Remark 2.3.1(2) that if δ comes from a grading deg, then δ_n can be chosen as $\delta_n(p) = \mathsf{deg}_n(p) \cdot p$, where deg_n is the induced grading at the corresponding level of the resolution.

LEMMA 3.3. *If* $\delta = \beta||a_n...a_1$ *corresponds to a shortcut, then we can choose* $\delta_n = 0$ *for* $n \geq 2$, *and* $\delta_1(1 \otimes v \otimes 1) = \sum_{i=1}^n a_n...a_{i+1} \otimes a_i \otimes a_{i-1}...a_1$ *if* $v = \beta$ *and* 0 *if not.*

PROOF. It follows by a direct computation. □

Recall that, being in characteristic zero, we can choose a basis of $\mathrm{HH}^6(A)$ consisting of the elements $[c^2||e]$, where c runs in the set of cycles of $Q_\mathbb{T}$ modulo rotation, and $e = s(c) = t(c)$.

PROPOSITION 3.4. *Given* $[\alpha||\alpha]$ *in the basis of* $\mathrm{HH}^1(A)$ *and* $[c^2||e]$ *in the basis of* $\mathrm{HH}^6(A)$, *the bracket* $[\alpha||\alpha, c^2||e]$ *is as follows:*

$$[\alpha||\alpha, c^2||e] = \begin{cases} -2c^2||e & \text{if } \alpha \text{ appears in } c, \\ 0 & \text{if not.} \end{cases}$$

This coincides with the degree of $c^2||e$ *with respect to the grading defined by* $\alpha||\alpha$*, that is,* α *is in degree* 1 *and all the other arrows are in degree* 0.

PROOF. It follows from the computation of the lifting of $\alpha||\alpha$. □

REMARK 3.4.1. Let us compute the necessary brackets:

- $[\mathrm{HH}^6(A), \mathrm{HH}^6(A)] = 0$, since the result lives in $\mathrm{HH}^{11}(A) = 0$.
- $[\mathrm{HH}^0(A), \mathrm{HH}^6(A)] = 0$, since the result lives in $\mathrm{HH}^5(A) = 0$.
- $[\mathrm{HH}^1(A), \mathrm{HH}^0(A)]$ is given by
  - $\left[\alpha||\alpha, \sum_{i=1}^{|Q_0|} e_i||e_i\right] = 0 = \left[\beta||a_n...a_1, \sum_{i=1}^{|Q_0|} e_i||e_i\right]$.
  - $[\alpha||\alpha, s(b_1)||b_m...b_1] = \#\{i|\alpha = b_i\} \cdot s(b_1)||b_m...b_1$ for each cycle in Q arising from a triangle of type 0.
  - the other brackets in $[\mathrm{HH}^1(A), \mathrm{HH}^0(A)]$ are 0.

3.3. A family of subalgebras of the Weyl algebra. The Weyl algebra may be presented as the quotient of the free $\Bbbk$-algebra generated by two variables x, y divided by the two-sided ideal generated by $yx - xy - 1$. It is isomorphic to the algebra of differential operators over the one dimensional affine space, where x acts by multiplication and y corresponds to the usual derivative $\frac{\partial}{\partial x}$. Replacing this action by $h \cdot \frac{\partial}{\partial x}$ for any non zero fixed polynomial $h(x) \in \Bbbk[x]$ also provides a derivation. The algebras we study in [**LS**] are Ore extensions of the polynomial algebra in one variable. It is well-known that the only other possible Ore extensions –here we allow $h = 0$– are quantum planes or quantum Weyl algebras. Explicitly, given a field $\Bbbk$ and a non zero polynomial $h(x) \in \Bbbk[x]$, let A_h be the unital associative $\Bbbk$-algebra with two generators x and $\hat{y}$, subject to the relation $\hat{y}x - x\hat{y} = h$. There is an embedding of A_h in A_1 given by $x \mapsto x$, $\hat{y} \mapsto yh$, as in [**BLO**, Lem. 3.1]. Taking $\hat{y} = yh$, we consider A_h as the unital subalgebra of the Weyl algebra A_1 generated by x and $\hat{y} = yh$, where $[y, x] = 1$ and $[\hat{y}, x] = h$.

The family A_h parametrizes many well-known algebras, which we study simultaneously. As previously said, for $h = 1$, we retrieve the first Weyl algebra A_1. Other particular cases that have attracted attention are A_x, which is the universal enveloping algebra of the two-dimensional non abelian Lie algebra and A_{x^2}, the *Jordan plane*, which is a Nichols algebra of non diagonal type. More generally, taking $h = x^n$ with $n \geq 3$ and setting x in degree 1 and $\hat{y}$ in degree $n-1$, A_{x^n} is Artin-Schelter regular of global dimension two, although it does not admit any regrading so that it becomes generated in degree one, as observed by Stephenson [**Ste**].

In [**LS**] we describe the Gerstenhaber structure of the Hochschild cohomology $\mathrm{HH}^*(A_h)$. This is especially interesting in case $char(\Bbbk) = 0$ and $\mathsf{gcd}(h, h') \neq 1$ since in this case the description of $\mathrm{HH}^1(A_h)$ is related to the Witt algebra, and the $\mathrm{HH}^1(A_h)$-Lie module structure of $\mathrm{HH}^2(A_h)$ can be described in terms of the representation theory of the Witt algebra. Since the minimal resolution of A_h as A_h-bimodule has length 2, $\mathrm{HH}^i(A_h)$ is zero for i greater than 2. In particular, this resolution allows us to give an explicit description of $\mathrm{HH}^2(A_h)$ in positive characteristic.

Our main results are the following. See [**LS**] for proofs and more details.

THEOREM 3.5. *Assume* $char(\Bbbk) = p > 0$ *and let* $\mathsf{Z}(A_h)$ *denote the centre of* A_h. *Then* $\mathrm{HH}^2(A_h)$ *is a free* $\mathsf{Z}(A_h)$*-module if and only if* $\mathsf{gcd}(h, h') = 1$. *In this case,* $\mathrm{HH}^2(A_h)$ *has rank one over* $\mathsf{Z}(A_h)$ *and, moreover,* $\mathrm{HH}^*(A_h)$ *is a free* $\mathsf{Z}(A_h)$*-module.*

We also give an explicit description of $\mathrm{HH}^2(A_h)$ in positive characteristic, although this is a bit involved. On the other hand, in characteristic zero, $\mathrm{HH}^2(A_h)$ can be presented as a space of polynomials as in the following statement.

THEOREM 3.6. *Assume char*$(\Bbbk) = 0$. *There are isomorphisms*

$$\mathrm{HH}^2(A_h) \cong A_h/\mathsf{gcd}(h, h')A_h \cong D[\hat{y}],$$

where $D = (\Bbbk[x]/\mathsf{gcd}(h, h')\Bbbk[x])$. *In particular,* $\mathrm{HH}^2(A_h) = 0$ *if and only if* h *is a separable polynomial; otherwise,* $\mathrm{HH}^2(A_h)$ *is infinite dimensional.*

Moreover, we describe in detail $\mathrm{HH}^2(A_h)$ as a $\mathrm{HH}^1(A_h)$-Lie module.

THEOREM 3.7. *Assume that char*$(\Bbbk) = 0$ *and* $\mathsf{gcd}(h, h') \neq 1$. *Let* $m_h + 1$ *be the largest exponent occurring in the decomposition of* h *in* $\Bbbk[x]$ *into irreducible factors. The structure of* $\mathrm{HH}^2(A_h)$ *as a Lie module, under the Gerstenhaber bracket, over the Lie algebra* $\mathrm{HH}^1(A_h)$ *is as follows:*

(1) *There is a filtration of length* m_h *by* $\mathrm{HH}^1(A_h)$*-submodules*

$$\mathrm{HH}^2(A_h) = P_0 \supsetneq P_1 \supsetneq \cdots \supsetneq P_{m_h-1} \supsetneq P_{m_h} = 0.$$

such that each factor P_i/P_{i+1} *is semisimple.*

(2) *The irreducible summands of each* P_i/P_{i+1} *can be naturally seen as obtained from intermediate series modules for the Witt algebra, under a suitable finite field extension of* $\Bbbk$.

(3) $\mathrm{HH}^2(A_h)$ *has finite composition length, equal to the number of irreducible factors of* $\mathsf{gcd}(h, h')$, *counted with multiplicity.*

(4) $\mathrm{HH}^2(A_h)$ *is a semisimple* $\mathrm{HH}^1(A_h)$*-module if and only if* h *is not divisible by the cube of any non-constant polynomial.*

Notice that, in case $\Bbbk$ is of characteristic 0 and algebraically closed, from this theorem and previous results obtained in [**BLO2**] we can recover the number of irreducible factors appearing in h and the corresponding multiplicities. More specifically, let $\lambda(h)$ denote the partition encoding the multiplicities of the irreducible factors of h. We can conclude that if $\lambda(h)$ and $\lambda(g)$ are different partitions, then A_h is not derived equivalent to A_g.

4. Final remarks

The explicit description of $\mathrm{HH}^1(A)$ as Lie algebra and of the other Hochschild cohomology spaces as $\mathrm{HH}^1(A)$-Lie modules, allows now to relate the structure obtained with the representation theory of the algebra A. In [**Skow**], the author gives an extensive but not complete classification of self-injective tame algebras up to derived equivalence. In [**RSS**] we proved that for almost all these algebras – with just two exceptions– $\mathrm{HH}^1(A)$ is a solvable Lie algebra in arbitrary characteristic. Also, we show that almost all tame blocks of finite groups of tame representation type have solvable first Hochschild cohomology. However, not all associative algebras have solvable HH^1: for example if A is the Kronecker algebra and $char(\Bbbk) \neq 2$, then $\mathrm{HH}^1(A) = \mathfrak{sl}(2, \Bbbk)$ while for $A = \Bbbk[x]/\langle x^p\rangle$ and $char(\Bbbk) = p$, $\mathrm{HH}^1(A)$ is isomorphic to the Jacobson-Witt algebra.

Acknowledgments

The author would like to thank my coauthors in articles related to the Gerstenhaber bracket: Dalia Artenstein, Cristian Chaparro, Marcelo Lanzilotta, Samuel Lopes, Joanna Meinel, Van Nguyen, Bregje Pauwels, María Julia Redondo, Sibylle Schroll, Mariano Suárez-Álvarez, and also Claude Cibils and Eduardo Marcos for discussions. They also want to thank the referee for a careful reading of the text. They are grateful to Jan Trlifaj and all the organisers of the ICRA conference in Prague.

References

[ABCP] Ibrahim Assem, Thomas Brüstle, Gabrielle Charbonneau-Jodoin, and Pierre-Guy Plamondon, *Gentle algebras arising from surface triangulations*, Algebra Number Theory **4** (2010), no. 2, 201–229, DOI 10.2140/ant.2010.4.201. MR2592019

[BLO] Georgia Benkart, Samuel A. Lopes, and Matthew Ondrus, *A parametric family of subalgebras of the Weyl algebra I. Structure and automorphisms*, Trans. Amer. Math. Soc. **367** (2015), no. 3, 1993–2021, DOI 10.1090/S0002-9947-2014-06144-8. MR3286506

[BLO2] Georgia Benkart, Samuel A. Lopes, and Matthew Ondrus, *Derivations of a parametric family of subalgebras of the Weyl algebra*, J. Algebra **424** (2015), 46–97, DOI 10.1016/j.jalgebra.2014.11.007. MR3293213

[CSS20] Cristian Chaparro, Sibylle Schroll, and Andrea Solotar, *On the Lie algebra structure of the first Hochschild cohomology of gentle algebras and Brauer graph algebras*, J. Algebra **558** (2020), 293–326, DOI 10.1016/j.jalgebra.2020.02.003. MR4102135

[CSSS] C. Chaparro, S. Schroll, A. Solotar, M. Suárez-Álvarez. The Hochschild cohomology Gerstenhaber algebra for quadratic monomial algebras. In preparation.

[dlPX] José A. de la Peña and Changchang Xi, *Hochschild cohomology of algebras with homological ideals*, Tsukuba J. Math. **30** (2006), no. 1, 61–79, DOI 10.21099/tkbjm/1496165029. MR2248284

[EG] K. Erdmann and R. M. Green, *On representations of affine Temperley-Lieb algebras. II*, Pacific J. Math. **191** (1999), no. 2, 243–273, DOI 10.2140/pjm.1999.191.243. MR1738191

[FST] Sergey Fomin, Michael Shapiro, and Dylan Thurston, *Cluster algebras and triangulated surfaces. I. Cluster complexes*, Acta Math. **201** (2008), no. 1, 83–146, DOI 10.1007/s11511-008-0030-7. MR2448067

[G] V. Ginzburg, Lectures on Noncommutative Geometry, preprint (2005), `arxiv.org/math/0506603`.

[Her] Reiner Hermann, *Monoidal categories and the Gerstenhaber bracket in Hochschild cohomology*, Mem. Amer. Math. Soc. **243** (2016), no. 1151, v+146, DOI 10.1090/memo/1151. MR3518219

[H] G. Hochschild, *On the cohomology groups of an associative algebra*, Ann. of Math. (2) **46** (1945), 58–67, DOI 10.2307/1969145. MR11076

[HKR] G. Hochschild, Bertram Kostant, and Alex Rosenberg, *Differential forms on regular affine algebras*, Trans. Amer. Math. Soc. **102** (1962), 383–408, DOI 10.2307/1993614. MR142598

[K] Bernhard Keller, *Hochschild cohomology and derived Picard groups*, J. Pure Appl. Algebra **190** (2004), no. 1-3, 177–196, DOI 10.1016/j.jpaa.2003.10.030. MR2043327

[KX] Steffen König and Changchang Xi, *Strong symmetry defined by twisting modules, applied to quasi-hereditary algebras with triangular decomposition and vanishing radical cube*, Comm. Math. Phys. **197** (1998), no. 2, 427–441, DOI 10.1007/s002200050458. MR1652759

[LF] Daniel Labardini-Fragoso, *Quivers with potentials associated to triangulated surfaces*, Proc. Lond. Math. Soc. (3) **98** (2009), no. 3, 797–839, DOI 10.1112/plms/pdn051. MR2500873

[LCX] Huanhuan Li, Yunge Xu, and Yuan Chen, *Hochschild cohomology rings of Temperley-Lieb algebras*, Chin. Ann. Math. Ser. B **36** (2015), no. 4, 613–624, DOI 10.1007/s11401-015-0903-y. MR3360819

[L] Markus Linckelmann, *On the Hochschild cohomology of commutative Hopf algebras*, Arch. Math. (Basel) **75** (2000), no. 6, 410–412, DOI 10.1007/s000130050523. MR1799425

[LS] S. Lopes, A. Solotar, Lie structure on the Hochschild cohomology of a family of subalgebras of the Weyl algebra, `arXiv:1903.01226`.

[MNPRS] J. Meinel, V. C. Nguyen, B. Pauwels, M. J. Redondo, and A. Solotar, The Gerstenhaber structure on the Hochschild cohomology of a class of special biserial algebras, `arXiv:1803.10310`.

[NW] Cris Negron and Sarah Witherspoon, *An alternate approach to the Lie bracket on Hochschild cohomology*, Homology Homotopy Appl. **18** (2016), no. 1, 265–285, DOI 10.4310/HHA.2016.v18.n1.a14. MR3498646

[RR] María Julia Redondo and Lucrecia Román, *Gerstenhaber algebra structure on the Hochschild cohomology of quadratic string algebras*, Algebr. Represent. Theory **21** (2018), no. 1, 61–86, DOI 10.1007/s10468-017-9704-1. MR3748354

[R] Jeremy Rickard, *Derived equivalences as derived functors*, J. London Math. Soc. (2) **43** (1991), no. 1, 37–48, DOI 10.1112/jlms/s2-43.1.37. MR1099084

[RSS] L. Rubio y Degrassi, S. Schroll, A. Solotar, The first Hochschild cohomology as a Lie algebra, `arXiv:1903.12145`.

[SF] Selene Sánchez-Flores, *The Lie structure on the Hochschild cohomology of a modular group algebra*, J. Pure Appl. Algebra **216** (2012), no. 3, 718–733, DOI 10.1016/j.jpaa.2011.08.007. MR2864773

[Sch] Stefan Schwede, *An exact sequence interpretation of the Lie bracket in Hochschild cohomology*, J. Reine Angew. Math. **498** (1998), 153–172, DOI 10.1515/crll.1998.048. MR1629858

[Skow] Andrzej Skowroński, *Selfinjective algebras: finite and tame type*, Trends in representation theory of algebras and related topics, Contemp. Math., vol. 406, Amer. Math. Soc., Providence, RI, 2006, pp. 169–238, DOI 10.1090/conm/406/07658. MR2258046

[St] Jim Stasheff, *The intrinsic bracket on the deformation complex of an associative algebra*, J. Pure Appl. Algebra **89** (1993), no. 1-2, 231–235, DOI 10.1016/0022-4049(93)90096-C. MR1239562

[Ste] D. R. Stephenson, *Artin-Schelter regular algebras of global dimension three*, J. Algebra **183** (1996), no. 1, 55–73, DOI 10.1006/jabr.1996.0207. MR1397387

[Str] Claudia Strametz, *The Lie algebra structure on the first Hochschild cohomology group of a monomial algebra*, J. Algebra Appl. **5** (2006), no. 3, 245–270, DOI 10.1142/S0219498806001120. MR2235810

[SA] Mariano Suárez-Álvarez, *A little bit of extra functoriality for Ext and the computation of the Gerstenhaber bracket*, J. Pure Appl. Algebra **221** (2017), no. 8, 1981–1998, DOI 10.1016/j.jpaa.2016.10.015. MR3623179

[Va] Y. Valdivieso-Díaz, Hochschild cohomology of Jacobian algebras from unpunctured surfaces: a geometric computation, `arxiv:1512.00738`.

[Va2] Yadira Valdivieso-Díaz, *A note on (co)homologies of algebras from unpunctured surfaces*, J. Algebra Appl. **17** (2018), no. 5, 1850088, 10, DOI 10.1142/S0219498818500883. MR3795383

[V] Yury Volkov, *Gerstenhaber bracket on the Hochschild cohomology via an arbitrary resolution*, Proc. Edinb. Math. Soc. (2) **62** (2019), no. 3, 817–836, DOI 10.1017/s0013091518000901. MR3974969

[Z] Alexander Zimmermann, *Fine Hochschild invariants of derived categories for symmetric algebras*, J. Algebra **308** (2007), no. 1, 350–367, DOI 10.1016/j.jalgebra.2006.08.021. MR2290926

IMAS and Dto de Matemática, Facultad de Ciencias Exactas y Naturales, Universidad de Buenos Aires, Ciudad Universitaria, Pabellón 1, (1428) Buenos Aires, Argentina

Email address: `asolotar@dm.uba.ar`

Selected Published Titles in This Series

758 **Jan Šťovíček and Jan Trlifaj, Editors,** Representation Theory and Beyond, 2020

756 **Joeri Van der Veken, Alfonso Carriazo, Ivko Dimitrić, Yun Myung Oh, Bogdan D. Suceavă, and Luc Vrancken, Editors,** Geometry of Submanifolds, 2020

755 **Marion Scheepers and Ondřej Zindulka, Editors,** Centenary of the Borel Conjecture, 2020

754 **Susanne C. Brenner, Igor Shparlinski, Chi-Wang Shu, and Daniel B. Szyld, Editors,** 75 Years of Mathematics of Computation, 2020

753 **Matthew Krauel, Michael Tuite, and Gaywalee Yamskulna, Editors,** Vertex Operator Algebras, Number Theory and Related Topics, 2020

752 **Samuel Coskey and Grigor Sargsyan, Editors,** Trends in Set Theory, 2020

751 **Ashish K. Srivastava, André Leroy, Ivo Herzog, and Pedro A. Guil Asensio, Editors,** Categorical, Homological and Combinatorial Methods in Algebra, 2020

750 **A. Bourhim, J. Mashreghi, L. Oubbi, and Z. Abdelali, Editors,** Linear and Multilinear Algebra and Function Spaces, 2020

749 **Guillermo Cortiñas and Charles A. Weibel, Editors,** K-theory in Algebra, Analysis and Topology, 2020

748 **Donatella Danielli and Irina Mitrea, Editors,** Advances in Harmonic Analysis and Partial Differential Equations, 2020

747 **Paul Bruillard, Carlos Ortiz Marrero, and Julia Plavnik, Editors,** Topological Phases of Matter and Quantum Computation, 2020

746 **Erica Flapan and Helen Wong, Editors,** Topology and Geometry of Biopolymers, 2020

745 **Federico Binda, Marc Levine, Manh Toan Nguyen, and Oliver Röndigs, Editors,** Motivic Homotopy Theory and Refined Enumerative Geometry, 2020

744 **Pieter Moree, Anke Pohl, L'ubomír Snoha, and Tom Ward, Editors,** Dynamics: Topology and Numbers, 2020

743 **H. Garth Dales, Dmitry Khavinson, and Javad Mashreghi, Editors,** Complex Analysis and Spectral Theory, 2020

742 **Francisco-Jesús Castro-Jiménez, David Bradley Massey, Bernard Teissier, and Meral Tosun, Editors,** A Panorama of Singularities, 2020

741 **Houssam Abdul-Rahman, Robert Sims, and Amanda Young, Editors,** Analytic Trends in Mathematical Physics, 2020

740 **Alina Bucur and David Zureick-Brown, Editors,** Analytic Methods in Arithmetic Geometry, 2019

739 **Yaiza Canzani, Linan Chen, and Dmitry Jakobson, Editors,** Probabilistic Methods in Geometry, Topology and Spectral Theory, 2019

738 **Shrikrishna G. Dani, Surender K. Jain, Jugal K. Verma, and Meenakshi P. Wasadikar, Editors,** Contributions in Algebra and Algebraic Geometry, 2019

737 **Fernanda Botelho, Editor,** Recent Trends in Operator Theory and Applications, 2019

736 **Jane Hawkins, Rachel L. Rossetti, and Jim Wiseman, Editors,** Dynamical Systems and Random Processes, 2019

735 **Yanir A. Rubinstein and Bernard Shiffman, Editors,** Advances in Complex Geometry, 2019

734 **Peter Kuchment and Evgeny Semenov, Editors,** Differential Equations, Mathematical Physics, and Applications, 2019

733 **Peter Kuchment and Evgeny Semenov, Editors,** Functional Analysis and Geometry, 2019

For a complete list of titles in this series, visit the
AMS Bookstore at **www.ams.org/bookstore/conmseries/**.